Analytical Mechanics

SEVENTH EDITION

Grant R. Fowles
University of Utah

George L. Cassiday
University of Utah

First mode
$N = 1$
$a_k = A \sin\left(\frac{k\pi}{4}\right)$

A

$A \sin \frac{3\pi}{4}$

a_1 a_2 a_3

Second mode
$N = 2$
$a_k = A \sin\left(\frac{2k\pi}{4}\right)$

A

$a_2 = 0$

$-A$

a_3

Third mode
$N = 3$
$a_k = A \sin\left(\frac{3k\pi}{4}\right)$

a_1

a_3

$A \sin \frac{3\pi}{4}$

$A \sin \frac{9\pi}{4}$

$-A$

a_2

THOMSON

BROOKS/COLE

Australia • Canada • Mexico • Singapore • Spain
United Kingdom • United States

THOMSON

BROOKS/COLE

Publisher: David Harris
Acquisitions Editor: Chris Hall
Development Editor: Rebecca Heider
Editorial Assistant: Seth Dobrin
Technology Project Manager: Sam Subity
Marketing Manager: Kelley McAllister
Advertising Project Manager: Stacey Purviance
Project Manager, Editorial
    Production: Belinda Krohmer
Art Director: Rob Hugel
Print/Media Buyer: Doreen Suruki
Permissions Editor: Sarah Harkrader

Production Service: Nesbitt Graphics, Inc.
Text Designer: John Edeen
Copy Editor: Nesbitt Graphics, Inc.
Illustrator: Nesbitt Graphics, Inc.
Cover Designer: Belinda Fernandez
Cover Image: Graham Watson
Cover Printer: Lehigh Press
Compositor: International Typesetting
    and Composition
Printer: The Maple-Vail Book
    Manufacturing Group

Printed in the United States of America
1 2 3 4 5 6 7 08 07 06 05 04

For more information about our products,
contact us at:
**Thomson Learning Academic
Resource Center
1-800-423-0563**

For permission to use material from this text or
product, submit a request online at:
http://www.thomsonrights.com

Any additional questions about permissions can
be submitted by email to
thomsonrights@thomson.com.

Library of Congress Control Number: 2003115137

ISBN 0-534-49492-7

International Student Edition: ISBN 0-534-40813-3
(Not for sale in the United States)

**Thomson Brooks/Cole
10 Davis Drive
Belmont, CA 94002
USA**

Asia
Thomson Learning
5 Shenton Way #01-01
UIC Building
Singapore 068808

Australia/New Zealand
Thomson Learning
102 Dodds Street
Southbank, Victoria 3006
Australia

Canada
Nelson
1120 Birchmount Road
Toronto, Ontario M1K 5G4
Canada

Europe/Middle East/Africa
Thomson Learning
High Holborn House
50/51 Bedford Row
London WC1R 4LR
United Kingdom

Latin America
Thomson Learning
Seneca, 53
Colonia Polanco
11560 Mexico D.F.
Mexico

Spain/Portugal
Paraninfo
Calle Magallanes, 25
28015 Madrid, Spain

Preface

This textbook is intended for an undergraduate course in classical mechanics taken by students majoring in physics, physical science, or engineering. We assume that the student has taken a year of calculus-based general physics and a year of differential/integral calculus. We recommend that a course in differential equations and matrix algebra be taken prior to, or concurrently with, this course in classical mechanics.

The seventh edition of this text adheres to the same general philosophy of the previous editions: it centers on the development and exposition of Newtonian mechanics with the more advanced Lagrangian and Hamiltonian formalism introduced and used only in the last two chapters. New material has been added to, and old material has been eliminated from, some of the chapters. We have expended much effort to stamp out annoying typographical errors, inadvertent mistakes, and unclear presentations. Explanations of some of the more difficult concepts have been expanded, and many figures and examples have been added with the intent of achieving greater clarification. Several sections have been greatly modified, and some new ones have been added. Those sections that are to be used with software tools such as *Mathcad* and *Mathematica* as part of the problem solving strategy have been streamlined with many of the details relegated to an appendix.

A brief synopsis of each chapter follows:

- Chapter 1. A brief introduction to dimensional analysis and vector algebra; concepts of velocity and acceleration.
- Chapter 2. Newton's laws of motion; motion in one dimension. Expansion of discussion of inertial frames of reference. Introduction to solving problems numerically using *Mathcad*: vertical fall through a fluid.
- Chapter 3. Harmonic motion, resonance, the driven oscillator. Numerical solution of non-linear oscillator problems.
- Chapter 4. Motion of a particle in three dimensions. Potential energy and conservative forces. Introduction to solving problems numerically using *Mathematica*; projectile motion in a resistive medium; Mickey Mantle's "tape measure" homerun.
- Chapter 5. The analysis of motion in a noninertial frame of reference and fictitious forces. Numerical solution of projectile motion in a rotating frame of reference.
- Chapter 6. Gravitation. Expanded discussion of central forces. Conic sections and orbits. Expanded discussion of orbital energy. Criteria for stable orbits. Rutherford scattering.
- Chapter 7. Many particle systems. The three-body problem: numerical solution. Lagrangian points. Conservation laws and collisions. Expanded presentation of rocket motion.
- Chapter 8. Rotation of a body about a fixed axis. Expanded discussion of laminar motion. Moments of inertia.

- Chapter 9. Rotation of a body in three dimensions. Numerical solutions of problems involving the rotation of bodies with differing principal moments of inertia. Motion of tops and gyroscopes. Stability of a rotating bicycle wheel (why Lance doesn't fall over).
- Chapter 10. Lagrangian and Hamiltonian mechanics. Hamilton's and D'Alembert's principles. Conservation laws.
- Chapter 11. Coupled oscillators. Normal coordinates and normal modes of oscillation. The eigenvalue problem. The loaded string and wave motion.

More worked examples have been added to this edition. Most worked examples can be found at the end of each section. The problems found in the first set at the end of each chapter can be solved analytically. A second set contains problems that require numerical techniques, typically by using *Mathcad, Mathematica,* or any other software tool favored by the student or required by the instructor.

The appendices contain information or reference material that should help the time-challenged student solve problems without resorting to time-consuming data searches elsewhere. Answers to a few selected odd-numbered problems are given at the end of the text.

An updated problem solutions manual is available to instructors who adopt the text. Brooks Cole/Thomson Learning may provide complementary aids and supplements to those qualified under our adoption policy. Please contact your sales representative for more information.

Acknowledgments

The authors wish to acknowledge Mathsoft, Inc. for supplying us with current versions of *Mathcad*. We also acknowledge Wolfram Assoc. for selling us a copy of their current version of *Mathematica 4* at a reduced cost. We would also like to thank those who aided in recommendations concerning the generation of this current edition. Linda McDonald of Northpark University and M. Anthony Reynolds of Embry-Riddle University provided in-depth reviews of certain chapters, and Zenaida Uy of Millersville University sent in comments as she taught with the book. Thanks also to the following individuals who answered an on-line survey about their Mechanics course: Charles Benesh, Wesleyan College; Mark S. Boley, Western Illinois University; Donald Bord, University of Michigan, Dearborn; Chris Burns, Swarthmore College; Steve Cederbloom, Mount Union College; Kelvin Chu, University of Vermont; Jim Crumley, St. John's University/College of St. Benedict; Vic DeCarlo, DePauw University; William Franz, Randolph-Macon College; Junichiro Fukai, Auburn; John G. Hardie, Christopher Newport University; Jim McCoy, Tarleton State University; Carl E. Mungan, U.S. Naval Academy; Rolfe G. Petschek, Case Western Reserve University; Brian P. Schwartz, Carthage College; C. Gregory Seab, University of New Orleans; Peter Skiff, Bard College; James Wheeler, Lock Haven University of Pennsylvania; E.J. Zita, The Evergreen State College.

I would also like to acknowledge my editor, Rebecca Heider, my children, Pat and Katie, and my wife, Nancy Cohn, for their strong support and encouragement during the preparation of this edition.

George L. Cassiday

Contents Overview

1 Fundamental Concepts: Vectors   1

2 Newtonian Mechanics: Rectilinear Motion of a Particle   47

3 Oscillations   82

4 General Motion of a Particle in Three Dimensions   144

5 Noninertial Reference Systems   184

6 Gravitation and Central Forces   218

7 Dynamics of Systems of Particles   275

8 Mechanics of Rigid Bodies: Planar Motion   323

9 Motion of Rigid Bodies in Three Dimensions   361

10 Lagrangian Mechanics   417

11 Dynamics of Oscillating Systems   465

Appendices   A-1

Answers to Selected Odd-Numbered Problems   ANS-1

Selected References   R-1

Index   I-1

Contents

1 Fundamental Concepts: Vectors   1

  1.1 Introduction   1
  1.2 Measure of Space and Time: Units and Dimensions   2
  1.3 Vectors   9
  1.4 The Scalar Product   15
  1.5 The Vector Product   19
  1.6 An Example of the Cross Product: Moment of a Force   22
  1.7 Triple Products   23
  1.8 Change of Coordinate System: The Transformation Matrix   25
  1.9 Derivative of a Vector   30
  1.10 Position Vector of a Particle: Velocity and Acceleration in
       Rectangular Coordinates   31
  1.11 Velocity and Acceleration in Plane Polar Coordinates   36
  1.12 Velocity and Acceleration in Cylindrical and Spherical Coordinates 39

**2 Newtonian Mechanics: Rectilinear
Motion of a Particle**   47

  2.1 Newton's Law of Motion: Historical Introduction   47
  2.2 Rectilinear Motion: Uniform Acceleration Under a
      Constant Force   60
  2.3 Forces that Depend on Position: The Concepts of Kinetic
      and Potential Energy   63
  2.4 Velocity-Dependent Forces: Fluid Resistance
      and Terminal Velocity   69
  *2.5 Vertical Fall Through a Fluid: Numerical Solution   75

3 Oscillations   82

  3.1 Introduction   82
  3.2 Linear Restoring Force: Harmonic Motion   84
  3.3 Energy Considerations in Harmonic Motion   93
  3.4 Damped Harmonic Motion   96
  *3.5 Phase Space   106
  3.6 Forced Harmonic Motion: Resonance   113

*3.7   The Nonlinear Oscillator: Method of Successive Approximations   125
*3.8   The Nonlinear Oscillator: Chaotic Motion   129
*3.9   Nonsinusoidal Driving Force: Fourier Series   135

4   General Motion of a Particle in Three Dimensions   144

4.1   Introduction: General Principles   144
4.2   The Potential Energy Function in Three-Dimensional Motion:
      The Del Operator   151
4.3   Forces of the Separable Type: Projectile Motion   156
4.4   The Harmonic Oscillator in Two and Three Dimensions   167
4.5   Motion of Charged Particles in Electric and Magnetic Fields   173
4.6   Constrained Motion of a Particle   176

5   Noninertial Reference Systems   184

5.1   Accelerated Coordinate Systems and Inertial Forces   184
5.2   Rotating Coordinate Systems   189
5.3   Dynamics of a Particle in a Rotating Coordinate System   196
5.4   Effects of Earth's Rotation   201
*5.5   Motion of a Projectile in a Rotating Cylinder   207
5.6   The Foucault Pendulum   212

6   Gravitation and Central Forces   218

6.1   Introduction   218
6.2   Gravitational Force between a Uniform Sphere and a Particle   223
6.3   Kepler's Laws of Planetary Motion   225
6.4   Kepler's Second Law: Equal Areas   226
6.5   Kepler's First Law: The Law of Ellipses   229
6.6   Kepler's Third Law: The Harmonic Law   238
6.7   Potential Energy in a Gravitational Field: Gravitational Potential   244
6.8   Potential Energy in a General Central Field   250
6.9   Energy Equation of an Orbit in a Central Field   251
6.10 Orbital Energies in an Inverse-Square Field   251
6.11 Limits of the Radial Motion: Effective Potential   257
6.12 Nearly Circular Orbits in Central Fields: Stability   260
6.13 Apsides and Apsidal Angles for Nearly Circular Orbits   262
6.14 Motion in an Inverse-Square Repulsive Field:
      Scattering of Alpha Particles   264

7   Dynamics of Systems of Particles   275

7.1   Introduction: Center of Mass and Linear Momentum of a System   275
7.2   Angular Momentum and Kinetic Energy of a System   278
7.3   Motion of Two Interacting Bodies: The Reduced Mass   283

*7.4   The Restricted Three-Body Problem   288
7.5   Collisions   303
7.6   Oblique Collisions and Scattering: Comparison of
      Laboratory and Center of Mass Coordinates   306
7.7   Motion of a Body with Variable Mass: Rocket Motion   312

8   Mechanics of Rigid Bodies: Planar Motion   323

8.1   Center of Mass of a Rigid Body   323
8.2   Rotation of a Rigid Body about a Fixed Axis: Moment of Inertia   328
8.3   Calculation of the Moment of Inertia   330
8.4   The Physical Pendulum   338
8.5   The Angular Momentum of a Rigid Body in Laminar Motion   344
8.6   Examples of the Laminar Motion of a Rigid Body   347
8.7   Impulse and Collisions Involving Rigid Bodies   354

9   Motion of Rigid Bodies in Three Dimensions   361

9.1   Rotation of a Rigid Body about an Arbitrary Axis: Moments and
      Products of Inertia—Angular Momentum and Kinetic Energy   361
9.2   Principal Axes of a Rigid Body   371
9.3   Euler's Equations of Motion of a Rigid Body   381
9.4   Free Rotation of a Rigid Body: Geometric
      Description of the Motion   383
9.5   Free Rotation of a Rigid Body with an Axis of Symmetry:
      Analytical Treatment   384
9.6   Description of the Rotation of a Rigid Body Relative to a Fixed
      Coordinate System: The Eulerian Angles   391
9.7   Motion of a Top   397
9.8   The Energy Equation and Nutation   401
9.9   The Gyrocompass   407
9.10 Why Lance Doesn't Fall Over (Mostly)   409

10 Lagrangian Mechanics   417

10.1 Hamilton's Variational Principle: An Example   419
10.2 Generalized Coordinates   423
10.3 Calculating Kinetic and Potential Energies in Terms of
      Generalized Coordinates: An Example   426
10.4 Lagrange's Equations of Motion for Conservative Systems   430
10.5 Some Applications of Lagrange's Equations   431
10.6 Generalized Momenta: Ignorable Coordinates   438
10.7 Forces of Constraint: Lagrange Multipliers   444
10.8 D'Alembert's Principle: Generalized Forces   449
10.9 The Hamiltonian Function: Hamilton's Equations   455

11 Dynamics of Oscillating Systems   **465**

11.1 Potential Energy and Equilibrium: Stability   465
11.2 Oscillation of a System with One Degree of
      Freedom about a Position of Stable Equilibrium   469
11.3 Coupled Harmonic Oscillators: Normal Coordinates   472
11.4 General Theory of Vibrating Systems   493
11.5 Vibration of a Loaded String or Linear Array of
      Coupled Harmonic Oscillators   498
11.6 Vibration of a Continuous System:
      The Wave Equation   505

Appendix A   Units   A-1
Appendix B   Complex Numbers and Identities   A-4
Appendix C   Conic Sections   A-7
Appendix D   Service Expansions   A-11
Appendix E   Special Functions   A-13
Appendix F   Curvilinear Coordinates   A-15
Appendix G   Fourier Series   A-17
Appendix H   Matrices   A-19
Appendix I   Software Tools: *Mathcad* and *Mathematica*   A-24

Answers to Selected Odd-Numbered Problems   **ANS-1**

Selected References   **R-1**

Index   **I-1**

Fundamental Concepts:
Vectors

"Let no one unversed in geometry enter these portals."

Plato's inscription over his academy in Athens

1.1 | Introduction

The science of classical mechanics deals with the motion of objects through absolute *space* and *time* in the Newtonian sense. Although central to the development of classical mechanics, the concepts of space and time would remain arguable for more than two and a half centuries following the publication of Sir Isaac Newton's *Philosophie naturalis principia mathematica* in 1687. As Newton put it in the first pages of the *Principia*, "Absolute, true and mathematical time, of itself, and from its own nature, flows equably, without relation to anything external, and by another name is called duration. Absolute space, in its own nature, without relation to anything external, remains always similar and immovable."

Ernst Mach (1838–1916), who was to have immeasurable influence on Albert Einstein, questioned the validity of these two Newtonian concepts in *The Science of Mechanics: A Critical and Historical Account of Its Development* (1907). There he claimed that Newton had acted contrary to his expressed intention of "framing no hypotheses," that is, accepting as fundamental premises of a scientific theory nothing that could not be inferred directly from "observable phenomena" or induced from them by argument. Indeed, although Newton was on the verge of overtly expressing this intent in Book III of the *Principia* as the fifth and last rule of his *Regulae Philosophandi* (rules of reasoning in philosophy), it is significant that he refrained from doing so.

Throughout his scientific career he exposed and rejected many hypotheses as false; he tolerated many as merely harmless; he put to use those that were verifiable. But he encountered a class of hypotheses that, neither "demonstrable from the phenomena nor following from them by argument based on induction," proved impossible to avoid. His concepts of space and time fell in this class. The acceptance of such hypotheses as fundamental was an embarrassing necessity; hence, he hesitated to adopt the frame-no-hypotheses rule. Newton certainly could be excused this sin of omission. After all, the adoption of these hypotheses and others of similar ilk (such as the "force" of gravitation) led to an elegant and comprehensive view of the world the likes of which had never been seen.

Not until the late 18th and early 19th centuries would experiments in electricity and magnetism yield observable phenomena that could be understood only from the vantage point of a new space–time paradigm arising from Albert Einstein's special relativity. Hermann Minkowski introduced this new paradigm in a semipopular lecture in Cologne, Germany in 1908 as follows:

> Gentlemen! The views of space and time which I wish to lay before you have sprung from the soil of experimental physics and therein lies their strength. They are radical. From now on, space by itself and time by itself are doomed to fade away into the shadows, and only a kind of union between the two will preserve an independent reality.

Thus, even though his own concepts of space and time were superceded, Newton most certainly would have taken great delight in seeing the emergence of a new space–time concept based upon observed "phenomena," which vindicated his unwritten frame-no-hypotheses rule.

1.2 | Measure of Space and Time: Units[1] and Dimensions

We shall assume that space and time are described strictly in the Newtonian sense. Three-dimensional space is Euclidian, and positions of points in that space are specified by a set of three numbers (x, y, z) relative to the origin $(0, 0, 0)$ of a rectangular Cartesian coordinate system. A length is the spatial separation of two points relative to some standard length.

Time is measured relative to the duration of reoccurrences of a given configuration of a cyclical system—say, a pendulum swinging to and fro, an Earth rotating about its axis, or electromagnetic waves from a cesium atom vibrating inside a metallic cavity. The time of occurrence of any event is specified by a number t, which represents the number of reoccurrences of a given configuration of a chosen cyclical standard. For example, if 1 vibration of a standard physical pendulum is used to define 1 s, then to say that some event occurred at $t = 2.3$ s means that the standard pendulum executed 2.3 vibrations after its "start" at $t = 0$, when the event occurred.

All this sounds simple enough, but a substantial difficulty has been swept under the rug: Just what are the standard units? The choice of standards has usually been made more for political reasons than for scientific ones. For example, to say that a person is 6 feet tall is to say that the distance between the top of his head and the bottom of his foot is six times the length of something, which is taken to be the standard unit of 1 foot.

---

[1] A delightful account of the history of the standardization of units can be found in H. A. Klein, *The Science of Measurement—A Historical Survey*, Dover Publ., Mineola, 1988, ISBN 0-486-25839-4 (pbk).

In an earlier era that standard might have been the length of an actual human foot or something that approximated that length, as per the writing of Leonardo da Vinci on the views of the Roman architect–engineer Vitruvius Pollio (first century B.C.E.):

> . . . Vitruvius declares that Nature has thus arranged the measurements of a man: four fingers make 1 palm and 4 palms make 1 foot; six palms make 1 cubit; 4 cubits make once a man's height; 4 cubits make a pace, and 24 palms make a man's height . . .

Clearly, the adoption of such a standard does not make for an accurately reproducible measure. An early homemaker might be excused her fit of anger upon being "short-footed" when purchasing a bolt of cloth measured to a length normalized to the foot of the current short-statured king.

The Unit of Length

The French Revolution, which ended with the Napoleanic *coup d'etat* of 1799, gave birth to (among other things) an extremely significant plan for reform in measurement. The product of that reform, the metric system, expanded in 1960 into the Système International d'Unités (SI).

In 1791, toward the end of the first French National Assembly, Charles Maurice de Talleyrand-Perigord (1754–1838) proposed that a task of weight and measure reform be undertaken by a "blue ribbon" panel with members selected from the French Academy of Sciences. This problem was not trivial. Metrologically, as well as politically, France was still absurdly divided, confused, and complicated. A given unit of length recognized in Paris was about 4% longer than that in Bordeaux, 2% longer than that in Marseilles, and 2% shorter than that in Lille. The Academy of Sciences panel was to change all that. Great Britain and the United States refused invitations to take part in the process of unit standardization. Thus was born the antipathy of English-speaking countries toward the metric system.

The panel chose 10 as the numerical base for all measure. The fundamental unit of length was taken to be one ten-millionth of a quadrant, or a quarter of a full meridian. A surveying operation, extending from Dunkirk on the English Channel to a site near Barcelona on the Mediterranean coast of Spain (a length equivalent to 10 degrees of latitude or one ninth of a quadrant), was carried out to determine this fundamental unit of length accurately. Ultimately, this monumental trek, which took from 1792 until 1799, changed the standard meter—estimated from previous, less ambitious surveys—by less than 0.3 mm, or about 3 parts in 10,000. We now know that this result, too, was in error by a similar factor. The length of a standard quadrant of meridian is 10,002,288.3 m, a little over 2 parts in 10,000 greater than the quadrant length established by the Dunkirk–Barcelona expedition.

Interestingly enough, in 1799, the year in which the Dunkirk–Barcelona survey was completed, the national legislature of France ratified new standards, among them the meter. The standard meter was now taken to be the distance between two fine scratches made on a bar of a dense alloy of platinum and iridium shaped in an X-like cross section to minimize sagging and distortion. The United States has two copies of this bar, numbers 21 and 27, stored at the Bureau of Standards in Gaithersburg, MD, just outside Washington, DC. Measurements based on this standard are accurate to about 1 part in 10^6. Thus, an object (a bar of platinum), rather than the concepts that led to it, was established as the standard meter. The Earth might alter its circumference if it so chose, but the standard meter would remain safe forever in a vault in Sevres, just outside Paris, France. This standard persisted until the 1960s.

The 11th General Conference of Weights and Measures, meeting in 1960, chose a reddish-orange radiation produced by atoms of krypton-86 as the next standard of length, with the meter defined in the following way:

> The meter is the length equal to 1,650,763.73 wavelengths in vacuum of the radiation corresponding to the transition between the levels $2\,p^{10}$ and $5\,d^5$ of the krypton-86 atom.

Krypton is all around us; it makes up about 1 part per million of the Earth's present atmosphere. Atmospheric krypton has an atomic weight of 83.8, being a mixture of six different isotopes that range in weight from 78 to 86. Krypton-86 composes about 60% of these. Thus, the meter was defined in terms of the "majority kind" of krypton. Standard lamps contained no more than 1% of the other isotopes. Measurements based on this standard were accurate to about 1 part in 10^8.

Since 1983 the meter standard has been specified in terms of the velocity of light. A meter is the distance light travels in 1/299,792,458 s in a vacuum. In other words, the velocity of light is defined to be 299,792,458 m/s. Clearly, this makes the standard of length dependent on the standard of time.

The Unit of Time

Astronomical motions provide us with three great "natural" time units: the day, the month, and the year. The day is based on the Earth's spin, the month on the moon's orbital motion about the Earth, and the year on the Earth's orbital motion about the Sun. Why do we have ratios of 60:1 and 24:1 connecting the day, hour, minute, and second? These relationships were born about 6000 years ago on the flat alluvial plains of Mesopotamia (now Iraq), where civilization and city-states first appeared on Earth. The Mesopotamian number system was based on 60, not on 10 as ours is. It seems likely that the ancient Mesopotamians were more influenced by the 360 days in a year, the 30 days in a month, and the 12 months in a year than by the number of fingers on their hands. It was in such an environment that sky watching and measurement of stellar positions first became precise and continuous. The movements of heavenly bodies across the sky were converted to clocks.

The second, the basic unit of time in SI, began as an arbitrary fraction (1/86,400) of a mean solar day ($24 \times 60 \times 60 = 86,400$). The trouble with astronomical clocks, though, is that they do not remain constant. The mean solar day is lengthening, and the lunar month, or time between consecutive full phases, is shortening. In 1956 a new second was defined to be 1/31,556,926 of one particular and carefully measured mean solar year, that of 1900. That second would not last for long! In 1967 it was redefined again, in terms of a specified number of oscillations of a cesium atomic clock.

A cesium atomic clock consists of a beam of cesium-133 atoms moving through an evacuated metal cavity and absorbing and emitting microwaves of a characteristic resonant frequency, 9,192,631,770 Hertz (Hz), or about 10^{10} cycles per second. This absorption and emission process occurs when a given cesium atom changes its atomic configuration and, in the process, either gains or loses a specific amount of energy in the form of microwave radiation. The two differing energy configurations correspond to situations in which the spins of the cesium nucleus and that of its single outer-shell electron are either opposed (lowest energy state) or aligned (highest energy state). This kind of a "spin-flip" atomic transition is called a *hyperfine transition*. The energy difference and, hence, the resonant frequency are precisely determined by the invariable structure of the cesium atom. It does not differ from one atom to another. A properly adjusted and maintained cesium clock can keep time with a stability of about 1 part in 10^{12}. Thus, in one year, its deviation from the right time should be no more than about 30 μs (30×10^{-6} s). When two different cesium clocks are compared, it is found that they maintain agreement to about 1 part in 10^{10}.

It was inevitable then that in 1967, because of such stability and reproducibility, the 13th General Conference on Weights and Measures would substitute the cesium-133 atom for any and all of the heavenly bodies as the primary basis for the unit of time. The conference established the new basis with the following historic words:

> The second is the duration of 9,192,631,770 periods of the radiation corresponding to the transition between two hyperfine levels of the cesium-133 atom.

So, just as the meter is no longer bound to the surface of the Earth, the second is no longer derived from the "ticking" of the heavens.

The Unit of Mass

This chapter began with the statement that the science of mechanics deals with the motion of objects. *Mass* is the final concept needed to specify completely any physical quantity.[2] The *kilogram* is its basic unit. This primary standard, too, is stored in a vault in Sevres, France, with secondaries owned and kept by most major governments of the world. Note that the units of length and time are based on atomic standards. They are *universally* reproducible and virtually indestructible. Unfortunately, the unit of mass is not yet quite so robust.

A concept involving mass, which we shall have occasion to use throughout this text, is that of the *particle*, or point mass, an entity that possesses mass but no spatial extent. Clearly, the particle is a nonexistent idealization. Nonetheless, the concept serves as a useful approximation of physical objects in a certain context, namely, in a situation where the dimension of the object is small compared to the dimensions of its environment. Examples include a bug on a phonograph record, a baseball in flight, and the Earth in orbit around the Sun.

The units (*kilogram, meter,* and *second*) constitute the basis of the SI system.[3] Other systems are commonly used also, for example, the cgs (*centimeter, gram, second*) and the fps (*foot, pound, second*) systems. These systems may be regarded as secondary because they are defined relative to the SI standard. See Appendix A.

Dimensions

Normally, we think of dimensions as the three mutually orthogonal directions in space along which an object can move. For example, the motion of an airplane can be described in terms of its movement along the directions: east–west, north–south, and up–down. However, in physics, the term has an analogous but more fundamental meaning.

---

[2] The concept of mass is treated in Chapter 2.

[3] Other basic and derived units are listed in Appendix A.

EXAMPLE 1.2.1

Converting Units

What is the length of a light year (LY) in meters?

Solution:

The speed of light is $c = 1$ LY/Y. The distance light travels in $T = 1$ Y is

$$D = cT = (1 \text{ LY/}\cancel{Y}) \times 1 \cancel{Y} = 1 \text{ LY}.$$

If we want to express one light year in terms of meters, we start with the speed of light expressed in those units. It is given by $c = 3.00 \times 10^8$ m/s. However, the time unit used in this value is expressed in seconds, while the interval of time T is expressed in years, so

$$1 \text{ LY} = (3.00 \times 10^8 \text{ m/s}) \times (1 \text{ Y}) = 3.00 \times 10^8 \text{ m} \times (1 \text{ Y/1 s})$$

The different times in the result must be expressed in the same unit to obtain a dimensionless ratio, leaving an answer in units of meters only. Converting 1 Y into its equivalent value in seconds achieves this.

$$1 \text{ LY} = (3.00 \times 10^8 \text{ m}) \times (1 \cancel{Y}/1 \text{ s}) \times (365 \cancel{day}/\cancel{Y}) \times (24 \cancel{hr}/\cancel{day}) \times (60 \cancel{min}/\cancel{hr}) \times (60 \text{ s/}\cancel{min})$$
$$= (3.00 \times 10^8 \text{ m}) \times (3.15 \times 10^7 \cancel{s}/1 \cancel{s}) = 9.46 \times 10^{15} \text{ m}$$

We have multiplied 1 year by a succession of ratios whose values each are intrinsically dimensionless and equal to one. For example, 365 days = 1 year, so (365 days/1 year) = (1 year/1 year) = 1. The multiplications have not changed the intrinsic value of the result. They merely convert the value (1 year) into its equivalent value in seconds to "cancel out" the seconds unit, leaving a result expressed in meters.

No more than three fundamental quantities are needed to completely describe or characterize the behavior of any physical system that we encounter in the study of classical mechanics: the space that bodies occupy, the matter of which they consist, and the time during which those bodies move. In other words, classical mechanics deals with the motion of physical objects through space and time. All measurements of that motion ultimately can be broken down into combinations of measurements of mass, length, and time. The acceleration a of a falling apple is measured as a change in speed per change in time and the change in speed is measured as a change in position (length) per change in time. Thus, the measurement of acceleration is completely characterized by measurements of length and time. The concepts of mass, length, and time are far more fundamental than are the arbitrary units we choose to provide a scale for their measurement. Mass, length, and time specify the three primary *dimensions* of all physical quantities. We use the symbols $[M]$, $[L]$, and $[T]$ to characterize these three primary dimensions. The dimension of any physical quantity is defined to be the algebraic combination of $[M]$, $[L]$, and $[T]$ that is needed to fully characterize a measurement of the physical quantity.

In other words, the dimension of any physical quantity can be written as $[M]^\alpha [L]^\beta [T]^\gamma$, where α, β, and γ are powers of their respective dimension. For example, the dimension of acceleration a is

$$[a] = \left[\frac{L/T}{T} \right] = [L][T]^{-2}$$

Be aware! Do not confuse the dimension of a quantity with the units chosen to express it. Acceleration can be expressed in units of feet per second per second, kilometers per hour per hour, or, if you were Galileo investigating a ball rolling down an inclined plane, in units of *punti* per beat per beat! All of these units are consistent with the dimension $[L] [T]^{-2}$.

Dimensional Analysis

Dimensional analysis of equations that express relationships between different physical quantities is a powerful tool that can be used to immediately determine whether the result of a calculation has even the possibility of being correct or not. All equations must have consistent dimensions. The dimension of a physical quantity on the left hand side of an equation must have the same dimension as the combination of dimensions of all physical quantities on the right hand side. For example, later on in Example 6.5.3, we calculate the speed of satellite in a circular orbit of radius R_c about the Earth (radius R_e) and obtain the result

$$v_c = \left(\frac{gR_e^2}{R_c} \right)^{1/2}$$

in which g is the acceleration due to gravity, which we introduce in Section 2.2. If this result is correct, the dimensions on both sides of the equation must be identical. Let's see. First, we write down the combination of dimensions on the right side of the equation and reduce them as far as possible

$$\left(\frac{([L][T]^{-2})[L]^2}{[L]} \right)^{1/2} = ([L]^2[T]^{-2})^{1/2} = [L][T]^{-1}$$

The dimensions of the speed v_c are also $[L] [T]^{-1}$. The dimensions match; thus, the answer *could* be correct. It could also be incorrect. Dimensional analysis does not tell us unequivocally that it is correct. It can only tell us unequivocally that it is incorrect in those cases in which the dimensions fail to match.

Determining Relationships by Dimensional Analysis

Dimensional analysis can also be used as a way to obtain relationships between physical quantities without going through the labor of a more detailed analysis based on the laws of physics. As an example, consider the simple pendulum, which we analyze in Example 3.2.2. It consists of a small bob of mass m attached to the end of a massless, rigid string of length l. When displaced from its equilibrium configuration, in which it hangs vertically with the mass at its lowest possible position, it swings to and fro because gravity tries to restore the mass to its minimum height above the ground. In the absence of friction, air resistance and all other dissipative forces, it continues to swing to and fro forever! The time it takes to return to any configuration and direction of motion is called its period, or the time τ it takes to execute one complete cycle of its motion. The question before us is: How does its period τ depend on any physical parameters that characterize the pendulum and its environment?

First, we list those parameters that could be relevant. Because we've postulated that the pendulum consists, in part, of an idealized string of zero mass and no flexibility, that it suffers no air resistance and no friction, we eliminate from consideration any factors that are derivable from them. That leaves only three: the mass m of the pendulum bob, the length l of the string, and the acceleration g due to gravity. The period of the pendulum has dimension $[T]$ and the combination of m, l, and g that equates to the period must have dimensions that reduce to $[T]$, also. In other words, the period of the pendulum τ depends on an algebraic combination of m, l, and g of the form

$$\tau \propto m^{\alpha} l^{\beta} g^{\gamma}$$

whose dimensional relationship must be

$$[T] = [M]^{\alpha} [L]^{\beta} ([L]^{\gamma}[T]^{-2\gamma})$$

Because there are no powers of $[M]$ on the left-hand side, $\alpha = 0$ and the mass of the pendulum bob is irrelevant. To match the dimension $[T]$ on both sides of the equation, $\gamma = -\frac{1}{2}$, and to match the dimension $[L]$, $\beta + \gamma = 0$, or $\beta = \frac{1}{2}$. Thus, we conclude that

$$\tau \propto \sqrt{\frac{l}{g}}$$

Dimensional analysis can be taken no further than this. It does not give us the constant of proportionality, but it does tell us how τ likely depends on l and g and it does tell us that the period is independent of the mass m of the bob. Moreover, a single measurement of the period of a pendulum of known length l, would give us the constant of proportionality.

We did leave out one other possible factor, the angle of the pendulum's swing. Could its value affect the period? Maybe, but dimensional analysis alone does not tell us. The angle of swing is a dimensionless quantity, and the period could conceivably depend on it in a myriad of ways. Indeed, we see in Example 3.7.1, that the angle does affect the period if the angular amplitude of the swing is large enough. Yet, what we have learned simply by applying dimensional analysis is quite remarkable. A more detailed analysis based on the laws of physics should yield a result that is consistent with the one obtained from simple dimensional analysis, or we should try to understand why it does not. Whenever we find ourselves faced with such a dilemma, we discover that there is a strong likelihood that we've fouled up the detailed analysis.

Dimensional analysis applied this way is not always so simple. Experience is usually required to zero in on the relevant variables and to make a guess of the relevant functional dependencies. In particular, when trigonometric functions are involved, their lack of dimensionality thwarts dimensional analysis. Be that as it may, it remains a valuable weapon of attack that all students should have in their arsenal.

1.3 | Vectors

The motion of dynamical systems is typically described in terms of two basic quantities: scalars and vectors. A *scalar* is a physical quantity that has magnitude only, such as the mass of an object. It is completely specified by a single number, in appropriate units. Its value is independent of any coordinates chosen to describe the motion of the system. Other familiar examples of scalars include density, volume, temperature, and energy. Mathematically, scalars are treated as real numbers. They obey all the normal algebraic rules of addition, subtraction, multiplication, division, and so on.

A *vector*, however, has both magnitude and direction, such as the displacement from one point in space to another. Unlike a scalar, a vector requires a set of numbers for its complete specification. The values of those numbers are, in general, coordinate system dependent. Besides displacement in space, other examples of vectors include velocity, acceleration, and force. Mathematically, vectors combine with each other according to the parallelogram rule of addition which we soon discuss.[4] The vector concept has led to the emergence of a branch of mathematics that has proved indispensable to the development of the subject of classical mechanics. Vectors provide a compact and elegant way of describing the behavior of even the most complicated physical systems. Furthermore, the use of vectors in the application of physical laws insures that the results we obtain are independent of our choice of coordinate system.

In most written work, a distinguishing mark, such as an arrow, customarily designates a vector, for example, $\vec{A}$. In this text, however, for the sake of simplicity, we denote vector quantities simply by boldface type, for example, **A.** We use ordinary italic type to represent scalars, for example, A.

A given vector **A** is specified by stating its magnitude and its direction relative to some arbitrarily chosen coordinate system. It is represented diagrammatically as a directed line segment, as shown in three-dimensional space in Figure 1.3.1.

A vector can also be specified as the set of its *components*, or projections onto the coordinate axes. For example, the set of three scalars, (A_x, A_y, A_z), shown in Figure 1.3.1, are the components of the vector **A** and are an equivalent representation. Thus, the equation

$$\mathbf{A} = (A_x, A_y, A_z) \tag{1.3.1}$$

implies that either the symbol **A** or the set of three components (A_x, A_y, A_z) referred to a particular coordinate system can be used to specify the vector. For example, if the vector **A** represents a displacement from a point P_1 (x_1, y_1, z_1) to the point P_2 (x_2, y_2, z_2), then its

---

[4] An example of a directed quantity that does not obey the rule for addition is a finite rotation of an object about a given axis. The reader can readily verify that two successive rotations about different axes do not produce the same result as a single rotation determined by the parallelogram rule. For the present, we shall not be concerned with such nonvector-directed quantities.

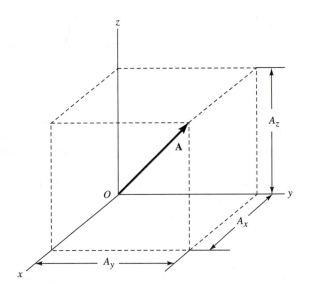

Figure 1.3.1   A vector **A** and its components in Cartesian coordinates.

three components are $A_x = x_2 - x_1$, $A_y = y_2 - y_1$, $A_z = z_2 - z_1$, and the equivalent representation of **A** is its set of three scalar components, $(x_2 - x_1, y_2 - y_1, z_2 - z_1)$. If **A** represents a force, then A_x is the x-component of the force, and so on.

If a particular discussion is limited to vectors in a plane, only two components are necessary for their specification. In general, one can define a mathematical space of any number of dimensions n. Thus, the set of n-numbers $(A_1, A_2, A_3, \ldots, A_n)$ represent a vector in an n-dimensional space. In this abstract sense, a vector is an ordered set of numbers.

We begin the study of vector algebra with some formal statements concerning vectors.

I. *Equality of Vectors*
The equation

$$\mathbf{A} = \mathbf{B} \tag{1.3.2}$$

or

$$(A_x, A_y, A_z) = (B_x, B_y, B_z)$$

is equivalent to the three equations

$$A_x = B_x \qquad A_y = B_y \qquad A_z = B_z$$

That is, two vectors are equal if, and only if, their respective components are equal. Geometrically, equal vectors are parallel and have the same length, but they do not necessarily have the same position. Equal vectors are shown in Figure 1.3.2. Though equal, they are physically separate. (Equal vectors are not necessarily equivalent in all respects. Thus, two vectorially equal forces acting at *different* points on an object may produce different mechanical effects.)

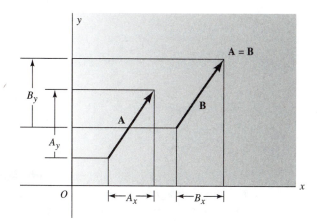

Figure 1.3.2 Illustration of equal vectors.

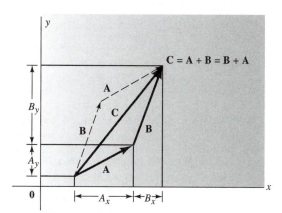

Figure 1.3.3 Addition of two vectors.

II. *Vector Addition*

The addition of two vectors is defined by the equation

$$\mathbf{A} + \mathbf{B} = (A_x, A_y, A_z) + (B_x, B_y, B_z) = (A_x + B_x, A_y + B_y, A_z + B_z) \qquad (1.3.3)$$

The sum of two vectors is a vector whose components are sums of the components of the given vectors. The geometric representation of the vector sum of two non-parallel vectors is the third side of a triangle, two sides of which are the given vectors. The vector sum is illustrated in Figure 1.3.3. The sum is also given by the parallelogram rule, as shown in the figure. The vector sum is defined, however, according to the above equation even if the vectors do not have a common point.

III. *Multiplication by a Scalar*

If c is a scalar and $\mathbf{A}$ is a vector,

$$c\mathbf{A} = c(A_x, A_y, A_z) = (cA_x, cA_y, cA_z) = \mathbf{A}c \qquad (1.3.4)$$

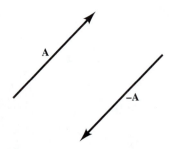

Figure 1.3.4 The negative of a vector.

The product $c\mathbf{A}$ is a vector whose components are c times those of $\mathbf{A}$. Geometrically, the vector $c\mathbf{A}$ is parallel to $\mathbf{A}$ and is c times the length of $\mathbf{A}$. When $c = -1$, the vector $-\mathbf{A}$ is one whose direction is the reverse of that of $\mathbf{A}$, as shown in Figure 13.4.

IV. *Vector Subtraction*
Subtraction is defined as follows:

$$\mathbf{A} - \mathbf{B} = \mathbf{A} + (-1)\mathbf{B} = (A_x - B_x, A_y - B_y, A_z - B_z) \tag{1.3.5}$$

That is, subtraction of a given vector $\mathbf{B}$ from the vector $\mathbf{A}$ is equivalent to adding $-\mathbf{B}$ to $\mathbf{A}$.

V. *The Null Vector*
The vector $\mathbf{O} = (0,0,0)$ is called the *null* vector. The direction of the null vector is undefined. From (IV) it follows that $\mathbf{A} - \mathbf{A} = \mathbf{O}$. Because there can be no confusion when the null vector is denoted by a zero, we shall hereafter use the notation $\mathbf{O} = 0$.

VI. *The Commutative Law of Addition*
This law holds for vectors; that is,

$$\mathbf{A} + \mathbf{B} = \mathbf{B} + \mathbf{A} \tag{1.3.6}$$

because $A_x + B_x = B_x + A_x$, and similarly for the y and z components.

VII. *The Associative Law*
The associative law is also true, because

$$\begin{aligned}
\mathbf{A} + (\mathbf{B} + \mathbf{C}) &= (A_x + (B_x + C_x), A_y + (B_y + C_y), A_z + (B_z + C_z)) \\
&= ((A_x + B_x) + C_x, (A_y + B_y) + C_y, (A_z + B_z) + C_z) \\
&= (\mathbf{A} + \mathbf{B}) + \mathbf{C}
\end{aligned} \tag{1.3.7}$$

VIII. *The Distributive Law*
Under multiplication by a scalar, the distributive law is valid because, from (II) and (III),

$$\begin{aligned}
c(\mathbf{A} + \mathbf{B}) &= c(A_x + B_x, A_y + B_y, A_z + B_z) \\
&= (c(A_x + B_x), c(A_y + B_y), c(A_z + B_z)) \\
&= (cA_z + cB_x, cA_y + cB_y, cA_z + cB_z) \\
&= c\mathbf{A}_x + c\mathbf{B}
\end{aligned} \tag{1.3.8}$$

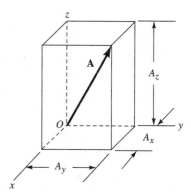

Figure 1.3.5 Magnitude of a vector **A**:
$A = (A_x^2 + A_y^2 + A_z^2)^{1/2}$.

Thus, vectors obey the rules of ordinary algebra as far as the above operations are concerned.

IX. *Magnitude of a Vector*

The magnitude of a vector **A,** denoted by $|\mathbf{A}|$ or by A, is defined as the square root of the sum of the squares of the components, namely,

$$A = |\mathbf{A}| = \left(A_x^2 + A_y^2 + A_z^2\right)^{1/2} \tag{1.3.9}$$

where the positive root is understood. Geometrically, the magnitude of a vector is its length, that is, the length of the diagonal of the rectangular parallelepiped whose sides are A_x, A_y, and A_z, expressed in appropriate units. See Figure 1.3.5.

X. *Unit Coordinate Vectors*

A *unit vector* is a vector whose magnitude is unity. Unit vectors are often designated by the symbol **e,** from the German word *Einheit*. The three unit vectors

$$\mathbf{e}_x = (1,0,0) \qquad \mathbf{e}_y = (0,1,0) \qquad \mathbf{e}_z = (0,0,1) \tag{1.3.10}$$

are called *unit coordinate vectors* or *basis vectors*. In terms of basis vectors, any vector can be expressed as a vector sum of components as follows:

$$\begin{aligned}
\mathbf{A} = (A_x, A_y, A_z) &= (A_x, 0, 0) + (0, A_y, 0) + (0, 0, A_z) \\
&= A_x(1,0,0) + A_y(0,1,0) + A_z(0,0,1) \\
&= \mathbf{e}_x A_x + \mathbf{e}_y A_y + \mathbf{e}_z A_z
\end{aligned} \tag{1.3.11}$$

A widely used notation for Cartesian unit vectors uses the letters **i, j,** and **k,** namely,

$$\mathbf{i} = \mathbf{e}_x \qquad \mathbf{j} = \mathbf{e}_y \qquad \mathbf{k} = \mathbf{e}_z \tag{1.3.12}$$

We shall usually employ this notation hereafter.

The directions of the Cartesian unit vectors are defined by the orthogonal coordinate axes, as shown in Figure 1.3.6. They form a right-handed or a left-handed triad,

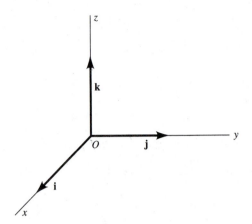

Figure 1.3.6   The unit vectors **ijk**.

depending on which type of coordinate system is used. It is customary to use right-handed coordinate systems. The system shown in Figure 1.3.6 is right-handed. (The handedness of coordinate systems is defined in Section 1.5.)

EXAMPLE 1.3.1

Find the sum and the magnitude of the sum of the two vectors $\mathbf{A} = (1,0,2)$ and $\mathbf{B} = (0,1,1)$.

Solution:

Adding components, we have $\mathbf{A} + \mathbf{B} = (1,0,2) + (0,1,1) = (1,1,3)$.

$$|\mathbf{A} + \mathbf{B}| = (1 + 1 + 9)^{1/2} = \sqrt{11}$$

EXAMPLE 1.3.2

For the above two vectors, express the difference in **ijk** form.

Solution:

Subtracting components, we have

$$\mathbf{A} - \mathbf{B} = (1,-1,1) = \mathbf{i} - \mathbf{j} + \mathbf{k}$$

EXAMPLE 1.3.3

A helicopter flies 100 m vertically upward, then 500 m horizontally east, then 1000 m horizontally north. How far is it from a second helicopter that started from the same point and flew 200 m upward, 100 m west, and 500 m north?

Solution:

Choosing up, east, and north as basis directions, the final position of the first helicopter is expressed vectorially as $\mathbf{A} = (100, 500, 1000)$ and the second as $\mathbf{B} = (200, -100, 500)$, in meters. Hence, the distance between the final positions is given by the expression

$$|\mathbf{A} - \mathbf{B}| = |((100 - 200), (500 + 100), (1000 - 500))| \text{ m}$$
$$= (100^2 + 600^2 + 500^2)^{1/2} \text{ m}$$
$$= 787.4 \text{ m}$$

1.4 | The Scalar Product

Given two vectors $\mathbf{A}$ and $\mathbf{B}$, the scalar product or "dot" product, $\mathbf{A} \cdot \mathbf{B}$, is the scalar defined by the equation

$$\mathbf{A} \cdot \mathbf{B} = A_x B_x + A_y B_y + A_z B_z \tag{1.4.1}$$

From the above definition, scalar multiplication is *commutative*,

$$\mathbf{A} \cdot \mathbf{B} = \mathbf{B} \cdot \mathbf{A} \tag{1.4.2}$$

because $A_x B_x = B_x A_x$, and so on. It is also *distributive*,

$$\mathbf{A} \cdot (\mathbf{B} + \mathbf{C}) = \mathbf{A} \cdot \mathbf{B} + \mathbf{A} \cdot \mathbf{C} \tag{1.4.3}$$

because if we apply the definition (1.4.1) in detail,

$$\mathbf{A} \cdot (\mathbf{B} + \mathbf{C}) = A_x(B_x + C_x) + A_y(B_y + C_y) + A_z(B_z + C_z)$$
$$= A_x B_x + A_y B_y + A_z B_z + A_x C_x + A_y C_y + A_z C_z \tag{1.4.4}$$
$$= \mathbf{A} \cdot \mathbf{B} + \mathbf{A} \cdot \mathbf{C}$$

The dot product $\mathbf{A} \cdot \mathbf{B}$ has a simple geometrical interpretation and can be used to calculate the angle θ between those two vectors. For example, shown in Figure 1.4.1 are the two vectors $\mathbf{A}$ and $\mathbf{B}$ separated by an angle θ, along with an x', y', z' coordinate system arbitrarily chosen as a basis for those vectors. However, because the quantity $\mathbf{A} \cdot \mathbf{B}$ is a scalar, its value is independent of choice of coordinates. With no loss of generality, we can rotate the x', y', z' system into an x, y, z coordinate system, such that the x-axis is aligned with the vector $\mathbf{A}$ and the z-axis is perpendicular to the plane defined by the two vectors. This coordinate system is also shown in Figure 1.4.1. The components of the vectors, and their dot product, are much simpler to evaluate in this system. The vector $\mathbf{A}$ is expressed as $(A, 0, 0)$ and the vector $\mathbf{B}$ as $(B_x, B_y, 0)$ or $(B \cos \theta, B \sin \theta, 0)$. Thus,

$$\mathbf{A} \cdot \mathbf{B} = A_x B_x = A(B \cos \theta) = |\mathbf{A}| \, |\mathbf{B}| \cos \theta \tag{1.4.5}$$

Geometrically, $B \cos \theta$ is simply the projection of $\mathbf{B}$ onto $\mathbf{A}$. If we had aligned the x-axis along $\mathbf{B}$, we would have obtained the same result but with the geometrical interpretation that $\mathbf{A} \cdot \mathbf{B}$ is now the projection of $\mathbf{A}$ onto $\mathbf{B}$ times the length of $\mathbf{B}$. Thus, $\mathbf{A} \cdot \mathbf{B}$ can be interpreted as either the projection of $\mathbf{A}$ onto $\mathbf{B}$ times the length of $\mathbf{B}$ or that of $\mathbf{B}$ onto $\mathbf{A}$ times the length of $\mathbf{A}$. Either interpretation is correct. Perhaps more importantly, we

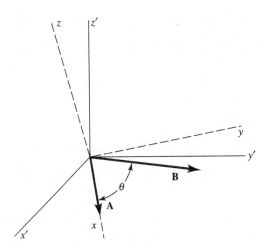

Figure 1.4.1 Evaluating a dot product between two vectors.

can see that we have just proved that the cosine of the angle between two line segments is given by

$$\cos\theta = \frac{\mathbf{A}\cdot\mathbf{B}}{|\mathbf{A}||\mathbf{B}|} = \frac{\mathbf{A}\cdot\mathbf{B}}{AB} \tag{1.4.6}$$

This last equation may be regarded as an alternative definition of the dot product.

(**Note:** *If* **A** · **B** *is equal to zero and neither* **A** *nor* **B** *is null, then* $\cos\theta$ *is zero and* **A** *is perpendicular to* **B**.)

The square of the magnitude of a vector **A** is given by the dot product of **A** with itself,

$$A^2 = |\mathbf{A}|^2 = \mathbf{A}\cdot\mathbf{A} \tag{1.4.7}$$

From the definitions of the unit coordinate vectors **i**, **j**, and **k**, it is clear that the following relations hold:

$$\mathbf{i}\cdot\mathbf{i} = \mathbf{j}\cdot\mathbf{j} = \mathbf{k}\cdot\mathbf{k} = 1 \tag{1.4.8}$$
$$\mathbf{i}\cdot\mathbf{j} = \mathbf{i}\cdot\mathbf{k} = \mathbf{j}\cdot\mathbf{k} = 0$$

Expressing Any Vector as the Product of Its Magnitude by a Unit Vector: Projection

Consider the equation

$$\mathbf{A} = \mathbf{i}A_x + \mathbf{j}A_y + \mathbf{k}A_z \tag{1.4.9}$$

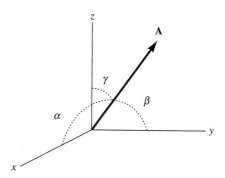

Figure 1.4.2   Direction angles α, β, γ of a
vector.

Multiply and divide on the right by the magnitude of **A**:

$$\mathbf{A} = A\left(\mathbf{i}\,\frac{A_x}{A} + \mathbf{j}\,\frac{A_y}{A} + \mathbf{k}\,\frac{A_z}{A}\right) \qquad (1.4.10)$$

Now $A_x/A = \cos\alpha$, $A_y/A = \cos\beta$, and $A_z/A = \cos\gamma$ are the *direction cosines* of the vector
A, and α, β, and γ are the *direction angles.* Thus, we can write

$$\mathbf{A} = A(\mathbf{i}\,\cos\alpha + \mathbf{j}\,\cos\beta + \mathbf{k}\,\cos\gamma) = A(\cos\alpha,\,\cos\beta,\,\cos\gamma) \qquad (1.4.11a)$$

or

$$\mathbf{A} = A\mathbf{n} \qquad (1.4.11b)$$

where **n** is a unit vector whose components are $\cos\alpha$, $\cos\beta$, and $\cos\gamma$. See Figure 1.4.2.
Consider any other vector **B**. Clearly, the projection of **B** on **A** is just

$$B \cos\theta = \frac{\mathbf{B} \cdot \mathbf{A}}{A} = \mathbf{B} \cdot \mathbf{n} \qquad (1.4.12)$$

where θ is the angle between **A** and **B**.

EXAMPLE 1.4.1

Component of a Vector: Work

As an example of the dot product, suppose that an object under the action of a constant
force[5] undergoes a linear displacement $\Delta\mathbf{s}$, as shown in Figure 1.4.3. By definition, the
work ΔW done by the force is given by the product of the component of the force **F** in
the direction of $\Delta\mathbf{s}$, multiplied by the magnitude Δs of the displacement; that is,

$$\Delta W = (F \cos\theta)\,\Delta s$$

---

[5] The concept of force is discussed in Chapter 2.

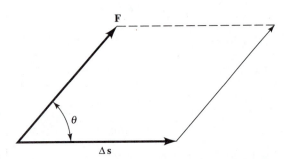

Figure 1.4.3 A force acting on a body undergoing a displacement.

where θ is the angle between $\mathbf{F}$ and $\Delta\mathbf{s}$. But the expression on the right is just the dot product of $\mathbf{F}$ and $\Delta\mathbf{s}$, that is,

$$\Delta W = \mathbf{F} \cdot \Delta\mathbf{s}$$

EXAMPLE 1.4.2

Law of Cosines

Consider the triangle whose sides are $\mathbf{A}$, $\mathbf{B}$, and $\mathbf{C}$, as shown in Figure 1.4.4. Then $\mathbf{C} = \mathbf{A} + \mathbf{B}$. Take the dot product of $\mathbf{C}$ with itself,

$$\mathbf{C} \cdot \mathbf{C} = (\mathbf{A} + \mathbf{B}) \cdot (\mathbf{A} + \mathbf{B})$$
$$= \mathbf{A} \cdot \mathbf{A} + 2\mathbf{A} \cdot \mathbf{B} + \mathbf{B} \cdot \mathbf{B}$$

The second step follows from the application of the rules in Equations 1.4.2 and 1.4.3. Replace $\mathbf{A} \cdot \mathbf{B}$ with $AB \cos \theta$ to obtain

$$C^2 = A^2 + 2AB \cos \theta + B^2$$

which is the familiar law of cosines.

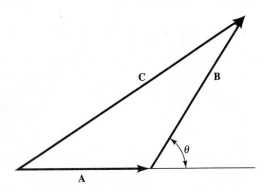

Figure 1.4.4 The law of cosines.

EXAMPLE 1.4.3

Find the cosine of the angle between a long diagonal and an adjacent face diagonal of a cube.

Solution:

We can represent the two diagonals in question by the vectors $\mathbf{A} = (1, 1, 1)$ and $\mathbf{B} = (1, 1, 0)$. Hence, from Equations 1.4.1 and 1.4.6,

$$\cos\theta = \frac{\mathbf{A}\cdot\mathbf{B}}{AB} = \frac{1+1+0}{\sqrt{3}\sqrt{2}} = \sqrt{\frac{2}{3}} = 0.8165$$

EXAMPLE 1.4.4

The vector $a\mathbf{i} + \mathbf{j} - \mathbf{k}$ is perpendicular to the vector $\mathbf{i} + 2\mathbf{j} - 3\mathbf{k}$. What is the value of a?

Solution:

If the vectors are perpendicular to each other, their dot product must vanish ($\cos 90° = 0$).

$$(a\mathbf{i} + \mathbf{j} - \mathbf{k}) \cdot (\mathbf{i} + 2\mathbf{j} - 3\mathbf{k}) = a + 2 + 3 = a + 5 = 0$$

Therefore,

$$a = -5$$

1.5 | The Vector Product

Given two vectors $\mathbf{A}$ and $\mathbf{B}$, the vector product or cross product, $\mathbf{A} \times \mathbf{B}$, is defined as the vector whose components are given by the equation

$$\mathbf{A} \times \mathbf{B} = (A_y B_z - A_z B_y, A_z B_x - A_x B_z, A_x B_y - A_y B_x) \tag{1.5.1}$$

It can be shown that the following rules hold for cross multiplication:

$$\mathbf{A} \times \mathbf{B} = -\mathbf{B} \times \mathbf{A} \tag{1.5.2}$$

$$\mathbf{A} \times (\mathbf{B} + \mathbf{C}) = \mathbf{A} \times \mathbf{B} + \mathbf{A} \times \mathbf{C} \tag{1.5.3}$$

$$n(\mathbf{A} \times \mathbf{B}) = (n\mathbf{A}) \times \mathbf{B} = \mathbf{A} \times (n\mathbf{B}) \tag{1.5.4}$$

The proofs of these follow directly from the definition and are left as an exercise.

(**Note:** *The first equation states that the cross product is anticommutative.*)

According to the definitions of the unit coordinate vectors (Section 1.3), it follows that

$$\mathbf{i} \times \mathbf{i} = \mathbf{j} \times \mathbf{j} = \mathbf{k} \times \mathbf{k} = 0$$
$$\mathbf{j} \times \mathbf{k} = \mathbf{i} = -\mathbf{k} \times \mathbf{j}$$
$$\mathbf{i} \times \mathbf{j} = \mathbf{k} = -\mathbf{j} \times \mathbf{i} \qquad (1.5.5)$$
$$\mathbf{k} \times \mathbf{i} = \mathbf{j} = -\mathbf{i} \times \mathbf{k}$$

These latter three relations define a right-handed triad. For example,

$$\mathbf{i} \times \mathbf{j} = (0 - 0, 0 - 0, 1 - 0) = (0,0,1) = \mathbf{k} \qquad (1.5.6)$$

The remaining equations are proved in a similar manner.
    The cross product expressed in **ijk** form is

$$\mathbf{A} \times \mathbf{B} = \mathbf{i}(A_y B_z - A_z B_y) + \mathbf{j}(A_z B_x - A_x B_z) + \mathbf{k}(A_x B_y - A_y B_x) \qquad (1.5.7)$$

Each term in parentheses is equal to a determinant,

$$\mathbf{A} \times \mathbf{B} = \mathbf{i} \begin{vmatrix} A_y & A_z \\ B_y & B_z \end{vmatrix} + \mathbf{j} \begin{vmatrix} A_z & A_x \\ B_z & B_x \end{vmatrix} + \mathbf{k} \begin{vmatrix} A_x & A_y \\ B_x & B_y \end{vmatrix} \qquad (1.5.8)$$

and finally

$$\mathbf{A} \times \mathbf{B} = \begin{vmatrix} \mathbf{i} & \mathbf{j} & \mathbf{k} \\ A_x & A_y & A_z \\ B_x & B_y & B_z \end{vmatrix} \qquad (1.5.9)$$

which is verified by expansion. The determinant form is a convenient aid for remembering the definition of the cross product. From the properties of determinants, if **A** is parallel to **B**—that is, if $\mathbf{A} = c\mathbf{B}$—then the two lower rows of the determinant are proportional and so the determinant is null. Thus, the cross product of two parallel vectors is null.
    Let us calculate the magnitude of the cross product. We have

$$|\mathbf{A} \times \mathbf{B}|^2 = (A_y B_z - A_z B_y)^2 + (A_z B_x - A_x B_z)^2 + (A_x B_y - A_y B_x)^2 \qquad (1.5.10)$$

This can be reduced to

$$|\mathbf{A} \times \mathbf{B}|^2 = \left(A_x^2 + A_y^2 + A_z^2\right)\left(B_x^2 + B_y^2 + B_z^2\right) - (A_x B_x + A_y B_y + A_z B_z)^2 \qquad (1.5.11)$$

or, from the definition of the dot product, the above equation may be written in the form

$$|\mathbf{A} \times \mathbf{B}|^2 = A^2 B^2 - (\mathbf{A} \cdot \mathbf{B})^2 \qquad (1.5.12)$$

Taking the square root of both sides of Equation 1.15.12 and using Equation 1.4.6, we can express the magnitude of the cross product as

$$|\mathbf{A} \times \mathbf{B}| = AB(1 - \cos^2 \theta)^{1/2} = AB \sin \theta \qquad (1.5.13)$$

where θ is the angle between $\mathbf{A}$ and $\mathbf{B}$.

To interpret the cross product geometrically, we observe that the vector $\mathbf{C} = \mathbf{A} \times \mathbf{B}$ is perpendicular to both $\mathbf{A}$ and to $\mathbf{B}$ because

$$\begin{aligned}
\mathbf{A} \cdot \mathbf{C} &= A_x C_x + A_y C_y + A_z C_z \\
&= A_x(A_y B_z - A_z B_y) + A_y(A_z B_x - A_x B_z) + A_z(A_x B_y - A_y B_x) \qquad (1.5.14) \\
&= 0
\end{aligned}$$

Similarly, $\mathbf{B} \cdot \mathbf{C} = 0$; thus, the vector $\mathbf{C}$ is perpendicular to the plane containing the vectors $\mathbf{A}$ and $\mathbf{B}$.

The sense of the vector $\mathbf{C} = \mathbf{A} \times \mathbf{B}$ is determined from the requirement that the three vectors $\mathbf{A}$, $\mathbf{B}$, and $\mathbf{C}$ form a right-handed triad, as shown in Figure 1.5.1. (This is consistent with the previously established result that in the right-handed triad $\mathbf{ijk}$ we have $\mathbf{i} \times \mathbf{j} = \mathbf{k}$.) Therefore, from Equation 1.5.13 we see that we can write

$$\mathbf{A} \times \mathbf{B} = (AB \sin \theta)\mathbf{n} \qquad (1.5.15)$$

where $\mathbf{n}$ is a unit vector normal to the plane of the two vectors $\mathbf{A}$ and $\mathbf{B}$. The sense of $\mathbf{n}$ is given by the *right-hand rule,* that is, the direction of advancement of a right-handed screw rotated from the positive direction of $\mathbf{A}$ to that of $\mathbf{B}$ through the smallest angle between them, as illustrated in Figure 1.5.1. Equation 1.5.15 may be regarded as an alternative definition of the cross product in a right-handed coordinate system.

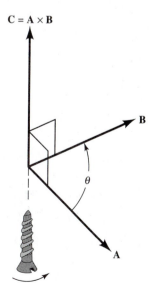

Figure 1.5.1   The cross product of two vectors.

EXAMPLE 1.5.1

Given the two vectors $\mathbf{A} = 2\mathbf{i} + \mathbf{j} - \mathbf{k}$, $\mathbf{B} = \mathbf{i} - \mathbf{j} + 2\mathbf{k}$, find $\mathbf{A} \times \mathbf{B}$.

Solution:

In this case it is convenient to use the determinant form

$$\mathbf{A} \times \mathbf{B} = \begin{vmatrix} \mathbf{i} & \mathbf{j} & \mathbf{k} \\ 2 & 1 & -1 \\ 1 & -1 & 2 \end{vmatrix} = \mathbf{i}(2-1) + \mathbf{j}(-1-4) + \mathbf{k}(-2-1)$$

$$= \mathbf{i} - 5\mathbf{j} - 3\mathbf{k}$$

EXAMPLE 1.5.2

Find a unit vector normal to the plane containing the two vectors $\mathbf{A}$ and $\mathbf{B}$ above.

Solution:

$$\mathbf{n} = \frac{\mathbf{A} \times \mathbf{B}}{|\mathbf{A} \times \mathbf{B}|} = \frac{\mathbf{i} - 5\mathbf{j} - 3\mathbf{k}}{[1^2 + 5^2 + 3^2]^{1/2}}$$

$$= \frac{\mathbf{i}}{\sqrt{35}} - \frac{5\mathbf{j}}{\sqrt{35}} - \frac{3\mathbf{k}}{\sqrt{35}}$$

EXAMPLE 1.5.3

Show by direct evaluation that $\mathbf{A} \times \mathbf{B}$ is a vector with direction perpendicular to $\mathbf{A}$ and $\mathbf{B}$ and magnitude $AB \sin\theta$.

Solution:

Use the frame of reference discussed for Figure 1.4.1 in which the vectors $\mathbf{A}$ and $\mathbf{B}$ are defined to be in the x, y plane; $\mathbf{A}$ is given by $(A, 0, 0)$ and $\mathbf{B}$ is given by $(B \cos\theta, B \sin\theta, 0)$. Then

$$\mathbf{A} \times \mathbf{B} = \begin{vmatrix} \mathbf{i} & \mathbf{j} & \mathbf{k} \\ A & 0 & 0 \\ B \cos\theta & B \sin\theta & 0 \end{vmatrix} = \mathbf{k} AB \sin\theta$$

1.6 | An Example of the Cross Product: Moment of a Force

Moments of force, or *torques*, are represented by cross products. Let a force $\mathbf{F}$ act at a point $P(x, y, z)$, as shown in Figure 1.6.1, and let the vector $\mathbf{OP}$ be designated by $\mathbf{r}$; that is,

$$\mathbf{OP} = \mathbf{r} = \mathbf{i}x + \mathbf{j}y + \mathbf{k}z \tag{1.6.1}$$

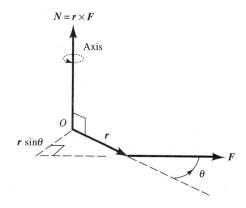

Figure 1.6.1 Illustration of the moment of
a force about a point O.

The moment $\mathbf{N}$ of force, or the *torque* $\mathbf{N}$, about a given point O is defined as the cross
product

$$\mathbf{N} = \mathbf{r} \times \mathbf{F} \tag{1.6.2}$$

Thus, the moment of a force about a point is a vector quantity having a magnitude and a
direction. If a single force is applied at a point P on a body that is initially at rest and is
free to turn about a fixed point O as a pivot, then the body tends to rotate. The axis of this
rotation is perpendicular to the force $\mathbf{F}$, and it is also perpendicular to the line OP; therefore, the direction of the torque vector $\mathbf{N}$ is along the axis of rotation.

The magnitude of the torque is given by

$$|\mathbf{N}| = |\mathbf{r} \times \mathbf{F}| = rF \sin \theta \tag{1.6.3}$$

in which θ is the angle between $\mathbf{r}$ and $\mathbf{F}$. Thus, $|\mathbf{N}|$ can be regarded as the product of the
magnitude of the force and the quantity $r \sin \theta$, which is just the perpendicular distance
from the line of action of the force to the point O.

When several forces are applied to a single body at different points, the moments add
vectorially. This follows from the distributive law of vector multiplication. The condition
for rotational equilibrium is that the vector sum of all the moments is zero:

$$\sum_i (\mathbf{r}_i \times \mathbf{F}_i) = \sum_i \mathbf{N}_i = 0 \tag{1.6.4}$$

A more complete discussion of force moments is given in Chapters 8 and 9.

1.7 | Triple Products

The expression

$$\mathbf{A} \cdot (\mathbf{B} \times \mathbf{C})$$

is called the *scalar triple product* of $\mathbf{A}$, $\mathbf{B}$, and $\mathbf{C}$. It is a scalar because it is the dot product of two vectors. Referring to the determinant expressions for the cross product,

Equations 1.5.8 and 1.5.9, we see that the scalar triple product may be written

$$\mathbf{A} \cdot (\mathbf{B} \times \mathbf{C}) = \begin{vmatrix} A_x & A_y & A_z \\ B_x & B_y & B_z \\ C_x & C_y & C_z \end{vmatrix} \tag{1.7.1}$$

Because the exchange of the terms of two rows or of two columns of a determinant changes its sign but not its absolute value, we can derive the following useful equation:

$$\mathbf{A} \cdot (\mathbf{B} \times \mathbf{C}) = (\mathbf{A} \times \mathbf{B}) \cdot \mathbf{C} \tag{1.7.2}$$

Thus, the dot and the cross may be interchanged in the scalar triple product.

The expression

$$\mathbf{A} \times (\mathbf{B} \times \mathbf{C})$$

is called the *vector triple product*. It is left for the student to prove that the following equation holds for the vector triple product:

$$\mathbf{A} \times (\mathbf{B} \times \mathbf{C}) = \mathbf{B}(\mathbf{A} \cdot \mathbf{C}) - \mathbf{C}(\mathbf{A} \cdot \mathbf{B}) \tag{1.7.3}$$

This last result can be remembered simply as the "back minus cab" rule.

Vector triple products are particularly useful in the study of rotating coordinate systems and rotations of rigid bodies, which we take up in later chapters. A geometric application is given in Problem 1.12 at the end of this chapter.

EXAMPLE 1.7.1

Given the three vectors $\mathbf{A} = \mathbf{i}$, $\mathbf{B} = \mathbf{i} - \mathbf{j}$, and $\mathbf{C} = \mathbf{k}$, find $\mathbf{A} \cdot (\mathbf{B} \times \mathbf{C})$.

Solution:

Using the determinant expression, Equation 1.7.1, we have

$$\mathbf{A} \cdot (\mathbf{B} \times \mathbf{C}) = \begin{vmatrix} 1 & 0 & 0 \\ 1 & -1 & 0 \\ 0 & 0 & 1 \end{vmatrix} = 1(-1 + 0) = -1$$

EXAMPLE 1.7.2

Find $\mathbf{A} \times (\mathbf{B} \times \mathbf{C})$ above.

Solution:

From Equation 1.7.3 we have

$$\mathbf{A} \times (\mathbf{B} \times \mathbf{C}) = \mathbf{B}(\mathbf{A} \cdot \mathbf{C}) - \mathbf{C}(\mathbf{A} \cdot \mathbf{B}) = (\mathbf{i} - \mathbf{j})0 - \mathbf{k}(1 - 0) = -\mathbf{k}$$

EXAMPLE 1.7.3

Show that the vector triple product is nonassociative.

Solution:

$$(\mathbf{a} \times \mathbf{b}) \times \mathbf{c} = -\mathbf{c} \times (\mathbf{a} \times \mathbf{b}) = -\mathbf{a}(\mathbf{c} \cdot \mathbf{b}) + \mathbf{b}(\mathbf{c} \cdot \mathbf{a})$$

$$\mathbf{a} \times (\mathbf{b} \times \mathbf{c}) - (\mathbf{a} \times \mathbf{b}) \times \mathbf{c} = \mathbf{a}(\mathbf{c} \cdot \mathbf{b}) - \mathbf{c}(\mathbf{a} \cdot \mathbf{b})$$

which is not necessarily zero.

1.8 | Change of Coordinate System: The Transformation Matrix

In this section we show how to represent a vector in different coordinate systems. Consider the vector **A** expressed relative to the triad **ijk**:

$$\mathbf{A} = \mathbf{i}A_x + \mathbf{j}A_y + \mathbf{k}A_z \tag{1.8.1}$$

Relative to a new triad **i′j′k′** having a different orientation from that of **ijk**, the same vector **A** is expressed as

$$\mathbf{A} = \mathbf{i}'A_{x'} + \mathbf{j}'A_{y'} + \mathbf{k}'A_{z'} \tag{1.8.2}$$

Now the dot product $\mathbf{A} \cdot \mathbf{i}'$ is just $A_{x'}$, that is, the projection of **A** on the unit vector **i′**. Thus, we may write

$$A_{x'} = \mathbf{A} \cdot \mathbf{i}' = (\mathbf{i} \cdot \mathbf{i}')A_x + (\mathbf{j} \cdot \mathbf{i}')A_y + (\mathbf{k} \cdot \mathbf{i}')A_z$$
$$A_{y'} = \mathbf{A} \cdot \mathbf{j}' = (\mathbf{i} \cdot \mathbf{j}')A_x + (\mathbf{j} \cdot \mathbf{j}')A_y + (\mathbf{k} \cdot \mathbf{j}')A_z \tag{1.8.3}$$
$$A_{z'} = \mathbf{A} \cdot \mathbf{k}' = (\mathbf{i} \cdot \mathbf{k}')A_x + (\mathbf{j} \cdot \mathbf{k}')A_y + (\mathbf{k} \cdot \mathbf{k}')A_z$$

The scalar products $(\mathbf{i} \cdot \mathbf{i}')$, $(\mathbf{i} \cdot \mathbf{j}')$, and so on are called the *coefficients of transformation*. They are equal to the direction cosines of the axes of the primed coordinate system relative to the unprimed system. The unprimed components are similarly expressed as

$$A_x = \mathbf{A} \cdot \mathbf{i} = (\mathbf{i}' \cdot \mathbf{i})A_{x'} + (\mathbf{j}' \cdot \mathbf{i})A_{y'} + (\mathbf{k}' \cdot \mathbf{i})A_{z'}$$
$$A_y = \mathbf{A} \cdot \mathbf{j} = (\mathbf{i}' \cdot \mathbf{j})A_{x'} + (\mathbf{j}' \cdot \mathbf{j})A_{y'} + (\mathbf{k}' \cdot \mathbf{j})A_{z'} \tag{1.8.4}$$
$$A_z = \mathbf{A} \cdot \mathbf{k} = (\mathbf{i}' \cdot \mathbf{k})A_{x'} + (\mathbf{j}' \cdot \mathbf{k})A_{y'} + (\mathbf{k}' \cdot \mathbf{k})A_{z'}$$

All the coefficients of transformation in Equation 1.8.4 also appear in Equation 1.8.3, because $\mathbf{i} \cdot \mathbf{i}' = \mathbf{i}' \cdot \mathbf{i}$ and so on, but those in the rows (equations) of Equation 1.8.4 appear in the columns of terms in Equation 1.8.3, and conversely. The transformation rules expressed in these two sets of equations are a general property of vectors. As a matter of fact, they constitute an alternative way of defining vectors.[6]

---

[6] See, for example, J. B. Marion and S. T. Thornton, *Classical Dynamics*, 5th ed., Brooks/Cole—Thomson Learning, Belmont, CA, 2004.

The equations of transformation are conveniently expressed in matrix notation.[7] Thus, Equation 1.8.3 is written

$$\begin{pmatrix} A_{x'} \\ A_{y'} \\ A_{z'} \end{pmatrix} = \begin{pmatrix} \mathbf{i} \cdot \mathbf{i'} & \mathbf{j} \cdot \mathbf{i'} & \mathbf{k} \cdot \mathbf{i'} \\ \mathbf{i} \cdot \mathbf{j'} & \mathbf{j} \cdot \mathbf{j'} & \mathbf{k} \cdot \mathbf{j'} \\ \mathbf{i} \cdot \mathbf{k'} & \mathbf{j} \cdot \mathbf{k'} & \mathbf{k} \cdot \mathbf{k'} \end{pmatrix} \begin{pmatrix} A_x \\ A_y \\ A_z \end{pmatrix} \tag{1.8.5}$$

The 3-by-3 matrix in Equation 1.8.5 is called the *transformation matrix*. One advantage of the matrix notation is that successive transformations are readily handled by means of matrix multiplication.

The application of a given transformation matrix to some vector $\mathbf{A}$ is also formally equivalent to rotating that vector within the unprimed (fixed) coordinate system, the components of the rotated vector being given by Equation 1.8.5. Thus, finite rotations can be represented by matrices. (Note that the sense of rotation of the vector in this context is opposite that of the rotation of the coordinate system in the previous context.)

From Example 1.8.2 the transformation matrix for a rotation about a different coordinate axis—say, the y-axis through an angle θ—is given by the matrix

$$\begin{pmatrix} \cos\theta & 0 & -\sin\theta \\ 0 & 1 & 0 \\ \sin\theta & 0 & \cos\theta \end{pmatrix}$$

Consequently, the matrix for the combination of two rotations, the first being about the z-axis (angle ϕ) and the second being about the new y'-axis (angle θ), is given by the matrix product

$$\begin{pmatrix} \cos\theta & 0 & -\sin\theta \\ 0 & 1 & 0 \\ \sin\theta & 0 & \cos\theta \end{pmatrix} \begin{pmatrix} \cos\phi & \sin\phi & 0 \\ -\sin\phi & \cos\phi & 0 \\ 0 & 0 & 1 \end{pmatrix} = \begin{pmatrix} \cos\theta\cos\phi & \cos\theta\sin\phi & -\sin\theta \\ -\sin\phi & \cos\phi & 0 \\ \sin\theta\cos\phi & \sin\theta\sin\phi & \cos\theta \end{pmatrix} \tag{1.8.6}$$

Now matrix multiplication is, in general, noncommutative; therefore, we might expect that the result would be different if the order of the rotations, and, therefore, the order of the matrix multiplication, were reversed. This turns out to be the case, which the reader can verify. This is in keeping with a remark made earlier, namely, that finite rotations do not obey the law of vector addition and, hence, are not vectors even though a single rotation has a direction (the axis) and a magnitude (the angle of rotation). However, we show later that infinitesimal rotations do obey the law of vector addition and can be represented by vectors.

---

[7] A brief review of matrices is given in Appendix H.

EXAMPLE 1.8.1

Express the vector $\mathbf{A} = 3\mathbf{i} + 2\mathbf{j} + \mathbf{k}$ in terms of the triad $\mathbf{i'j'k'}$, where the $x'y'$–axes are rotated 45° around the z-axis, with the z- and z´-axes coinciding, as shown in Figure 1.8.1. Referring to the figure, we have for the coefficients of transformation $\mathbf{i} \cdot \mathbf{i'} = \cos 45°$ and so on; hence,

$$\mathbf{i} \cdot \mathbf{i'} = 1/\sqrt{2} \qquad \mathbf{j} \cdot \mathbf{i'} = 1/\sqrt{2} \qquad \mathbf{k} \cdot \mathbf{i'} = 0$$
$$\mathbf{i} \cdot \mathbf{j'} = -1/\sqrt{2} \qquad \mathbf{j} \cdot \mathbf{j'} = 1/\sqrt{2} \qquad \mathbf{k} \cdot \mathbf{j'} = 0$$
$$\mathbf{i} \cdot \mathbf{k'} = 0 \qquad \mathbf{j} \cdot \mathbf{k'} = 0 \qquad \mathbf{k} \cdot \mathbf{k'} = 1$$

These give

$$A_{x'} = \frac{3}{\sqrt{2}} + \frac{2}{\sqrt{2}} = \frac{5}{\sqrt{2}} \qquad A_{y'} = \frac{-3}{\sqrt{2}} + \frac{2}{\sqrt{2}} = \frac{-1}{\sqrt{2}} \qquad A_{z'} = 1$$

so that, in the primed system, the vector $\mathbf{A}$ is given by

$$\mathbf{A} = \frac{5}{\sqrt{2}}\mathbf{i'} - \frac{1}{\sqrt{2}}\mathbf{j'} + \mathbf{k'}$$

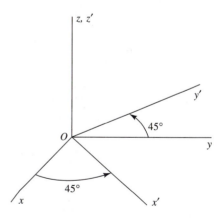

Figure 1.8.1   Rotated axes.

EXAMPLE 1.8.2

Find the transformation matrix for a rotation of the primed coordinate system through an angle ϕ about the z-axis. (Example 1.8.1 is a special case of this.) We have

$$\mathbf{i} \cdot \mathbf{i'} = \mathbf{j} \cdot \mathbf{j'} = \cos\phi$$
$$\mathbf{j} \cdot \mathbf{i'} = -\mathbf{i} \cdot \mathbf{j'} = \sin\phi$$
$$\mathbf{k} \cdot \mathbf{k'} = 1$$

and all other dot products are zero; hence, the transformation matrix is

$$\begin{pmatrix} \cos\phi & \sin\phi & 0 \\ -\sin\phi & \cos\phi & 0 \\ 0 & 0 & 1 \end{pmatrix}$$

EXAMPLE 1.8.3

Orthogonal Transformations

In more advanced texts, vectors are defined as quantities whose components change according to the rules of *orthogonal* transformations. The development of this subject lies outside the scope of this text; however, we give a simple example of such a transformation that the student may gain some appreciation for the elegance of this more abstract definition of vectors. The rotation of a Cartesian coordinate system is an example of an orthogonal transformation. Here we show how the components of a vector transform when the Cartesian coordinate system in which its components are expressed is rotated through some angle θ and then back again.

Let us take the velocity $\mathbf{v}$ of a projectile of mass m traveling through space along a parabolic trajectory as an example of the vector.[8] In Figure 1.8.2, we show the position and velocity of the projectile at some instant of time t. The direction of $\mathbf{v}$ is tangent to the trajectory of the projectile and designates its instantaneous direction of travel. Because the motion takes place in two dimensions only, we can specify the velocity in terms of its components along the x- and y-axes of a two-dimensional Cartesian coordinate system. We can also specify the velocity of the projectile in terms of components referred to an $x'y'$ coordinate system obtained by rotating the xy system through the angle θ. We choose an angle of rotation θ that aligns the x'-axis with the direction of the velocity vector.

We express the coordinate rotation in terms of the transformation matrix, defined in Equation 1.8.5. We write all vectors as column matrices; thus, the vector $\mathbf{v} = (v_x, v_y)$ is

$$\mathbf{v} = \begin{pmatrix} v_x \\ v_y \end{pmatrix} = \begin{pmatrix} v\cos\theta \\ v\sin\theta \end{pmatrix}$$

Given the components in one coordinate system, we can calculate them in the other using the transformation matrix of Equation 1.8.5. We represent this matrix by the symbol $\mathbf{R}$.[9]

$$\mathbf{R} = \begin{pmatrix} i \cdot i' & j \cdot i' \\ i \cdot j' & j \cdot j' \end{pmatrix} = \begin{pmatrix} \cos\theta & \sin\theta \\ -\sin\theta & \cos\theta \end{pmatrix}$$

---

[8] Galileo demonstrated back in 1609 that the trajectory of such a projectile is a parabola. See for example: (1) Stillman Drake, *Galileo at Work—His Scientific Biography,* Dover Publications, New York 1978. (2) *Galileo Manuscripts,* Folio 116v, vol. 72, Biblioteca Nationale Centrale, Florence, Italy.

[9] We also denote matrices in this text with boldface type symbols. Whether the symbol represents a vector or a matrix should be clear from the context.

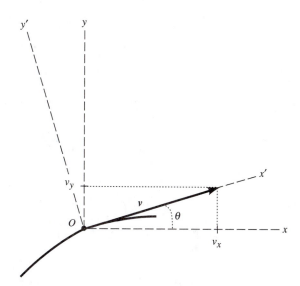

Figure 1.8.2   Velocity of a moving particle referred to two different two-dimensional coordinate systems.

The components of $\mathbf{v}'$ in the $x'y'$ coordinate system are

$$\mathbf{v}' = \begin{pmatrix} v \\ 0 \end{pmatrix} = \begin{pmatrix} \cos\theta & \sin\theta \\ -\sin\theta & \cos\theta \end{pmatrix} \begin{pmatrix} v\cos\theta \\ v\sin\theta \end{pmatrix}$$

or symbolically, $\mathbf{v}' = \mathbf{R}\mathbf{v}$. Here we have denoted the vector in the primed coordinate system by $\mathbf{v}'$. Bear in mind, though, that $\mathbf{v}$ and $\mathbf{v}'$ represent the same vector. The velocity vector points along the direction of the x'-axis in the rotated $x'y'$ coordinate system and, consistent with the figure, $v_{x'} = v$ and $v_{y'} = 0$. The components of a vector change values when we express the vector in coordinate systems rotated with respect to each other.

The square of the magnitude of $\mathbf{v}$ is

$$(\mathbf{v}\cdot\mathbf{v}) = \tilde{\mathbf{v}}\mathbf{v} = (v\cos\theta \quad v\sin\theta)\begin{pmatrix} v\cos\theta \\ v\sin\theta \end{pmatrix} = v^2\cos^2\theta + v^2\sin^2\theta = v^2$$

($\tilde{\mathbf{v}}$ is the transpose of the column vector $\mathbf{v}$—the transpose $\tilde{\mathbf{A}}$ of any matrix $\mathbf{A}$ is obtained by interchanging its columns with its rows.)

Similarly, the square of the magnitude of $\mathbf{v}'$ is

$$(\mathbf{v}'\cdot\mathbf{v}') = \tilde{\mathbf{v}}'\mathbf{v}' = (v \quad 0)\begin{pmatrix} v \\ 0 \end{pmatrix} = v^2 + 0^2 = v^2$$

In each case, the magnitude of the vector is a scalar v whose value is independent of our choice of coordinate system. The same is true of the mass of the projectile. If its mass is one kilogram in the xy coordinate system, then its mass is one kilogram in the $x'y'$ coordinate system. Scalar quantities are *invariant* under a rotation of coordinates.

Suppose we transform back to the xy coordinate system. We should obtain the original components of $\mathbf{v}$. The transformation back is obtained by rotating the $x'y'$ coordinate system through the angle $-\theta$. The transformation matrix that accomplishes this can be obtained by changing the sign of θ in the matrix $\mathbf{R}$.

$$\mathbf{R}(-\theta) = \begin{pmatrix} \cos(-\theta) & \sin(-\theta) \\ -\sin(-\theta) & \cos(-\theta) \end{pmatrix} = \begin{pmatrix} \cos\theta & -\sin\theta \\ \sin\theta & \cos\theta \end{pmatrix} \equiv \tilde{\mathbf{R}}$$

We see that the rotation back is generated by the *transpose* of the matrix $\mathbf{R}$, or $\tilde{\mathbf{R}}$.

If we now operate on $\mathbf{v}'$ with $\tilde{\mathbf{R}}$, we obtain $\tilde{\mathbf{R}}\mathbf{v}' = \tilde{\mathbf{R}}\mathbf{R}\mathbf{v} = \mathbf{v}$ or in matrix notation

$$\begin{pmatrix} \cos\theta & -\sin\theta \\ \sin\theta & \cos\theta \end{pmatrix}\begin{pmatrix} v \\ 0 \end{pmatrix} = \begin{pmatrix} \cos\theta & -\sin\theta \\ \sin\theta & \cos\theta \end{pmatrix}\begin{pmatrix} \cos\theta & \sin\theta \\ -\sin\theta & \cos\theta \end{pmatrix}\begin{pmatrix} v\cos\theta \\ v\sin\theta \end{pmatrix}$$

$$= \begin{pmatrix} 1 & 0 \\ 0 & 1 \end{pmatrix}\begin{pmatrix} v\cos\theta \\ v\sin\theta \end{pmatrix} = \begin{pmatrix} v\cos\theta \\ v\sin\theta \end{pmatrix}$$

In other words, $\tilde{\mathbf{R}}\mathbf{R} = \mathbf{I}$, the *identity* operator, or $\tilde{\mathbf{R}} = \mathbf{R}^{-1}$, the inverse of $\mathbf{R}$. Transformations that exhibit this characteristic are called orthogonal transformations. Rotations of coordinate systems are examples of such a transformation.

1.9 | Derivative of a Vector

Up to this point we have been concerned mainly with vector algebra. We now begin the study of the calculus of vectors and its use in the description of the motion of particles.

Consider a vector $\mathbf{A}$, whose components are functions of a single variable u. The vector may represent position, velocity, and so on. The parameter u is usually the time t, but it can be any quantity that determines the components of $\mathbf{A}$:

$$\mathbf{A}(u) = \mathbf{i}A_x(u) + \mathbf{j}A_y(u) + \mathbf{k}A_z(u) \tag{1.9.1}$$

The derivative of $\mathbf{A}$ with respect to u is defined, quite analogously to the ordinary derivative of a scalar function, by the limit

$$\frac{d\mathbf{A}}{du} = \lim_{\Delta u \to 0} \frac{\Delta\mathbf{A}}{\Delta u} = \lim_{\Delta u \to 0}\left(\mathbf{i}\frac{\Delta A_x}{\Delta u} + \mathbf{j}\frac{\Delta A_y}{\Delta u} + \mathbf{k}\frac{\Delta A_z}{\Delta u} \right)$$

where $\Delta A_x = A_x(u + \Delta u) - A_x(u)$ and so on. Hence,

$$\frac{d\mathbf{A}}{du} = \mathbf{i}\frac{dA_x}{du} + \mathbf{j}\frac{dA_y}{du} + \mathbf{k}\frac{dA_z}{du} \tag{1.9.2}$$

The derivative of a vector is a vector whose Cartesian components are ordinary derivatives.

It follows from Equation 1.9.2 that the derivative of the sum of two vectors is equal to the sum of the derivatives, namely,

$$\frac{d}{du}(\mathbf{A} + \mathbf{B}) = \frac{d\mathbf{A}}{du} + \frac{d\mathbf{B}}{du} \tag{1.9.3}$$

The rules for differentiating vector products obey similar rules of vector calculus. For example,

$$\frac{d(n\mathbf{A})}{du} = \frac{dn}{du}\mathbf{A} + n\frac{d\mathbf{A}}{du} \tag{1.9.4}$$

$$\frac{d(\mathbf{A} \cdot \mathbf{B})}{du} = \frac{d\mathbf{A}}{du} \cdot \mathbf{B} + \mathbf{A} \cdot \frac{d\mathbf{B}}{du} \tag{1.9.5}$$

$$\frac{d(\mathbf{A} \times \mathbf{B})}{du} = \frac{d\mathbf{A}}{du} \times \mathbf{B} + \mathbf{A} \times \frac{d\mathbf{B}}{du} \tag{1.9.6}$$

Notice that it is necessary to preserve the order of the terms in the derivative of the cross product. The proofs are left as an exercise for the student.

1.10 | Position Vector of a Particle: Velocity and Acceleration in Rectangular Coordinates

In a given reference system, the position of a particle can be specified by a single vector, namely, the displacement of the particle relative to the origin of the coordinate system. This vector is called the *position vector* of the particle. In rectangular coordinates (Figure 1.10.1), the position vector is simply

$$\mathbf{r} = \mathbf{i}x + \mathbf{j}y + \mathbf{k}z \tag{1.10.1}$$

The components of the position vector of a moving particle are functions of the time, namely,

$$x = x(t) \qquad y = y(t) \qquad z = z(t) \tag{1.10.2}$$

In Equation 1.9.2 we gave the formal definition of the derivative of any vector with respect to some parameter. In particular, if the vector is the position vector $\mathbf{r}$ of a moving

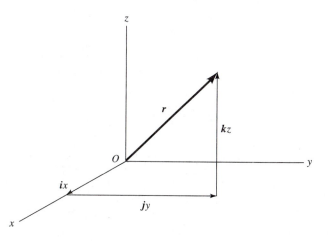

Figure 1.10.1 The position vector **r** and its components in a Cartesian coordinate system.

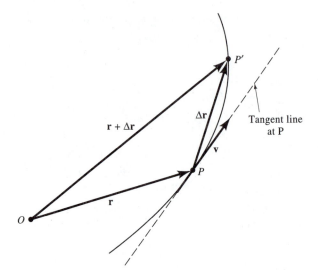

Figure 1.10.2   The velocity vector of a moving particle as the limit of the ratio $\Delta \mathbf{r}/\Delta t$.

particle and the parameter is the time t, the derivative of $\mathbf{r}$ with respect to t is called the *velocity*, which we shall denote by $\mathbf{v}$:

$$\mathbf{v} = \frac{d\mathbf{r}}{dt} = \mathbf{i}\dot{x} + \mathbf{j}\dot{y} + \mathbf{k}\dot{z} \tag{1.10.3}$$

where the dots indicate differentiation with respect to t. (This convention is standard and is used throughout the book.) Let us examine the geometric significance of the velocity vector. Suppose a particle is at a certain position at time t. At a time Δt later, the particle will have moved from the position $\mathbf{r}(t)$ to the position $\mathbf{r}(t + \Delta t)$. The vector displacement during the time interval Δt is

$$\Delta \mathbf{r} = \mathbf{r}(t + \Delta t) - \mathbf{r}(t) \tag{1.10.4}$$

so the quotient $\Delta \mathbf{r}/\Delta t$ is a *vector* that is parallel to the displacement. As we consider smaller and smaller time intervals, the quotient $\Delta \mathbf{r}/\Delta t$ approaches a limit $d\mathbf{r}/dt$, which we call the *velocity*. The vector $d\mathbf{r}/dt$ expresses both the direction of motion and the rate. This is shown graphically in Figure 1.10.2. In the time interval Δt, the particle moves along the path from P to P'. As Δt approaches zero, the point P' approaches P, and the direction of the vector $\Delta \mathbf{r}/\Delta t$ approaches the direction of the tangent to the path at P. The velocity vector, therefore, is always tangent to the path of motion.

The magnitude of the velocity is called the *speed*. In rectangular components the speed is just

$$v = |\mathbf{v}| = (\dot{x}^2 + \dot{y}^2 + \dot{z}^2)^{1/2} \tag{1.10.5}$$

If we denote the cumulative scalar distance along the path with s, then we can express the speed alternatively as

$$v = \frac{ds}{dt} = \lim_{\Delta t \to 0} \frac{\Delta s}{\Delta t} = \lim_{\Delta t \to 0} \frac{[(\Delta x)^2 + (\Delta y)^2 + (\Delta z)^2]^{1/2}}{\Delta t} \tag{1.10.6}$$

which reduces to the expression on the right of Equation 1.10.5.

The time derivative of the velocity is called the *acceleration*. Denoting the acceleration with **a**, we have

$$\mathbf{a} = \frac{d\mathbf{v}}{dt} = \frac{d^2\mathbf{r}}{dt^2} \tag{1.10.7}$$

In rectangular components,

$$\mathbf{a} = \mathbf{i}\ddot{x} + \mathbf{j}\ddot{y} + \mathbf{k}\ddot{z} \tag{1.10.8}$$

Thus, acceleration is a vector quantity whose components, in rectangular coordinates, are the second derivatives of the positional coordinates of a moving particle.

EXAMPLE 1.10.1

Projectile Motion

Let us examine the motion represented by the equation

$$\mathbf{r}(t) = \mathbf{i}bt + \mathbf{j}\left(ct - \frac{gt^2}{2}\right) + \mathbf{k}0$$

This represents motion in the xy plane, because the z component is constant and equal to zero. The velocity **v** is obtained by differentiating with respect to t, namely,

$$\mathbf{v} = \frac{d\mathbf{r}}{dt} = \mathbf{i}b + \mathbf{j}(c - gt)$$

The acceleration, likewise, is given by

$$\mathbf{a} = \frac{d\mathbf{v}}{dt} = -\mathbf{j}g$$

Thus, **a** is in the negative y direction and has the constant magnitude g. The path of motion is a parabola, as shown in Figure 1.10.3. The speed v varies with t according to the equation

$$v = [b^2 + (c - gt)^2]^{1/2}$$

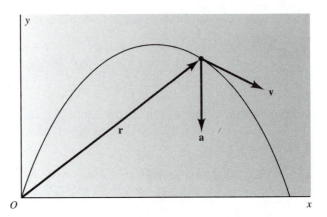

Figure 1.10.3 Position, velocity, and acceleration vectors of a particle (projectile) moving in a parabolic path.

EXAMPLE 1.10.2

Circular Motion

Suppose the position vector of a particle is given by

$$\mathbf{r} = \mathbf{i}b \sin \omega t + \mathbf{j}b \cos \omega t$$

where ω is a constant.

Let us analyze the motion. The distance from the origin remains constant:

$$|\mathbf{r}| = r = (b^2 \sin^2 \omega t + b^2 \cos^2 \omega t)^{1/2} = b$$

So the path is a circle of radius b centered at the origin. Differentiating $\mathbf{r}$, we find the velocity vector

$$\mathbf{v} = \frac{d\mathbf{r}}{dt} = \mathbf{i}b\omega \cos \omega t - \mathbf{j}b\omega \sin \omega t$$

The particle traverses its path with constant speed:

$$v = |\mathbf{v}| = (b^2\omega^2 \cos^2 \omega t + b^2\omega^2 \sin^2 \omega t)^{1/2} = b\omega$$

The acceleration is

$$\mathbf{a} = \frac{d\mathbf{v}}{dt} = -\mathbf{i}b\omega^2 \sin \omega t - \mathbf{j}b\omega^2 \cos \omega t$$

In this case the acceleration is perpendicular to the velocity, because the dot product of $\mathbf{v}$ and $\mathbf{a}$ vanishes:

$$\mathbf{v} \cdot \mathbf{a} = (b\omega \cos \omega t)(-b\omega^2 \sin \omega t) + (-b\omega \sin \omega t)(-b\omega^2 \cos \omega t) = 0$$

Comparing the two expressions for $\mathbf{a}$ and $\mathbf{r}$, we see that we can write

$$\mathbf{a} = -\omega^2 \mathbf{r}$$

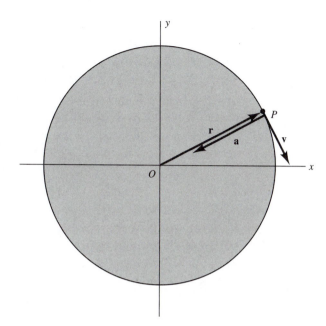

Figure 1.10.4 A particle
moving in a circular path
with constant speed.

so **a** and **r** are oppositely directed; that is, **a** always points toward the center of the circular path (Fig. 1.10.4).

<hr>

EXAMPLE 1.10.3

Rolling Wheel

Let us consider the following position vector of a particle P:

$$\mathbf{r} = \mathbf{r}_1 + \mathbf{r}_2$$

in which

$$\mathbf{r}_1 = \mathbf{i}b\omega t + \mathbf{j}b$$
$$\mathbf{r}_2 = \mathbf{i}b \sin \omega t + \mathbf{j}b \cos \omega t$$

Now $\mathbf{r}_1$ by itself represents a point moving along the line $y = b$ at constant velocity, provided ω is constant; namely,

$$\mathbf{v}_1 = \frac{d\mathbf{r}_1}{dt} = \mathbf{i}b\omega$$

The second part, $\mathbf{r}_2$, is just the position vector for circular motion, as discussed in Example 1.10.2. Hence, the vector sum $\mathbf{r}_1 + \mathbf{r}_2$ represents a point that describes a circle of radius b about a moving center. This is precisely what occurs for a particle on the rim of a rolling wheel, $\mathbf{r}_1$ being the position vector of the center of the wheel and $\mathbf{r}_2$ being

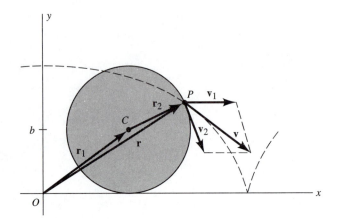

Figure 1.10.5 The cycloidal path of a particle on a rolling wheel.

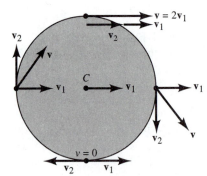

Figure 1.10.6 Velocity vectors for various points on a rolling wheel.

the position vector of the particle P *relative* to the moving center. The actual path is a *cycloid*, as shown in Figure 1.10.5. The velocity of P is

$$\mathbf{v} = \mathbf{v}_1 + \mathbf{v}_2 = \mathbf{i}(b\omega + b\omega \cos \omega t) - \mathbf{j}b\omega \sin \omega t$$

In particular, for $\omega t = 0, 2\pi, 4\pi, \ldots$, we find that $\mathbf{v} = \mathbf{i}2b\omega$, which is just twice the velocity of the center C. At these points the particle is at the uppermost part of its path. Furthermore, for $\omega t = \pi, 3\pi, 5\pi, \ldots$, we obtain $\mathbf{v} = 0$. At these points the particle is at its lowest point and is instantaneously in contact with the ground. See Figure 1.10.6.

1.11 | Velocity and Acceleration in Plane Polar Coordinates

It is often convenient to employ polar coordinates r, θ to express the position of a particle moving in a plane. Vectorially, the position of the particle can be written as the product of the radial distance r by a unit radial vector $\mathbf{e}_r$:

$$\mathbf{r} = r\mathbf{e}_r \tag{1.11.1}$$

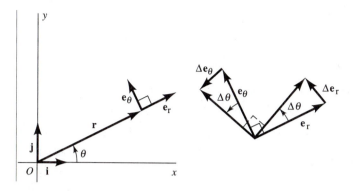

Figure 1.11.1   Unit vectors for plane polar coordinates.

As the particle moves, both r and $\mathbf{e}_r$ vary; thus, they are both functions of the time. Hence, if we differentiate with respect to t, we have

$$\mathbf{v} = \frac{d\mathbf{r}}{dt} = \dot{r}\mathbf{e}_r + r\frac{d\mathbf{e}_r}{dt} \tag{1.11.2}$$

To calculate the derivative $d\mathbf{e}_r/dt$, let us consider the vector diagram shown in Figure 1.11.1. A study of the figure shows that when the direction of $\mathbf{r}$ changes by an amount $\Delta\theta$, the corresponding change $\Delta\mathbf{e}_r$ of the unit radial vector is as follows: The magnitude $|\Delta\mathbf{e}_r|$ is approximately equal to $\Delta\theta$ and the direction of $\Delta\mathbf{e}_r$ is very nearly perpendicular to $\mathbf{e}_r$. Let us introduce another unit vector, $\mathbf{e}_\theta$, whose direction is perpendicular to $\mathbf{e}_r$. Then we have

$$\Delta\mathbf{e}_r \simeq \mathbf{e}_\theta \Delta\theta \tag{1.11.3}$$

If we divide by Δt and take the limit, we get

$$\frac{d\mathbf{e}_r}{dt} = \mathbf{e}_\theta \frac{d\theta}{dt} \tag{1.11.4}$$

for the time derivative of the unit radial vector. In a precisely similar way, we can argue that the change in the unit vector $\mathbf{e}_\theta$ is given by the approximation

$$\Delta\mathbf{e}_\theta \simeq -\mathbf{e}_r \Delta\theta \tag{1.11.5}$$

Here the minus sign is inserted to indicate that the direction of the change $\Delta\mathbf{e}_\theta$ is opposite to the direction of $\mathbf{e}_r$, as can be seen from Figure 1.11.1. Consequently, the time derivative is given by

$$\frac{d\mathbf{e}_\theta}{dt} = -\mathbf{e}_r \frac{d\theta}{dt} \tag{1.11.6}$$

By using Equation 1.11.4 for the derivative of the unit radial vector, we can finally write the equation for the velocity as

$$\mathbf{v} = \dot{r}\mathbf{e}_r + r\dot{\theta}\mathbf{e}_\theta \tag{1.11.7}$$

Thus, $\dot{r}$ is the radial component of the velocity vector, and $r\dot{\theta}$ is the transverse component.

To find the acceleration vector, we take the derivative of the velocity with respect to time. This gives

$$\mathbf{a} = \frac{d\mathbf{v}}{dt} = \ddot{r}\mathbf{e}_r + \dot{r}\frac{d\mathbf{e}_r}{dt} + (\dot{r}\dot{\theta} + r\ddot{\theta})\mathbf{e}_\theta + r\dot{\theta}\frac{d\mathbf{e}_\theta}{dt} \tag{1.11.8}$$

The values of $d\mathbf{e}_r/dt$ and $d\mathbf{e}_\theta/dt$ are given by Equations 1.11.4 and 1.11.6 and yield the following equation for the acceleration vector in plane polar coordinates:

$$\mathbf{a} = (\ddot{r} - r\dot{\theta}^2)\mathbf{e}_r + (r\ddot{\theta} + 2\dot{r}\dot{\theta})\mathbf{e}_\theta \tag{1.11.9}$$

Thus, the radial component of the acceleration vector is

$$a_r = \ddot{r} - r\dot{\theta}^2 \tag{1.11.10}$$

and the transverse component is

$$a_\theta = r\ddot{\theta} + 2\dot{r}\dot{\theta} = \frac{1}{r}\frac{d}{dt}(r^2\dot{\theta}) \tag{1.11.11}$$

The above results show, for instance, that if a particle moves on a circle of constant radius b, so that $\dot{r} = 0$, then the radial component of the acceleration is of magnitude $b\dot{\theta}^2$ and is directed inward toward the center of the circular path. The transverse component in this case is $b\ddot{\theta}$. On the other hand, if the particle moves along a fixed radial line—that is, if θ is constant—then the radial component is just $\ddot{r}$ and the transverse component is zero. If r and θ both vary, then the general expression (1.11.9) gives the acceleration.

EXAMPLE 1.11.1

A honeybee hones in on its hive in a spiral path in such a way that the radial distance decreases at a constant rate, $r = b - ct$, while the angular speed increases at a constant rate, $\dot{\theta} = kt$. Find the speed as a function of time.

Solution:

We have $\dot{r} = -c$ and $\ddot{r} = 0$. Thus, from Equation 1.11.7,

$$\mathbf{v} = -c\mathbf{e}_r + (b - ct)kt\mathbf{e}_\theta$$

so

$$v = [c^2 + (b - ct)^2 k^2 t^2]^{1/2}$$

which is valid for $t \leq b/c$. Note that $v = c$ both for $t = 0$, $r = b$ and for $t = b/c$, $r = 0$.

EXAMPLE 1.11.2

On a horizontal turntable that is rotating at constant angular speed, a bug is crawling outward on a radial line such that its distance from the center increases quadratically with time: $r = bt^2$, $\theta = \omega t$, where b and ω are constants. Find the acceleration of the bug.

Solution:

We have $\dot{r} = 2bt$, $\ddot{r} = 2b$, $\dot{\theta} = \omega$, $\ddot{\theta} = 0$. Substituting into Equation 1.11.9, we find

$$\mathbf{a} = \mathbf{e}_r(2b - bt^2\omega^2) + \mathbf{e}_\theta[0 + 2(2bt)\omega]$$
$$= b(2 - t^2\omega^2)\mathbf{e}_r + 4b\omega t\mathbf{e}_\theta$$

Note that the radial component of the acceleration becomes negative for large t in this example, although the radius is always increasing monotonically with time.

1.12 | Velocity and Acceleration in Cylindrical and Spherical Coordinates

Cylindrical Coordinates

In the case of three-dimensional motion, the position of a particle can be described in cylindrical coordinates R, ϕ, z. The position vector is then written as

$$\mathbf{r} = R\mathbf{e}_R + z\mathbf{e}_z \tag{1.12.1}$$

where $\mathbf{e}_R$ is a unit radial vector in the xy plane and $\mathbf{e}_z$ is the unit vector in the z direction. A third unit vector $\mathbf{e}_\phi$ is needed so that the three vectors $\mathbf{e}_R\mathbf{e}_\phi\mathbf{e}_z$ constitute a right-handed triad, as illustrated in Figure 1.12.1. We note that $\mathbf{k} = \mathbf{e}_z$.

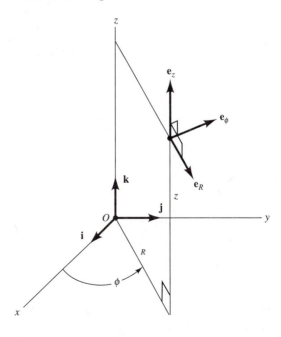

Figure 1.12.1   Unit vectors for cylindrical coordinates.

The velocity and acceleration vectors are found by differentiating, as before. This again involves derivatives of the unit vectors. An argument similar to that used for the plane case shows that $d\mathbf{e}_R/dt = \mathbf{e}_\phi \dot{\phi}$ and $d\mathbf{e}_\phi/dt = -\mathbf{e}_R \dot{\phi}$. The unit vector $\mathbf{e}_z$ does not change in direction, so its time derivative is zero.

In view of these facts, the velocity and acceleration vectors are easily seen to be given by the following equations:

$$\mathbf{v} = \dot{R}\,\mathbf{e}_R + R\dot{\phi}\,\mathbf{e}_\phi + \dot{z}\,\mathbf{e}_z \tag{1.12.2}$$

$$\mathbf{a} = (\ddot{R} - R\dot{\phi}^2)\mathbf{e}_R + (2\dot{R}\dot{\phi} + R\ddot{\phi})\mathbf{e}_\phi + \ddot{z}\,\mathbf{e}_z \tag{1.12.3}$$

These give the values of $\mathbf{v}$ and $\mathbf{a}$ in terms of their components in the *rotated* triad $\mathbf{e}_R\,\mathbf{e}_\phi\,\mathbf{e}_z$.

An alternative way of obtaining the derivatives of the unit vectors is to differentiate the following equations, which are the relationships between the fixed unit triad **ijk** and the rotated triad:

$$\begin{aligned}
\mathbf{e}_R &= \mathbf{i}\cos\phi + \mathbf{j}\sin\phi \\
\mathbf{e}_\phi &= -\mathbf{i}\sin\phi + \mathbf{j}\cos\phi \\
\mathbf{e}_z &= \mathbf{k}
\end{aligned} \tag{1.12.4}$$

The steps are left as an exercise. The result can also be found by use of the rotation matrix, as given in Example 1.8.2.

Spherical Coordinates

When spherical coordinates r, θ, ϕ are employed to describe the position of a particle, the position vector is written as the product of the radial distance r and the unit radial vector $\mathbf{e}_r$, as with plane polar coordinates. Thus,

$$\mathbf{r} = r\mathbf{e}_r \tag{1.12.5}$$

The direction of $\mathbf{e}_r$ is now specified by the two angles ϕ and θ. We introduce two more unit vectors, $\mathbf{e}_\phi$ and $\mathbf{e}_\theta$, as shown in Figure 1.12.2.

The velocity is

$$\mathbf{v} = \frac{d\mathbf{r}}{dt} = \dot{r}\mathbf{e}_r + r\frac{d\mathbf{e}_r}{dt} \tag{1.12.6}$$

Our next problem is how to express the derivative $d\mathbf{e}_r/dt$ in terms of the unit vectors in the rotated triad.

Referring to Figure 1.12.2, we can derive relationships between the **ijk** and $\mathbf{e}_r\mathbf{e}_\theta\mathbf{e}_\phi$ triads. For example, because any vector can be expressed in terms of its projections on to the x, y, z, coordinate axes

$$\mathbf{e}_r = \mathbf{i}(\mathbf{e}_r \cdot \mathbf{i}) + \mathbf{j}(\mathbf{e}_r \cdot \mathbf{j}) + \mathbf{k}(\mathbf{e}_r \cdot \mathbf{k}) \tag{1.12.7}$$

$\mathbf{e}_r \cdot \mathbf{i}$ is the projection of the unit vector $\mathbf{e}_r$ directly onto the unit vector $\mathbf{i}$. According to Equation 1.4.11a, it is equal to $\cos\alpha$, the cosine of the angle between those two unit vectors. We need to express this dot product in terms of θ and ϕ, not α. We can obtain the

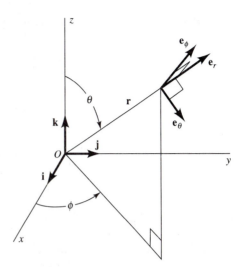

Figure 1.12.2 Unit vectors for spherical coordinates.

desired relation by making two successive projections to get to the x-axis. First project $\mathbf{e}_r$ onto the xy plane, and then project from there onto the x-axis. The first projection gives us a factor of $\sin \theta$, while the second yields a factor of $\cos \phi$. The magnitude of the projection obtained in this way is the desired dot product:

$$\mathbf{e}_r \cdot \mathbf{i} = \sin \theta \cos \phi \qquad (1.12.8a)$$

The remaining dot products can be evaluated in a similar way,

$$\mathbf{e}_r \cdot \mathbf{j} = \sin \theta \sin \phi \qquad \text{and} \qquad \mathbf{e}_r \cdot \mathbf{k} = \cos \theta \qquad (1.12.8b)$$

The relationships for $\mathbf{e}_\theta$ and $\mathbf{e}_\phi$ can be obtained as above, yielding the desired relations

$$\mathbf{e}_r = \mathbf{i} \sin \theta \cos \phi + \mathbf{j} \sin \theta \sin \phi + \mathbf{k} \cos \theta$$
$$\mathbf{e}_\theta = \mathbf{i} \cos \theta \cos \phi + \mathbf{j} \cos \theta \sin \phi - \mathbf{k} \sin \theta \qquad (1.12.9)$$
$$\mathbf{e}_\phi = -\mathbf{i} \sin \phi + \mathbf{j} \cos \phi$$

which express the unit vectors of the rotated triad in terms of the fixed triad **ijk.** We note the similarity between this transformation and that of the second part of Example 1.8.2. The two are, in fact, identical if the correct identification of rotations is made. Let us differentiate the first equation with respect to time. The result is

$$\frac{d\mathbf{e}_r}{dt} = \mathbf{i}(\dot{\theta} \cos \theta \cos \phi - \dot{\phi} \sin \theta \sin \phi) + \mathbf{j}(\dot{\theta} \cos \theta \sin \phi + \dot{\phi} \sin \theta \cos \phi) - \mathbf{k}\dot{\theta} \sin \theta \qquad (1.12.10)$$

Next, by using the expressions for $\mathbf{e}_\phi$ and $\mathbf{e}_\theta$ in Equation 1.12.9, we find that the above equation reduces to

$$\frac{d\mathbf{e}_r}{dt} = \mathbf{e}_\phi \dot{\phi} \sin \theta + \mathbf{e}_\theta \dot{\theta} \qquad (1.12.11a)$$

The other two derivatives are found through a similar procedure. The results are

$$\frac{d\mathbf{e}_\theta}{dt} = -\mathbf{e}_r\dot{\theta} + \mathbf{e}_\phi\dot{\phi}\cos\theta \tag{1.12.11b}$$

$$\frac{d\mathbf{e}_\phi}{dt} = -\mathbf{e}_r\dot{\phi}\sin\theta - \mathbf{e}_\theta\dot{\phi}\cos\theta \tag{1.12.11c}$$

The steps are left as an exercise. Returning now to the problem of finding **v**, we insert the expression for $d\mathbf{e}_r/dt$ given by Equation 1.12.11a into Equation 1.12.6. The final result is

$$\mathbf{v} = \mathbf{e}_r\dot{r} + \mathbf{e}_\phi r\dot{\phi}\sin\theta + \mathbf{e}_\theta r\dot{\theta} \tag{1.12.12}$$

giving the velocity vector in terms of its components in the rotated triad.

To find the acceleration, we differentiate the above expression with respect to time. This gives

$$\mathbf{a} = \frac{d\mathbf{v}}{dt}$$

$$= \mathbf{e}_r\ddot{r} + \dot{r}\frac{d\mathbf{e}_r}{dt} + \mathbf{e}_\phi\frac{d(r\dot{\phi}\sin\theta)}{dt} + r\dot{\phi}\sin\theta\frac{d\mathbf{e}_\phi}{dt} + \mathbf{e}_\theta\frac{d(r\dot{\theta})}{dt} + r\dot{\theta}\frac{d\mathbf{e}_\theta}{dt} \tag{1.12.13}$$

Upon using the previous formulas for the derivatives of the unit vectors, the above expression for the acceleration reduces to

$$\mathbf{a} = (\ddot{r} - r\dot{\phi}^2\sin^2\theta - r\dot{\theta}^2)\mathbf{e}_r + (r\ddot{\theta} + 2\dot{r}\dot{\theta} - r\dot{\phi}^2\sin\theta\cos\theta)\mathbf{e}_\theta$$
$$+ (r\ddot{\phi}\sin\theta + 2\dot{r}\dot{\phi}\sin\theta + 2r\dot{\theta}\dot{\phi}\cos\theta)\mathbf{e}_\phi \tag{1.12.14}$$

giving the acceleration vector in terms of its components in the triad $\mathbf{e}_r\mathbf{e}_\theta\mathbf{e}_\phi$.

EXAMPLE 1.12.1

A bead slides on a wire bent into the form of a helix, the motion of the bead being given in cylindrical coordinates by $R = b$, $\phi = \omega t$, $z = ct$. Find the velocity and acceleration vectors as functions of time.

Solution:

Differentiating, we find $\dot{R} = \ddot{R} = 0$, $\dot{\phi} = \omega$, $\ddot{\phi} = 0$, $\dot{z} = c$, $\ddot{z} = 0$. So, from Equations 1.12.2 and 1.12.3, we have

$$\mathbf{v} = b\omega\mathbf{e}_\phi + c\mathbf{e}_z$$
$$\mathbf{a} = -b\omega^2\mathbf{e}_R$$

Thus, in this case both velocity and acceleration are constant in magnitude, but they vary in direction because both $\mathbf{e}_\phi$ and $\mathbf{e}_r$ change with time as the bead moves.

EXAMPLE 1.12.2

A wheel of radius b is placed in a gimbal mount and is made to rotate as follows. The wheel spins with constant angular speed ω_1 about its own axis, which in turn rotates with constant angular speed ω_2 about a vertical axis in such a way that the axis of the wheel stays in a horizontal plane and the center of the wheel is motionless. Use spherical coordinates to find the acceleration of any point on the rim of the wheel. In particular, find the acceleration of the highest point on the wheel.

Solution:

We can use the fact that spherical coordinates can be chosen such that $r = b$, $\theta = \omega_1 t$, and $\phi = \omega_2 t$ (Fig. 1.12.3). Then we have $\dot{r} = \quad = 0$, $\dot{\theta} = \omega_1$, $\ddot{\theta} = 0$, $\dot{\phi} = \omega_2$, $\ddot{\phi} = 0$. Equation 1.12.14 gives directly

$$\mathbf{a} = (-b\omega_2^2 \sin^2\theta - b\omega_1^2)\mathbf{e}_r - b\omega_2^2 \sin\theta\cos\theta\,\mathbf{e}_\theta + 2b\omega_1\omega_2 \cos\theta\,\mathbf{e}_\phi$$

The point at the top has coordinate $\theta = 0$, so at that point

$$\mathbf{a} = -b\omega_1^2 \mathbf{e}_r + 2b\omega_1\omega_2 \mathbf{e}_\phi$$

The first term on the right is the centripetal acceleration, and the last term is a transverse acceleration normal to the plane of the wheel.

Figure 1.12.3   A rotating wheel on a rotating mount.

Problems

1.1   Given the two vectors $\mathbf{A} = \mathbf{i} + \mathbf{j}$ and $\mathbf{B} = \mathbf{j} + \mathbf{k}$, find the following:
(a) $\mathbf{A} + \mathbf{B}$ and $|\mathbf{A} + \mathbf{B}|$
(b) $3\mathbf{A} - 2\mathbf{B}$
(c) $\mathbf{A} \cdot \mathbf{B}$
(d) $\mathbf{A} \times \mathbf{B}$ and $|\mathbf{A} \times \mathbf{B}|$

1.2   Given the three vectors $\mathbf{A} = 2\mathbf{i} + \mathbf{j}$, $\mathbf{B} = \mathbf{i} + \mathbf{k}$, and $\mathbf{C} = 4\mathbf{j}$, find the following:
(a) $\mathbf{A} \cdot (\mathbf{B} + \mathbf{C})$ and $(\mathbf{A} + \mathbf{B}) \cdot \mathbf{C}$
(b) $\mathbf{A} \cdot (\mathbf{B} \times \mathbf{C})$ and $(\mathbf{A} \times \mathbf{B}) \cdot \mathbf{C}$
(c) $\mathbf{A} \times (\mathbf{B} \times \mathbf{C})$ and $(\mathbf{A} \times \mathbf{B}) \times \mathbf{C}$

1.3   Find the angle between the vectors $\mathbf{A} = a\mathbf{i} + 2a\mathbf{j}$ and $\mathbf{B} = a\mathbf{i} + 2a\mathbf{j} + 3a\mathbf{k}$. (*Note:* These two vectors define a face diagonal and a body diagonal of a rectangular block of sides a, $2a$, and $3a$.)

1.4   Consider a cube whose edges are each of unit length. One corner coincides with the origin of an xyz Cartesian coordinate system. Three of the cube's edges extend from the origin along the positive direction of each coordinate axis. Find the vector that begins at the origin and extends
(a) along a major diagonal of the cube;
(b) along the diagonal of the lower face of the cube.
(c) Calling these vectors $\mathbf{A}$ and $\mathbf{B}$, find $\mathbf{C} = \mathbf{A} \times \mathbf{B}$.
(d) Find the angle between $\mathbf{A}$ and $\mathbf{B}$.

1.5   Assume that two vectors $\mathbf{A}$ and $\mathbf{B}$ are known. Let $\mathbf{C}$ be an unknown vector such that $\mathbf{A} \cdot \mathbf{C} = u$ is a known quantity and $\mathbf{A} \times \mathbf{C} = \mathbf{B}$. Find $\mathbf{C}$ in terms of $\mathbf{A}$, $\mathbf{B}$, u, and the magnitude of $\mathbf{A}$.

1.6   Given the time-varying vector

$$\mathbf{A} = \mathbf{i}\alpha t + \mathbf{j}\beta t^2 + \mathbf{k}\gamma t^3$$

where α, β, and γ are constants, find the first and second time derivatives $d\mathbf{A}/dt$ and $d^2\mathbf{A}/dt^2$.

1.7   For what value (or values) of q is the vector $\mathbf{A} = \mathbf{i}q + 3\mathbf{j} + \mathbf{k}$ perpendicular to the vector $\mathbf{B} = \mathbf{i}q - q\mathbf{j} + 2\mathbf{k}$?

1.8   Give an algebraic proof and a geometric proof of the following relations:

$$|\mathbf{A} + \mathbf{B}| \leq |\mathbf{A}| + |\mathbf{B}|$$
$$|\mathbf{A} \cdot \mathbf{B}| \leq |\mathbf{A}||\mathbf{B}|$$

1.9   Prove the vector identity $\mathbf{A} \times (\mathbf{B} \times \mathbf{C}) = \mathbf{B}(\mathbf{A} \cdot \mathbf{C}) - \mathbf{C}(\mathbf{A} \cdot \mathbf{B})$.

1.10   Two vectors $\mathbf{A}$ and $\mathbf{B}$ represent concurrent sides of a parallelogram. Show that the area of the parallelogram is equal to $|\mathbf{A} \times \mathbf{B}|$.

1.11   Show that $\mathbf{A} \cdot (\mathbf{B} \times \mathbf{C})$ is not equal to $\mathbf{B} \cdot (\mathbf{A} \times \mathbf{C})$.

1.12   Three vectors $\mathbf{A}$, $\mathbf{B}$, and $\mathbf{C}$ represent three concurrent edges of a parallelepiped. Show that the volume of the parallelepiped is equal to $|\mathbf{A} \cdot (\mathbf{B} \times \mathbf{C})|$.

1.13   Verify the transformation matrix for a rotation about the z-axis through an angle ϕ followed by a rotation about the y'-axis through an angle θ, as given in Example 1.8.2.

1.14   Express the vector $2\mathbf{i} + 3\mathbf{j} - \mathbf{k}$ in the primed triad $\mathbf{i'j'k'}$ in which the $x'y'$-axes are rotated about the z-axis (which coincides with the z'-axis) through an angle of $30°$.

1.15   Consider two Cartesian coordinate systems xyz and $x'\ y'\ z'$ that initially coincide. The $x'\ y'\ z'$ undergoes three successive counterclockwise $45°$ rotations about the following axes: first, about the fixed z-axis; second, about its own x'-axis (which has now been rotated); finally, about its own z'-axis (which has also been rotated). Find the components of a unit vector $\mathbf{X}$ in the xyz coordinate system that points along the direction of the x'-axis in the rotated $x'\ y'\ z'$ system. (*Hint: It would be useful to find three transformation matrices that depict each of the above rotations. The resulting transformation matrix is simply their product.*)

1.16   A racing car moves on a circle of constant radius b. If the speed of the car varies with time t according to the equation $v = ct$, where c is a positive constant, show that that the angle between the velocity vector and the acceleration vector is $45°$ at time $t = \sqrt{b/c}$. (*Hint: At this time the tangential and normal components of the acceleration are equal in magnitude.*)

1.17   A small ball is fastened to a long rubber band and twirled around in such a way that the ball moves in an elliptical path given by the equation

$$\mathbf{r}(t) = \mathbf{i}b \cos \omega t + \mathbf{j}2b \sin \omega t$$

where b and ω are constants. Find the speed of the ball as a function of t. In particular, find v at $t = 0$ and at $t = \pi/2\omega$, at which times the ball is, respectively, at its minimum and maximum distances from the origin.

1.18   A buzzing fly moves in a helical path given by the equation

$$\mathbf{r}(t) = \mathbf{i}b \sin \omega t + \mathbf{j}b \cos \omega t + \mathbf{k}ct^2$$

Show that the magnitude of the acceleration of the fly is constant, provided b, ω, and c are constant.

1.19   A bee goes out from its hive in a spiral path given in plane polar coordinates by

$$r = be^{kt} \qquad \theta = ct$$

where b, k, and c are positive constants. Show that the angle between the velocity vector and the acceleration vector remains constant as the bee moves outward. (*Hint: Find $\mathbf{v} \cdot \mathbf{a}/va$.*)

1.20   Work Problem 1.18 using cylindrical coordinates where $R = b$, $\phi = \omega t$, and $z = ct^2$.

1.21   The position of a particle as a function of time is given by

$$\mathbf{r}(t) = \mathbf{i}(1 - e^{-kt}) + \mathbf{j}e^{kt}$$

where k is a positive constant. Find the velocity and acceleration of the particle. Sketch its trajectory.

1.22   An ant crawls on the surface of a ball of radius b in such a manner that the ant's motion is given in spherical coordinates by the equations

$$r = b \qquad \phi = \omega t \qquad \theta = \frac{\pi}{2}\left[1 + \frac{1}{4}\cos(4\omega t)\right]$$

Find the speed of the ant as a function of the time t. What sort of path is represented by the above equations?

1.23   Prove that $\mathbf{v} \cdot \mathbf{a} = v\dot{v}$ and, hence, that for a moving particle $\mathbf{v}$ and $\mathbf{a}$ are perpendicular to each other if the speed v is constant. (*Hint: Differentiate both sides of the equation* $\mathbf{v} \cdot \mathbf{v} = v^2$ *with respect to t. Note,* $\dot{v}$ *is not the same as* $|\mathbf{a}|$. *It is the magnitude of the acceleration of the particle along its instantaneous direction of motion.*)

1.24   Prove that

$$\frac{d}{dt}[\mathbf{r} \cdot (\mathbf{v} \times \mathbf{a})] = \mathbf{r} \cdot (\mathbf{v} \times \dot{\mathbf{a}})$$

1.25   Show that the tangential component of the acceleration of a moving particle is given by the expression

$$a_\tau = \frac{\mathbf{v} \cdot \mathbf{a}}{v}$$

and the normal component is therefore

$$a_n = \left(a^2 - a_\tau^2\right)^{1/2} = \left[a^2 - \frac{(\mathbf{v} \cdot \mathbf{a})^2}{v^2}\right]^{1/2}$$

1.26   Use the above result to find the tangential and normal components of the acceleration as functions of time in Problems 1.18 and 1.19.

1.27   Prove that $|\mathbf{v} \times \mathbf{a}| = v^3/\rho$, where ρ is the radius of curvature of the path of a moving particle.

1.28   A wheel of radius b rolls along the ground with constant forward acceleration a_0. Show that, at any given instant, the magnitude of the acceleration of any point on the wheel is $(a_0^2 + v^4/b^2)^{1/2}$ relative to the center of the wheel and is also $a_0[2 + 2\cos\theta + v^4/a_0^2b^2 - (2v^2/a_0b)\sin\theta]^{1/2}$ relative to the ground. Here v is the instantaneous forward speed, and θ defines the location of the point on the wheel, measured forward from the highest point. Which point has the greatest acceleration relative to the ground?

1.29   What is the value of x that makes of following transformation $\mathbf{R}$ orthogonal?

$$\mathbf{R} = \begin{pmatrix} x & x & 0 \\ -x & x & 0 \\ 0 & 0 & 1 \end{pmatrix}$$

What transformation is represented by $\mathbf{R}$?

1.30   Use vector algebra to derive the following trigonometric identities
(a) $\cos(\theta - \phi) = \cos\theta \cos\phi + \sin\theta \sin\phi$
(b) $\sin(\theta - \phi) = \sin\theta \cos\phi - \cos\theta \sin\phi$

$V(x)$

Allowed region

Turning points

2

Newtonian Mechanics:
Rectilinear Motion of a Particle

"Salviati: But if this is true, and if a large stone moves with a speed of, say, eight while a smaller one moves with a speed of four, then when they are united, the system will move with a speed less than eight; but the two stones when tied together make a stone larger than that which before moved with a speed of eight. Hence the heavier body moves with less speed than the lighter; an effect which is contrary to your supposition. Thus you see how, from your supposition that the heavier body moves more rapidly than the lighter one, I infer that the heavier body moves more slowly."

Galileo—*Dialogues Concerning Two New Sciences*

2.1 | Newton's Laws of Motion: Historical Introduction

In his *Principia* of 1687, Isaac Newton laid down three fundamental laws of motion, which would forever change mankind's perception of the world:

I. Every body continues in its state of rest, or of uniform motion in a straight line, unless it is compelled to change that state by forces impressed upon it.

II. The change of motion is proportional to the motive force impressed and is made in the direction of the line in which that force is impressed.

III. To every action there is always imposed an equal reaction; or, the mutual actions of two bodies upon each other are always equal and directed to contrary parts.

These three laws of motion are now known collectively as Newton's laws of motion or, more simply, as Newton's laws. It is arguable whether or not these are indeed all his laws. However, no one before Newton stated them quite so precisely, and certainly no one before him had such a clear understanding of the overall implication and power of these laws. The behavior of natural phenomena that they imply seems to fly in the face of common experience. As any beginning student of physics soon discovers, Newton's laws become "reasonable" only with the expenditure of great effort in attempting to understand thoroughly the apparent vagaries of physical systems.

Aristotle (384–322 B.C.E.) had frozen the notion of the way the world works for almost 20 centuries by invoking powerfully logical arguments that led to a physics in which all moving, earthbound objects ultimately acquired a state of rest unless acted upon by some motive force. In his view, a force was required to keep earthly things moving, even at constant speed—a law in distinct contradiction with Newton's first and second laws. On the other hand, heavenly bodies dwelt in a more perfect realm where perpetual circular motion was the norm and no forces were required to keep this celestial clockwork ticking.

Modern scientists heap scorn upon Aristotle for burdening us with such obviously flawed doctrine. He is particularly criticized for his failure to carry out even the most modest experiment that would have shown him the error of his ways. At that time, though, it was a commonly held belief that experiment was not a suitable enterprise for any self-respecting philosopher, and thus Aristotle, raised with that belief, failed to acquire a true picture of nature. This viewpoint is a bit misleading, however. Although he did no experiments in natural philosophy, Aristotle was a keen observer of nature, one of the first. If he was guilty of anything, it was less a failure to observe nature than a failure to follow through with a process of abstraction based upon observation. Indeed, bodies falling through air accelerate initially, but ultimately they attain a nearly constant velocity of fall. Heavy objects, in general, fall faster than lighter ones. It takes a sizable force to haul a ship through water, and the greater the force, the greater the ship's speed. A spear thrown vertically upward from a moving chariot will land behind the charioteer, not on top of him. And the motion of heavenly bodies does go on and on, apparently following a curved path forever without any visible motive means. Of course, nowadays we can understand these things if we pay close attention to all the variables that affect the motion of objects and then apply Newton's laws correctly.

That Aristotle failed to extract Newton's laws from such observations of the real world is a consequence only of the fact that he observed the world and interpreted its workings in a rather superficial way. He was basically unaware of the then-subtle effects of air resistance, friction, and the like. It was only with the advent of the ability and motivation to carry out precise experiments followed by a process of abstraction that led to the revolutionary point of view of nature represented by the Newtonian paradigm. Even today, the workings of that paradigm are most easily visualized in the artificial realm of our own minds, emptied of the real world's imperfections of friction and air resistance (look at any elementary physics book and see how often one encounters the phrase "neglecting friction"). Aristotle's physics, much more than Newton's, reflects the workings of a nature quite coincident with the common misconception of modern people in general (including the typical college student who chooses a curriculum curiously devoid of courses in physics).

There is no question that the first law, the so-called law of inertia, had already been set forth prior to the time of Newton. This law, commonly attributed to Galileo (1564–1642), was actually first formulated by René Descartes (1596–1650). According to Descartes, "inertia" made bodies persist in motion forever, not in perfect

Aristotelian circles but in a straight line. Descartes came to this conclusion not by experiment but by pure thought. In contrast to belief in traditional authority (which at that time meant belief in the teachings of Aristotle), Descartes believed that only one's own thinking could be trusted. It was his intent to "explain effects by their causes, and not causes by their effects." For Descartes, pure reasoning served as the sole basis of certainty. Such a paradigm would aid the transition from an Aristotelian worldview to a Newtonian one, but it contained within itself the seeds of its own destruction.

It was not too surprising that Descartes failed to grasp the implication of his law of inertia regarding planetary motion. Planets certainly did not move in straight lines. Descartes, more ruthless in his methods of thought than any of his predecessors, reasoned that some physical thing had to "drive" the planets along in their curved paths. Descartes rebelled in horror at the notion that the required physical force was some invisible entity reaching out across the void to grab the planets and hold them in their orbits. Moreover, having no knowledge of the second law, Descartes never realized that the required force was not a "driving" force but a force that had to be directed "inward" toward the Sun. He, along with many others of that era, was certain that the planets had to be pushed along in their paths around the Sun (or Earth). Thus, he concocted the notion of an all-pervading, ether-like fluid made of untold numbers of unseen particles, rotating in vortices, within which the planets were driven round and round—an erroneous conclusion that arose from the fancies of a mind engaged only in pure thought, minimally constrained by experimental or observational data.

Galileo, on the other hand, mainly by clear argument based on actual experimental results, had gradually commandeered a fairly clear understanding of what would come to be the first of Newton's laws, as well as the second. A necessary prelude to the final synthesis of a correct system of mechanics was his observation that a pendulum undergoing small oscillations was isochronous; that is, its period of oscillation was independent of its amplitude. This discovery led to the first clocks capable of making accurate measurements of small time intervals, a capability that Aristotle did not have. Galileo would soon exploit this capability in carrying out experiments of unprecedented precision with objects either freely falling or sliding down inclined planes. Generalizing from the results of his experiments, Galileo came very close to formulating Newton's first two laws.

For example, concerning the first law, Galileo noted, as had Aristotle, that an object sliding along a level surface indeed came to rest. But here Galileo made a wonderful mental leap that took him far past the dialectics of Aristotle. He imagined a second surface, more slippery than the first. An object given a push along the second surface would travel farther before stopping than it would if given a similar push along the first surface. Carrying this process of abstraction to its ultimate conclusion, Galileo reasoned that an object given a push along a surface of "infinite slipperiness" (i.e., "neglecting friction") would, in fact, go on forever, never coming to rest. Thus, contrary to Aristotle's physics, he reasoned that a force is not required merely to keep an object in motion. In fact, some force must be applied to stop it. This is very close to Newton's law of inertia but, astonishingly, Galileo did not argue that motion, in the absence of forces, would continue forever in a straight line!

For Galileo and his contemporaries, the world was not an impersonal one ruled by mechanical laws. Instead, it was a cosmos that marched to the tune of an infinitely intelligent craftsman. Following the Aristotelian tradition, Galileo saw a world ordered according to the perfect figure, the circle. Rectilinear motion implied disorder. Objects that found themselves in such a state of affairs would not continue to fly in a straight line for-ever but would ultimately lapse into their more natural state of perfect circular motion. The experiments necessary to discriminate between straight-line motion forever and straight-line motion ultimately evolving to pure circular motion obviously could not be performed in practice, but only within the confines of one's own mind, and only if that mind had been properly freed from the conditioning of centuries of ill-founded dogma. Galileo, brilliant though he was, still did battle with the ghosts of the past and had not yet reached that required state of mind.

Galileo's experiments with falling bodies led him to the brink of Newton's second law. Again, as Aristotle had known, Galileo saw that heavy objects, such as stones, did fall faster than lighter ones, such as feathers. However, by carefully timing similarly shaped objects, albeit of different weights, Galileo discovered that such objects accelerated as they fell and all reached the ground at more or less the same time! Indeed, very heavy objects, even though themselves differing greatly in weight, fell at almost identical rates, with a speed that increased about 10 m/s each second. (Incidentally, the famous experiment of dropping cannonballs from the Leaning Tower of Pisa might not have been carried out by Galileo but by one of his chief Aristotelian antagonists at Pisa, Giorgio Coressio, and in hopes not of refuting but of confirming the Aristotelian view that larger bodies must fall more quickly than small ones!)[1] It was again through a process of brilliant abstraction that Galileo realized that if the effects of air resistance could be eliminated, all objects would fall with the same acceleration, regardless of weight or shape. Thus, even more of Aristotle's edifice was torn apart; a heavier weight does not fall faster than a light one, and a force causes objects to accelerate, not to move at constant speed.

Galileo's notions of mechanics on Earth were more closely on target with Newton's laws than the conjectures of any of his predecessors had been. He sometimes applied them bril-liantly in defense of the Copernican viewpoint, that is, a heliocentric model of the solar system. In particular, even though his notion of the law of inertia was somewhat flawed, he applied it correctly in arguing that terrestrial-based experiments could not be used to demonstrate that the Earth could not be in motion around the Sun. He pointed out that a stone dropped from the mast of a moving ship would not "be left behind" since the stone would share the ship's horizontal speed. By analogy, in contrast to Aristotelian argument, a stone dropped from a tall tower would not be left behind by an Earth in motion. This powerful argument implied that no such observation could be used to demonstrate whether or not the Earth was rotating. The argument contained the seeds of relativity theory.

Unfortunately, as mentioned above, Galileo could not entirely break loose from the Aristotelian dogma of circular motion. In strict contradiction to the law of inertia, he postulated that a body left to itself will continue to move forever, not in a straight line but in a circular orbit. His reasoning was as follows:

> . . . straight motion being by nature infinite (because a straight line is infinite and inde-terminate), it is impossible that anything should have by nature the principle of moving in a straight line; or, in other words, towards a place where it is impossible to arrive, there being no finite end. For that which cannot be done, nor endeavors to move whither it is impossible to arrive.

---

[1] *Aristotle, Galileo, and the Tower of Pisa*, L. Cooper, Cornell University Press, Ithaca, 1935.

This statement also contradicted his intimate knowledge of centrifugal forces, that is, the tendency of an object moving in a circle to fly off on a tangent in a straight line. He knew that earthbound objects could travel in circles only if this centrifugal force was either balanced or overwhelmed by some other offsetting force. Indeed, one of the Aristotelian arguments against a rotating Earth was that objects on the Earth's surface would be flung off it. Galileo argued that this conclusion was not valid, because the Earth's "gravity" overwhelmed this centrifugal tendency! Yet somehow he failed to make the mental leap that some similar effect must keep the planets in circular orbit about the Sun!

So ultimately it was Newton who pulled together all the fragmentary knowledge that had been accumulated about the motion of earthbound objects into the brilliant synthesis of the three laws and then demonstrated that the motion of heavenly objects obeyed those laws as well.

Newton's laws of motion can be thought of as a prescription for calculating or predicting the subsequent motion of a particle (or system of particles), given a knowledge of its position and velocity at some instant in time. These laws, in and of themselves, say nothing about the reason why a given physical system behaves the way it does. Newton was quite explicit about that shortcoming. He refused to speculate (at least in print) why objects move the way they do. Whatever "mechanism" lay behind the workings of physical systems remained forever hidden from Newton's eyes. He simply stated that, for whatever reason, this is the way things work, as demonstrated by the power of his calculational prescription to predict, with astonishing accuracy, the evolution of physical systems set in motion. Much has been learned since the time of Newton, but a basic fact of physical law persists: the laws of motion are mathematical prescriptions that allow us to predict accurately the future motion of physical systems, given a knowledge of their current state. The laws describe how things work. They do not tell us why.

Newton's First Law: Inertial Reference Systems

The first law describes a common property of matter, namely, *inertia*. Loosely speaking, inertia is the resistance of all matter to having its motion changed. If a particle is at rest, it resists being moved; that is, a force is required to move it. If the particle is in motion, it resists being brought to rest. Again, a force is required to bring it to rest. It almost seems as though matter has been endowed with an innate abhorrence of acceleration. Be that as it may, for whatever reason, it takes a force to accelerate matter; in the absence of applied forces, matter simply persists in its current velocity state—forever.

A mathematical description of the motion of a particle requires the selection of a *frame of reference*, or a set of coordinates in configuration space that can be used to specify the position, velocity, and acceleration of the particle at any instant of time. A frame of reference in which Newton's first law of motion is valid is called an *inertial frame of reference*. This law rules out accelerated frames of reference as inertial, because an object "really" at rest or moving at constant velocity, seen from an accelerated frame of reference, would appear to be accelerated. Moreover, an object seen to be at rest in such a frame would be seen to be accelerated with respect to the inertial frame. So strong is our belief in the concept of inertia and the validity of Newton's laws of motion that we would be forced to invent "fictitious" forces to account for the apparent lack of acceleration of an object at rest in an accelerated frame of reference.

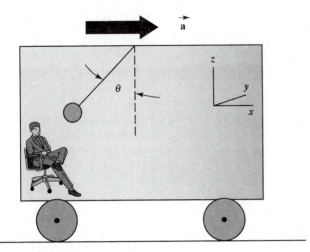

Figure 2.1.1   A plumb bob hangs at an angle θ in an accelerating frame of reference.

A simple example of a noninertial frame of reference should help clarify the situation. Consider an observer inside a railroad boxcar accelerating down the track with an acceleration **a**. Suppose a plumb bob were suspended from the ceiling of the boxcar. How would it appear to the observer? Take a look at Figure 2.1.1. The point here is that the observer in the boxcar is in a noninertial frame of reference and is at rest with respect to it. He sees the plumb bob, also apparently at rest, hanging at an angle θ with respect to the vertical. He knows that, in the absence of any forces other than gravity and tension in the plumb line, such a device should align itself vertically. It does not, and he concludes that some unknown force must be pushing or pulling the plumb bob toward the back of the car. (Indeed, he too feels such a force, as anyone who has ever been in an accelerating vehicle knows from first-hand experience.)

A question that naturally arises is how is it possible to determine whether or not a given frame of reference constitutes an inertial frame? The answer is nontrivial! (For example, if the boxcar were sealed off from the outside world, how would the observer know that the apparent force causing the plumb bob to hang off-vertical was not due to the fact that the whole boxcar was "misaligned" with the direction of gravity—that is, the force due to gravity was actually in the direction indicated by the angle θ?) Observers would have to know that *all* external forces on a body had been eliminated before checking to see whether or not objects in their frame of reference obeyed Newton's first law. It would be necessary to isolate a body completely to eliminate all forces acting upon it. This is impossible, because there would always be some gravitational forces acting unless the body were removed to an infinite distance from all other matter.

Is there a perfect inertial frame of reference? For most practical purposes, a coordinate system attached to the Earth's surface is approximately inertial. For example, a billiard ball seems to move in a straight line with constant speed as long as it does not collide with other balls or hit the cushion. If its motion were measured with very high precision, however, we would see that its path is slightly curved. This is due to the fact that the Earth is rotating and its surface is therefore accelerating toward its axis. Hence, a coordinate system attached to the Earth's surface is not inertial. A better system would be one that uses the center of the Earth as coordinate origin, with the Sun and a star as reference points. But even this system would not be inertial because of the Earth's orbital motion around the Sun.

Suppose, then, we pick a coordinate system whose origin is centered on the Sun. Strictly speaking, this is not a perfect inertial frame either, because the Sun partakes of the general rotational motion of the Milky Way galaxy. So, we try the center of the Milky Way, but to our chagrin, it is part of a local group, or small cluster, of some 20 galaxies that all rotate about their common center of mass. Continuing on, we see that the local group lies on the edge of the Virgo supercluster, which contains dozens of clusters of galaxies centered on the 2000-member-rich Virgo cluster, 60 million light years away, all rotating about their common center of mass! As a final step in this continuing saga of seeming futility, we might attempt to find a frame of reference that is at rest with respect to the observed relative motion of all the matter in the universe; however, we cannot observe all the matter. Some of the potentially visible matter is too dim to be seen, and some matter isn't even potentially visible, the so-called *dark matter*, whose existence we can only infer by indirect means. Furthermore, the universe appears to have a large supply of *dark energy*, also invisible, which nonetheless makes its presence known by accelerating the expansion of the universe.

However, all is not lost. The universe began with the *Big Bang* about 12.7 billion years ago and has been expanding ever since. Some of the evidence for this is the observation of the <u>C</u>osmic <u>M</u>icrowave <u>B</u>ackground radiation (CMB), a relic of the primeval fireball that emerged from that singular event.[2] Its existence provides us with a novel means of actually measuring the Earth's "true" velocity through space, without reference to neighboring galaxies, clusters, or superclusters. If we were precisely at rest with respect to the universal expansion,[3] then we would see the CMB as perfectly *isotropic*, that is, the distribution of the radiation would be the same in all directions in the sky. The reason for this is that initially, the universe was extremely hot and the radiation and matter that sprang forth from the Big Bang interacted fairly strongly and were tightly coupled together. But 380,000 years later, the expanding universe cooled down to a temperature of about 3000 K and matter, which up to that point consisted mostly of electrically charged protons and electrons, then combined to form neutral hydrogen atoms and the radiation *decoupled* from it. Since then, the universe has expanded even more, by a factor of about 1000, and has cooled to a temperature of about 2.73 K. The spectral distribution of the left over CMB has changed accordingly. Indeed, the radiation is remarkably, though not perfectly,

---

[2] For the most up-to-date information about the CMB, dark matter, and dark energy, visit the NASA Goddard Space Flight Center at http://map.gsfc.nasa.gov and look for articles discussing the Wilkinson Microwave Anisotropy Project (WMAP). For a general discussion of the CMB and its implications, the reader is referred to almost any current astronomy text, such as *The Universe*, 6th ed., Kaufmann and Freedman, Wiley Publishing, Indianapolis, 2001.

[3] A common analog of this situation is an inflating balloon on whose surface is attached a random distribution of buttons. Each button is fixed and, therefore, "at rest" relative to the expanding two-dimensional surface. Any frame of reference attached to any button would be a valid inertial frame of reference.

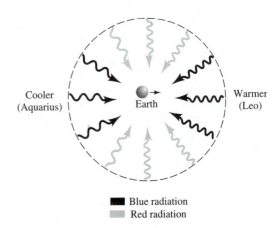

Cooler
(Aquarius)

Earth

Warmer
(Leo)

Figure 2.1.2   Motion of Earth through
the cosmic microwave background.

▬ Blue radiation
▬ Red radiation

isotropic. Radiation arriving at Earth from the direction of the constellation Leo appears
to be coming from a slightly warmer region of the universe and, thus, has a slightly shorter
or "bluer" wavelength than radiation arriving from the opposite direction in the constellation Aquarius (Figure 2.1.2). This small spectral difference occurs because the Earth
moves about 400 km/s towards Leo, which causes a small *Doppler shift* in the observed
spectral distribution.[4] Observers in a frame of reference moving from Leo toward Aquarius
at 400 km/s *relative to Earth* would see a perfectly isotropic distribution (except for some
variations that originated when the radiation decoupled from matter in localized regions
of space of slightly different matter densities). These observers would be at rest with
respect to the overall expansion of the universe! It is generally agreed that such a frame
of reference comes closest to a perfect inertial frame.

However, do not think that we are implying that there is such a thing as an *absolute*
inertial frame of reference. In part, the theory of relativity resulted from the failure of
attempts to find an absolute frame of reference in which all of the fundamental laws of
physics, not just Newton's first law of motion, were supposed to be valid. This led Einstein
to the conclusion that the failure to find an absolute frame was because of the simple
reason that none exists. Consequently, he proposed as a cornerstone of the theory of relativity that the fundamental laws of physics are the same in all inertial frames of reference and that there is no single preferred inertial frame.

Interestingly, Galileo, who predated Einstein by 300 years, had arrived at a very
similar conclusion. Consider the words that one of his characters, Salviati, speaks to
another, Sagredo, in his infamous *Dialogue Concerning the Two Chief World Systems*,[5]
which poetically expresses the gist of *Galilean relativity*.

---

[4] Relative motion toward a source of light decreases the observed wavelength of the light. Relative motion away
from the source increases the observed wavelength. This change in observed wavelength is called the Doppler
Effect. A shortening is called a blueshift and a lengthening is called a redshift.

[5] *Dialogue Concerning the Two Chief World Systems*, Galileo Galilei (1632), *The Second Day*, 2nd printing,
p. 186, translated by Stillman Drake, University of California Press, Berkeley, 1970.

"Shut yourself up with some friend in the main cabin below decks on some large ship, and have with you there some flies, butterflies, and other small flying animals. Have a large bowl of water with some fish in it; hang up a bottle that empties drop by drop into a wide vessel beneath it. With the ship standing still, observe carefully how the little animals fly with equal speed to all sides of the cabin. The fish swim indifferently in all directions; the drops fall into the vessel beneath; and, in throwing something to your friend, you need throw it no more strongly in one direction than another, the distances being equal; jumping with your feet together, you pass equal spaces in every direction. When you have observed all these things carefully (though there is no doubt that when the ship is standing still, everything must happen in this way), have the ship proceed with any speed you like, so long as the motion is uniform and not fluctuating this way and that. You will discover not the least change in all the effects named, nor could you tell from any of them whether the ship was moving or standing still. In jumping, you will pass on the floor the same spaces as before, nor will you make larger jumps toward the stern than toward the prow even though the ship is moving quite rapidly, despite the fact that during the time you are in the air, the floor under you will be going in a direction opposite to your jump. In throwing something to your companion, you will need no more force to get it to him whether he is in the direction of the bow or the stern, with yourself situated opposite. The droplets will fall as before into the vessel, without dropping toward the stern, although while the drops are in the air the ship runs many spans. The fish in their water will swim toward the front of their bowl with no more effort than toward the back, and will go with equal ease toward bait placed anywhere around the edges of the bowl. Finally the butterflies and flies will continue their flights indifferently toward every side, nor will it ever happen that they are concentrated toward the stern, as if tired out from keeping up with the course of the ship, from which they will have been separated during long intervals by keeping themselves in the air. . . . "

EXAMPLE 2.1.1

Is the Earth a Good Inertial Reference Frame?

Calculate the centripetal acceleration (see Example 1.12.2), relative to the acceleration due to gravity g, of

(a) a point on the surface of the Earth's equator (the radius of the Earth is $R_E = 6.4 \times 10^3$ km)
(b) the Earth in its orbit about the Sun (the radius of the Earth's orbit is $a_E = 150 \times 10^6$ km)
(c) the Sun in its rotation about the center of the galaxy (the radius of the Sun's orbit about the center of the galaxy is $R_G = 2.8 \times 10^4$ LY. Its orbital speed is $v_G = 220$ km/s)

Solution:

The centripetal acceleration of a point rotating in a circle of radius R is given by

$$a_c = \omega^2 R = \left(\frac{2\pi}{T}\right)^2 R = \frac{4\pi^2 R}{T^2}$$

where T is period of one complete rotation. Thus, relative to g we have

$$\frac{a_c}{g} = \frac{4\pi^2 R}{gT^2}$$

(a) $\dfrac{a_c}{g} = \dfrac{4\pi^2 (6.4 \times 10^6 \, \text{m})}{9.8 \, \text{m} \cdot \text{s}^{-2} (3.16 \times 10^7 \, \text{s})^2} = 3.4 \times 10^{-3}$

(b) 6×10^{-4}

(c) 1.5×10^{-12}

Question for Discussion

Suppose that you step inside an express elevator on the 120th floor of a tall skyscraper. The elevator starts its descent, but as in your worst nightmare, the support elevator cable snaps and you find yourself suddenly in freefall. Realizing that your goose is cooked— or soon will be—you decide to conduct some physics experiments during the little time you have left on Earth—or above it! First, you take your wallet out of your pocket and remove a dollar bill. You hold it in front of your face and let it go. Wonder of wonders— it does nothing! It just hangs there seemingly suspended in front of your face (Figure 2.1.3)! Being an educated person with a reasonably good understanding of Newton's first law of motion, you conclude that there is no force acting on the dollar bill. Being a skeptical person, however, you decide to subject this conclusion to a second test. You take a piece of string from your pocket, tie one end to a light fixture on the ceiling of the falling elevator, attach your wallet to the other end, having thus fashioned a crude

Figure 2.1.3   Person in falling elevator.

plumb bob. You know that a hanging plumb bob aligns itself in the direction of gravity, which you anticipate is perpendicular to the plane of the ceiling. However, you discover that no matter how you initially align the plumb bob relative to the ceiling, it simply hangs in that orientation. There appears to be no gravitational force acting on the plumb bob, either. Indeed, there appears to be no force of any kind acting on any object inside the elevator. You now wonder why your physics instructor had such difficulty trying to find a perfect inertial frame of reference, because you appear to have discovered one quite easily—just get into a freely falling elevator. Unfortunately, you realize that within a few moments, you will not be able to share the joy of your discovery with anyone else.

So—is an elevator in free fall a perfect inertial frame of reference, or not?

Hint: Consider this quotation by Albert Einstein.

At that moment there came to me the happiest moment of my life . . . for an observer falling freely from the roof of a house no gravitational force exists during his fall—at least not in his immediate vicinity. That is, if the observer releases any objects, they remain in a state of rest or uniform motion relative to him, respectively, independent of their unique chemical and physical nature. Therefore, no observer is entitled to interpret his state as that of "rest."

For a more detailed discussion of inertial frames of reference and their relationship to gravity, read the delightful book, *Spacetime Physics*, 2nd ed., by Taylor and Wheeler, W. H. Freeman & Co., New York, 1992.

Mass and Force: Newton's Second and Third Laws

The quantitative measure of inertia is called *mass*. We are all familiar with the notion that the more massive an object is, the more resistive it is to acceleration. Go push a bike to get it rolling, and then try the same thing with a car. Compare the efforts. The car is much more massive and a much larger force is required to accelerate it than the bike. A more quantitative definition may be constructed by considering two masses, m_1 and m_2, attached by a spring and initially at rest in an inertial frame of reference. For example, we could imagine the two masses to be on a frictionless surface, almost achieved in practice by two carts on an air track, commonly seen in elementary physics class demonstrations. Now imagine someone pushing the two masses together, compressing the spring, and then suddenly releasing them so that they fly apart, attaining speeds v_1 and v_2. We *define* the ratio of the two masses to be

$$\frac{m_2}{m_1} = \left|\frac{\mathbf{v}_1}{\mathbf{v}_2}\right|$$

(2.1.1)

If we let m_1 be the standard of mass, then all other masses can be operationally defined in the above way relative to the standard. This operational definition of mass is consistent with Newton's second and third laws of motion, as we shall soon see. Equation 2.1.1 is equivalent to

$$\Delta(m_1\mathbf{v}_1) = -\Delta(m_2\mathbf{v}_2)$$

(2.1.2)

because the initial velocities of each mass are zero and the final velocities $\mathbf{v}_1$ and $\mathbf{v}_2$ are in opposite directions. If we divide by Δt and take limits as $\Delta t \to 0$, we obtain

$$\frac{d}{dt}(m_1\mathbf{v}_1) = -\frac{d}{dt}(m_2\mathbf{v}_2) \qquad (2.1.3)$$

The product of mass and velocity, $m\mathbf{v}$, is called *linear momentum*. The "change of motion" stated in the second law of motion was rigorously defined by Newton to be the time rate of change of the linear momentum of an object, and so the second law can be rephrased as follows: *The time rate of change of an object's linear momentum is proportional to the impressed force,* $\mathbf{F}$. Thus, the second law can be written as

$$\mathbf{F} = k\frac{d(m\mathbf{v})}{dt} \qquad (2.1.4)$$

where k is a constant of proportionality. Considering the mass to be a constant, independent of velocity (which is not true of objects moving at "relativistic" speeds or speeds approaching the speed of light, 3×10^8 m/s, a situation that we do not consider in this book), we can write

$$\mathbf{F} = km\frac{d\mathbf{v}}{dt} = km\mathbf{a} \qquad (2.1.5)$$

where $\mathbf{a}$ is the resultant acceleration of a mass m subjected to a force $\mathbf{F}$. The constant of proportionality can be taken to be $k = 1$ by defining the unit of force in the SI system to be that which causes a 1-kg mass to be accelerated 1 m/s^2. This force unit is called 1 newton.

Thus, we finally express Newton's second law in the familiar form

$$\mathbf{F} = \frac{d(m\mathbf{v})}{dt} = m\mathbf{a} \qquad (2.1.6)$$

The force $\mathbf{F}$ on the left side of Equation 2.1.6 is the *net* force acting upon the mass m; that is, it is the vector sum of all of the individual forces acting upon m.

We note that Equation 2.1.3 is equivalent to

$$\mathbf{F}_1 = -\mathbf{F}_2 \qquad (2.1.7)$$

or Newton's third law, namely, that two interacting bodies exert equal and opposite forces upon one another. Thus, our definition of mass is consistent with both Newton's second and third laws.

Linear Momentum

Linear momentum proves to be such a useful notion that it is given its own symbol:

$$\mathbf{p} = m\mathbf{v} \qquad (2.1.8)$$

Newton's second law may be written as

$$\mathbf{F} = \frac{d\mathbf{p}}{dt} \qquad (2.1.9)$$

Thus, Equation 2.1.3, which describes the behavior of two mutually interacting masses, is equivalent to

$$\frac{d}{dt}(\mathbf{p}_1 + \mathbf{p}_2) = 0 \qquad (2.1.10)$$

or

$$\mathbf{p}_1 + \mathbf{p}_2 = \text{constant} \qquad (2.1.11)$$

In other words, Newton's third law implies that the total momentum of two mutually interacting bodies is a constant. This constancy is a special case of the more general situation in which the total linear momentum of an isolated system (a system subject to no net externally applied forces) is a conserved quantity. The law of linear momentum conservation is one of the most fundamental laws of physics and is valid even in situations in which Newtonian mechanics fails.

EXAMPLE 2.1.2

A spaceship of mass M is traveling in deep space with velocity $v_i = 20$ km/s relative to the Sun. It ejects a rear stage of mass $0.2\,M$ with a relative speed $u = 5$ km/s (Figure 2.1.4). What then is the velocity of the spaceship?

Solution:

The system of spaceship plus rear stage is a closed system upon which no external forces act (neglecting the gravitational force of the Sun); therefore, the total linear momentum is conserved. Thus

$$\mathbf{P}_f = \mathbf{P}_i$$

where the subscripts i and f refer to initial and final values respectively. Taking velocities in the direction of the spaceship's travel to be positive, before ejection of the rear stage, we have

$$P_i = Mv_i$$

Let U be the velocity of the ejected rear stage and v_f be the velocity of the ship after ejection. The total momentum of the system after ejection is then

$$P_f = 0.20\,MU + 0.80\,Mv_f$$

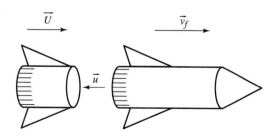

Figure 2.1.4   Spaceship ejecting a rear stage.

The speed u of the ejected stage relative to the spaceship is the difference in velocities of the spaceship and stage

$$u = v_f - U$$

or

$$U = v_f - u$$

Substituting this latter expression into the equation above and using the conservation of momentum condition, we find

$$0.20\, M(v_f - u) + 0.8\, Mv_f = Mv_i$$

which gives us

$$v_f = v_i + 0.2\,u = 20 \text{ km/s} + 0.20\,(5 \text{ km/s}) = 21 \text{ km/s}$$

Motion of a Particle

Equation 2.1.6 is the fundamental equation of motion for a particle subject to the influence of a *net* force, $\mathbf{F}$. We emphasize this point by writing $\mathbf{F}$ as $\mathbf{F}_{net}$, the vector sum of all the forces acting on the particle.

$$\mathbf{F_{net}} = \Sigma\, \mathbf{F}_i = m\frac{d^2\mathbf{r}}{dt^2} = m\mathbf{a} \qquad (2.1.12)$$

The usual problem of dynamics can be expressed in the following way: Given a knowledge of the forces acting on a particle (or system of particles), calculate the acceleration of the particle. Knowing the acceleration, calculate the velocity and position as functions of time. This process involves solving the second-order differential equation of motion represented by Equation 2.1.12. A complete solution requires a knowledge of the initial conditions of the problem, such as the values of the position and velocity of the particle at time $t = 0$. The initial conditions plus the dynamics dictated by the differential equation of motion of Newton's second law completely determine the subsequent motion of the particle. In some cases this procedure cannot be carried to completion in an analytic way. The solution of a complex problem will, in general, have to be carried out using numerical approximation techniques on a digital computer.

2.2 | Rectilinear Motion: Uniform Acceleration Under a Constant Force

When a moving particle remains on a single straight line, the motion is said to be *rectilinear*. In this case, without loss of generality we can choose the x-axis as the line of motion. The general equation of motion is then

$$F_x(x,\dot{x},t) = m\ddot{x} \qquad (2.2.1)$$

(**Note:** *In the rest of this chapter, we usually use the single variable* x *to repre-sent the position of a particle. To avoid excessive and unnecessary use of sub-scripts, we often use the symbols* v *and* a *for* $\dot{x}$ *and* $\ddot{x}$*, respectively, rather than* v_x *and* a_x*, and* F *rather than* F_x*.*)

The simplest situation is that in which the force is constant. In this case we have con-stant acceleration

$$\ddot{x} = \frac{dv}{dt} = \frac{F}{m} = \text{constant} = a \qquad (2.2.2a)$$

and the solution is readily obtained by direct integration with respect to time:

$$\dot{x} = v = at + v_0 \qquad (2.2.2b)$$

$$x = \frac{1}{2}at^2 + v_0 t + x_0 \qquad (2.2.2c)$$

where v_0 is the velocity and x_0 is the position at $t = 0$. By eliminating the time t between Equations 2.2.2b and 2.2.2c, we obtain

$$2a(x - x_0) = v^2 - v_0^2 \qquad (2.2.2d)$$

The student will recall the above familiar equations of uniformly accelerated motion. There are a number of fundamental applications. For example, in the case of a body falling freely near the surface of the Earth, neglecting air resistance, the acceleration is very nearly constant. We denote the acceleration of a freely falling body with **g**. Its mag-nitude is $g = 9.8$ m/s^2. The downward force of gravity (the *weight*) is, accordingly, equal to *m***g**. The gravitational force is always present, regardless of the motion of the body, and is independent of any other forces that may be acting.[6] We henceforth call it *m***g**.

EXAMPLE 2.2.1

Consider a block that is free to slide down a smooth, frictionless plane that is inclined at an angle θ to the horizontal, as shown in Figure 2.2.1(a). If the height of the plane is h and the block is released from rest at the top, what will be its speed when it reaches the bottom?

Solution:

We choose a coordinate system whose positive x-axis points down the plane and whose y-axis points "upward," perpendicular to the plane, as shown in the figure. The only force along the x direction is the component of gravitational force, $mg \sin \theta$, as shown in Figure 2.2.1(b). It is constant. Thus, Equations 2.2.2a–d are the equations of motion, where

$$\ddot{x} = a = \frac{F_x}{m} = g \sin \theta$$

---

[6] Effects of the Earth's rotation are studied in Chapter 5.

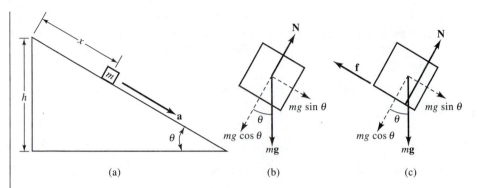

Figure 2.2.1   (a) A block sliding down an inclined plane. (b) Force diagram (no friction). (c) Force diagram (friction $f = \mu_\kappa N$).

and

$$x - x_0 = \frac{h}{\sin\theta}$$

Thus,

$$v^2 = 2(g\sin\theta)\left(\frac{h}{\sin\theta}\right) = 2gh$$

Suppose that, instead of being smooth, the plane is rough; that is, it exerts a frictional force **f** on the particle. Then the net force in the x direction, (see Figure 2.2.1(c)), is equal to $mg\sin\theta - f$. Now, for sliding contact it is found that the magnitude of the frictional force is proportional to the magnitude of the normal force N; that is,

$$f = \mu_\kappa N$$

where the constant of proportionality μ_κ is known as the *coefficient of sliding* or *kinetic friction*.[7] In the example under discussion, the normal force, as shown in the figure, is equal to $mg\cos\theta$; hence,

$$f = \mu_\kappa mg\cos\theta$$

Consequently, the net force in the x direction is equal to

$$mg\sin\theta - \mu_\kappa mg\cos\theta$$

Again the force is constant, and Equations 2.2.2a–d apply where

$$\ddot{x} = \frac{F_x}{m} = g(\sin\theta - \mu_k\cos\theta)$$

---

[7]There is another coefficient of friction called the *static* coefficient μ_s, which, when multiplied by the normal force, gives the maximum frictional force under static contact, that is, the force required to barely start an object to move when it is initially at rest. In general, $\mu_s > \mu_\kappa$.

The speed of the particle increases if the expression in parentheses is positive—that is, if $\theta > \tan^{-1} \mu_\kappa$. The angle, $\tan^{-1} \mu_\kappa$, usually denoted by ϵ, is called the *angle of kinetic friction*. If $\theta = \epsilon$, then $a = 0$, and the particle slides down the plane with constant speed. If $\theta < \epsilon$, a is negative, and so the particle eventually comes to rest. For motion *up* the plane, the direction of the frictional force is reversed; that is, it is in the positive x direction. The acceleration (actually deceleration) is then $\ddot{x} = g(\sin\theta + \mu_\kappa \cos\theta)$.

2.3 | Forces that Depend on Position: The Concepts of Kinetic and Potential Energy

It is often true that the force a particle experiences depends on the particle's position with respect to other bodies. This is the case, for example, with electrostatic and gravitational forces. It also applies to forces of elastic tension or compression. If the force is independent of velocity or time, then the differential equation for rectilinear motion is simply

$$F(x) = m\ddot{x} \qquad (2.3.1)$$

It is usually possible to solve this type of differential equation by one of several methods, such as using the chain rule to write the acceleration in the following way:

$$\ddot{x} = \frac{d\dot{x}}{dt} = \frac{dx}{dt}\frac{d\dot{x}}{dx} = v\frac{dv}{dx} \qquad (2.3.2)$$

so the differential equation of motion may be written

$$F(x) = mv\frac{dv}{dx} = \frac{m}{2}\frac{d(v^2)}{dx} = \frac{dT}{dx} \qquad (2.3.3)$$

The quantity $T = \frac{1}{2}mv^2$ is called the *kinetic energy* of the particle. We can now express Equation 2.3.3 in integral form:

$$W = \int_{x_0}^{x} F(x)\,dx = T - T_0 \qquad (2.3.4)$$

The integral $\int F(x)\,dx$ is the *work* W done on the particle by the impressed force $F(x)$. *The work is equal to the change in the kinetic energy of the particle.* Let us *define* a function $V(x)$ such that

$$-\frac{dV(x)}{dx} = F(x) \qquad (2.3.5)$$

The function $V(x)$ is called the *potential energy*; it is defined only to within an arbitrary additive constant. In terms of $V(x)$, the work integral is

$$W = \int_{x_0}^{x} F(x)\,dx = -\int_{x_0}^{x} dV = -V(x) + V(x_0) = T - T_0 \qquad (2.3.6)$$

Notice that Equation 2.3.6 remains unaltered if $V(x)$ is changed by adding *any* constant C, because

$$-[V(x) + C] + [V(x_0) + C] = -V(x) + V(x_0) \qquad (2.3.7)$$

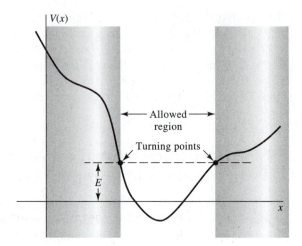

Figure 2.3.1 Graph of a one-dimensional potential energy function $V(x)$ showing the allowed region of motion and the turning points for a given value of the total energy E.

We now transpose terms and write Equation 2.3.6 in the following form:

$$T_0 + V(x_0) = \text{constant} = T + V(x) \equiv E \tag{2.3.8}$$

This is the *energy equation*. E is defined to be the *total energy* of the particle (technically, it's the total *mechanical* energy). It is equal to the sum of the kinetic and potential energies and is constant throughout the motion of the particle. This constancy results from the fact that the impressed force is a function only of the position x (of the particle and consequently can be derived from a corresponding potential energy) function $V(x)$. Such a force is said to be *conservative*.[8] Nonconservative forces—that is, those for which no potential energy function exists—are usually of a dissipational nature, such as friction.

The motion of the particle can be obtained by solving the energy equation (Equation 2.3.8) for v,

$$v = \frac{dx}{dt} = \pm\sqrt{\frac{2}{m}[E - V(x)]} \tag{2.3.9}$$

which can be written in integral form,

$$\int_{x_0}^{x} \frac{dx}{\pm\sqrt{\frac{2}{m}[E - V(x)]}} = t - t_0 \tag{2.3.10}$$

thus giving t as a function of x.

In view of Equation 2.3.9, we see that the expression for v is real only for those values of x such that $V(x)$ is less than or equal to the total energy E. Physically, this means that the particle is confined to the region or regions for which the condition $V(x) \leq E$ is satisfied. Furthermore, v goes to zero when $V(x) = E$. This means that the particle must come to rest and reverse its motion at points for which the equality holds. These points are called the *turning points* of the motion. The above facts are illustrated in Figure 2.3.1.

---

[8] A more complete discussion of conservative forces is found in Chapter 4.

EXAMPLE 2.3.1

Free Fall

The motion of a freely falling body (discussed above under the case of constant acceleration) is an example of conservative motion. If we choose the x direction to be positive upward, then the gravitational force is equal to $-mg$. Therefore, $-dV/dx = -mg$, and $V = mgx + C$. The constant of integration C is arbitrary and merely depends on the choice of the reference level for measuring V. We can choose $C = 0$, which means that $V = 0$ when $x = 0$. The energy equation is then

$$\tfrac{1}{2}mv^2 + mgx = E$$

The energy constant E is determined from the initial conditions. For instance, let the body be projected upward with initial speed v_0 from the origin $x = 0$. These values give $E = mv_0^2/2 = mv^2/2 + mgx$, so

$$v^2 = v_0^2 - 2gx$$

The turning point of the motion, which is in this case the maximum height, is given by setting $v = 0$. This gives $0 = v_0^2 - 2gx_{max}$, or

$$h = x_{max} = \frac{v_0^2}{2g}$$

EXAMPLE 2.3.2

Variation of Gravity with Height

In Example 2.3.1 we assumed that g was constant. Actually, the force of gravity between two particles is inversely proportional to the square of the distance between them (Newton's law of gravity).[9] Thus, the gravitational force that the Earth exerts on a body of mass m is given by

$$F_r = -\frac{GMm}{r^2}$$

in which G is Newton's constant of gravitation, M is the mass of the Earth, and r is the distance from the center of the Earth to the body. By definition, this force is equal to the quantity $-mg$ when the body is at the surface of the Earth, so $mg = GMm/r_e^2$. Thus, $g = GM/r_e^2$ is the acceleration of gravity at the Earth's surface. Here r_e is the radius of the Earth (assumed to be spherical). Let x be the distance above the surface, so that $r = r_e + x$. Then, neglecting any other forces such as air resistance, we can write

$$F(x) = -mg\frac{r_e^2}{(r_e + x)^2} = m\ddot{x}$$

---

[9] We study Newton's law of gravity in more detail in Chapter 6.

for the differential equation of motion of a vertically falling (or rising) body with the variation of gravity taken into account. To integrate, we set $\ddot{x} = v\,dv/dx$. Then

$$-mgr_e^2 \int_{x_0}^{x} \frac{dx}{(r_e + x)^2} = \int_{v_0}^{v} mv\,dv$$

$$mgr_e^2 \left(\frac{1}{r_e + x} - \frac{1}{r_e + x_0} \right) = \frac{1}{2}mv^2 - \frac{1}{2}mv_0^2$$

This is just the *energy equation* in the form of Equation 2.3.6. The potential energy is $V(x) = -mg[r_e^2/(r_e + x)]$ rather than mgx.

Maximum Height: Escape Speed

Suppose a body is projected upward with initial speed v_0 at the surface of the Earth, $x_0 = 0$. The energy equation then yields, upon solving for v^2, the following result:

$$v^2 = v_0^2 - 2gx\left(1 + \frac{x}{r_e}\right)^{-1}$$

This reduces to the result for a uniform gravitational field of Example 2.2.1, if x is very small compared to r_e so that the term x/r_e can be neglected. The turning point (maximum height) is found by setting $v = 0$ and solving for x. The result is

$$x_{max} = h = \frac{v_0^2}{2g}\left(1 - \frac{v_0^2}{2gr_e}\right)^{-1}$$

Again we get the formula of Example 2.2.1 if the second term in the parentheses can be ignored, that is, if v_0^2 is much smaller than $2gr_e$.

Using this last, exact expression, we solve for the value of v_0 that gives an infinite value for h. This is called the *escape speed,* and it is found by setting the quantity in parentheses equal to zero. The result is

$$v_e = (2gr_e)^{1/2}$$

This gives, for $g = 9.8$ m/s^2 and $r_e = 6.4 \times 10^6$ m,

$$v_e \approx 11 \text{ km/s} \approx 7 \text{ mi/s}$$

for the numerical value of the escape speed from the surface of the Earth.

In the Earth's atmosphere, the average speed of air molecules (O_2 and N_2) is about 0.5 km/s, which is considerably less than the escape speed, so the Earth retains its atmosphere. The moon, on the other hand, has no atmosphere; because the escape speed at the moon's surface, owing to the moon's small mass, is considerably smaller than that at the Earth's surface, any oxygen or nitrogen would eventually disappear. The Earth's atmosphere, however, contains no significant amount of hydrogen, even though hydrogen is the most abundant element in the universe as a whole. A hydrogen atmosphere would have escaped from the Earth long ago, because the molecular speed of hydrogen is large enough (owing to the small mass of the hydrogen molecule) that at any instant a significant number of hydrogen molecules would have speeds exceeding the escape speed.

EXAMPLE 2.3.3

The *Morse function* $V(x)$ approximates the potential energy of a vibrating diatomic molecule as a function of x, the distance of separation of its constituent atoms, and is given by

$$V(x) = V_0\left[1 - e^{-(x-x_0)/\delta}\right]^2 - V_0$$

where V_0, x_0, and δ are parameters chosen to describe the observed behavior of a particular pair of atoms. The force that each atom exerts on the other is given by the derivative of this function with respect to x. Show that x_0 is the separation of the two atoms when the potential energy function is a minimum and that its value for that distance of separation is $V(x_0) = -V_0$. (When the molecule is in this configuration, we say that it is in equilibrium.)

Solution:

The potential energy of the diatomic molecule is a minimum when its derivative with respect to x, the distance of separation, is zero. Thus,

$$F(x) = -\frac{dV(x)}{dx} = 0 =$$

$$2\frac{V_0}{\delta}\left(1 - e^{-(x-x_0)/\delta}\right)\left(e^{-(x-x_0)/\delta}\right) = 0$$

$$1 - e^{-(x-x_0)/\delta} = 0$$

$$\ln(1) = -(x - x_0)/\delta = 0$$

$$\therefore x = x_0$$

The value of the potential energy at the minimum can be found by setting $x = x_0$ in the expression for $V(x)$. This gives $V(x_0) = -V_0$.

EXAMPLE 2.3.4

Shown in Figure 2.3.2 is the potential energy function for a diatomic molecule. Show that, for separation distances x close to x_0, the potential energy function is parabolic and the resultant force on each atom of the pair is linear, always directed toward the equilibrium position.

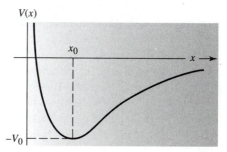

Figure 2.3.2   Potential energy function for a diatomic molecule.

Solution:

All we need do here is expand the potential energy function near the equilibrium position.

$$V(x) \approx V_0\left[1-\left(1-\left(\frac{x-x_0}{\delta}\right)\right)\right]^2 - V_0$$

$$\approx \frac{V_0}{\delta^2}(x-x_0)^2 - V_0$$

$$F(x) = -\frac{dV(x)}{dx} = -\frac{2V_0}{\delta^2}(x-x_0)$$

(**Note:** The force is linear and is directed in such a way as to restore the diatomic molecule to its equilibrium position.)

EXAMPLE 2.3.5

The binding energy $(-V_0)$ of the diatomic hydrogen molecule H_2 is -4.52 eV (1 eV = 1.6×10^{-19} joules; 1 joule = 1 N · m). The values of the constants x_0 and δ are .074 and .036 nm, respectively (1 nm = 10^{-9} m). Assume that at room temperature the total energy of the hydrogen molecule is about $\Delta E = 1/40$ eV higher than its binding energy. Calculate the maximum separation of the two atoms in the diatomic hydrogen molecule.

Solution:

Because the molecule has a little more energy than its minimum possible value, the two atoms will vibrate between two values of x, where their kinetic energy is zero. At these turning points, all the energy is potential; hence,

$$V(x) = -V_0 + \Delta E \approx \frac{V_0}{\delta^2}(x-x_0)^2 - V_0$$

$$x = x_0 \pm \delta\sqrt{\frac{\Delta E}{V_0}}$$

Putting in numbers, we see that the hydrogen molecule vibrates at room temperature a distance of about ±4% of its equilibrium separation.

For this situation where the oscillation is small, the two atoms undergo a symmetrical displacement about their equilibrium position. This arises from approximating the potential function as a parabola near equilibrium. Note from Figure 2.3.2 that, farther away from the equilibrium position, the potential energy function is not symmetrical, being steeper at smaller distances of separation. Thus, as the diatomic molecule is "heated up," on the average it spends an increasingly greater fraction of its time separated by a distance greater than their separation at equilibrium. This is why most substances tend to expand when heated.

2.4 | Velocity-Dependent Forces: Fluid Resistance and Terminal Velocity

It often happens that the force that acts on a body is a function of the velocity of the body. This is true, for example, in the case of viscous resistance exerted on a body moving through a fluid. If the force can be expressed as a function of v only, the differential equation of motion may be written in either of the two forms

$$F_0 + F(v) = m\frac{dv}{dt} \tag{2.4.1}$$

$$F_0 + F(v) = mv\frac{dv}{dx} \tag{2.4.2}$$

Here F_0 is any constant force that does not depend on v. Upon separating variables, integration yields either t or x as a function of v. A second integration can then yield a functional relationship between x and t.

For normal fluid resistance, including air resistance, $F(v)$ is not a simple function and generally must be found through experimental measurements. However, a fair approximation for many cases is given by the equation

$$F(v) = -c_1 v - c_2 v |v| = -v (c_1 + c_2 |v|) \tag{2.4.3}$$

in which c_1 and c_2 are constants whose values depend on the size and shape of the body. (The absolute-value sign is necessary on the last term because the force of fluid resistance is always opposite to the direction of v.) If the above form for $F(v)$ is used to find the motion by solving Equation 2.4.1 or 2.4.2, the resulting integrals are somewhat messy. But for the limiting cases of small v and large v, respectively, the linear or the quadratic term in $F(v)$ dominates, and the differential equations become somewhat more manageable.

For spheres in air, approximate values for the constants in the equation for $F(v)$ are, in SI units,

$$c_1 = 1.55 \times 10^{-4} D$$
$$c_2 = 0.22 D^2$$

where D is the diameter of the sphere in meters. The ratio of the quadratic term $c_2 v |v|$ to the linear term $c_1 v$ is, thus,

$$\frac{0.22 v |v| D^2}{1.55 \times 10^{-4} vD} = 1.4 \times 10^3 |v| D$$

This means that, for instance, with objects of baseball size ($D \sim 0.07$ m), the quadratic term dominates for speeds in excess of 0.01 m/s (1 cm/s), and the linear term dominates for speeds less than this value. For speeds around this value, *both* terms must be taken into account. (See Problem 2.15.)

EXAMPLE 2.4.1

Horizontal Motion with Linear Resistance

Suppose a block is projected with initial velocity v_0 on a smooth horizontal surface and that there is air resistance such that the linear term dominates. Then, in the direction of the motion, $F_0 = 0$ in Equations 2.4.1 and 2.4.2, and $F(v) = -c_1 v$. The differential equation of motion is then

$$-c_1 v = m \frac{dv}{dt}$$

which gives, upon integrating,

$$t = \int_{v_0}^{v} -\frac{m \, dv}{c_1 v} = -\frac{m}{c_1} \ln \left(\frac{v}{v_0} \right)$$

Solution:

We can easily solve for v as a function of t by multiplying by $-c_1/m$ and taking the exponential of both sides. The result is

$$v = v_0 e^{-c_1 t/m}$$

Thus, the velocity decreases exponentially with time. A second integration gives

$$x = \int_0^t v_0 e^{-c_1 t/m} \, dt$$

$$= \frac{m v_0}{c_1} (1 - e^{-c_1 t/m})$$

showing that the block approaches a limiting position given by $x_{lim} = m v_0/c_1$.

EXAMPLE 2.4.2

Horizontal Motion with Quadratic Resistance

If the parameters are such that the quadratic term dominates, then for positive v we can write

$$-c_2 v^2 = m \frac{dv}{dt}$$

which gives

$$t = \int_{v_0}^{v} \frac{-m \, dv}{c_2 v^2} = \frac{m}{c_2} \left(\frac{1}{v} - \frac{1}{v_0} \right)$$

Solution:

Solving for v, we get

$$v = \frac{v_0}{1 + kt}$$

where $k = c_2 v_0/m$. A second integration gives us the position as a function of time:

$$x(t) = \int_0^t \frac{v_0 dt}{1 + kt} = \frac{v_0}{k} \ln(1 + kt)$$

Thus, as $t \to \infty$, v decreases as $1/t$, but the position does not approach a limit as was obtained in the case of a linear retarding force. Why might this be? You might guess that a quadratic retardation should be more effective in stopping the block than is a linear one. This is certainly true at large velocities, but as the velocity approaches zero, the quadratic retarding force goes to zero much faster than the linear one—enough to allow the block to continue on its merry way, albeit at a very slow speed.

Vertical Fall Through a Fluid: Terminal Velocity

(a) *Linear case*. For an object falling vertically in a resisting fluid, the force F_0 in Equations 2.4.1 and 2.4.2 is the weight of the object, namely, $-mg$ for the x-axis positive in the upward direction. For the linear case of fluid resistance, we then have for the differential equation of motion

$$-mg - c_1 v = m \frac{dv}{dt} \tag{2.4.4}$$

Separating variables and integrating, we find

$$t = \int_{v_0}^v \frac{m \, dv}{-mg - c_1 v} = -\frac{m}{c_1} \ln \frac{mg + c_1 v}{mg + c_1 v_0} \tag{2.4.5}$$

in which v_0 is the initial velocity at $t = 0$. Upon multiplying by $-c_1/m$ and taking the exponential, we can solve for v:

$$v = -\frac{mg}{c_1} + \left(\frac{mg}{c_1} + v_0 \right) e^{-c_1 t/m} \tag{2.4.6}$$

The exponential term drops to a negligible value after a sufficient time ($t \gg m/c_1$), and the velocity approaches the limiting value $-mg/c_1$. The limiting velocity of a falling body is called the *terminal velocity*; it is that velocity at which the force of resistance is just equal and opposite to the weight of the body so that the total force is zero, and so the acceleration is zero. The magnitude of the terminal velocity is the *terminal speed*.

Let us designate the terminal speed mg/c_1 by v_t, and let us write τ (which we may call the *characteristic time*) for m/c_1. Equation 2.4.6 may then be written in the more significant form

$$v = -v_t(1 - e^{-t/\tau}) + v_0 e^{-t/\tau} \tag{2.4.7}$$

These two terms represent two velocities: the terminal velocity v_t, which exponentially "fades in," and the initial velocity v_0, which exponentially "fades out" due to the action of the viscous drag force.

In particular, for an object dropped from rest at time $t = 0$, $v_0 = 0$, we find

$$v = -v_t(1 - e^{-t/\tau}) \tag{2.4.8}$$

Thus, after one characteristic time the speed is $1 - e^{-1}$ times the terminal speed, after two characteristic times it is the factor $1 - e^{-2}$ of v_t, and so on. After an interval of 5τ, the speed is within 1% of the terminal value, namely, $(1 - e^{-5})v_t = 0.993\, v_t$.

(b) *Quadratic case.* In this case, the magnitude of $F(v)$ is proportional to v^2. To ensure that the force remains resistive, we must remember that the sign preceding the $F(v)$ term depends on whether or not the motion of the object is upward or downward. This is the case for any resistive force proportional to an *even* power of velocity. A general solution usually involves treating the upward and downward motions separately. Here, we simplify things somewhat by considering only the situation in which the body is either dropped from rest or projected downward with an initial velocity v_0. We leave it as an exercise for the student to treat the upward-going case. We take the downward direction to be the positive y direction. The differential equation of motion is

$$m\frac{dv}{dt} = mg - c_2 v^2 = mg\left(1 - \frac{c_2}{mg}v^2\right)$$

$$= mg\left(1 - \frac{v^2}{v_t^2}\right)$$

$$\frac{dv}{dt} = g\left(1 - \frac{v^2}{v_t^2}\right)$$

(2.4.9)

where

$$v_t = \sqrt{\frac{mg}{c_2}} \qquad (terminal\ speed)$$

(2.4.10)

Integrating Equation 2.4.9 gives t as a function of v,

$$t - t_0 = \int_{v_0}^{v} \frac{dv}{g\left(1 - \frac{v^2}{v_t^2}\right)} = \tau\left(\tanh^{-1}\frac{v}{v_t} - \tanh^{-1}\frac{v_0}{v_t}\right)$$

(2.4.11)

where

$$\tau = \frac{v_t}{g} = \sqrt{\frac{m}{c_2 g}} \qquad (characteristic\ time)$$

(2.4.12)

Solving for v, we obtain

$$v = v_t \tanh\left(\frac{t - t_0}{\tau} - \tanh^{-1}\frac{v_0}{v_t}\right)$$

(2.4.13)

If the body is released from rest at time $t = 0$,

$$v = v_t \tanh\frac{t}{\tau} = v_t\left(\frac{e^{2t/\tau} - 1}{e^{2t/\tau} + 1}\right)$$

(2.4.14)

The terminal speed is attained after the lapse of a few characteristic times; for example, at $t = 5\tau$, the speed is $0.99991\, v_t$. Graphs of speed versus time of fall for the linear and quadratic cases are shown in Figure 2.4.1.

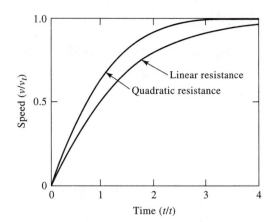

Figure 2.4.1 Graphs of speed (units of terminal speed) versus time (units of time constant τ) for a falling body.

In many instances we would like to know the speed attained upon falling a given distance. We could find this out by integrating Equation 2.4.13, obtaining y as a function of time, and then eliminating the time parameter to find speed versus distance. A more direct solution can be obtained by direct modification of the fundamental differential equation of motion so that the independent variable is distance instead of time. For example, because

$$\frac{dv}{dt} = \frac{dv}{dy}\frac{dy}{dt} = \frac{1}{2}\frac{dv^2}{dy} \tag{2.4.15}$$

Equation 2.4.9 can be rewritten with y as the independent variable:

$$\frac{dv^2}{dy} = 2g\left(1 - \frac{v^2}{v_t^2}\right) \tag{2.4.16}$$

We solve this equation as follows:

$$u = 1 - \frac{v^2}{v_t^2} \quad \text{so} \quad \frac{du}{dy} = -\frac{1}{v_t^2}\frac{dv^2}{dy} = -\left(\frac{2g}{v_t^2}\right)u$$

$$u = u(y = 0)e^{-2gy/v_t^2} \quad \text{but} \quad u(y = 0) = 1 - \frac{v_0^2}{v_t^2}$$

$$u = \left(1 - \frac{v_0^2}{v_t^2}\right)e^{-2gy/v_t^2} = 1 - \frac{v^2}{v_t^2}$$

$$\therefore v^2 = v_t^2\left(1 - e^{-2gy/v_t^2}\right) + v_0^2 e^{-2gy/v_t^2} \tag{2.4.17}$$

Thus, we see that the squares of the initial velocity and terminal velocity exponentially fade in and out within a characteristic length of $v_t^2/2g$.

EXAMPLE 2.4.3

Falling Raindrops and Basketballs

Calculate the terminal speed in air and the characteristic time for (a) a very tiny spherical raindrop of diameter 0.1 mm = 10^{-4} m and (b) a basketball of diameter 0.25 m and mass 0.6 kg.

Solution:

To decide which type of force law to use, quadratic or linear, we recall the expression that gives the ratio of the quadratic to the linear force for air resistance, namely, $1.4 \times 10^3 |v| D$. For the raindrop this is $0.14v$, and for the basketball it is $350v$, numerically, where v is in meters per second. Thus, for the raindrop, v must exceed $1/0.14 = 7.1$ m/s for the quadratic force to dominate. In the case of the basketball, v must exceed only $1/350 = 0.0029$ m/s for the quadratic force to dominate. We conclude that the linear case should hold for the falling raindrop, while the quadratic case should be correct for the basketball. (See also Problem 2.15.)

The volume of the raindrop is $\pi D^3/6 = 0.52 \times 10^{-12}$ m^3, so, multiplying by the density of water, 10^3 kg/m^3, gives the mass $m = 0.52 \times 10^{-9}$ kg. For the drag coefficient we get $c_1 = 1.55 \times 10^{-4} D = 1.55 \times 10^{-8}$ N · s/m. This gives a terminal speed

$$v_t = \frac{mg}{c_1} = \frac{0.52 \times 10^{-9} \times 9.8}{1.55 \times 10^{-8}} \text{ m/s} = 0.33 \text{ m/s}$$

The characteristic time is

$$\tau = \frac{v_t}{g} = \frac{0.33 \text{ m/s}}{9.8 \text{ m/s}^2} = 0.034 \text{ s}$$

For the basketball the drag constant is $c_2 = 0.22 D^2 = 0.22 \times (0.25)^2 = 0.0138$ N · s^2/m^3, and so the terminal speed is

$$v_t = \left(\frac{mg}{c_2}\right)^{1/2} = \left(\frac{0.6 \times 9.8}{0.0138}\right)^{1/2} \text{ m/s} = 20.6 \text{ m/s}$$

and the characteristic time is

$$\tau = \frac{v_t}{g} = \frac{20.6 \text{ m/s}}{9.8 \text{ m/s}^2} = 2.1 \text{ s}$$

Thus, the raindrop practically attains its terminal speed in less than 1 s when starting from rest, whereas it takes several seconds for the basketball to come to within 1% of the terminal value.

For more information on aerodynamic drag, the reader is referred to an article by C. Frohlich in *Am. J. Phys.*, **52**, 325 (1984) and the extensive list of references cited therein.

*2.5 | Vertical Fall Through a Fluid: Numerical Solution

Many problems in classical mechanics are described by fairly complicated equations of motion that cannot be solved analytically in closed form. When one encounters such a problem, the only available alternative is to try to solve the problem numerically. Once one decides that such a course of action is necessary, many alternatives open up. The widespread use of personal computers (PCs) with large amounts of memory and hard-disk storage capacity has made it possible to implement a wide variety of problem-solving tools in high-level languages without the tedium of programming. The tools in most widespread use among physicists include the software packages *Mathcad, Mathematica* (see Appendix I), and *Maple,* which are designed specifically to solve mathematical problems numerically (and symbolically).

As we proceed through the remaining chapters in this text, we use one or another of these tools, usually at the end of the chapter, to solve a problem for which no closed-form solution exists. Here we have used *Mathcad* to solve the problem of an object falling vertically through a fluid. The problem was solved analytically in the preceding section, and we use the solution we obtained there as a check on the numerical result we obtain here, in hopes of illustrating the power and ease of the numerical problem-solving technique.

Linear and quadratic cases revisited.   The first-order differential equation of motion for an object falling vertically through a fluid in which the retarding force is linear was given by Equation 2.4.4:

$$mg - c_1 v = m \frac{dv}{dt} \qquad (2.5.1a)$$

Here, though, we have chosen the downward y direction to be positive, because we consider only the situation in which the object is dropped from rest. The equation can be put into a much simpler form by expressing it in terms of the characteristic time $\tau = m/c_1$ and terminal velocity $v_t = mg/c_1$.

$$\frac{dv/v_t}{dt/\tau} = 1 - \frac{v}{v_t} \qquad (2.5.1b)$$

Now, in the above equation, we "scale" the velocity v and the time of fall t in units of v_t and τ, respectively; that is, we let $u = v/v_t$ and $T = t/\tau$. The preceding equation becomes

$$Linear: \quad \frac{du}{dT} = u' = 1 - u \qquad (2.5.1c)$$

where we denote the first derivative of u by u'.

---

*Sections in the text marked with * may be skipped with impunity.

An analysis similar to the one above leads to the following "scaled" first-order differential equation of motion for the case in which the retarding force is quadratic (see Equation 2.4.9).

$$\text{Quadratic:} \quad \frac{du}{dT} = u' = 1 - u^2 \qquad (2.5.2)$$

The *Mathcad* software package comes with the *rkfixed* function, a general-purpose Runge–Kutta solver that can be used on *n*th-order differential equations or on systems of differential equations whose initial conditions are known. This is the situation that faces us in both of the preceding cases. All we need do, it turns out, to solve these two differential equations is to "supply" them to the *rkfixed* function in *Mathcad*. This function uses the fourth-order Runge–Kutta method[10] to solve the equations. When called in *Mathcad*, it returns a two-column matrix in which

- the left-hand (or 0th) column contains the data points at which the solution to the differential equation is evaluated (in the case here, the data points are the times T_i);
- the right-hand (or first) column contains the corresponding values of the solution (the values u_i).

The syntax of the call to the function and the arguments of the function is:
rkfixed(**y**, x_0, x_f, *npoints*, **D**)

y	= a vector of *n* initial values, where *n* is the order of the differential equation or the size of the system of equations you're solving. For a single first-order differential equation, like the one in this case, the vector degenerates to a single initial value, $y(0) = y(x_0)$.
x_0, x_f	= the endpoints of the interval within which the solutions to the differential equation are to be evaluated. The initial values of **y** are the values at x_0.
npoints	= the number of points beyond the initial point at which the solution is to be evaluated. This value sets the number of rows to (1 + *npoints*) in the matrix *rkfixed*.
D(x, y)	= an *n*-element vector function containing the first derivatives of the unknown functions **y**. Again, for a single first-order differential equation, this vector function degenerates to a single function equal to the first derivative of the single function *y*.

We show on the next two pages an example of a *Mathcad* worksheet in which we obtained a numerical solution for the above first-order differential equations (2.5.1c and 2.5.2). The worksheet was imported to this text directly from *Mathcad*. What is shown there should be self-explanatory, but exactly how to implement the solution might not be. We discuss the details of how to do it in Appendix I. The important thing here is to note the simplicity of the solution (as evidenced by the brevity of the worksheet) and its accuracy (as can be seen by comparing the numerical solutions shown in Figure 2.5.1 with the analytic solutions shown in Figure 2.4.1). The accuracy is further detailed in Figure 2.5.2, where we have

---

[10] See, for example, R. L. Burden and J. Douglas Faires, *Numerical Analysis*, 6th ed, Brooks/Cole, Pacific Grove, ITP, 1977.

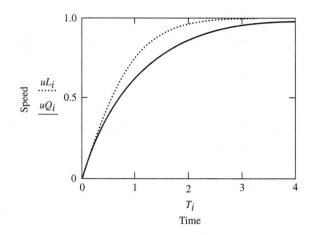

Figure 2.5.1 Numerical solution of speed versus time for a falling body. uL, linear case; uQ, quadratic case.

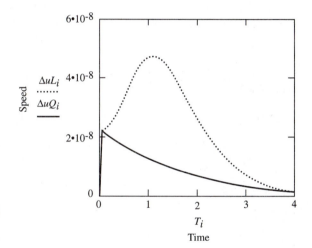

Figure 2.5.2 Difference between analytic and numerical solutions for the speed of a falling object. ΔuL, linear case; ΔuQ, quadratic case.

plotted the percent difference between the numerical and analytic solutions. The worst error, about 5×10^{-8}, occurs in the quadratic solution. Even greater accuracy could be achieved by dividing the time interval $(0-4)$ into even more data points than the 100 chosen here.

Mathcad Solution for Speed of Falling Object: Linear Retarding Force.

$u_0 := 0$ ← Define initial value (use [to make the subscript)

$D(T, u) := 1 - u$ ← Define function for first derivative u'

$Y := rkfixed(u, 0, 4, 100, D)$ ← Evaluates solution at 100 points between 0 and 4 using fourth-order Runge–Kutta.

$i := 0 .. \text{rows}(Y) - 1$

← i denotes each element pair in the matrix Y (a 101×2 matrix). First column contains data points (time T) where solution (velocity u) is evaluated. Second column contains u values.

$uL_i := (Y^{\langle 1 \rangle})_i$

← Rename normalized velocity, linear case

Mathcad Solution for Speed of Falling Object: Quadratic Retarding Force.

$u_0 := 0$

← Define initial value (use [to make the subscript)

$D(T, u) := 1 - u^2$

← Define function for first derivative u'

$Z := rkfixed\,(u, 0, 4, 100, D)$

← Evaluates solution at 100 points between 0 and 4 using fourth-order Runge–Kutta.

$T_i := 0.04i$

← Define time in terms of array element

$uQ_i := (Z^{\langle 1 \rangle})_i$

← Rename normalized velocity, quadratic case

Difference Between Analytic and Numerical Solutions.

$v_i := 1 - e^{-T_i}$

← Analytic solution for linear retarding force

$$u_i := \frac{(e^{2 \cdot T_i} - 1)}{(e^{2 \cdot T_i} + 1)}$$

← Analytic solution for quadratic retarding force

$$\Delta uL_i := \frac{(v_i - uL_i)}{v_i}$$

← Difference, linear case

$$\Delta uQ_i := \frac{(u_i - uQ_i)}{u_i}$$

← Difference, quadratic case

Problems

2.1   Find the velocity $\dot{x}$ and the position x as functions of the time t for a particle of mass m, which starts from rest at $x = 0$ and $t = 0$, subject to the following force functions:
(a) $F_x = F_0 + ct$
(b) $F_x = F_0 \sin ct$
(c) $F_x = F_0 e^{ct}$
where F_0 and c are positive constants.

2.2   Find the velocity $\dot{x}$ as a function of the displacement x for a particle of mass m, which starts from rest at $x = 0$, subject to the following force functions:
(a) $F_x = F_0 + cx$
(b) $F_x = F_0 e^{-cx}$
(c) $F_x = F_0 \cos cx$
where F_0 and c are positive constants.

2.3   Find the potential energy function $V(x)$ for each of the forces in Problem 2.2.

2.4   A particle of mass m is constrained to lie along a frictionless, horizontal plane subject to a force given by the expression $F(x) = -kx$. It is projected from $x = 0$ to the right along the positive x direction with initial kinetic energy $T_0 = 1/2 \, kA^2$. k and A are positive constants. Find **(a)** the potential energy function $V(x)$ for this force; **(b)** the kinetic energy, and **(c)** the total energy of the particle as a function of its position. **(d)** Find the turning points of the motion. **(e)** Sketch the potential, kinetic, and total energy functions. (*Optional*: Use *Mathcad* or *Mathematica* to plot these functions. Set k and A each equal to 1.)

2.5   As in the problem above, the particle is projected to the right with initial kinetic energy T_0 but subject to a force $F(x) = -kx + kx^3/A^2$, where k and A are positive constants. Find **(a)** the potential energy function $V(x)$ for this force; **(b)** the kinetic energy, and **(c)** the total energy of the particle as a function of its position. **(d)** Find the turning points of the motion and the condition the total energy of the particle must satisfy if its motion is to exhibit turning points. **(e)** Sketch the potential, kinetic, and total energy functions. (*Optional*: Use *Mathcad* or *Mathematica* to plot these functions. Set k and A each equal to 1.)

2.6   A particle of mass m moves along a frictionless, horizontal plane with a speed given by $v(x) = \alpha/x$, where x is its distance from the origin and α is a positive constant. Find the force $F(x)$ to which the particle is subject.

2.7   A block of mass M has a string of mass m attached to it. A force $\mathbf{F}$ is applied to the string, and it pulls the block up a frictionless plane that is inclined at an angle θ to the horizontal. Find the force that the string exerts on the block.

2.8   Given that the velocity of a particle in rectilinear motion varies with the displacement x according to the equation

$$\dot{x} = bx^{-3}$$

where b is a positive constant, find the force acting on the particle as a function of x. (*Hint: $F = m\ddot{x} = m\dot{x}\, d\dot{x}/dx$.*)

2.9   A baseball (radius = .0366 m, mass = .145 kg) is dropped from rest at the top of the Empire State Building (height = 1250 ft). Calculate **(a)** the initial potential energy of the baseball, **(b)** its final kinetic energy, and **(c)** the total energy dissipated by the falling baseball by computing the line integral of the force of air resistance along the baseball's total distance of fall. Compare this last result to the difference between the baseball's initial potential energy and its final kinetic energy. (*Hint: In part (c) make approximations when evaluating the hyperbolic functions obtained in carrying out the line integral.*)

2.10   A block of wood is projected up an inclined plane with initial speed v_0. If the inclination of the plane is 30° and the coefficient of sliding friction $\mu_x = 0.1$, find the total time for the block to return to the point of projection.

2.11   A metal block of mass m slides on a horizontal surface that has been lubricated with a heavy oil so that the block suffers a viscous resistance that varies as the $\frac{3}{2}$ power of the speed:

$$F(v) = -cv^{3/2}$$

If the initial speed of the block is v_0 at $x = 0$, show that the block cannot travel farther than $2mv_0^{1/2}/c$.

2.12   A gun is fired straight up. Assuming that the air drag on the bullet varies quadratically with speed, show that the speed varies with height according to the equations

$$v^2 = Ae^{-2kx} - \frac{g}{k} \quad \text{(upward motion)}$$

$$v^2 = \frac{g}{k} - Be^{2kx} \quad \text{(downward motion)}$$

in which A and B are constants of integration, g is the acceleration of gravity, and $k = c_2/m$ where c_2 is the drag constant and m is the mass of the bullet. (*Note:* x is measured positive upward, and the gravitational force is assumed to be constant.)

2.13   Use the above result to show that, when the bullet hits the ground on its return, the speed is equal to the expression

$$\frac{v_0 v_t}{\left(v_0^2 + v_t^2\right)^{1/2}}$$

in which v_0 is the initial upward speed and

$$v_t = (mg/c_2)^{1/2} = \text{terminal speed} = (g/k)^{1/2}$$

(This result allows one to find the fraction of the initial kinetic energy lost through air friction.)

2.14   A particle of mass m is released from rest a distance b from a fixed origin of force that attracts the particle according to the inverse square law:

$$F(x) = -kx^{-2}$$

Show that the time required for the particle to reach the origin is

$$\pi \left(\frac{mb^3}{8k}\right)^{1/2}$$

2.15   Show that the terminal speed of a falling spherical object is given by

$$v_t = [(mg/c_2) + (c_1/2c_2)^2]^{1/2} - (c_1/2c_2)$$

when *both* the linear and the quadratic terms in the drag force are taken into account.

2.16   Use the above result to calculate the terminal speed of a soap bubble of mass 10^{-7} kg and diameter 10^{-2} m. Compare your value with the value obtained by using Equation 2.4.10.

2.17   Given: The force acting on a particle is the product of a function of the distance and a function of the velocity: $F(x, v) = f(x)g(v)$. Show that the differential equation of motion can be solved by integration. If the force is a product of a function of distance and a function of time, can the equation of motion be solved by simple integration? Can it be solved if the force is a product of a function of time and a function of velocity?

2.18   The force acting on a particle of mass m is given by

$$F = kvx$$

in which k is a positive constant. The particle passes through the origin with speed v_0 at time $t = 0$. Find x as a function of t.

2.19   A surface-going projectile is launched horizontally on the ocean from a stationary war-ship, with initial speed v_0. Assume that its propulsion system has failed and it is slowed by a retarding force given by $F(v) = -Ae^{\alpha v}$. **(a)** Find its speed as a function of time, $v(t)$. Find **(b)** the time elapsed and **(c)** the distance traveled when the projectile finally comes to rest. A and α are positive constants.

2.20   Assume that a water droplet falling though a humid atmosphere gathers up mass at a rate that is proportional to its cross-sectional area A. Assume that the droplet starts from rest and that its initial radius R_0 is so small that it suffers no resistive force. Show that **(a)** its radius and **(b)** its speed increase linearly with time.

Computer Problems

C 2.1   A parachutist of mass 70 kg jumps from a plane at an altitude of 32 km above the surface of the Earth. Unfortunately, the parachute fails to open. (In the following parts, neglect horizontal motion and assume that the initial velocity is zero.)
   (a) Calculate the time of fall (accurate to 1 s) until ground impact, given no air resistance and a constant value of g.
   (b) Calculate the time of fall (accurate to 1 s) until ground impact, given constant g and a force of air resistance given by

$$F(v) = -c_2 v |v|$$

where c_2 is 0.5 in SI units for a falling man and is constant.
   (c) Calculate the time of fall (accurate to 1 s) until ground impact, given c_2 scales with atmospheric density as

$$c_2 = 0.5 e^{-y/H}$$

where $H = 8$ km is the scale height of the atmosphere and y is the height above ground. Furthermore, assume that g is no longer constant but is given by

$$g = \frac{9.8}{\left(1 + \dfrac{y}{R_e}\right)^2} \ \text{ms}^{-2}$$

where R_e is the radius of the Earth and is 6370 km.
   (d) For case (c), plot the acceleration, velocity, and altitude of the parachutist as a function of time. Explain why the acceleration becomes positive as the parachutist falls.

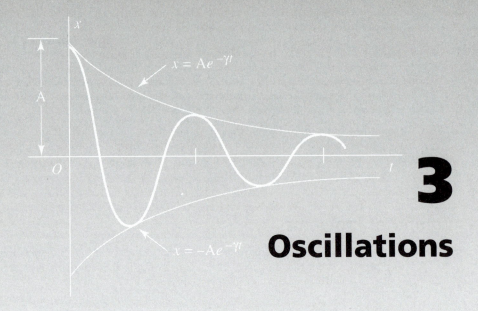

3

Oscillations

"These are the Phenomena of Springs and springy bodies, which as they have not hitherto been by any that I know reduced to Rules—It is very evident that the Rule or Law of Nature in every springing body is, that the force or power thereof to restore itself to its natural position is always proportionate to the distance or space it is removed therefrom—"

Robert Hooke—*De Potentia Restitutiva,* 1678

3.1 | Introduction

The solar system was the most fascinating and intensively studied mechanical system known to early humans. It is a marvelous example of periodic motion. It is not clear how long people would have toiled in mechanical ignorance were it not for this periodicity or had our planet been the singular observable member of the solar system. Everywhere around us we see systems engaged in a periodic dance: the small oscillations of a pendulum clock, a child playing on a swing, the rise and fall of the tides, the swaying of a tree in the wind, the vibrations of the strings on a violin. Even things that we cannot see march to the tune of a periodic beat: the vibrations of the air molecules in the woodwind instruments of a symphony, the hum of the electrons in the wires of our modern civilization, the vibrations of the atoms and molecules that make up our bodies. It is fitting that we cannot even say the word *vibration* properly without the tip of the tongue oscillating.

The essential feature that all these phenomena have in common is *periodicity,* a pattern of movement or displacement that repeats itself over and over again. The pattern may be simple or it may be complex. For example, Figure 3.1.1(a) shows a record of the horizontal displacement of a supine human body resting on a nearly frictionless surface, such as a thin layer of air. The body oscillates horizontally back and forth due to the mechanical action of the heart, pumping blood through and around the aortic

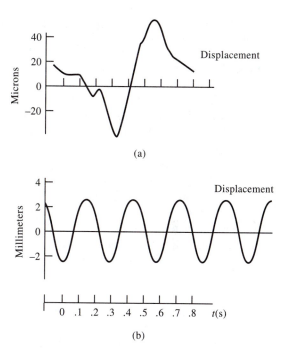

Figure 3.1.1 (a) Recoil vibrations of a human subject resting on a frictionless surface (in response to the pumping action of the heart). (b) Horizontal displacement of a simple pendulum about equilibrium.

arch. Such a recording is called a *ballistocardiogram*.[1] Figure 3.1.1(b) shows the almost perfect sine curve representing the horizontal displacement of a simple pendulum executing small oscillations about its equilibrium position. In both cases, the horizontal axis represents the steady advance of time. The period of the motion is readily identified as the time required for one complete cycle of the motion to occur.

It is with the hope of being able to describe all the complicated forms of periodic motion Mother Nature exhibits, such as that shown in Figure 3.1.1(a), that we undertake an analysis of her simplest form—*simple harmonic motion* (exemplified in Fig. 3.1.1(b)).

Simple harmonic motion exhibits two essential characteristics. (1) It is described by a second-order, linear differential equation with constant coefficients. Thus, the *superposition principle* holds; that is, if two particular solutions are found, their sum is also a solution. We will see evidence of this in the examples to come. (2) The period of the motion, or the time required for a particular configuration (not only position, but velocity as well) to repeat itself, is independent of the maximum displacement from equilibrium. We have already remarked that Galileo was the first to exploit this essential feature of the pendulum by using it as a clock. These features are true only if the displacements from equilibrium are "small." "Large" displacements result in the appearance of nonlinear terms in the differential equations of motion, and the resulting oscillatory solutions no longer obey the principle of superposition or exhibit amplitude-independent periods. We briefly consider this situation toward the end of this chapter.

---

[1] George B. Benedek and Felix M. H. Villars, *Physics —with Illustrative Examples from Medicine and Biology*, Addison-Wesley, New York, 1974.

3.2 | Linear Restoring Force: Harmonic Motion

One of the simplest models of a system executing simple harmonic motion is a mass on a frictionless surface attached to a wall by means of a spring. Such a system is shown in Figure 3.2.1. If X_e is the unstretched length of the spring, the mass will sit at that position, undisturbed, if initially placed there at rest. This position represents the equilibrium configuration of the mass, that is, the one in which its potential energy is a minimum or, equivalently, where the net force on it vanishes. If the mass is pushed or pulled away from this position, the spring will be either compressed or stretched. It will then exert a force on the mass, which will always attempt to restore it to its equilibrium configuration.

We need an expression for this restoring force if we are to calculate the motion of the mass. We can estimate the mathematical form of this force by appealing to arguments based on the presumed nature of the potential energy of this system. Recall from Example 2.3.3 that the Morse potential—the potential energy function of the diatomic hydrogen molecule, a bound system of two particles—has the shape of a well or a cup. Mathematically, it was given by the following expression:

$$V(x) = V_0(1 - \exp(-x/\delta))^2 - V_0 \tag{3.2.1}$$

We showed that this function exhibited quadratic behavior near its minimum and that the resulting force between the two atoms was linear, always acting to restore them to their equilibrium configuration. In general, any potential energy function can be described approximately by a polynomial function of the displacement x for displacements not too far from equilibrium

$$V(x) = a_0 + a_1 x + a_2 x^2 + a_3 x^3 + \cdots \tag{3.2.2a}$$

Furthermore, because only *differences* in potential energies are relevant for the behavior of physical systems, the constant term in each of the above expressions may be taken to be zero; this amounts to a simple reassignment of the value of the potential energy at some reference point. We also argue that the linear term in the above expression must be identically zero. This condition follows from the fact that the first derivative of any function must vanish at its minimum, presuming that the function and its derivatives are continuous, as they must be if the function is to describe the behavior of a real, physical system. Thus, the approximating polynomial takes the form

$$V(x) = a_2 x^2 + a_3 x^3 + \cdots \tag{3.2.2b}$$

For example, Figure 3.2.2(a) is a plot of the Morse potential along with an approximating eighth-order polynomial "best fit." The width δ of the potential and its depth

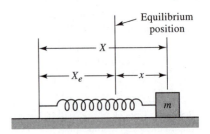

Figure 3.2.1 A model of the simple harmonic oscillator.

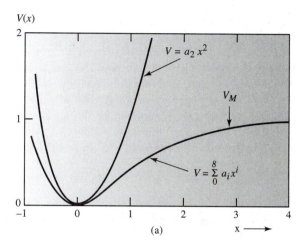

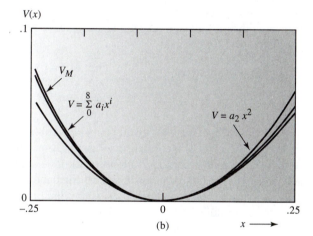

Figure 3.2.2 (a) The Morse potential, its eighth-order approximating polynomial and the quadratic term only. (b) Same as (a) but magnified in scale around $x = 0$.

(the V_0 coefficient) were both set equal to 1.0 (the bare constant V_0 was set equal to 0). The fit was made over the rather sizable range $\Delta x = [-1, 4] = 5\delta$. The result is

$$V(x) = \sum_{i=0}^{8} a_i x^i \qquad (3.2.2c)$$

$a_0 = 1.015 \cdot 10^{-4}$	$a_1 = 0.007$	$a_2 = 0.995$
$a_3 = -1.025$	$a_4 = 0.611$	$a_5 = -0.243$
$a_6 = 0.061$	$a_7 = -0.009$	$a_8 = 5.249 \cdot 10^{-4}$

The polynomial function fits the Morse potential quite well throughout the quoted displacement range. If one examines closely the coefficients of the eighth-order fit, one sees that the first two terms are essentially zero, as we have argued they should be. Therefore, we also show plotted only the quadratic term $V(x) \approx a_2 \cdot x^2$. It seems as though this term does not agree very well with the Morse potential. However, if we "explode" the plot around $x = 0$ (see Fig. 3.2.2(b)), we see that for small displacements—say, $-0.1\delta \le x \le +0.1\delta$—there is virtually no difference among the purely quadratic term, the eighth-order polynomial fit, and the actual Morse potential. For such small displacements, the potential function is, indeed, purely quadratic. One might argue that this example was contrived; however, it is fairly representative of many physical systems.

The potential energy function for the system of spring and mass must exhibit similar behavior near the equilibrium position at X_e, dominated by a purely quadratic term. The spring's restoring force is thus given by the familiar Hooke's law,

$$F(x) = -\frac{dV(x)}{dx} = -(2a_2)x = -kx \qquad (3.2.3)$$

where $k = 2a_2$ is the *spring constant*. In fact, this is how we define small displacements from equilibrium, that is, those for which Hooke's law is valid or the restoring force is linear. That the derived force must be a restoring one is a consequence of the fact that the derivative of the potential energy function must be negative for positive displacements from equilibrium and vice versa for negative ones. Newton's second law of motion for the mass can now be written as

$$m\ddot{x} + kx = 0 \qquad (3.2.4a)$$

$$\ddot{x} + \frac{k}{m}x = 0 \qquad (3.2.4b)$$

Equation 3.2.4b can be solved in a wide variety of ways. It is a second-order, linear differential equation with constant coefficients. As previously stated, the principle of superposition holds for its solutions. Before solving the equation here, we point out those characteristics we expect the solution to exhibit. First, the motion is both periodic and bounded. The mass vibrates back and forth between two limiting positions. Suppose we pull the mass out to some position x_{m1} and then release it from rest. The restoring force, initially equal to $-kx_{m1}$, pulls the mass toward the left in Figure 3.2.1, where it vanishes at $x = 0$, the equilibrium position. The mass now finds itself moving to the left with some velocity v, and so it passes on through equilibrium. Then the restoring force begins to build up strength as the spring compresses, but now directed toward the right. It slows the mass down until it stops, just for an instant, at some position, $-x_{m2}$. The spring, now fully compressed, starts to shove the mass back toward the right. But again momentum carries it through the equilibrium position until the now-stretching spring finally manages to stop it— we might guess—at x_{m1}, the initial configuration of the system. This completes one cycle of the motion—a cycle that repeats itself, apparently forever! Clearly, the resultant functional dependence of x upon t must be represented by a periodic and bounded function. Sine and/or cosine functions come to mind, because they exhibit the sort of behavior we are describing here. In fact, sines and cosines are the real solutions of Equation 3.2.4b. Later on, we show that other functions, imaginary exponentials, are actually equivalent to sines and cosines and are easier to use in describing the more complicated systems soon to be discussed.

A solution is given by

$$x = A\sin(\omega_0 t + \phi_0) \qquad (3.2.5)$$

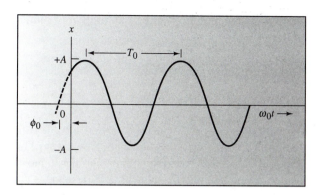

Figure 3.2.3 Displacement versus $\omega_0 t$ for the simple harmonic oscillator.

which can be verified by substituting it into Equation 3.2.4b

$$\omega_0 = \sqrt{\frac{k}{m}} \qquad (3.2.6)$$

is the *angular frequency* of the system. The motion represented by Equation 3.2.5 is a sinusoidal oscillation about equilibrium. A graph of the displacement x versus $\omega_0 t$ is shown in Figure 3.2.3. The motion exhibits the following features. (1) It is characterized by a single angular frequency ω_0. The motion repeats itself after the angular argument of the sine function ($\omega_0 t + \phi_0$) advances by 2π or after one cycle has occurred (hence, the name *angular frequency* for ω_0). The time required for a phase advance of 2π is given by

$$\omega_0(t + T_0) + \phi_0 = \omega_0 t + \phi_0 + 2\pi$$

$$\therefore T_0 = \frac{2\pi}{\omega_0} \qquad (3.2.7)$$

T_0 is called the period of the motion. (2) The motion is bounded; that is, it is confined within the limits $-A \le x \le +A$. A, the maximum displacement from equilibrium, is called the amplitude of the motion. It is independent of the angular frequency ω_0. (3) The phase angle ϕ_0 is the initial value of the angular argument of the sine function. It determines the value of the displacement x at time $t = 0$. For example, at $t = 0$ we have

$$x(t = 0) = A \sin(\phi_0) \qquad (3.2.8)$$

The maximum displacement from equilibrium occurs at a time t_m given by the condition that the angular argument of the sine function is equal to $\pi/2$, or

$$\omega_0 t_m = \frac{\pi}{2} - \phi_0 \qquad (3.2.9)$$

One commonly uses the term *frequency* to refer to the reciprocal of the period of the oscillation or

$$f_0 = \frac{1}{T_0} \qquad (3.2.10)$$

where f_0 is the number of cycles of vibration per unit time. It is related to the angular frequency ω_0 by

$$2\pi f_0 = \omega_0 \tag{3.2.11a}$$

$$f_0 = \frac{1}{T_0} = \frac{1}{2\pi}\sqrt{\frac{k}{m}} \tag{3.2.11b}$$

The unit of frequency (cycles per second, or s^{-1}) is called the *hertz* (Hz) in honor of Heinrich Hertz, who is credited with the discovery of radio waves. Note that $1\,\text{Hz} = 1\,s^{-1}$. The word *frequency* is used sloppily sometimes to mean either cycles per second or radians per second (angular frequency). The meaning is usually clear from the context.

Constants of the Motion and Initial Conditions

Equation 3.2.5, the solution for simple harmonic motion, contains two arbitrary constants, A and ϕ_0. The value of each constant can be determined from knowledge of the initial conditions of the specific problem at hand. As an example of the simplest and most commonly described initial condition, consider a mass initially displaced from equilibrium to a position x_m, where it is then released from rest. The displacement at $t = 0$ is a maximum. Therefore, $A = x_m$ and $\phi_0 = \pi/2$.

As an example of another simple situation, suppose the oscillator is at rest at $x = 0$, and at time $t = 0$ it receives a sharp blow that imparts to it an initial velocity v_0 in the positive x direction. In such a case the initial phase is given by $\phi_0 = 0$. This automatically ensures that the solution yields $x = 0$ at $t = 0$. The amplitude can be found by differentiating x to get the velocity of the oscillator as a function of time and then demanding that the velocity equal v_0 at $t = 0$. Thus,

$$v(t) = \dot{x}(t) = \omega_0 A \cos(\omega_0 t + \phi_0) \tag{3.2.12a}$$

$$v(0) = v_0 = \omega_0 A \tag{3.2.12b}$$

$$\therefore A = \frac{v_0}{\omega_0} \tag{3.2.12c}$$

For a more general scenario, consider a mass initially displaced to some position x_0 and given an initial velocity v_0. The constants can then be determined as follows:

$$x(0) = A \sin\phi_0 = x_0 \tag{3.2.13a}$$

$$\dot{x}(0) = \omega_0 A \cos\phi_0 = v_0 \tag{3.2.13b}$$

$$\therefore \tan\phi_0 = \frac{\omega_0 x_0}{v_0} \tag{3.2.13c}$$

$$A^2 = x_0^2 + \frac{v_0^2}{\omega_0^2} \tag{3.2.13d}$$

This more general solution reduces to either of those described above, as can easily be seen by setting v_0 or x_0 equal to zero.

Simple Harmonic Motion as the Projection of a Rotating Vector

Imagine a vector **A** rotating at a constant angular velocity ω_0. Let this vector denote the position of a point P in uniform circular motion. The projection of the vector onto a line (which we call the x-axis) in the same plane as the circle traces out simple harmonic motion. Suppose the vector **A** makes an angle θ with the x-axis at some time t, as shown in Figure 3.2.4. Because $\dot{\theta} = \omega_0$, the angle θ increases with time according to

$$\theta = \omega_0 t + \theta_0 \tag{3.2.14}$$

where θ_0 is the value of θ at $t = 0$. The projection of P onto the x-axis is given by

$$x = A \cos\theta = A \cos(\omega_0 t + \theta_0) \tag{3.2.15}$$

This point oscillates in simple harmonic motion as P goes around the circle in uniform angular motion.

Our picture describes x as a cosine function of t. We can show the equivalence of this expression to the sine function given by Equation 3.2.5 by measuring angles to the vector **A** from the y-axis, instead of the x-axis as shown in Figure 3.2.4. If we do this, the projection of **A** onto the x-axis is given by

$$x = A \sin(\omega_0 t + \phi_0) \tag{3.2.16}$$

We can see this equivalence in another way. We set the phase difference between ϕ_0 and θ_0 to $\pi/2$ and then substitute into the above equation, obtaining

$$\phi_0 - \theta_0 = \frac{\pi}{2} \tag{3.2.17a}$$

$$\cos(\omega_0 t + \theta_0) = \cos\left(\omega_0 t + \phi_0 - \frac{\pi}{2}\right)$$
$$= \sin(\omega_0 t + \phi_0) \tag{3.2.17b}$$

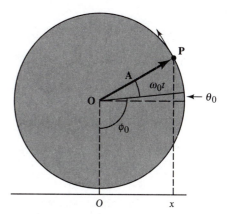

Figure 3.2.4 Simple harmonic motion as a projection of uniform circular motion.

We now see that simple harmonic motion can be described equally well by a sine function or a cosine function. The one we choose is largely a matter of taste; it depends upon our choice of initial phase angle to within an arbitrary constant.

You might guess from the above commentary that we could use a sum of sine and cosine functions to represent the general solution for harmonic motion. For example, we can convert the sine solution of Equation 3.2.5 directly to such a form, using the trigonometric identity for the sine of a sum of angles:

$$x(t) = A \sin(\omega_0 t + \phi_0) = A \sin\phi_0 \cos\omega_0 t + A \cos\phi_0 \sin\omega_0 t$$
$$= C \cos\omega_0 t + D \sin\omega_0 t \qquad\qquad (3.2.18)$$

Neither A nor ϕ_0 appears explicitly in the solution. They are there implicitly; that is,

$$\tan\phi_0 = \frac{C}{D} \qquad\qquad A^2 = C^2 + D^2 \qquad\qquad (3.2.19)$$

There are occasions when this form may be the preferred one.

Effect of a Constant External Force on a Harmonic Oscillator

Suppose the same spring shown in Figure 3.2.1 is held in a vertical position, supporting the same mass m (Fig. 3.2.5). The total force acting is now given by adding the weight mg to the restoring force,

$$F = -k(X - X_e) + mg \qquad\qquad (3.2.20)$$

where the positive direction is down. This equation could be written $F = -kx + mg$ by defining x to be $X - X_e$, as previously. However, it is more convenient to define the variable x in a different way, namely, as the displacement from the *new* equilibrium position

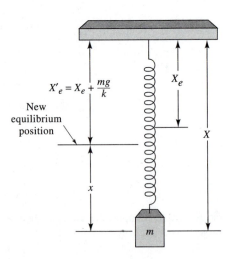

$$X'_e = X_e + \frac{mg}{k}$$

New
equilibrium
position

Figure 3.2.5   The vertical case for the harmonic oscillator.

X'_e obtained by setting $F = 0$ in Equation 3.2.20: $0 = -k(X'_e - X_e) + mg$, which gives $X'_e = X_e + mg/k$. We now define the displacement as

$$x = X - X'_e = X - X_e - \frac{mg}{k} \tag{3.2.21}$$

Putting this into Equation 3.2.20 gives, after a very little algebra,

$$F = -kx \tag{3.2.22}$$

so the differential equation of motion is again

$$m\ddot{x} + kx = 0 \tag{3.2.23}$$

and our solution in terms of our newly defined x is identical to that of the horizontal case. It should now be evident that *any* constant external force applied to a harmonic oscillator merely shifts the equilibrium position. The equation of motion remains unchanged if we measure the displacement x from the new equilibrium position.

EXAMPLE 3.2.1

When a light spring supports a block of mass m in a vertical position, the spring is found to stretch by an amount D_1 over its unstretched length. If the block is furthermore pulled downward a distance D_2 from the equilibrium position and released—say, at time $t = 0$—find (a) the resulting motion, (b) the velocity of the block when it passes back upward through the equilibrium position, and (c) the acceleration of the block at the top of its oscillatory motion.

Solution:

First, for the equilibrium position we have

$$F_x = 0 = -kD_1 + mg$$

where x is chosen positive downward. This gives us the value of the stiffness constant:

$$k = \frac{mg}{D_1}$$

From this we can find the angular frequency of oscillation:

$$\omega_0 = \sqrt{\frac{k}{m}} = \sqrt{\frac{g}{D_1}}$$

We shall express the motion in the form $x(t) = A \cos \omega_0 t + B \sin \omega_0 t$. Then

$$\dot{x} = -A\omega_0 \sin \omega_0 t + B\omega_0 \cos \omega_0 t.$$

From the initial conditions we find

$$x_0 = D_2 = A \qquad \dot{x}_0 = 0 = B\omega_0 \qquad B = 0$$

The motion is, therefore, given by

(a)
$$x(t) = D_2 \cos\left(\sqrt{\frac{g}{D_1}}\,t\right)$$

in terms of the given quantities. Note that the mass m does not appear in the final expression. The velocity is then

$$\dot{x}(t) = -D_2 \sqrt{\frac{g}{D_1}}\, \sin\left(\sqrt{\frac{g}{D_1}}\,t\right)$$

and the acceleration

$$\ddot{x}(t) = -D_2 \frac{g}{D_1} \cos\left(\sqrt{\frac{g}{D_1}}\,t\right)$$

As the block passes upward through the equilibrium position, the argument of the sine term is $\pi/2$ (one-quarter period), so

(b)
$$\dot{x} = -D_2 \sqrt{\frac{g}{D_1}} \qquad \text{(center)}$$

At the top of the swing the argument of the cosine term is π (one-half period), which gives

(c)
$$\ddot{x} = D_2 \frac{g}{D_1} \qquad \text{(top)}$$

In the case $D_1 = D_2$, the downward acceleration at the top of the swing is just g. This means that the block, at that particular instant, is in *free fall*; that is, the spring is exerting zero force on the block.

EXAMPLE 3.2.2

The Simple Pendulum

The so-called simple pendulum consists of a small plumb bob of mass m swinging at the end of a light, inextensible string of length l, Figure 3.2.6. The motion is along a circular arc defined by the angle θ, as shown. The restoring force is the component of the weight mg acting in the direction of increasing θ along the path of motion: $F_s = -mg \sin\theta$. If we treat the bob as a particle, the differential equation of motion is, therefore,

$$m\ddot{s} = -mg \sin\theta$$

Now $s = l\theta$, and, for small θ, $\sin\theta = \theta$ to a fair approximation. So, after canceling the m's and rearranging terms, we can write the differential equation of motion in terms of either θ or s as follows:

$$\ddot{\theta} + \frac{g}{l}\theta = 0 \qquad\qquad \ddot{s} + \frac{g}{l}s = 0$$

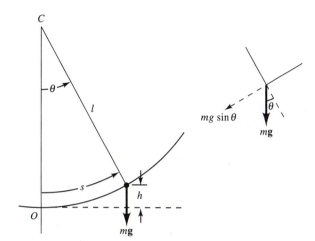

Figure 3.2.6 The simple pendulum.

Although the motion is along a curved path rather than a straight line, the differential equation is mathematically identical to that of the linear harmonic oscillator, Equation 3.2.4b, with the quantity g/l replacing k/m. Thus, to the extent that the approximation $\sin \theta = \theta$ is valid, we can conclude that the motion is simple harmonic with angular frequency

$$\omega_0 = \sqrt{\frac{g}{l}}$$

and period

$$T_0 = \frac{2\pi}{\omega_0} = 2\pi \sqrt{\frac{l}{g}}$$

This formula gives a period of very nearly 2 s, or a half-period of 1 s, when the length l is 1 m. More accurately, for a half-period of 1 s, known as the "seconds pendulum," the precise length is obtained by setting $T_0 = 2$ s and solving for l. This gives $l = g/\pi^2$ numerically, when g is expressed in m/s². At sea level at a latitude of 45°, the value of the acceleration of gravity is $g = 9.8062$ m/s². Accordingly, the length of a seconds pendulum at that location is $9.8062/9.8696 = 0.9936$ m.

3.3 | Energy Considerations in Harmonic Motion

Consider a particle under the action of a linear restoring force $F_x = -kx$. Let us calculate the work done by an external force F_{ext} in moving the particle from the equilibrium position ($x = 0$) to some position x. Assume that we move the particle very slowly so that it does not gain any kinetic energy; that is, the applied external force is barely greater in magnitude than the restoring force $-kx$; hence, $F_{ext} = -F_x = kx$, so

$$W = \int_0^x F_{ext}\, dx = \int_0^x kx\, dx = \frac{k}{2}x^2 \tag{3.3.1}$$

In the case of a spring obeying Hooke's law, the work is stored in the spring as potential energy: $W = V(x)$, where

$$V(x) = \frac{1}{2}kx^2 \tag{3.3.2}$$

Thus, $F_x = -dV/dx = -kx$, as required by the definition of V. The total energy, when the particle is undergoing harmonic motion, is given by the sum of the kinetic and potential energies, namely,

$$E = \frac{1}{2}m\dot{x}^2 + \frac{1}{2}kx^2 \tag{3.3.3}$$

This equation epitomizes the harmonic oscillator in a rather fundamental way: The kinetic energy is quadratic in the velocity variable, and the potential energy is quadratic in the displacement variable. The total energy is constant if there are no other forces except the restoring force acting on the particle.

The motion of the particle can be found by starting with the energy equation (3.3.3). Solving for the velocity gives

$$\dot{x} = \pm\left(\frac{2E}{m} - \frac{kx^2}{m}\right)^{1/2} \tag{3.3.4}$$

which can be integrated to give t as a function of x as follows:

$$t = \int \frac{dx}{\pm[(2E/m) - (k/m)x^2]^{1/2}} = \mp(m/k)^{1/2} \cos^{-1}(x/A) + C \tag{3.3.5}$$

in which C is a constant of integration and A is the amplitude given by

$$A = \left(\frac{2E}{k}\right)^{1/2} \tag{3.3.6}$$

Upon solving the integrated equation for x as a function of t, we find the same relationship as in the preceding section, with the addition that we now have an explicit value for the amplitude. We can also obtain the amplitude directly from the energy equation (3.3.3) by finding the turning points of the motion where $\dot{x} = 0$: The value of x must lie between $\pm A$ in order for $\dot{x}$ to be real. This is illustrated in Figure 3.3.1.

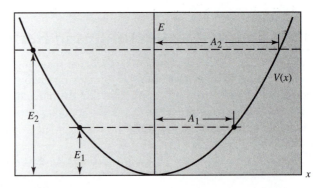

Figure 3.3.1 Graph of the parabolic potential energy function of the harmonic oscillator. The turning points defining the amplitude are indicated for two different values of the total energy.

We also see from the energy equation that the maximum value of the speed, which we call v_{max}, occurs at $x = 0$. Accordingly, we can write

$$E = \tfrac{1}{2}mv_{max}^2 = \tfrac{1}{2}kA^2 \tag{3.3.7}$$

As the particle oscillates, the kinetic and potential energies continually change. The constant total energy is entirely in the form of kinetic energy at the center, where $x = 0$ and $\dot{x} = \pm v_{max}$, and it is all potential energy at the extrema, where $\dot{x} = 0$ and $x = \pm A$.

EXAMPLE 3.3.1

The Energy Function of the Simple Pendulum

The potential energy of the simple pendulum (Fig. 3.2.6) is given by the expression

$$V = mgh$$

where h is the vertical distance from the reference level (which we choose to be the level of the equilibrium position). For a displacement through an angle θ (Fig. 3.2.6), we see that $h = l - l\cos\theta$, so

$$V(\theta) = mgl(1 - \cos\theta)$$

Now the series expansion for the cosine is $\cos\theta = 1 - \theta^2/2! + \theta^4/4! - \cdots$, so for small θ we have approximately $\cos\theta = 1 - \theta^2/2$. This gives

$$V(\theta) = \tfrac{1}{2}mgl\,\theta^2$$

or, equivalently, because $s = l\theta$,

$$V(s) = \tfrac{1}{2}\frac{mg}{l}s^2$$

Thus, to a first approximation, the potential energy function is quadratic in the displacement variable. In terms of s, the total energy is given by

$$E = \tfrac{1}{2}m\dot{s}^2 + \tfrac{1}{2}\frac{mg}{l}s^2$$

in accordance with the general statement concerning the energy of the harmonic oscillator discussed above.

EXAMPLE 3.3.2

Calculate the average kinetic, potential, and total energies of the harmonic oscillator. (Here we use the symbol K for kinetic energy and T_0 for the period of the motion.)

Solution:

$$\langle K \rangle = \frac{1}{T_0}\int_0^{T_0} K(t)\,dt = \frac{1}{T_0}\int_0^{T_0}\tfrac{1}{2}m\dot{x}^2\,dt$$

but

$$x = A \sin(\omega_0 t + \phi_0)$$
$$\dot{x} = \omega_0 A \cos(\omega_0 t + \phi_0)$$

Setting $\phi_0 = 0$ and letting $u = \omega_0 t = (2\pi/T_0) \cdot t$, we obtain

$$\langle K \rangle = \frac{1}{T_0} \left[\frac{1}{2} m\omega_0^2 A^2 \int_0^{T_0} \cos^2(\omega_0 t)\, dt \right]$$
$$= \frac{1}{2\pi} \left[\frac{1}{2} m\omega_0^2 A^2 \int_0^{2\pi} \cos^2 u\, du \right]$$

We can make use of the fact that

$$\frac{1}{2\pi} \int_0^{2\pi} (\sin^2 u + \cos^2 u)\, du = \frac{1}{2\pi} \int_0^{2\pi} du = 1$$

to obtain

$$\frac{1}{2\pi} \int_0^{2\pi} \cos^2 u\, du = \frac{1}{2}$$

because the areas under the $\cos^2$ and $\sin^2$ terms throughout one cycle are identical. Thus,

$$\langle K \rangle = \frac{1}{4} m\omega_0^2 A^2$$

The calculation of the average potential energy proceeds along similar lines.

$$V = \frac{1}{2} kx^2 = \frac{1}{2} kA^2 \sin^2 \omega_0 t$$
$$\langle V \rangle = \frac{1}{2} kA^2 \frac{1}{T_0} \int_0^{T_0} \sin^2 \omega_0 t\, dt$$
$$= \frac{1}{2} kA^2 \frac{1}{2\pi} \int_0^{2\pi} \sin^2 u\, du$$
$$= \frac{1}{4} kA^2$$

Now, because $k/m = \omega_0^2$ or $k = m\omega_0^2$, we obtain

$$\langle V \rangle = \frac{1}{4} kA^2 = \frac{1}{4} m\omega_0^2 A^2 = \langle K \rangle$$
$$\langle E \rangle = \langle K \rangle + \langle V \rangle = \frac{1}{2} m\omega_0^2 A^2 = \frac{1}{2} kA^2 = E$$

The average kinetic energies and potential energies are equal; therefore, the average energy of the oscillator is equal to its total instantaneous energy.

3.4 | Damped Harmonic Motion

The foregoing analysis of the harmonic oscillator is somewhat idealized in that we have failed to take into account frictional forces. These are always present in a mechanical system to some extent. Analogously, there is always a certain amount of resistance in an electrical circuit. For a specific model, let us consider an object of mass m that is supported by

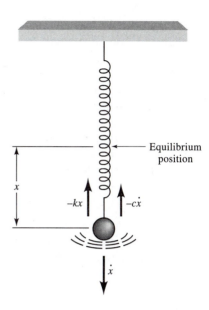

Figure 3.4.1   A model for the damped harmonic oscillator.

a light spring of stiffness k. We assume that there is a viscous retarding force that is a linear function of the velocity, such as is produced by air drag at low speeds.[2] The forces are indicated in Figure 3.4.1.

If x is the displacement from equilibrium, then the restoring force is $-kx$, and the retarding force is $-c\dot{x}$, where c is a constant of proportionality. The differential equation of motion is, therefore, $m\ddot{x} = -kx - c\dot{x}$, or

$$m\ddot{x} + c\dot{x} + kx = 0 \qquad (3.4.1)$$

As with the undamped case, we divide Equation 3.4.1 by m to obtain

$$\ddot{x} + \frac{c}{m}\dot{x} + \frac{k}{m}x = 0 \qquad (3.4.2)$$

If we substitute the *damping factor* γ, defined as

$$\gamma \equiv \frac{c}{2m} \qquad (3.4.3)$$

and $\omega_0^2 \,(=k/m)$ into Equation 3.4.2, it assumes the simpler form

$$\ddot{x} + 2\gamma\dot{x} + \omega_0^2 x = 0 \qquad (3.4.4)$$

The presence of the velocity-dependent term $2\gamma\dot{x}$ complicates the problem; simple sine or cosine solutions do not work, as can be verified by trying them. We introduce a method of solution that works rather well for second-order differential equations with constant

---

[2] Nonlinear drag is more realistic in many situations; however, the equations of motion are much more difficult to solve and are not treated here.

coefficients. Let D be the differential operator d/dt. We "operate" on x with a quadratic function of D chosen in such a way that we generate Equation 3.4.4:

$$\left[D^2 + 2\gamma D + \omega_0^2\right]x = 0 \tag{3.4.5a}$$

We interpret this equation as an "operation" by the term in brackets on x. The operation by D^2 means first operate on x with D and then operate on the result of that operation with D again. This procedure yields $\ddot{x}$, the first term in Equation 3.4.4. The operator equation (Equation 3.4.5a) is, therefore, equivalent to the differential equation (Equation 3.4.4). The simplification that we get by writing the equation this way arises when we factor the operator term, using the binomial theorem, to obtain

$$\left[D + \gamma - \sqrt{\gamma^2 - \omega_0^2}\right]\left[D + \gamma + \sqrt{\gamma^2 - \omega_0^2}\right]x = 0 \tag{3.4.5b}$$

The operation in Equation 3.4.5b is identical to that in Equation 3.4.5a, but we have reduced the operation from second-order to a product of two first-order ones. Because the order of operation is arbitrary, the general solution is a sum of solutions obtained by setting the result of each first-order operation on x equal to zero. Thus, we obtain

$$x(t) = A_1 e^{-(\gamma - q)t} + A_2 e^{-(\gamma + q)t} \tag{3.4.6}$$

where

$$q = \sqrt{\gamma^2 - \omega_0^2} \tag{3.4.7}$$

The student can verify that this is a solution by direct substitution into Equation 3.4.4. A problem that we soon encounter, though, is that the above exponents may be real or complex, because the factor q could be imaginary. We see what this means in just a minute.

There are three possible scenarios:

I. q real > 0 *Overdamping*
II. q real $= 0$ *Critical damping*
III. q imaginary   *Underdamping*

I. *Overdamped.* Both exponents in Equation 3.4.6 are real. The constants A_1 and A_2 are determined by the initial conditions. The motion is an exponential decay with two different decay constants, $(\gamma - q)$ and $(\gamma + q)$. A mass, given some initial displacement and released from rest, returns slowly to equilibrium, prevented from oscillating by the strong damping force. This situation is depicted in Figure 3.4.2.

II. *Critical damping.* Here $q = 0$. The two exponents in Equation 3.4.6 are each equal to γ. The two constants A_1 and A_2 are no longer independent. Their sum forms a single constant A. The solution degenerates to a single exponential decay function. A completely general solution requires two different functions and independent constants to satisfy the boundary conditions specified by an initial position and velocity. To find a solution with two independent constants, we return to Equation 3.4.5b:

$$(D + \gamma)(D + \gamma)x = 0 \tag{3.4.8a}$$

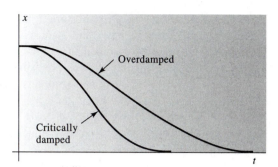

Figure 3.4.2 Displacement versus time for critically damped and overdamped oscillators released from rest after an initial displacement.

Switching the order of operation does not work here, because the operators are the same. We have to carry out the entire operation on x before setting the result to zero. To do this, we make the substitution $u = (D + \gamma)x$, which gives

$$(D + \gamma)u = 0$$
$$u = Ae^{-\gamma t}$$

(3.4.8b)

Equating this to $(D + \gamma)x$, the final solution is obtained as follows:

$$Ae^{-\gamma t} = (D + \gamma)x$$
$$A = e^{\gamma t}(D + \gamma)x = D(xe^{\gamma t})$$
$$\therefore \ xe^{\gamma t} = At + B$$
$$x(t) = Ate^{-\gamma t} + Be^{-\gamma t}$$

(3.4.9)

The solution consists of two different functions, $te^{-\gamma t}$ and $e^{-\gamma t}$, and two constants of integration, A and B, as required. As in case I, if a mass is released from rest after an initial displacement, the motion is nonoscillatory, returning asymptotically to equilibrium. This case is also shown in Figure 3.4.2. Critical damping is highly desirable in many systems, such as the mechanical suspension systems of motor vehicles.

III. *Underdamping.* If the constant γ is small enough that $\gamma^2 - \omega_0^2 < 0$, the factor q in Equation 3.4.7 is imaginary. A mass initially displaced and then released from rest oscillates, not unlike the situation described earlier for no damping force at all. The only difference is the presence of the real factor $-\gamma$ in the exponent of the solution that leads to the ultimate death of the oscillatory motion. Let us now reverse the factors under the square root sign in Equation 3.4.7 and write q as $i\omega_d$. Thus,

$$\omega_d = \sqrt{\omega_0^2 - \gamma^2} = \sqrt{\frac{k}{m} - \frac{c^2}{4m^2}}$$

(3.4.10)

where ω_0 and ω_d are the angular frequencies of the undamped and underdamped harmonic oscillators, respectively. We now rewrite the general solution represented by Equation 3.4.6 in terms of the factors described here,

$$x(t) = C_+ e^{-(\gamma - i\omega_d)t} + C_- e^{-(\gamma + i\omega_d)t}$$
$$= e^{-\gamma t}(C_+ e^{i\omega_d t} + C_- e^{-i\omega_d t})$$

(3.4.11)

where the constants of integration are C_+ and C_-. The solution contains a sum of imaginary exponentials. But the solution must be real—it is supposed to describe the real world! This reality demands that C_+ and C_- be complex conjugates of each other, a condition that ultimately allows us to express the solution in terms of sines and/or cosines. Thus, taking the complex conjugate of Equation 3.4.11,

$$x^*(t) = e^{-\gamma t}(C_+^* e^{-i\omega_d t} + C_-^* e^{+i\omega_d t}) = x(t) \qquad (3.4.12a)$$

Because $x(t)$ is real, $x^*(t) = x(t)$, and, therefore,

$$\therefore \ C_+^* = C_- = C \\ C_-^* = C_+ = C^* \qquad (3.4.12b) \\ \therefore \ x(t) = e^{-\gamma t}(C^* e^{+i\omega_d t} + C e^{-i\omega_d t})$$

It looks as though we have a solution that now has only a single constant of integration. In fact, C is a complex number. It is composed of two constants. We can express C and C^* in terms of two real constants, A and θ_0, in the following way.

$$C_- = C = \frac{A}{2} e^{-i\theta_0} \\ C_+ = C^* = \frac{A}{2} e^{+i\theta_0} \qquad (3.4.13)$$

We soon see that A is the maximum displacement and θ_0 is the initial phase angle of the motion. Thus, Equation 3.4.12b becomes

$$x(t) = e^{-\gamma t}\left(\frac{A}{2} e^{+i(\omega_d t + \theta_0)} + \frac{A}{2} e^{-i(\omega_d t + \theta_0)}\right) \qquad (3.4.14)$$

We now apply Euler's identity[3] to the above expressions, thus obtaining

$$\frac{A}{2} e^{+i(\omega_d t + \theta_0)} = \frac{A}{2} \cos(\omega_d t + \theta_0) + i\frac{A}{2} \sin(\omega_d t + \theta_0) \\ \frac{A}{2} e^{-i(\omega_d t + \theta_0)} = \frac{A}{2} \cos(\omega_d t + \theta_0) - i\frac{A}{2} \sin(\omega_d t + \theta_0) \qquad (3.4.15) \\ \therefore \ x(t) = e^{-\gamma t}(A \cos(\omega_d t + \theta_0))$$

Following our discussion in Section 3.2 concerning the rotating vector construct, we see that we can express the solution equally well as a sine function:

$$x(t) = e^{-\gamma t}(A \sin(\omega_d t + \phi_0)) \qquad (3.4.16)$$

The constants A, θ_0, and ϕ_0 have the same interpretation as those of Section 3.2. In fact, we see that the solution for the underdamped oscillator is nearly identical to that of the undamped oscillator. There are two differences: (1) The presence of the real exponential factor $e^{-\gamma t}$ leads to a gradual death of the oscillations, and (2) the underdamped oscillator's angular frequency is ω_d, not ω_0, because of the presence of the damping force.

---

[3] Euler's identity relates imaginary exponentials to sines and cosines. It is given by the expression $e^{iu} = \cos u + i \sin u$. This equality is demonstrated in Appendix D.

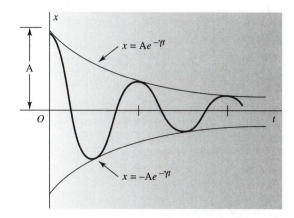

Figure 3.4.3 Graph of displacement versus time for the underdamped harmonic oscillator.

The underdamped oscillator vibrates a little more slowly than does the undamped oscillator. The period of the underdamped oscillator is given by

$$T_d = \frac{2\pi}{\omega_d} = \frac{2\pi}{\left(\omega_0^2 - \gamma^2\right)^{1/2}} \tag{3.4.17}$$

Figure 3.4.3 is a plot of the motion. Equation 3.4.15a shows that the two curves given by $x = Ae^{-\gamma t}$ and $x = -Ae^{-\gamma t}$ form an envelope of the curve of motion because the cosine factor takes on values between +1 and −1, including +1 and −1, at which points the curve of motion touches the envelope. Accordingly, the points of contact are separated by a time interval of one-half period, $T_d/2$. These points, however, are not quite the maxima and minima of the displacement. It is left to the student to show that the actual maxima and minima are also separated in time by the same amount. In one complete period the amplitude diminishes by a factor $e^{-\gamma T_d}$; also, in a time $\gamma^{-1} = 2m/c$ the amplitude decays by a factor $e^{-1} = 0.3679$.

In summary, our analysis of the freely running harmonic oscillator has shown that the presence of damping of the linear type causes the oscillator, given an initial motion, to eventually return to a state of rest at the equilibrium position. The return to equilibrium is either oscillatory or not, depending on the amount of damping. The critical condition, given by $\gamma = \omega_0$, characterizes the limiting case of the nonoscillatory mode of return.

Energy Considerations

The total energy of the damped harmonic oscillator is given by the sum of the kinetic and potential energies:

$$E = \tfrac{1}{2} m\dot{x}^2 + \tfrac{1}{2} kx^2 \tag{3.4.18}$$

This is constant for the undamped oscillator, as stated previously. Let us differentiate the above expression with respect to t:

$$\frac{dE}{dt} = m\dot{x}\ddot{x} + kx\dot{x} = (m\ddot{x} + kx)\dot{x} \tag{3.4.19}$$

Now the differential equation of motion is $m\ddot{x} + c\dot{x} + kx = 0$, or $m\ddot{x} + kx = -c\dot{x}$. Thus, we can write

$$\frac{dE}{dt} = -c\dot{x}^2 \tag{3.4.20}$$

for the time rate of change of total energy. We see that it is given by the product of the damping force and the velocity. Because this is always either zero or negative, the total energy continually decreases and, like the amplitude, eventually becomes negligibly small. The energy is dissipated as frictional heat by virtue of the viscous resistance to the motion.

Quality Factor

The rate of energy loss of a weakly damped harmonic oscillator is best characterized by a single parameter Q, called the *quality factor* of the oscillator. It is defined to be 2π times the energy stored in the oscillator divided by the energy lost in a single period of oscillation T_d. If the oscillator is weakly damped, the energy lost per cycle is small and Q is, therefore, large. We calculate Q in terms of parameters already derived and show that this is true.

The average rate of energy dissipation for the damped oscillator is given by Equation 3.4.20, $\dot{E} = -c\dot{x}^2$, so we need to calculate $\dot{x}$. Equation 3.4.16 gives $x(t)$:

$$x = Ae^{-\gamma t}\sin(\omega_d t + \phi_0) \tag{3.4.21a}$$

Differentiating it, we obtain

$$\dot{x} = -Ae^{-\gamma t}(\gamma \sin(\omega_d t + \phi_0) - \omega_d \cos(\omega_d t + \phi_0)) \tag{3.4.21b}$$

The energy lost during a single cycle of period $T_d = 2\pi/\omega_d$ is

$$\Delta E = \int_0^{T_d} \dot{E}\, dt \tag{3.4.22a}$$

If we change the variable of integration to $\theta = \omega_d t + \phi_0$, then $dt = d\theta/\omega_d$ and the integral over the period T_d transforms to an integral, from ϕ_0 to $\phi_0 + 2\pi$. The value of the integral over a full cycle doesn't depend on the initial phase ϕ_0 of the motion, so, for the sake of simplicity, we drop it from the limits of integration:

$$\Delta E = \frac{1}{\omega_d}\int_0^{2\pi} \dot{E}\, d\theta$$
$$= -\frac{cA^2}{\omega_d}\int_0^{2\pi} e^{-2\gamma t}\left[\gamma^2 \sin^2\theta - 2\gamma\omega_d \sin\theta\, \cos\theta + \omega_d^2 \cos^2\theta\right]d\theta \tag{3.4.22b}$$

Now we can extract the exponential factor $e^{-2\gamma t}$ from inside the integral, because in the case of weak damping ($\gamma \ll \omega_d$) its value does not change very much during a single cycle of oscillation:

$$\Delta E = \frac{-cA^2}{\omega_d}e^{-2\gamma t}\int_0^{2\pi}\left(\gamma^2 \sin^2\theta - 2\gamma\omega_d \sin\theta\, \cos\theta + \omega_d^2 \cos^2\theta\right)d\theta \tag{3.4.22c}$$

The integral of both $\sin^2\theta$ and $\cos^2\theta$ over one cycle is π, while the integral of the $\sin\theta$ $\cos\theta$ product vanishes. Thus, we have

$$\Delta E = \frac{-cA^2}{\omega_d}\pi e^{-2\gamma t}\left(\gamma^2 + \omega_d^2\right) = -cA^2 e^{-2\gamma t}\omega_0^2\left(\frac{\pi}{\omega_d}\right)$$
$$= -\gamma m\omega_0^2 A^2 e^{-2\gamma t}T_d$$

(3.4.22d)

where we have made use of the relations $\omega_0^2 = \omega_d^2 + \gamma^2$ and $\gamma = c/2m$. Now, if we identify the damping factor γ with a time constant τ, such that $\gamma = (2\tau)^{-1}$, we obtain for the *magnitude* of the energy loss in one cycle

$$\Delta E = \left(\tfrac{1}{2}mA^2\omega_0^2 e^{-t/\tau}\right)\frac{T_d}{\tau}$$
$$\frac{\Delta E}{E} = \frac{T_d}{\tau}$$

(3.4.22e)

where the energy stored in the oscillator (see Example 3.3.2) at any time t is

$$E(t) = \tfrac{1}{2}m\omega_0^2 A^2 e^{-t/\tau}$$

(3.4.23)

Clearly, the energy remaining in the oscillator during any cycle dies away exponentially with time constant τ. We, therefore, see that the quality factor Q is just 2π times the inverse of the ratios given in the expression above, or

$$Q = \frac{2\pi}{(T_d/\tau)} = \frac{2\pi\tau}{(2\pi/\omega_d)} = \omega_d\tau = \frac{\omega_d}{2\gamma}$$

(3.4.24)

For weak damping, the period of oscillation T_d is much less than the time constant τ, which characterizes the energy loss rate of the oscillator. Q is large under such circumstances. Table 3.4.1 gives some values of Q for several different kinds of oscillators.

TABLE 3.4.1	Values of Q for Several Physical Systems
Earth (for earthquake)	250–1400
Piano string	3000
Crystal in digital watch	10^4
Microwave cavity	10^4
Excited atom	10^7
Neutron star	10^{12}
Excited Fe^{57} nucleus	3×10^{12}

EXAMPLE 3.4.1

An automobile suspension system is critically damped, and its period of free oscillation with no damping is 1 s. If the system is initially displaced by an amount x_0 and released with zero initial velocity, find the displacement at $t = 1$ s.

Solution:

For critical damping we have $\gamma = c/2m = (k/m)^{1/2} = \omega_0 = 2\pi/T_0$. Hence, $\gamma = 2\pi$ s^{-1} in our case, because $T_0 = 1$ s. Now the general expression for the displacement in the critically damped case (Equation 3.4.9) is $x(t) = (At + B)e^{-\gamma t}$, so, for $t = 0$, $x_0 = B$. Differentiating, we have $\dot{x}(t) = (A - \gamma B - \gamma At)e^{-\gamma t}$, which gives $\dot{x}_0 = A - \gamma B = 0$, so $A = \gamma B = \gamma x_0$ in our problem. Accordingly,

$$x(t) = x_0(1 + \gamma t)e^{-\gamma t} = x_0(1 + 2\pi t)e^{-2\pi t}$$

is the displacement as a function of time. For $t = 1$ s, we obtain

$$x_0(1 + 2\pi)e^{-2\pi} = x_0(7.28)e^{-6.28} = 0.0136\, x_0$$

The system has practically returned to equilibrium.

EXAMPLE 3.4.2

The frequency of a damped harmonic oscillator is one-half the frequency of the same oscillator with no damping. Find the ratio of the maxima of successive oscillations.

Solution:

We have $\omega_d = \frac{1}{2}\omega_0 = (\omega_0^2 - \gamma^2)^{1/2}$, which gives $\omega_0^2/4 = \omega_0^2 - \gamma^2$, so $\gamma = \omega_0(3/4)^{1/2}$. Consequently,

$$\gamma T_d = \omega_0(3/4)^{1/2}\,[2\pi/(\omega_0/2)] = 10.88$$

Thus, the amplitude ratio is

$$e^{-\gamma T_d} = e^{-10.88} = 0.00002$$

This is a *highly damped* oscillator.

EXAMPLE 3.4.3

Given: The terminal speed of a baseball in free fall is 30 m/s. Assuming a linear air drag, calculate the effect of air resistance on a simple pendulum, using a baseball as the plumb bob.

Solution:

In Chapter 2 we found the terminal speed for the case of linear air drag to be given by $v_t = mg/c_1$, where c_1 is the linear drag coefficient. This gives

$$\gamma = \frac{c_1}{2m} = \frac{(mg/v_t)}{2m} = \frac{g}{2v_t} = \frac{9.8\ \text{ms}^{-2}}{60\ \text{ms}^{-1}} = 0.163\ \text{s}^{-1}$$

for the exponential damping constant. Consequently, the baseball pendulum's amplitude drops off by a factor e^{-1} in a time $\gamma^{-1} = 6.13$ s. This is independent of the length of the pendulum. Earlier, in Example 3.2.2, we showed that the angular frequency of oscillation of the simple pendulum of length l is given by $\omega_0 = (g/l)^{1/2}$ for *small* amplitude. Therefore, from Equation 3.4.17, the period of our pendulum is

$$T_d = 2\pi\left(\omega_0^2 - \gamma^2\right)^{-1/2} = 2\pi\left(\frac{g}{l} - 0.0265\ \text{s}^{-2}\right)^{-1/2}$$

In particular, for a baseball "seconds pendulum" for which the half-period is 1 s in the absence of damping, we have $g/l = \pi^2$, so the half-period with damping in our case is

$$\frac{T_d}{2} = \pi(\pi^2 - 0.0265)^{-1/2} \text{ s} = 1.00134 \text{ s}$$

Our solution somewhat exaggerates the effect of air resistance, because the drag function for a baseball is more nearly quadratic than linear in the velocity except at very low velocities, as discussed in Section 2.4.

EXAMPLE 3.4.4

A spherical ball of radius 0.00265 m and mass 5×10^{-4} kg is attached to a spring of force constant $k = .05$ N/m underwater. The mass is set to oscillate under the action of the spring. The coefficient of viscosity η for water is 10^{-3} Ns/m^2. (a) Find the number of oscillations that the ball will execute in the time it takes for the amplitude of the oscillation to drop by a factor of 2 from its initial value. (b) Calculate the Q of the oscillator.

Solution:

Stokes' law for objects moving in a viscous medium can be used to find c, the constant of proportionality of the $\dot{x}$ term, in the equation of motion (Equation 3.4.1) for the damped oscillator. The relationship is

$$c = 6\pi\eta r = 5 \cdot 10^{-5} \text{ Ns/m}$$

The energy of the oscillator dies away exponentially with time constant τ, and the amplitude dies away as $A = A_0 e^{-t/2\tau}$. Thus,

$$\frac{A}{A_0} = \frac{1}{2} = e^{-t/2\tau}$$

$$\therefore t = 2\tau \ln 2$$

Consequently, the number of oscillations during this time is

$$n = \omega_d t/2\pi$$
$$= \omega_d \tau (\ln 2)/\pi$$
$$= Q(\ln 2)/\pi$$

Because $\omega_0^2 = k/m = 100$ s^{-2}, $\tau = m/c = 10$ s, and $\gamma = 1/2\tau = 0.05$ s^{-1}, we obtain

$$Q = \left(\omega_0^2 - \gamma^2\right)^{1/2} \tau = (100 - 0.0025)^{1/2} \, 10 = 100$$

$$n = Q(\ln 2)/\pi = 22$$

If we had asked how many oscillations would occur in the time it takes for the amplitude to drop to $e^{-1/2}$, or about 0.606 times its initial value, the answer would have been $Q/2\pi$. Clearly Q is a measure of the rate at which an oscillator loses energy.

*3.5 | Phase Space

A physical system in motion that does not dissipate energy remains in motion. One that dissipates energy eventually comes to rest. An oscillating or rotating system that does not dissipate energy repeats its configuration each cycle. One that dissipates energy never does. The evolution of such a physical system can be graphically illustrated by examining its motion in a special space called *phase space*, rather than real space. The phase space for a single particle whose motion is restricted to lie along a single spatial coordinate consists of all the possible points in a "plane" whose horizontal coordinate is its position x and whose vertical coordinate is its velocity $\dot{x}$. Thus, the "position" of a particle on the phase-space plane is given by its "coordinates" $(x, \dot{x})$.[4] The future state of motion of such a particle is completely specified if its position and velocity are known simultaneously—say, its initial conditions $x(t_0)$ and $\dot{x}(t_0)$. We can, thus, picture the evolution of the motion of the particle from that point on by plotting its coordinates in phase space. Each point in such a plot can be thought of as a precursor for the next point. The trajectory of these points in phase space represents the complete time history of the particle.

Simple Harmonic Oscillator: No Damping Force

The simple harmonic oscillator that we discuss in this section is an example of a particle whose motion is restricted to a single dimension. Let's examine the phase-space motion of a simple harmonic oscillator that is not subject to any damping force. The solutions for its position and velocity as functions of time were given previously by Equations 3.2.5 and 3.2.12a:

$$x(t) = A\,\sin(\omega_0 t + \phi_0) \tag{3.5.1a}$$

$$\dot{x}(t) = A\omega_0\,\cos(\omega_0 t + \phi_0) \tag{3.5.1b}$$

Letting $y = \dot{x}$ we eliminate t from these two parametric equations to find the equation of the trajectory of the oscillator in phase space:

$$x^2(t) + \frac{y^2(t)}{\omega_0^2} = A^2(\sin^2(\omega_0 t + \phi_0) + \cos^2(\omega_0 t + \phi_0)) = A^2$$

$$\therefore \frac{x^2}{A^2} + \frac{y^2}{A^2\omega_0^2} = 1 \tag{3.5.2}$$

Equation 3.5.2 is the equation of an ellipse whose semimajor axis is A and whose semiminor axis is $\omega_0 A$. Shown in Figure 3.5.1 are several phase-space trajectories for the harmonic oscillator. The trajectories differ only in the amplitude A of the oscillation.

Note that the phase-path trajectories never intersect. The existence of a point common to two different trajectories would imply that two different future motions could evolve

---

*Again, as noted in Chapter 2, sections in the text marked with an asterisk may be skipped with impunity.

[4]Strictly speaking, phase space is defined as the ensemble of points (x, p) where x and p are the position and momentum of the particle. Because momentum is directly proportional to velocity, the space defined here is essentially a phase space.

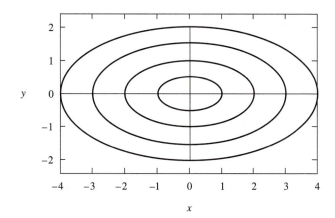

Figure 3.5.1 Phase-space plot for the simple harmonic oscillator ($\omega_0 = 0.5$ s^{-1}). No damping force ($\gamma = 0$ s^{-1}).

from a single set of conditions $(x(t_i), \dot{x}(t_i))$ at some time t_i. This cannot happen because, starting with specific values of $x(t_i)$ and $\dot{x}(t_i)$, Newton's laws of motion completely determine a unique future state of motion for the system.

Also note that the trajectories in this case form closed paths. In other words, the motion repeats itself, a consequence of the conservation of the total energy of the harmonic oscillator. In fact, the equation of the phase-space trajectory (Equation 3.5.2) is nothing more than a statement that the total energy is conserved. We can show this by substituting $E = \frac{1}{2}kA^2$ and $\omega_0^2 = k/m$ into Equation 3.5.2, obtaining

$$\frac{x^2}{2E/k} + \frac{y^2}{2E/m} = 1 \tag{3.5.3a}$$

which is equivalent to (replacing y with $\dot{x}$)

$$\tfrac{1}{2}kx^2 + \tfrac{1}{2}m\dot{x}^2 = V + T = E \tag{3.5.3b}$$

the energy equation (Equation 3.3.3) for the harmonic oscillator. Each closed phase-space trajectory, thus, corresponds to some definite, conserved total energy.

EXAMPLE 3.5.1

Consider a particle of mass m subject to a force of strength $+kx$, where x is the displacement of the particle from equilibrium. Calculate the phase space trajectories of the particle.

Solution:

The equation of motion of the particle is $m\ddot{x} = kx$. Letting $\omega^2 = k/m$ we have $\ddot{x} - \omega^2 x = 0$. Letting $y = \dot{x}$ and $y' = dy/dx$ we have $\dot{y} = \dot{x}y' = yy' = \omega^2 x$ or $ydy = \omega^2 x dx$. The solution is $y^2 - \omega^2 x^2 = C$ in which C is a constant of integration. The phase space trajectories are branches of a hyperbola whose asymptotes are $y = \pm\omega x$. The resulting phase space plot is shown in Figure 3.5.2. The trajectories are open ended, radiating away from the origin, which is an unstable equilibrium point.

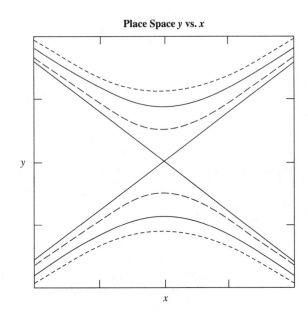

Place Space y vs. x

Figure 3.5.2 Phase space plot for $\ddot{x} - \omega^2 x = 0$.

The Underdamped Harmonic Oscillator

The phase-space trajectories for the harmonic oscillator subject to a weak damping force can be calculated in the same way as before. We anticipate, though, that the trajectories will not be closed. The motion does not repeat itself, because energy is constantly being dissipated. For the sake of illustration, we assume that the oscillator is started from rest at position x_0. The solutions for x and $\dot{x}$ are given by Equations 3.4.21a and b:

$$x = Ae^{-\gamma t} \sin(\omega_d t + \phi_0) \tag{3.5.4a}$$

$$\dot{x} = -Ae^{-\gamma t}(\gamma \sin(\omega_d t + \phi_0) - \omega_d \cos(\omega_d t + \phi_0)) \tag{3.5.4b}$$

Remember that because the initial phase angle ϕ_0 is given by the condition that $\dot{x}_0 = 0$, its value for the damped oscillator is not $\pi/2$ but $\phi_0 = \tan^{-1}\omega_d/\gamma$. It is difficult to eliminate t by brute force in the above parametric equations. Instead, we can illuminate the motion in phase space by applying a sequence of substitutions and linear transformations of the phase-space coordinates that simplifies the above expressions, leading to the form we've already discussed for the harmonic oscillator. First, substitute $\rho = Ae^{-\gamma t}$ and

$$\theta = \omega_d t + \phi_0$$

into the above equations, obtaining

$$x = \rho \sin\theta \tag{3.5.4c}$$

$$\dot{x} = -\rho(\gamma \sin\theta - \omega_d \cos\theta) \tag{3.5.4d}$$

Next, we apply the linear transformation $y = \dot{x} + \gamma x$ to Equation 3.5.4d, obtaining

$$y = \omega_d \rho \cos\theta \tag{3.5.5}$$

We then square this equation and carry out some algebra to obtain

$$y^2 = \omega_d^2 \rho^2 (1 - \sin^2 \theta)$$
$$y^2 = \omega_d^2 (\rho^2 - x^2)$$
$$\frac{x^2}{\rho^2} + \frac{y^2}{\omega_d^2 \rho^2} = 1$$

(3.5.6)

Voila! Equation 3.5.6 is identical *in form* to Equation 3.5.2. But here the variable y is a linear combination of x and $\dot{x}$ so the ensemble of points (x, y) represents a *modified* phase space. The trajectory of the oscillator in this space is an ellipse whose major and minor axes, characterized by ρ and $\omega_d \rho$, decrease exponentially with time. The trajectory starts off with a maximum value of $x_0 (= A \sin \phi_0)$ and then spirals inward toward the origin. The result is shown in Figure 3.5.3(a). The behavior of the trajectory in the x–$\dot{x}$ plane is similar and is shown in Figure 3.5.3(b). Two trajectories are shown in the plots for the cases of strong and weak damping. Which is which should be obvious.

As before, Equation 3.5.6 is none other than the energy equation for the damped harmonic oscillator. We can compare it to the results we obtained in our discussion in Section 3.4 for the rate of energy dissipation in the weakly damped oscillator. In the case of weak damping, the damping factor γ is small compared to ω_0, the undamped oscillator angular frequency (see Equation 3.4.10), and, thus, we have

$$\omega_d \approx \omega_0 \qquad y \approx \dot{x}$$

(3.5.7)

Hence, Equation 3.5.6 becomes

$$\frac{x^2}{\rho^2} + \frac{\dot{x}^2}{\rho^2 \omega_0^2} = 1$$

(3.5.8)

Note that this equation is identical in form to Equation 3.5.6, and consequently the trajectory seen in the x–$\dot{x}$ plane of Figure 3.5.3(b) for the case of weak damping is virtually identical to the modified phase-space trajectory of the weakly damped oscillator shown in Figure 3.5.3(a). Finally, upon substituting k/m for ω_0^2 and $A^2 e^{-2\gamma t}$ for ρ^2, we obtain

$$\tfrac{1}{2} kx^2 + \tfrac{1}{2} m\dot{x}^2 = \tfrac{1}{2} kA^2 e^{-2\gamma t}$$
$$= \tfrac{1}{2} m\omega_0^2 A^2 e^{-2\gamma t}$$

(3.5.9)

If we compare this result with Equation 3.4.23, we see that it represents the total energy remaining in the oscillator at any subsequent time t:

$$V(t) + T(t) = E(t)$$

(3.5.10)

The energy of the weakly damped harmonic oscillator dies away exponentially with a time constant $\tau = (2\gamma)^{-1}$. The spiral nature of its phase-space trajectory reflects this fact.

The Critically Damped Harmonic Oscillator

Equation 3.4.9 gave the solution for the critically damped oscillator:

$$x = (At + B)e^{-\gamma t}$$

(3.5.11)

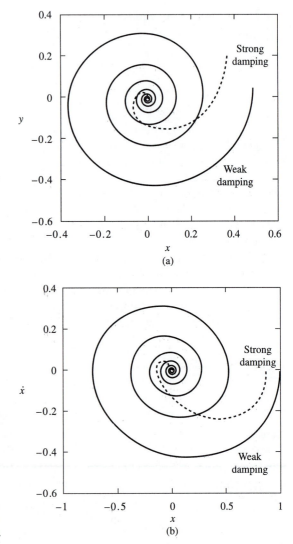

Figure 3.5.3 (a) Modified phase-space plot (see text) for the simple harmonic oscillator. (b) Phase-space plot ($\omega_0 = 0.5$ s^{-1}). Underdamped case: (1) weak damping ($\gamma = 0.05$ s^{-1}) and (2) strong damping ($\gamma = 0.25$ s^{-1}).

Taking the derivative of this equation, we obtain

$$\dot{x} = -\gamma(At + B)e^{-\gamma t} + Ae^{-\gamma t} \tag{3.5.12}$$

or

$$\dot{x} + \gamma x = Ae^{-\gamma t} \tag{3.5.13}$$

This last equation indicates that the phase-space trajectory should approach a straight line whose intercept is zero and whose slope is equal to $-\gamma$. The phase-space plot is shown in Figure 3.5.4 for motion starting off with the conditions $(x_0, \dot{x}_0) = (1, 0)$.

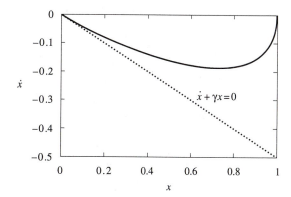

Figure 3.5.4 Phase-space plot for the simple harmonic oscillator ($\omega_0 = 0.5$ s^{-1}). Critical damping ($\gamma = 0.5$ s^{-1}).

The Overdamped Oscillator

Overdamping occurs when the damping parameter γ is larger than the angular frequency ω_0. Equation 3.4.6 then gives the solution for the motion:

$$x(t) = A_1 e^{-(\gamma-q)t} + A_2 e^{-(\gamma+q)t} \tag{3.5.14}$$

in which all the exponents are real. Taking the derivative of this equation, we find

$$\dot{x}(t) = -\gamma x + q e^{-\gamma t}(A_1 e^{qt} - A_2 e^{-qt}) \tag{3.5.15}$$

As in the case of critical damping, the phase path approaches zero along a straight line. However, approaches along two different lines are possible. To see what they are, it is convenient to let the motion start from rest at some displacement x_0. Given these conditions, a little algebra yields the following values for A_1 and A_2:

$$A_1 = \frac{(\gamma+q)}{2q}x_0 \qquad A_2 = -\frac{(\gamma-q)}{2q}x_0 \tag{3.5.16}$$

Some more algebra yields the following for two different linear combinations of x and $\dot{x}$:

$$\dot{x} + (\gamma - q)x = (\gamma - q)x_0 e^{-(\gamma+q)t} \tag{3.5.17a}$$
$$\dot{x} + (\gamma + q)x = (\gamma + q)x_0 e^{-(\gamma-q)t} \tag{3.5.17b}$$

The term on the right-hand side of each of the above equations dies out with time, and, thus, the phase-space asymptotes are given by the pairs of straight lines:

$$\dot{x} = -(\gamma - q)x \tag{3.5.18a}$$
$$\dot{x} = -(\gamma + q)x \tag{3.5.18b}$$

Except for special cases, phase-space paths of the motion always approach zero along the asymptote whose slope is $-(\gamma - q)$. That asymptote invariably "springs into existence" much faster than the other, because its exponential decay factor is $(\gamma + q)$ (Equations 3.5.17), the larger of the two.

Figure 3.5.5 shows the phase-space plot for an overdamped oscillator whose motion starts off with the values $(x_0, \dot{x}_0) = (1,0)$, along with the asymptote whose slope is $-(\gamma - q)$. Note how rapidly the trajectory locks in on the asymptote, unlike the case of critical damping, where it reaches the asymptote only toward the end of its motion. Obviously, overdamping is the most efficient way to knock the oscillation out of oscillatory motion!

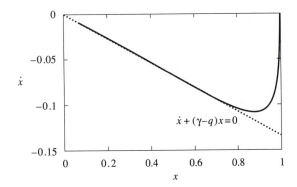

Figure 3.5.5 Phase-space plot for the simple harmonic oscillator ($\omega_0 = 0.5$ s^{-1}). Overdamping ($\gamma = 1$ s^{-1}).

EXAMPLE 3.5.2

A particle of unit mass is subject to a damping force $-\dot{x}$ and a force that depends on its displacement x from the origin that varies as $+x - x^3$. (a) Find the points of equilibrium of the particle and specify whether or not they are stable or unstable. (b) Use *Mathcad* to plot phase-space trajectories for the particle for three sets of starting conditions: $(x, y) =$ (i) $(-1, 1.40)$ (ii) $(-1, 1.45)$ (iii) $(0.01, 0)$ and describe the resulting motion.

Solution:

(a) The equation of motion is

$$\ddot{x} + \dot{x} - x + x^3 = 0$$

Let $y = \dot{x}$. Then

$$\dot{y} = -y + x - x^3$$

At equilibrium, both $y = 0$ and $\dot{y} = 0$. This is satisfied if

$$x - x^3 = x(1 - x^2) = x(1 - x)(1 + x) = 0$$

Thus, there are three equilibrium points $x = 0$ and $x = \pm 1$.

We can determine whether or not they are stable by *linearizing* the equation of motion for small excursions away from those points. Let u represent a small excursion of the particle away from an equilibrium point, which we designated by x_0. Thus, $x = x_0 + u$ and the equation of motion becomes

$$y = \dot{u} \qquad \text{and} \qquad \dot{y} = -y + (x_0 + u) - (x_0 + u)^3$$

Carrying out the expansion and dropping all terms non-linear in u, we get

$$\dot{y} = -y + \left(1 - 3x_0^2\right)u + x_0\left(1 - x_0^2\right)$$

The last term is zero, so

$$\dot{y} = -y + \left(1 - 3x_0^2\right)u$$

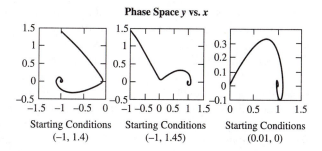

Figure 3.5.6 Phase space plots for $\ddot{x} + \dot{x} - x + x^3 = 0$.

If $(1 - 3x_0^2) < 0$ the motion is a stable, damped oscillation that eventually ceases at $x = x_0$. If $(1 - 3x_0^2) > 0$ the particle moves away from x_0 and the equilibrium is unstable. Thus, $x = \pm 1$ are points of stable equilibrium and x_0 is an unstable point.

(b) The three graphs in Figure 3.5.6 were generated by using *Mathcad's rkfixed* equation solver to solve the complete nonlinear equation of motion numerically. In all cases, no matter how the motion is started, the particle veers away from $x = 0$ and ultimately terminates at $x = \pm 1$. The motion for the third set of starting conditions is particularly illuminating. The particle is started at rest near, but not precisely at, $x = 0$. The particle is repelled away from that point, goes into damped oscillation about $x = 1$, and eventually comes to rest there. The points $x = \pm 1$ are called *attractors* and the point $x = 0$ is called a *repellor*.

3.6 | Forced Harmonic Motion: Resonance

In this section we study the motion of a damped harmonic oscillator that is subjected to a periodic driving force by an external agent. Suppose a force of the form $F_0 \cos \omega t$ is exerted upon such an oscillator. The equation of motion is

$$m\ddot{x} = -kx - c\dot{x} + F_0 \cos \omega t \qquad (3.6.1)$$

The most striking feature of such an oscillator is the way in which it responds as a function of the driving frequency even when the driving force is of fixed amplitude. A remarkable phenomenon occurs when the driving frequency is close in value to the natural frequency ω_0 of the oscillator. It is called *resonance*. Anyone who has ever pushed a child on a swing knows that the amplitude of oscillation can be made quite large if even the smallest push is made at just the right time. Small, periodic forces exerted on oscillators at frequencies well above or below the natural frequency are much less effective; the amplitude remains small. We initiate our discussion of forced harmonic motion with a qualitative description of the behavior that we might expect. Then we carry out a detailed analysis of the equation of motion (Equation 3.6.1), with our eyes peeled for the appearance of the phenomenon of resonance.

We already know that the undamped harmonic oscillator, subjected to any sort of disturbance that displaces it from its equilibrium position, oscillates at its natural frequency, $\omega_0 = \sqrt{(k/m)}$. The dissipative forces inevitably present in any real system changes the frequency of the oscillator slightly, from ω_0 to ω_d, and cause the free oscillation to die out. This motion is represented by a solution to the *homogeneous* differential equation (Equation 3.4.1, which is Equation 3.6.1 without the driving force present). A periodic driving force does two things to the oscillator: (1) It initiates a "free" oscillation at its natural frequency, and (2) it forces the oscillator to vibrate eventually at the driving frequency ω. For a short time the actual motion is a linear superposition of oscillations at these two frequencies, but with one dying away and the other persisting. The motion that dies away is called the *transient*. The final surviving motion, an oscillation at the driving frequency, is called the *steady-state* motion. It represents a solution to the inhomogeneous equation (Equation 3.6.1). Here we focus only upon the steady-state motion, whose anticipated features we describe below. To aid in the descriptive process, we assume for the moment that the damping term $-c\dot{x}$ is vanishingly small. Unfortunately, this approximation leads to the physical absurdity that the transient term never dies out—a rather paradoxical situation for a phenomenon described by the word *transient*! We just ignore this difficulty and focus totally upon the steady-state description, in hopes that the simplicity gained by this approximation gives us insight that helps when we finally solve the problem of the driven, damped oscillator.

In the absence of damping, Equation 3.6.1 can be written as

$$m\ddot{x} + kx = F_0 \cos \omega t \qquad (3.6.2)$$

The most dramatic feature of the resulting motion of this driven, undamped oscillator is a catastrophically large response at $\omega = \omega_0$. This we shall soon see, but what response might we anticipate at both extremely low ($\omega \ll \omega_0$) and high ($\omega \gg \omega_0$) frequencies? At low frequencies, we might expect the inertial term $m\ddot{x}$ to be negligible compared to the spring force $-kx$. The spring should appear to be quite stiff, compressing and relaxing very slowly, with the oscillator moving pretty much in phase with the driving force. Thus, we might guess that

$$x \approx A \cos \omega t$$

$$A = \frac{F_0}{k}$$

At high frequencies the acceleration should be large, so we might guess that $m\ddot{x}$ should dominate the spring force $-kx$. The response, in this case, is controlled by the mass of the oscillator. Its displacement should be small and 180° out of phase with the driving force, because the acceleration of a harmonic oscillator is 180° out of phase with the displacement. The veracity of these preliminary considerations emerge during the process of obtaining an actual solution.

First, let us solve Equation 3.6.2, representing the driven, undamped oscillator. In keeping with our previous descriptions of harmonic motion, we try a solution of the form

$$x(t) = A \cos(\omega t - \phi)$$

Thus, we assume that the steady-state motion is harmonic and that in the steady state it ought to respond at the driving frequency ω. We note, though, that its response might differ in phase from that of the driving force by an amount ϕ. ϕ is not the result of some initial condition! (It does not make any sense to talk about initial conditions for a steady-state solution.) To see if this assumed solution works, we substitute it into Equation 3.6.2, obtaining

$$-m\omega^2 A \cos(\omega t - \phi) + kA \cos(\omega t - \phi) = F_0 \cos \omega t$$

This works if ϕ can take on only two values, 0 and π. Let us see what is implied by this requirement. Solving the above equation for $\phi = 0$ and π, respectively, yields

$$A = \frac{F_0/m}{\left(\omega_0^2 - \omega^2\right)} \quad \phi = 0 \quad \omega < \omega_0$$

$$= \frac{F_0/m}{\left(\omega^2 - \omega_0^2\right)} \quad \phi = \pi \quad \omega > \omega_0$$

We plot the amplitude A and phase angle ϕ as functions of ω in Figure 3.6.1. Indeed, as can be seen from the plots, as ω passes through ω_0, the amplitude becomes catastrophically large, and, perhaps even more surprisingly, the displacement shifts discontinuously from being in phase with the driving force to being 180° out of phase. True, these results are not physically possible. However, they are idealizations of real situations. As we shall soon see, if we throw in just a little damping, at ω close to ω_0 the amplitude becomes large but finite. The phase shift "smooths out"; it is no longer discontinuous, although the shift is still quite abrupt.

(**Note:** *The behavior of the system mimics our description of the low-frequency and high-frequency limits.*)

The 0° and 180° phase differences between the displacement and driving force can be simply and vividly demonstrated. Hold the lighter end of a pencil or a pair of scissors (closed) or a spoon delicately between forefinger and thumb, squeezing just hard enough that it does not drop. To demonstrate the 0° phase difference, slowly move your hand back and forth horizontally in a direction parallel to the line formed between your forefinger and thumb. The bottom of this makeshift pendulum swings back and forth in phase with the hand motion and with a larger amplitude than the hand motion. To see the 180°

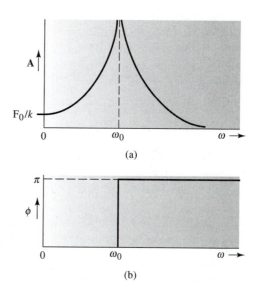

Figure 3.6.1   (a) The amplitude of a driven oscillator versus ω with no damping. (b) The phase lag of the displacement relative to the driving force versus ω.

phase shift, move your hand back and forth rather rapidly (high frequency). The bottom of the pendulum hardly moves at all, but what little motion it does undergo is 180° out of phase with the hand motion.

The Driven, Damped Harmonic Oscillator

We now seek the steady-state solution to Equation 3.6.1, representing the driven, damped harmonic oscillator. It is fairly straightforward to solve this equation directly, but it is algebraically simpler to use complex exponentials instead of sines and/or cosines. First, we represent the driving force as

$$F = F_0 e^{i\omega t} \tag{3.6.3}$$

so that Equation 3.6.1 becomes

$$m\ddot{x} + c\dot{x} + kx = F_0 e^{i\omega t} \tag{3.6.4}$$

The variable x is now complex, as is the applied force F. Remember, though, that by Euler's identity the real part of F is $F_0 \cos \omega t$.[5] If we solve Equation 3.6.4 for x, its real part will be a solution to Equation 3.6.1. In fact, when we find a solution to the above complex equation (Equation 3.6.4), we can be sure that the real parts of both sides are equal (as are the imaginary parts). It is the real parts that are equivalent to Equation 3.6.1 and, thus, the real, physical situation.

For the steady-state solution, let us, therefore, try the complex exponential

$$x(t) = Ae^{i(\omega t - \phi)} \tag{3.6.5}$$

where the amplitude A and phase difference ϕ are constants to be determined. If this "guess" is correct, we must have

$$m\frac{d^2}{dt^2} Ae^{i(\omega t - \phi)} + c\frac{d}{dt} Ae^{i(\omega t - \phi)} + kAe^{i(\omega t - \phi)} = F_0 e^{i\omega t} \tag{3.6.6a}$$

be true for all values of t. Upon performing the indicated operations and canceling the common factor $e^{i\omega t}$, we find

$$-m\omega^2 A + i\omega cA + kA = F_0 e^{i\phi} = F_0(\cos \phi + i \sin \phi) \tag{3.6.6b}$$

Equating the real and imaginary parts yields the two equations

$$A(k - m\omega^2) = F_0 \cos \phi \tag{3.6.7a}$$
$$c\omega A = F_0 \sin \phi$$

Upon dividing the second by the first and using the identity $\tan \phi = \sin \phi / \cos \phi$, we obtain the following relation for the phase angle:

$$\tan \phi = \frac{c\omega}{k - m\omega^2} \tag{3.6.7b}$$

---

[5] For a proof of Euler's identity, see Appendix D.

By squaring both sides of Equations 3.6.7a and adding and employing the identity $\sin^2 \phi + \cos^2 \phi = 1$, we find

$$A^2(k - m\omega^2)^2 + c^2\omega^2 A^2 = F_0^2 \tag{3.6.7c}$$

We can then solve for A, the amplitude of the steady-state oscillation, as a function of the driving frequency:

$$A(\omega) = \frac{F_0}{[(k - m\omega^2)^2 + c^2\omega^2]^{1/2}} \tag{3.6.7d}$$

In terms of our previous abbreviations $\omega_0^2 = k/m$ and $\gamma = c/2m$, we can write the expressions in another form, as follows:

$$\tan\phi = \frac{2\gamma\omega}{\omega_0^2 - \omega^2} \tag{3.6.8}$$

$$A(\omega) = \frac{F_0/m}{\left[\left(\omega_0^2 - \omega^2\right)^2 + 4\gamma^2\omega^2\right]^{1/2}} \tag{3.6.9}$$

A plot of the above amplitude A and phase difference ϕ versus driving frequency ω (Fig. 3.6.2) reveals a fetching similarity to the plots of Figure 3.6.1 for the case of the undamped oscillator. As can be seen from the plots, as the damping term approaches 0, the resonant peak gets larger and narrower, and the phase shift sharpens up, ultimately approaching infinity and discontinuity, respectively, at ω_0. What is not so obvious from these plots is that the amplitude resonant frequency is not ω_0 when damping is present (although the phase shift always passes through $\pi/2$ at ω_0)! Amplitude resonance occurs at some other value ω_r, which can be calculated by differentiating $A(\omega)$ and setting the result equal to zero. Upon solving the resultant equation for ω, we obtain

$$\omega_r^2 = \omega_0^2 - 2\gamma^2 \tag{3.6.10}$$

ω_r approaches ω_0 as γ, the damping term, goes to zero. Because the angular frequency of the freely running damped oscillator is given by $\omega_d = (\omega_0^2 - \gamma^2)^{1/2}$, we have

$$\omega_r^2 = \omega_d^2 - \gamma^2 \tag{3.6.11}$$

When the damping is weak, and only under this condition, the resonant frequency ω_r, the freely running, damped oscillator frequency ω_d, and the natural frequency ω_0 of the undamped oscillator are essentially identical.

At the extreme of strong damping, no amplitude resonance occurs if $\gamma > \omega_0/\sqrt{2}$, because the amplitude then becomes a monotonically decreasing function of ω. To see this, consider the limiting case $\gamma^2 = \omega_0^2/2$. Equation 3.6.9 then gives

$$A(\omega) = \frac{F_0/m}{\left[\left(\omega_0^2 - \omega^2\right)^2 + 2\omega_0^2\omega^2\right]^{1/2}} = \frac{F_0/m}{\left(\omega_0^4 + \omega^4\right)^{1/2}} \tag{3.6.12}$$

which clearly decreases with increasing values of ω, starting with $\omega = 0$.

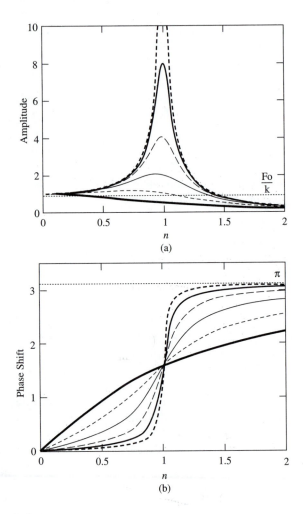

Figure 3.6.2 (a) Amplitude $A/(F_0/k)$ and (b) phase shift ϕ vs. driving frequency ($n = \omega/\omega_0$) for values of the damping constant γ given by $\gamma = 2^{-i}\omega_0$ ($i = 0, 1 \ldots 5$). Larger values of A and more abrupt phase shifts correspond to decreasing values of γ.

EXAMPLE 3.6.1

A seismograph may be modeled as a mass suspended by springs and a *dashpot* from a platform attached to the Earth (Figure 3.6.3). Oscillations of the Earth are passed through the platform to the suspended mass, which has a "pointer" to record its displacement relative to the platform. The dashpot provides a damping force. Ideally, the displacement A of the mass relative to the platform should closely mimic the displacement of the Earth D. Find the equation of motion of the mass m and choose parameters ω_0 and γ to insure that A lies within 10% of D. Assume during a ground tremor that the Earth oscillates with simple harmonic motion at $f = 10$ Hz.

Solution:

First we calculate the equation of motion of the mass m. Suppose the platform moves downward a distance z relative to its initial position and that m moves downward to a position y relative to the platform. The plunger in the dashpot is moving downward with

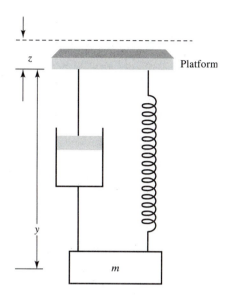

Figure 3.6.3 Seismograph model.

speed $\dot{z}$ while the pot containing the damping fluid is moving downward with speed $\dot{y} + \dot{z}$; therefore, the retarding, damping force is given by $c\dot{y}$. If l is the natural length of the spring, then

$$F = mg - c\dot{y} - k(y - l) = m(\ddot{y} + \ddot{z})$$

We let $y = x + mg/k + l$, so that x is the displacement of the mass from its equilibrium position (see Figure 3.2.5), and, in terms of x, the equation of motion becomes

$$m\ddot{x} + c\dot{x} + kx = -m\ddot{z}$$

During the tremor, as the platform oscillates with simple harmonic motion of amplitude D and angular frequency $\omega = 2\pi f$, we have $z = De^{i\omega t}$. Thus,

$$m\ddot{x} + c\dot{x} + kx = mD\omega^2 e^{i\omega t}$$

Comparing with Equation 3.6.4, and associating F_0/m with $D\omega^2$, the solution for the amplitude of oscillation given by Equation 3.6.9 can be expressed here as

$$A = D\omega^2 \left[\left(\omega_0^2 - \omega^2 \right)^2 + 4\gamma^2 \omega^2 \right]^{-1/2}$$

Dividing numerator and denominator by ω^2 we obtain

$$A = D \left[\left(\frac{\omega_0^2}{\omega^2} - 1 \right)^2 + 4\frac{\gamma^2}{\omega^2} \right]^{-1/2}$$

Expanding the term in the denominator gives

$$A = D \left[1 + \frac{\omega_0^4}{\omega^4} + \frac{2}{\omega^2} \left(2\gamma^2 - \omega_0^2 \right) \right]^{-1/2}$$

We can insure that $A \approx D$ for reasonable values of ω by setting $2\gamma^2 - \omega_0^2 = 0$ and $\omega_0/\omega < 0$. For example, for a fractional difference between A and D of 10%, we require that

$$\frac{D-A}{D} = 1 - \left(1 + \frac{\omega_0^4}{\omega^4}\right)^{-1/2} \approx \frac{1}{2}\frac{\omega_0^4}{\omega^4} < \frac{1}{10} \quad \text{or} \quad \omega_0 < 0.84\omega$$

This means that the free-running frequency of the oscillator is

$$f_0 = \omega_0/2\pi \le 8\,\text{Hz}$$

The damping parameter should be

$$\gamma = \omega_0/\sqrt{2} = 36.$$

Typically, this requires the use of "soft" springs and a heavy mass.

Amplitude of Oscillation at the Resonance Peak

The steady-state amplitude at the resonant frequency, which we call A_{max}, is obtained from Equations 3.6.9 and 3.6.10. The result is

$$A_{max} = \frac{F_0/m}{2\gamma\sqrt{\omega_0^2 - \gamma^2}} \tag{3.6.13a}$$

In the case of weak damping, we can neglect γ^2 and write

$$A_{max} \simeq \frac{F_0}{2\gamma m \omega_0} \tag{3.6.13b}$$

Thus, the amplitude of the induced oscillation at the resonant condition becomes very large if the damping factor γ is very small, and conversely. In mechanical systems large resonant amplitudes may or may not be desirable. In the case of electric motors, for example, rubber or spring mounts are used to minimize the transmission of vibration. The stiffness of these mounts is chosen so as to ensure that the resulting resonant frequency is far from the running frequency of the motor.

Sharpness of the Resonance: Quality Factor

The sharpness of the resonance peak is frequently of interest. Let us consider the case of weak damping $\gamma \ll \omega_0$. Then, in the expression for steady-state amplitude (Equation 3.6.9), we can make the following substitutions:

$$\omega_0^2 - \omega^2 = (\omega_0 + \omega)(\omega_0 - \omega) \tag{3.6.14a}$$
$$\simeq 2\omega_0(\omega_0 - \omega)$$
$$4\gamma^2\omega^2 \simeq 4\gamma^2\omega_0^2 \tag{3.6.14b}$$

These, together with the expression for A_{max}, allow us to write the amplitude equation in the following approximate form:

$$A(\omega) \approx \frac{A_{max}\gamma}{\sqrt{(\omega_0 - \omega)^2 + \gamma^2}}$$

(3.6.15)

The above equation shows that when $|\omega_0 - \omega| = \gamma$ or, equivalently, if

$$\omega = \omega_0 \pm \gamma$$

(3.6.16)

then

$$A^2 = \frac{1}{2}A_{max}^2$$

(3.6.17)

This means that γ is a measure of the width of the resonance curve. Thus, 2γ is the frequency difference between the points for which the energy is down by a factor of $\frac{1}{2}$ from the energy at resonance, because the energy is proportional to A^2.

The quality factor Q defined in Equation 3.4.24, which characterizes the rate of energy loss in the undriven, damped harmonic oscillator, also characterizes the sharpness of the resonance peak for the driven oscillator. In the case of weak damping, Q can be expressed as

$$Q = \frac{\omega_d}{2\gamma} \approx \frac{\omega_0}{2\gamma}$$

(3.6.18)

Thus, the total width $\Delta\omega$ at the half-energy points is approximately

$$\Delta\omega = 2\gamma \approx \frac{\omega_0}{Q}$$

(3.6.19a)

or, because $\omega = 2\pi f$,

$$\frac{\Delta\omega}{\omega_0} = \frac{\Delta f}{f_0} \approx \frac{1}{Q}$$

(3.6.19b)

giving the fractional width of the resonance peak.

This last expression for Q, so innocuous-looking, represents a key feature of feedback and control in electrical systems. Many electrical systems require the existence of a well-defined and precisely maintained frequency. High Q (of order 10^5) quartz oscillators, vibrating at their resonant frequency, are commonly employed as the control element in feedback circuits to provide frequency stability. A high Q results in a sharp resonance. If the frequency of the circuit under control by the quartz oscillator starts to wander or drift by some amount δf away from the resonance peak, feedback circuitry, exploiting the sharpness of the resonance, drives the circuit vigorously back toward the resonant frequency. The higher the Q of the oscillator and, thus, the narrower δf, the more stable the output of the frequency of the circuit.

The Phase Difference ϕ

Equation 3.6.8 gives the difference in phase ϕ between the applied driving force and the steady-state response:

$$\phi = \tan^{-1}\left[\frac{2\gamma\omega}{\left(\omega_0^2 - \omega^2\right)}\right] \tag{3.6.20}$$

The phase difference is plotted in Figure 3.6.2(b). We saw that for the driven, undamped oscillator, ϕ was 0° for $\omega < \omega_0$ and 180° for $\omega > \omega_0$. These values are the low- and high-frequency limits of the real motion. Furthermore, ϕ changed discontinuously at $\omega = \omega_0$. This, too, is an idealization of the real motion where the transition between the two limits is smooth, although for very small damping it is quite abrupt, changing essentially from one limit to the other as ω passes through a region within $\pm\gamma$ about ω_0.

At low driving frequencies $\omega \ll \omega_0$, we see that $\phi \to 0$ and the response is nearly in phase with the driving force. That this is reasonable can be seen upon examination of the amplitude of the oscillation (Equation 3.6.9). In the low-frequency limit, it becomes

$$A(\omega \to 0) \approx \frac{F_0/m}{\omega_0^2} = \frac{F_0/m}{k/m} = \frac{F_0}{k} \tag{3.6.21}$$

In other words, just as we claimed during our preliminary discussion of the driven oscillator, the spring, and not the mass or the friction, controls the response; the mass is slowly pushed back and forth by a force acting against the retarding force of the spring.

At resonance the response can be enormous. Physically, how can this be? Perhaps some insight can be gained by thinking about pushing a child on a swing. How is it done? Clearly, anyone who has experience pushing a swing does not stand behind the child and push when the swing is on the backswing. One pushes in the same direction the swing is *moving*, essentially in phase with its velocity, regardless of its position. To push a small child, we usually stand somewhat to the side and give a very small shove forward as the swing passes through the equilibrium position, when its speed is a maximum and the displacement is zero! In fact, this is the optimum way to achieve a resonance condition; a rather gentle force, judiciously applied, can lead to a large amplitude of oscillation. The maximum amplitude at resonance is given by Equation 3.6.13a and, in the case of weak damping, by Equation 3.6.13b, $A_{max} \approx F_0/2\gamma m\omega_0$. But from the expression above for the amplitude as $\omega \to 0$, we have $A(\omega \to 0) \approx F_0/m\omega_0^2$. Hence, the ratio is

$$\frac{A_{max}}{A(\omega \to 0)} = \frac{F_0/(2\gamma m\omega_0)}{F_0/\left(m\omega_0^2\right)} = \frac{\omega_0}{2\gamma} = \omega_0\tau = Q \tag{3.6.22}$$

The result is simply the Q of the oscillator. Imagine what would happen to the child on the swing if there were no frictional losses! We would continue to pump little bits of energy into the swing on a cycle-by-cycle basis, and with no energy loss per cycle, the amplitude would soon grow to a catastrophic dimension.

Now let us look at the phase difference. At $\omega = \omega_0$, $\phi = \pi/2$. Hence, the displacement "lags," or is behind, the driving force by 90°. In view of the foregoing discussion, this should make sense. The optimum time to dump energy into the oscillator is when it swings through zero at maximum velocity, that is, when the power input $\mathbf{F} \cdot \mathbf{v}$ is a maximum. For example, the real part of Equation 3.6.5 gives the displacement of the oscillator:

$$x(t) = A(\omega)Re(e^{i(\omega t - \phi)}) = A(\omega) \cos(\omega t - \phi) \tag{3.6.23}$$

and at resonance, for small damping, this becomes

$$
\begin{aligned}
x(t) &= A(\omega_0) \cos(\omega_0 t - \pi/2) \\
&= A(\omega_0) \sin \omega_0 t
\end{aligned}
\tag{3.6.24}
$$

The velocity, in general, is

$$\dot{x}(t) = -\omega A(\omega) \sin(\omega t - \phi) \tag{3.6.25}$$

which at resonance becomes

$$\dot{x}(t) = \omega_0 A(\omega_0) \cos \omega_0 t \tag{3.6.26}$$

Because the driving force at resonance is given by

$$F = F_0 \, Re(e^{i\omega_0 t}) = F_0 \, \cos \omega_0 t \tag{3.6.27}$$

we can see that the driving force is indeed in phase with the velocity of the oscillator, or 90° ahead of the displacement.

Finally, for large values of ω, $\omega \gg \omega_0$, $\phi \to \pi$, and the displacement is 180° out of phase with the driving force. The amplitude of the displacement becomes

$$A(\omega \gg \omega_0) \approx \frac{F_0}{m\omega^2} \tag{3.6.28}$$

In this case, the amplitude falls off as $1/\omega^2$. The mass responds essentially like a free object, being rapidly shaken back and forth by the applied force. The main effect of the spring is to cause the displacement to lag behind the driving force by 180°.

Electrical–Mechanical Analogs

When an electric current flows in a circuit comprising inductive, capacitive, and resistive elements, there is a precise analogy with a moving mechanical system of masses and springs with frictional forces of the type studied previously. Thus, if a current $i = dq/dt$ (q being the charge) flows through an inductance L, the potential difference across the inductance is $L\ddot{q}$, and the stored energy is $\frac{1}{2}L\dot{q}^2$. Hence, inductance and charge are analogous to mass and displacement, respectively, and potential difference is analogous to force. Similarly, if a capacitance C carries a charge q, the potential difference is $C^{-1}q$, and the stored energy is $\frac{1}{2}C^{-1}q^2$. Consequently, we see that the reciprocal of C is analogous to the stiffness constant of a spring. Finally, for an electric current i flowing through a resistance R, the potential difference is $iR = \dot{q}R$, and the rate of energy dissipation is $i^2R = \dot{q}^2R$ in analogy with the quantity $c\dot{x}^2$ for a mechanical system. Table 3.6.1 summarizes the situation.

TABLE 3.6.1	Electrical–Mechanical Analogs	
Mechanical		**Electrical**
x	Displacement	q	Charge
$\dot{x}$	Velocity	$\dot{q}=i$	Current
m	Mass	L	Inductance
k	Stiffness	C^{-1}	Reciprocal of capacitance
c	Damping resistance	R	Resistance
F	Force	V	Potential difference

EXAMPLE 3.6.2

The exponential damping factor γ of a spring suspension system is one-tenth the critical value. If the undamped frequency is ω_0, find (a) the resonant frequency, (b) the quality factor, (c) the phase angle ϕ when the system is driven at a frequency $\omega = \omega_0/2$, and (d) the steady-state amplitude at this frequency.

Solution:

(a) We have $\gamma = \gamma_{crit}/10 = \omega_0/10$, from Equation 3.4.7, so from Equation 3.6.10,

$$\omega_r = \left[\omega_0^2 - 2(\omega_0/10)^2\right]^{1/2} = \omega_0(0.98)^{1/2} = 0.99\,\omega_0$$

(b) The system can be regarded as weakly damped, so, from Equation 3.6.18,

$$Q \approx \frac{\omega_0}{2\gamma} = \frac{\omega_0}{2(\omega_0/10)} = 5$$

(c) From Equation 3.6.8 we have

$$\phi = \tan^{-1}\left(\frac{2\gamma\omega}{\omega_0^2 - \omega^2}\right) = \tan^{-1}\left[\frac{2(\omega_0/10)(\omega_0/2)}{\omega_0^2 - (\omega_0/2)^2}\right]$$

$$= \tan^{-1} 0.133 = 7.6°$$

(d) From Equation 3.6.9 we first calculate the value of the resonance denominator:

$$D(\omega = \omega_0/2) = \left[\left(\omega_0^2 - \omega_0^2/4\right)^2 + 4(\omega_0/10)^2(\omega_0/2)^2\right]^{1/2}$$

$$= [(9/16) + (1/100)]^{1/2}\,\omega_0^2 = 0.7566\omega_0^2$$

From this, the amplitude is

$$A(\omega = \omega_0/2) = \frac{F_0/m}{0.7566\omega_0^2} = 1.322\frac{F_0}{m\omega_0^2}$$

Notice that the factor $(F_0/m\omega_0^2) = F_0/k$ is the steady-state amplitude for zero driving frequency.

*3.7 | The Nonlinear Oscillator: Method of Successive Approximations

When a system is displaced from its equilibrium position, the restoring force may vary in a manner other than in direct proportion to the displacement. For example, a spring may not obey Hooke's law exactly; also, in many physical cases the restoring force function is inherently nonlinear, as is the case with the simple pendulum discussed in the example to follow.

In the nonlinear case the restoring force can be expressed as

$$F(x) = -kx + \epsilon(x) \tag{3.7.1}$$

in which the function $\epsilon(x)$ represents the departure from linearity. It is necessarily quadratic, or higher order, in the displacement variable x. The differential equation of motion under such a force, assuming no external forces are acting, can be written in the form

$$m\ddot{x} + kx = \epsilon(x) = \epsilon_2 x^2 + \epsilon_3 x^3 + \cdots \tag{3.7.2}$$

Here we have expanded $\epsilon(x)$ as a power series.

Solving the above type of equation usually requires some method of approximation. To illustrate one method, we take a particular case in which only the cubic term in $\epsilon(x)$ is of importance. Then we have

$$m\ddot{x} + kx = \epsilon_3 x^3 \tag{3.7.3}$$

Upon division by m and introduction of the abbreviations $\omega_0^2 = k/m$ and $\epsilon_3/m = \lambda$, we can write

$$\ddot{x} + \omega_0^2 x = \lambda x^3 \tag{3.7.4}$$

We find the solution by the *method of successive approximations*.

Now we know that for $\lambda = 0$ a solution is $x = A \cos \omega_0 t$. Suppose we try a *first* approximation of the same form,

$$x = A \cos \omega t \tag{3.7.5}$$

where, as we see, ω is not quite equal to ω_0. Inserting our trial solution into the differential equation gives

$$-A\omega^2 \cos\omega t + A\omega_0^2 \cos\omega t = \lambda A^3 \cos^3\omega t = \lambda A^3 \left(\tfrac{3}{4}\cos\omega t + \tfrac{1}{4}\cos 3\omega t\right) \tag{3.7.6a}$$

In the last step we have used the trigonometric identity $\cos^3 u = \tfrac{3}{4}\cos u + \tfrac{1}{4}\cos 3u$, which is easily derived by use of the relation $\cos^3 u = [(e^{iu} + e^{-iu})/2]^3$. Upon transposing and collecting terms, we get

$$\left(-\omega^2 + \omega_0^2 - \tfrac{3}{4}\lambda A^2\right)A \cos\omega t - \tfrac{1}{4}\lambda A^3 \cos 3\omega t = 0 \tag{3.7.6b}$$

Excluding the trivial case $A = 0$, we see that our trial solution does not exactly satisfy the differential equation. However, an approximation to the value of ω, *which is valid for small*

λ, is obtained by setting the quantity in parentheses equal to zero. This yields

$$\omega^2 = \omega_0^2 - \tfrac{3}{4}\lambda A^2 \tag{3.7.7a}$$

$$\omega = \omega_0\left(1 - \frac{3\lambda A^2}{4\omega_0^2}\right)^{1/2} \tag{3.7.7b}$$

for the frequency of our freely running nonlinear oscillator. As we can see, it is a function of the amplitude A.

To obtain a better solution, we must take into account the dangling term in Equation 3.7.6b involving the third harmonic, $\cos 3\omega t$. Accordingly, we take a *second* trial solution of the form

$$x = A\cos\omega t + B\cos 3\omega t \tag{3.7.8}$$

Putting this into the differential equation, we find, after collecting terms,

$$\left(-\omega^2 + \omega_0^2 - \tfrac{3}{4}\lambda A^2\right)A\cos\omega t + \left(-9B\omega^2 + \omega_0^2 B - \tfrac{1}{4}\lambda A^3\right)\cos 3\omega t \tag{3.7.9a}$$
$$+ (\text{terms involving } B\lambda \text{ and higher multiples of } \omega t) = 0$$

Setting the first quantity in parentheses equal to zero gives the same value for ω found in Equations 3.7.7. Equating the second to zero gives a value for the coefficient B, namely,

$$B = \frac{\tfrac{1}{4}\lambda A^3}{-9\omega^2 + \omega_0^2} = \frac{\lambda A^3}{-32\omega_0^2 + 27\lambda A^2} \approx -\frac{\lambda A^3}{32\omega_0^2} \tag{3.7.9b}$$

where we have assumed that the term in the denominator involving λA^2 is small enough to neglect. Our second approximation can be expressed as

$$x = A\cos\omega t - \frac{\lambda A^3}{32\omega_0^2}\cos 3\omega t \tag{3.7.10}$$

We stop at this point, but the process could be repeated to find yet a third approximation, and so on.

The above analysis, although it is admittedly very crude, brings out two essential features of free oscillation under a nonlinear restoring force; that is, the period of oscillation is a function of the amplitude of vibration, and the oscillation is not strictly sinusoidal but can be considered as the superposition of a mixture of harmonics. The vibration of a nonlinear system driven by a purely sinusoidal driving force is also distorted; that is, it contains harmonics. The loudspeaker of a stereo system, for example, may introduce distortion (harmonics) over and above that introduced by the electronic amplifying system.

EXAMPLE 3.7.1

The Simple Pendulum as a Nonlinear Oscillator

In Example 3.2.2 we treated the simple pendulum as a linear harmonic oscillator by using the approximation $\sin\theta \simeq \theta$. Actually, the sine can be expanded as a power series,

$$\sin\theta = \theta - \frac{\theta^3}{3!} + \frac{\theta^5}{5!} - \cdots$$

so the differential equation for the simple pendulum, $\ddot{\theta} + (g/l)\sin\theta = 0$, may be written in the form of Equation 3.7.2, and, by retaining only the linear and the cubic terms in the expansion for the sine, the differential equation becomes

$$\ddot{\theta} + \omega_0^2 \theta = \frac{\omega_0^2}{3!}\theta^3$$

in which $\omega_0^2 = g/l$. This is mathematically identical to Equation 3.7.4 with the constant $\lambda = \omega_0^2/3! = \omega_0^2/6$. The improved expression for the angular frequency, Equation 3.7.7b, then gives

$$\omega = \omega_0\left[1 - \frac{3(\omega_0^2/6)A^2}{4\omega_0^2}\right]^{1/2} = \omega_0\left(1 - \frac{A^2}{8}\right)^{1/2}$$

and

$$T = \frac{2\pi}{\omega} = 2\pi\sqrt{\frac{l}{g}}\left(1 - \frac{A^2}{8}\right)^{-1/2} = T_0\left(1 - \frac{A^2}{8}\right)^{-1/2}$$

for the period of the simple pendulum. Here A is the amplitude of oscillation expressed in radians. Our method of approximation shows that the period for nonzero amplitude is longer by the factor $(1 - A^2/8)^{-1/2}$ than that calculated earlier, assuming $\sin\theta = \theta$. For instance, if the pendulum is swinging with an amplitude of $90° = \pi/2$ radians (a fairly large amplitude), the factor is $(1 - \pi^2/32)^{-1/2} = 1.2025$, so the period is about 20% longer than the period for small amplitude. This is considerably greater than the increase due to damping of the baseball pendulum, treated in Example 3.4.3.

*The Self-Limiting Oscillator: Numerical Solution

Certain nonlinear oscillators exhibit an effect that cannot be generated by any linear oscillator—the limit cycle, that is, its oscillations are self-limiting. Examples of nonlinear oscillators that exhibit self-limiting behavior are the van der Pol oscillator, intensively studied by van der Pol[6] in his investigation of vacuum tube circuits, and the simple mechanical oscillator subject to dry friction (see Computer Problem 3.5), studied by Lord Rayleigh in his investigation of the vibrations of violin strings driven by bow strings.[7] Here we discuss a variant of the van der Pol equation of motion that describes a nonlinear oscillator, exhibiting self-limiting behavior whose limit cycle we can calculate explicitly rather than numerically. Consider an oscillator subject to a nonlinear damping force, whose overall equation of motion is

$$\ddot{x} - \gamma\left(A^2 - x^2 - \frac{\dot{x}^2}{\beta^2}\right)\dot{x} + \omega_0^2 x = 0 \qquad (3.7.11)$$

---

[6]B. van der Pol, *Phil. Mag.* 2, 978 (1926). Also see T. L. Chow, *Classical Mechanics,* New York, NY Wiley, 1995.

[7]P. Smith and R. Smith, *Mechanics,* Chichester, England Wiley, 1990.

Van der Pol's equation is identical to Equation 3.7.11 without the third term in parentheses, the velocity-dependent damping factor, $\dot{x}^2/\beta^2$ (see Computer Problem 3.3). The limit cycle becomes apparent with a slight rearrangement of the above terms and a substitution of the phase-space variable y for $\dot{x}$:

$$\dot{y} - \gamma A^2 \left[1 - \left(\frac{x^2}{A^2} + \frac{y^2}{A^2 \beta^2} \right) \right] y + \omega_0^2 x = 0 \qquad (3.7.12)$$

The nonlinear damping term is *negative* for all points (x, y) inside the ellipse given by

$$\frac{x^2}{A^2} + \frac{y^2}{A^2 \beta^2} = 1 \qquad (3.7.13)$$

It is zero for points on the ellipse and positive for points outside the ellipse. Therefore, no matter the state of the oscillator (described by its current position in phase space), it is driven toward states whose phase-space points lie along the ellipse. In other words, no matter how the motion is started, the oscillator ultimately vibrates with simple harmonic motion of amplitude A; its behavior is said to be "self-limiting," and this ellipse in phase space is called its limit cycle. The van der Pol oscillator behaves this way, but its limit cycle cannot be seen quite so transparently.

A complete solution can only be carried out numerically. We have used *Mathcad* to do this. For ease of calculation, we have set the factors A, β, and ω_0 equal to one. This amounts to transforming the elliptical limit cycle into a circular one of unit radius and scaling angular frequencies of vibration to ω_0. Thus, Equation 3.7.12 takes on the simple form

$$\dot{y} - \gamma(1 - x^2 - y^2)y + x = 0 \qquad (3.7.14)$$

A classic way to solve a single second-order differential equation is to turn it into an equivalent system of first-order ones and then use Runge–Kutta or some equivalent technique to solve them (see Appendix I). With the substitution of y for $\dot{x}$, we obtain the following two first-order differential equations:

$$\begin{aligned} \dot{x} &= y \\ \dot{y} &= -x + \gamma(1 - x^2 - y^2)y \end{aligned} \qquad (3.7.15)$$

In fact, these equations do not have to be solved numerically. One can easily verify that they have analytic solutions $x = \cos t$ and $y = -\sin t$, which represent the final limiting motion on the unit circle $x^2 + y^2 = 1$. It is captivating, however, to let the motion start from arbitrary values that lie both within and without the limit cycle, and watch the system evolve toward its limit cycle. This behavior can be observed only by solving the equations numerically—for example, using *Mathcad*.

As in the preceding chapter, we use the *Mathcad* equation solver, *rkfixed*, which employs the fourth-order Runge–Kutta technique to numerically solve first-order differential equations. We represent the variables x and y in *Mathcad* as x_1 and x_2, the components of a two-dimensional vector $\mathbf{x} = (x_1, x_2)$.

Mathcad Procedure

- Define a two-dimensional vector $\mathbf{x} = (x_1, x_2)$ containing initial values (x_0, y_0); that is,

$$\mathbf{x} = \begin{pmatrix} -0.5 \\ 0 \end{pmatrix}$$

(This starts motion off at $(x_0, y_0) = (-0.5, 0)$.)
- Define a vector-valued function $D(t, \mathbf{x})$ containing the first derivatives of the unknown functions $x(t)$ and $y(t)$ (Equations 3.7.15):

$$D(t, \mathbf{x}) = \begin{pmatrix} x_2 \\ -x_1 + \gamma\left(1 - x_1^2 - x_2^2\right)x_2 \end{pmatrix}$$

- Decide on time interval $[0, T]$ and the number of points, $npts$, within that interval where solutions are to be evaluated.
- Pass this information to the function $rkfixed$ (or $Rkadapt$ if the motion changes too rapidly within small time intervals somewhere within the time interval $[0, T]$ that you have selected); that is,

$$Z = rkfixed(\mathbf{x}, 0, T, npts, D)$$

or

$$Z = Rkadapt(\mathbf{x}, 0, T, npts, D)$$

The function $rkfixed$ (or $Rkadapt$) returns a matrix Z (in this case, two rows and three columns) whose first column contains the times t_i where the solution was evaluated and whose remaining two columns contain the values of $x(t_i)$ and $y(t_i)$. *Mathcad*'s graphing feature can then be used to generate the resulting phase-space plot, a two-dimensional scatter plot of $y(t_i)$ versus $x(t_i)$.

Figure 3.7.1 shows the result of a numerical solution to the above equation of motion. Indeed, as advertised, the system either spirals in or spirals out, finally settling on the limit cycle in which the damping force disappears. Once the oscillator "locks in" on its limit cycle, its motion is simply that of the simple harmonic oscillator, repetitive and completely predictable.

*3.8 | The Nonlinear Oscillator: Chaotic Motion

When do nonlinear oscillations occur in nature? We answer that question with a tautology: They occur when the equations of motion are nonlinear. This means that if there are two (or more) solutions, $x_1(t)$ and $x_2(t)$, to a nonlinear equation of motion, any *arbitrary* linear combination of them, $\alpha x_1(t) + \beta x_2(t)$ is, in general, not linear. We can illustrate this with a simple example. The first nonlinear oscillator discussed in Section 3.7 was described by Equation 3.7.4:

$$\ddot{x} + \omega_0^2 x = \lambda x^3 \tag{3.8.1}$$

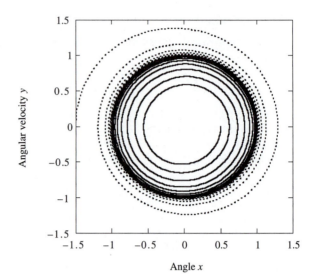

Figure 3.7.1 Phase-space plot for the self-limiting oscillator, with damping factor $\gamma = 0.1$ and starting values (x_0, y_0) of (a) $(0.5, 0)$, solid curve, and (b) $(-1.5, 0)$, dashed curve.

Assume that x_1 and x_2 individually satisfy the above equation. First, substitute their linear combination into the left side,

$$\alpha \ddot{x}_1 + \beta \ddot{x}_2 + \omega_0^2(\alpha x_1 + \beta x_2) = \alpha\left(\ddot{x}_1 + \omega_0^2 x_1\right) + \beta\left(\ddot{x}_2 + \omega_0^2 x_2\right)$$
$$= \alpha\left(\lambda x_1^3\right) + \beta\left(\lambda x_2^3\right) \tag{3.8.2}$$

where the last step follows from the fact that x_1 and x_2 are assumed to be solutions to Equation 3.8.1. Now substitute the linear combination into the right side of Equation 3.8.1 and equate it to the result of Equation 3.8.2:

$$(\alpha x_1 + \beta x_2)^3 = \left(\alpha x_1^3 + \beta x_2^3\right) \tag{3.8.3a}$$

With a little algebra, Equation 3.8.3a can be rewritten as

$$\alpha(\alpha^2 - 1)x_1^3 + 3\alpha^2\beta x_1^2 x_2 + 3\alpha\beta^2 x_1 x_2^2 + \beta(\beta^2 - 1)x_2^3 = 0 \tag{3.8.3b}$$

x_1 and x_2 are solutions to the equation of motion that vary with time t. Thus, the only way Equation 3.8.3b can be satisfied at all times is if α and β are identically zero, which violates the postulate that they are arbitrary factors. Clearly, if x_1 and x_2 are solutions to the nonlinear equation of motion, any linear combination of them is not. It is this nonlinearity that gives rise to the fascinating behavior of chaotic motion.

The essence of the chaotic motion of a nonlinear system is erratic and unpredictable behavior. It occurs in simple mechanical oscillators, such as pendula or vibrating objects, that are "overdriven" beyond their linear regime where their potential energy function is a quadratic function of distance from equilibrium (see Section 3.2). It occurs in the weather, in the convective motion of heated fluids, in the motion of objects bound to our solar system, in laser cavities, in electronic circuits, and even in some chemical reactions. Chaotic oscillation in such systems manifests itself as nonrepetitive behavior. The oscillation is bounded, but each "cycle" of oscillation is like none in the past or future. The oscillation seems to exhibit all the vagaries of purely random motion. Do not be confused

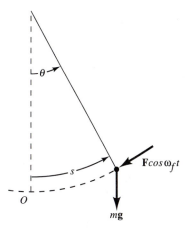

Figure 3.8.1   A simple pendulum driven in a resistive medium by a sinusoidally varying force, $F \cos \omega_f t$. The force is applied tangential to the arc path of the pendulum.

by this statement. "Chaotic" behavior of classical systems does not mean that they do not obey deterministic laws of nature. They do. Given initial conditions and the forces to which they are subject, classical systems do evolve in time in a way that is completely determined. We just may not be able to calculate that evolution with any degree of certainty.

We do not treat chaotic motion in great detail. Such treatment is beyond the mission of this text. The reader who wishes to remedy this deficiency is referred to many fine treatments of chaotic motion elsewhere.[8] Here we are content to introduce the phenomenon of chaos with an analysis of the damped simple pendulum that also can be driven into a chaotic state. We show that slight changes in the driving parameter can lead to wide divergences in the resulting motion, thus rendering prediction of its long-term evolution virtually impossible.

The Driven, Damped Harmonic Oscillator

We developed the equation of motion for the simple pendulum in Example 3.2.2. With the addition of a damping term and a forcing term, it becomes

$$m\ddot{s} = -c\dot{s} - mg\,\sin\theta + F\,\cos\,\omega_f t \tag{3.8.4}$$

where we have assumed that the driving force, $F \cos \omega_f t$, is applied tangent to the path of the pendulum whose arc distance from equilibrium is s (see Fig. 3.8.1).[9]

Let $s = l\theta$, $\gamma = c/m$, $\omega_c^2 = g/l$, and $\alpha = F/ml$, and apply a little algebra to obtain

$$\ddot{\theta} + \gamma\dot{\theta} + \omega_0^2\,\sin\theta = \alpha\,\cos\,\omega_f t \tag{3.8.5}$$

---

[8]J. B. Marion and S. T. Thornton, *Classical Dynamics*, 5th ed., Brooks/Cole—Thomson Learning, Belmont, CA, 2004.

[9]The equation of motion of the simple pendulum in terms of the angular variable θ can be derived most directly using the notion of applied torques and resulting rates of change of angular momentum. These concepts are not fully developed until Chapter 7.

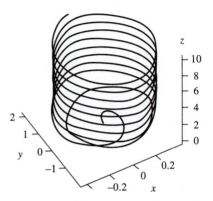

Figure 3.8.2 Three-dimensional phase-space plot of a driven, damped simple pendulum. The driving parameter is $\alpha = 0.9$. The driving angular frequency ω and damping parameter γ are $\frac{2}{3}$ and $\frac{1}{2}$ respectively. Coordinates plotted are $x = \theta/2\pi$, $y = \dot{\theta}$, $z = \omega t/2\pi$.

In our earlier discussion of the simple pendulum, we restricted the analysis of its motion to the regime of small oscillations where the approximation $\sin\theta \approx \theta$ could be used. We do not do that here. It is precisely when the pendulum is driven out of the small-angle regime that the nonlinear effect of the $\sin\theta$ term manifests itself, sometimes in the form of chaotic motion.

We simplify our analysis by scaling angular frequencies in units of ω_0 (in essence, let $\omega_0 = 1$), and we simplify notation by letting $x = \theta$ and $\omega = \omega_f$. The above equation becomes

$$\ddot{x} + \gamma\dot{x} + \sin x = \alpha \cos \omega t \qquad (3.8.6)$$

Exactly as before, we transform this second-order differential equation into three first-order ones by letting $y = \dot{x}$ and $z = \omega t$:

$$\begin{aligned} \dot{x} &= y \\ \dot{y} &= -\sin x - \gamma y + \alpha \cos z \\ \dot{z} &= \omega \end{aligned} \qquad (3.8.7)$$

Remember, these equations are dimensionless, and the driving angular frequency ω is a multiple of ω_0.

We use *Mathcad* as in the preceding example to solve these equations under a variety of conditions. For the descriptions that follow, we vary the driving "force" α and hold fixed both the driving frequency ω and the damping parameter γ at $\frac{2}{3}$ and $\frac{1}{2}$, respectively. The starting coordinates (x_0, y_0, z_0) of the motion are $(0, 0, 0)$ unless otherwise noted.

- *Driving parameter:* $\alpha = 0.9$
  These conditions lead to periodic motion. The future behavior of the pendulum is predictable. We have allowed the motion to evolve for a duration T equivalent to 10 driving cycles.[10] We show in Figure 3.8.2 a three-dimensional phase-space trajectory of the motion. The vertical axis represents the z-coordinate, or the flow of time, while the horizontal axes represent the two phase-space coordinates x and y. The trajectory starts at coordinates $(0, 0, 0)$ and spirals outward and upward in corkscrew-like fashion with the flow of time. There are 10 spirals corresponding to

---

[10] The duration of one driving cycle is $\tau = 2\pi/\omega$.

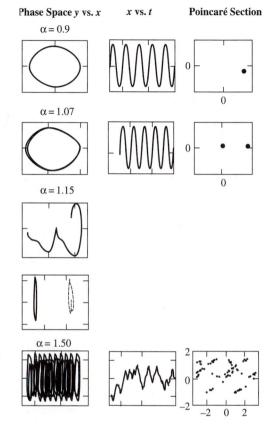

Figure 3.8.3 Damped, driven pendulum for different driving force parameters α. (i) Phase-space plots on the left (ii) angle vs. time in the center (iii) Poincaré sections on the right. Only phase-space plots of first two and last two cycles shown for $\alpha = 1.15$. Each plot represents two sets of starting conditions in which the initial angular velocities differ by only 1 part in 10^5.

the evolution of the motion over 10 driving cycle periods. The transient behavior dies out after the first few cycles as the pendulum attains a state of stable, steady-state oscillation. This is evident upon examination of the first of the top row of graphs in Figure 3.8.3. It is a two-dimensional projection of the three-dimensional, phase-space plot during the last 5 of the 10 total driving cycles. The resulting closed curve actually consists of five superimposed projected curves. The perfect superposition and closure demonstrates the stability and exact repeatability of the oscillation.

The second graph in the first row of Figure 3.8.3 is a plot of the angular position of the pendulum x as a function of the number of elapsed driving cycles n ($=\omega t/2\pi$). The repeatability of the oscillation, cycle after cycle, is evident here as well.

The third graph in the top row is a *Poincaré section* plot of the motion. Think of it as a stroboscopic snapshot of the three-dimensional, phase-space trajectory taken every drive cycle period. The times at which the snapshots are taken can be envisioned as a series of two-dimensional planes parallel to the x–y plane separated by a single drive cycle period. The intersection of the trajectory with any of these horizontal planes, or "slices," is a single point whose (x, y) phase-space coordinates represent the current state of the motion. The single point shown in the plot for this driving cycle parameter is actually five different superimposed

Poincaré sections taken during the last five cycles of the motion. This means that after the initial transient effects die out, the (x, y) phase-space coordinates repeat exactly every subsequent drive cycle period. In other words, the pendulum is oscillating at a single frequency, the drive cycle frequency, as one might expect.

- *Driving parameter:* $\alpha = 1.07$

  This value leads to an interesting effect shown in the second row of graphs in Figure 3.8.3, known as *period doubling*, in which the motion repeats itself exactly every other drive cycle. Close examination of the phase-space plot reveals two closed loops, one for each drive cycle. You might need glasses to see the effect in the second plot (angle vs. time), but close scrutiny reveals that there is a slight vertical displacement between adjacent cycles and that every other cycle is identical. The Poincaré section shows the effect best: two discrete points can be seen indicating that the motion consists of two different but repetitive oscillations.

- *Driving parameter:* $\alpha = 1.15$

  starting coordinates: $(-0.9, 0.54660, 0)$ and $(-0.9, 0.54661, 0)$

  This particular value of driving parameter leads to *chaotic motion* and allows us to graphically illustrate one of its defining characteristics: unpredictable behavior. Take a look at the two phase-space plots defined by $\alpha = 1.15$ in the third and fourth row of Figure 3.8.3. The first one is a phase-space plot of the first two cycles of the motion, for two different trials, each one started with the two slightly different set of starting coordinates given above. In each trial, the pendulum was started from the position, $x = -0.9$ at time $t = 0$, but with slightly different angular velocities y that differed by only 1 part in 10^5. The trajectory shown in the first plot is thus two trajectories, one for each trial. The two trajectories are identical indicating that the motion for the two trials during the first two cycles are virtually indistinguishable. Note that in each case, the pendulum has moved one cycle to the left in the phase-space coordinate x when it reaches its maximum speed y in the negative direction. This tells us that the pendulum swung through one complete revolution in the clockwise direction.

  However, the second graph, where we have plotted the two phase-space trajectories for the 99th and 100th cycles, shows that the motion of the pendulum has diverged dramatically between the two trials. The trajectory for the first trial is centered on $x = -2$ on the left-hand side of the graph, indicating that after 98 drive cycle periods have elapsed, the pendulum has made two more complete 2π clockwise revolutions than it did counterclockwise. The trajectory for the second trial is centered about $x = 6$ on the right-hand side of the graph, indicating that its slightly different starting angular velocity resulted in the pendulum making six more counterclockwise revolutions than it did clockwise. Furthermore, the phase-space trajectories for the two trials now have dramatically different shapes, indicating that the oscillation of the pendulum is quite different at this point in the two trials. An effect such as this invariably occurs if the parameters α, γ, and ω are set for chaotic motion. In the case here, if we were trying to predict the future motion of the pendulum by integrating the equations of motion numerically with a precision no better than 10^{-5}, we would fail miserably. Because any numerical solution has some precisional limit, even a completely deterministic system, such as we have in Newtonian dynamics, ultimately behaves in an unpredictable fashion—in other words, in a chaotic way.

- *Driving parameter:* $\alpha = 1.5$
  This value for the driving parameter also leads to chaotic motion. The three graphs in the last row of Figure 3.8.3 illustrate a second defining characteristic of chaotic motion, namely, nonrepeatability. Two hundred drive cycles have been plotted and during no single cycle is the motion identical to that of any other. If we had plotted y vs. x *modulo* 2π, as is done in many treatments of chaotic motion (thus, discounting all full revolutions by restricting the angular variable to the interval $[-\pi, \pi]$), the entire allowed area on the phase-space plot would be filled up, a clear signature of chaotic motion. The signature is still obvious in the Poincaré section plot, which actually consists of 200 distinct points, indicating that the motion never repeats itself during any drive cycle.

Finally, the richness of the motion of the driven, damped pendulum discussed here was elicited by simply varying the driving parameter within the interval $[0.9, 1.5]$. We saw that one value led to periodic behavior, one led to period doubling and two led to chaotic motion. Apparently, when one deals with driven, nonlinear oscillators, chaotic motion lurks just around the corner from the rather mundane periodic behavior that we and our predecessors have beat into the ground in textbooks throughout the past several hundred years. We urge each student to investigate these motions for him- or herself using a computer. It is remarkable how the slightest change in the parameters governing the equations of motion either leads to or terminates chaotic behavior, but, of course, that's what chaos is all about.

*3.9 | Nonsinusoidal Driving Force: Fourier Series

To determine the motion of a harmonic oscillator that is driven by an external periodic force that is *other* than "pure" sinusoidal, it is necessary to employ a somewhat more involved method than that of the previous sections. In this more general case it is convenient to use the *principle of superposition*. The principle is applicable to any system governed by a linear differential equation. In our application, the principle states that if the external driving force acting on a damped harmonic oscillator is given by a superposition of force functions

$$F_{ext} = \sum_n F_n(t) \tag{3.9.1}$$

such that the differential equation

$$m\ddot{x}_n + c\dot{x}_n + kx_n = F_n(t) \tag{3.9.2}$$

is individually satisfied by the functions $x_n(t)$, then the solution of the differential equation of motion

$$m\ddot{x} + c\dot{x} + kx = F_{ext} \tag{3.9.3}$$

is given by the superposition

$$x(t) = \sum_n x_n(t) \tag{3.9.4}$$

The validity of the principle is easily verified by substitution:

$$m\ddot{x} + c\dot{x} + kx = \sum_n (m\ddot{x}_n + c\dot{x}_n + kx_n) = \sum_n F_n(t) = F_{ext} \tag{3.9.5}$$

In particular, when the driving force is periodic—that is, if for any value of the time t

$$F_{ext}(t) = F_{ext}(t + T) \tag{3.9.6}$$

where T is the period—then the force function can be expressed as a superposition of harmonic terms according to *Fourier's theorem*. This theorem states that any periodic function $f(t)$ can be expanded as a sum as follows:

$$f(t) = \frac{1}{2}a_0 + \sum_{n=1}^{\infty} [a_n \cos(n\omega t) + b_n \sin(n\omega t)] \tag{3.9.7}$$

The coefficients are given by the following formulas (derived in Appendix G):

$$a_n = \frac{2}{T} \int_{-T/2}^{T/2} f(t) \cos(n\omega t)\, dt \qquad n = 0,1,2,\dots \tag{3.9.8a}$$

$$b_n = \frac{2}{T} \int_{-T/2}^{T/2} f(t) \sin(n\omega t)\, dt \qquad n = 1,2,\dots \tag{3.9.8b}$$

Here T is the period and $\omega = 2\pi/T$ is the fundamental frequency. If the function $f(t)$ is an *even* function—that is, if $f(t) = f(-t)$—then the coefficients $b_n = 0$ for all n. The series expansion is then known as a *Fourier cosine series*. Similarly, if we have an *odd* function so that $f(t) = -f(-t)$, then the a_n vanish, and the series is called a *Fourier sine series*. By use of the relation $e^{iu} = \cos u + i \sin u$, it is straightforward to verify that Equations 3.9.7 and 3.9.8a and b may also be expressed in complex exponential form as follows:

$$f(t) = \sum_n c_n e^{in\omega t} \qquad n = 0, \pm 1, \pm 2, \dots \tag{3.9.9}$$

$$c_n = \frac{1}{T} \int_{-T/2}^{T/2} f(t) e^{-in\omega t}\, dt \tag{3.9.10}$$

Thus, to find the steady-state motion of our harmonic oscillator subject to a given periodic driving force, we express the force as a Fourier series of the form of Equation 3.9.7 or 3.9.9, using Equations 3.9.8a and b or 3.9.10 to determine the Fourier coefficients a_n and b_n, or c_n. For each value of n, corresponding to a given harmonic $n\omega$ of the fundamental driving frequency ω, there is a response function $x_n(t)$. This function is the steady-state solution of the driven oscillator treated in Section 3.6. The superposition of all the $x_n(t)$ gives the actual motion. In the event that one of the harmonics of the driving frequency coincides, or nearly coincides, with the resonance frequency ω_r, then the response at that harmonic dominates the motion. As a result, if the damping constant γ is very small, the resulting oscillation may be very nearly sinusoidal even if a highly nonsinusoidal driving force is applied.

EXAMPLE 3.9.1

Periodic Pulse

To illustrate the above theory, we analyze the motion of a harmonic oscillator that is driven by an external force consisting of a succession of rectangular pulses:

$$F_{ext}(t) = F_0 \qquad NT - \tfrac{1}{2}\Delta T \le t \le NT + \tfrac{1}{2}\Delta T$$
$$F_{ext}(t) = 0 \qquad \text{Otherwise}$$

where $N = 0, \pm 1, \pm 2, \dots$, T is the time from one pulse to the next, and ΔT is the width of each pulse as shown in Figure 3.9.1. In this case, $F_{ext}(t)$ is an even function of t, so it can be expressed as a Fourier cosine series. Equation 3.9.8a gives the coefficients a_n,

$$a_n = \frac{2}{T} \int_{-\Delta T/2}^{+\Delta T/2} F_0 \cos(n\omega t)\,dt$$

$$= \frac{2}{T} F_0 \left[\frac{\sin(n\omega t)}{n\omega} \right]_{-\Delta T/2}^{+\Delta T/2} \qquad (3.9.11a)$$

$$= F_0 \frac{2\sin(n\pi \Delta T/T)}{n\pi}$$

where in the last step we use the fact that $\omega = 2\pi/T$. We see also that

$$a_0 = \frac{2}{T} \int_{-\Delta T/2}^{+\Delta T/2} F_0\,dt = F_0 \frac{2\,\Delta T}{T} \qquad (3.9.11b)$$

Thus, for our periodic pulse force we can write

$$F_{ext}(t) = F_0 \left[\frac{\Delta T}{T} + \frac{2}{\pi}\sin\left(\pi \frac{\Delta T}{T}\right)\cos(\omega t) + \frac{2}{2\pi}\sin\left(2\pi \frac{\Delta T}{T}\right)\cos(2\omega t) \right.$$

$$\left. + \frac{2}{3\pi}\sin\left(3\pi \frac{\Delta T}{T}\right)\cos(3\omega t) + \cdots \right] \qquad (3.9.12)$$

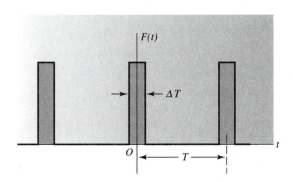

Figure 3.9.1 Rectangular-pulse driving force.

The first term in the above series expansion is just the *average* value of the external force: $F_{avg} = F_0(\Delta T/T)$. The second term is the Fourier component at the fundamental frequency ω. The remaining terms are harmonics of the fundamental: 2ω, 3ω, and so on.

Referring to Equations 3.6.5 and 3.9.4, we can now write the final expression for the motion of our pulse-driven oscillator. It is given by the superposition principle,

$$x(t) = \sum_n x_n(t) = \sum_n A_n \cos(n\omega t - \phi_n) \tag{3.9.13}$$

in which the respective amplitudes are (Equation 3.6.9)

$$A_n = \frac{a_n/m}{D_n(\omega)} = \frac{(F_0/m)(2/n\pi)\sin(n\pi\Delta T/T)}{\left[\left(\omega_0^2 - n^2\omega^2\right)^2 + 4\gamma^2 n^2\omega^2\right]^{1/2}} \tag{3.9.14}$$

and the phase angles (Equation 3.6.8)

$$\phi_n = \tan^{-1}\left(\frac{2\gamma n\omega}{\omega_0^2 - n^2\omega^2}\right) \tag{3.9.15}$$

Here m is the mass, γ is the decay constant, and ω_0 is the frequency of the freely running oscillator with no damping.

As a specific numerical example, let us consider the spring suspension system of Example 3.6.1 under the action of a periodic pulse for which the pulse width is one tenth the pulse period: $\Delta T/T = 0.1$. As before, we shall take the damping constant to be one-tenth critical, $\gamma = 0.1\,\omega_0$, and the pulse frequency to be one-half the undamped frequency of the system: $\omega = \omega_0/2$. The Fourier series for the driving force (Equation 3.9.12) is then

$$F_{ext}(t) = F_0\left[0.1 + \frac{2}{\pi}\sin(0.1\pi)\,\cos(\omega t) + \frac{2}{2\pi}\sin(0.2\pi)\,\cos(2\omega t)\right.$$

$$\left. + \frac{2}{3\pi}\sin(0.3\pi)\,\cos(3\omega t) + \cdots\right]$$

$$= F_0[0.1 + 0.197\,\cos(\omega t) + 0.187\,\cos(2\omega t) + 0.172\,\cos(3\omega t) + \cdots]$$

The resonance denominators in Equation 3.9.14 are given by

$$D_n = \left[\left(\omega_0^2 - n^2\frac{\omega_0^2}{4}\right)^2 + 4(0.1)^2\omega_0^2 n^2\frac{\omega_0^2}{4}\right]^{1/2} = \left[\left(1 - \frac{n^2}{4}\right)^2 + 0.01n^2\right]^{1/2}\omega_0^2$$

Thus,

$$D_0 = \omega_0^2 \qquad D_1 = 0.757\omega_0^2 \qquad D_2 = 0.2\omega_0^2 \qquad D_3 = 1.285\,\omega_0^2$$

The phase angles (Equation 3.9.15) are

$$\phi_n = \tan^{-1}\left(\frac{0.2n\omega_0^2/2}{\omega_0^2 - n^2\omega_0^2/4}\right) = \tan^{-1}\left(\frac{0.4n}{4 - n^2}\right)$$

which gives

$$\phi_0 = 0 \qquad \phi_1 = \tan^{-1}(0.133) = 0.132$$

$$\phi_2 = \tan^{-1}\infty = \pi/2 \qquad \phi_3 = \tan^{-1}(-0.24) = -0.236$$

The steady-state motion of the system is, therefore, given by the following series (Equation 3.9.13):

$$x(t) = \frac{F_0}{m\omega_0^2}[0.1 + 0.26\cos(\omega t - 0.132) + 0.935\sin(2\omega t) + 0.134\cos(3\omega t + 0.236) + \cdots]$$

The dominant term is the one involving the second harmonic $2\omega = \omega_0$, because ω_0 is close to the resonant frequency. Note also the phase of this term:

$$\cos(2\omega t - \pi/2) = \sin(2\omega t).$$

Problems

3.1 A guitar string vibrates harmonically with a frequency of 512 Hz (one octave above middle C on the musical scale). If the amplitude of oscillation of the centerpoint of the string is 0.002 m (2 mm), what are the maximum speed and the maximum acceleration at that point?

3.2 A piston executes simple harmonic motion with an amplitude of 0.1 m. If it passes through the center of its motion with a speed of 0.5 m/s, what is the period of oscillation?

3.3 A particle undergoes simple harmonic motion with a frequency of 10 Hz. Find the displacement x at any time t for the following initial condition:

$$t = 0 \qquad x = 0.25\,\text{m} \qquad \dot{x} = 0.1\,\text{m/s}$$

3.4 Verify the relations among the four quantities C, D, ϕ_0, and A given by Equation 3.2.19.

3.5 A particle undergoing simple harmonic motion has a velocity $\dot{x}_1$ when the displacement is x_1 and a velocity $\dot{x}_2$ when the displacement is x_2. Find the angular frequency and the amplitude of the motion in terms of the given quantities.

3.6 On the surface of the moon, the acceleration of gravity is about one-sixth that on the Earth. What is the half-period of a simple pendulum of length 1 m on the moon?

3.7 Two springs having stiffness k_1 and k_2, respectively, are used in a vertical position to support a single object of mass m. Show that the angular frequency of oscillation is $[(k_1 + k_2)/m]^{1/2}$ if the springs are tied in parallel, and $[k_1 k_2/(k_1 + k_2)m]^{1/2}$ if the springs are tied in series.

3.8 A spring of stiffness k supports a box of mass M in which is placed a block of mass m. If the system is pulled downward a distance d from the equilibrium position and then released, find the force of reaction between the block and the bottom of the box as a function of time. For what value of d does the block just begin to leave the bottom of the box at the top of the vertical oscillations? Neglect any air resistance.

3.9 Show that the ratio of two successive maxima in the displacement of a damped harmonic oscillator is constant. (*Note:* The maxima do not occur at the points of contact of the displacement curve with the curve $Ae^{-\gamma t}$.)

3.10 A damped harmonic oscillator with $m = 10$ kg, $k = 250$ N/m, and $c = 60$ kg/s is subject to a driving force given by $F_0 \cos \omega t$, where $F_0 = 48$ N.
(a) What value of ω results in steady-state oscillations with maximum amplitude? Under this condition:

(b) What is the maximum amplitude?

(c) What is the phase shift?

3.11 A mass m moves along the x-axis subject to an attractive force given by $17\beta^2 mx/2$ and a retarding force given by $3\beta m\dot{x}$, where x is its distance from the origin and β is a constant. A driving force given by $mA \cos \omega t$, where A is a constant, is applied to the particle along the x-axis.

(a) What value of ω results in steady-state oscillations about the origin with maximum amplitude?

(b) What is the maximum amplitude?

3.12 The frequency f_d of a damped harmonic oscillator is 100 Hz, and the ratio of the amplitude of two successive maxima is one half.

(a) What is the undamped frequency f_0 of this oscillator?

(b) What is the resonant frequency f_r?

3.13 Given: The amplitude of a damped harmonic oscillator drops to $1/e$ of its initial value after n complete cycles. Show that the ratio of period of the oscillation to the period of the same oscillator with no damping is given by

$$\frac{T_d}{T_0} = \left(1 + \frac{1}{4\pi^2 n^2}\right)^{1/2} \simeq 1 + \frac{1}{8\pi^2 n^2}$$

where the approximation in the last expression is valid if n is large. (See the approximation formulas in Appendix D.)

3.14 Work all parts of Example 3.6.2 for the case in which the exponential damping factor γ is one-half the critical value and the driving frequency is equal to $2\omega_0$.

3.15 For a lightly damped harmonic oscillator $\gamma \ll \omega_0$, show that the driving frequency for which the steady-state amplitude is one-half the steady-state amplitude at the resonant frequency is given by $\omega \simeq \omega_0 \pm \gamma\sqrt{3}$.

3.16 If a series LCR circuit is connected across the terminals of an electric generator that produces a voltage $V = V_0 e^{i\omega t}$, the flow of electrical charge q through the circuit is given by the following second-order differential equation:

$$L\frac{d^2 q}{dt^2} + R\frac{dq}{dt} + \frac{1}{C}q = V_0 e^{i\omega t}$$

(a) Verify the correspondence shown in Table 3.6.1 between the parameters of a driven mechanical oscillator and the above driven electrical oscillator.

(b) Calculate the Q of the electrical circuit in terms of the coefficients of the above differential equation.

(c) Show that, in the case of small damping, Q can be written as $Q = R_0/R$, where $R_0 = \sqrt{L/C}$ is the *characteristic impedance* of the circuit.

3.17 A damped harmonic oscillator is driven by an external force of the form

$$F_{ext} = F_0 \sin \omega t$$

Show that the steady-state solution is given by

$$x(t) = A(\omega) \sin(\omega t - \phi)$$

where $A(\omega)$ and ϕ are identical to the expressions given by Equations 3.6.9 and 3.6.8.

3.18 Solve the differential equation of motion of the damped harmonic oscillator driven by a damped harmonic force:

$$F_{ext}(t) = F_0 e^{-\alpha t} \cos \omega t$$

(*Hint:* $e^{-\alpha t} \cos \omega t = Re(e^{-\alpha t + i\omega t}) = Re(e^{\beta t})$, *where* $\beta = -\alpha + i\omega$. *Assume a solution of the form* $Ae^{\beta t - i\phi}$.)

3.19 A simple pendulum of length l oscillates with an amplitude of 45°.
(a) What is the period?
(b) If this pendulum is used as a laboratory experiment to determine the value of g, find the error included in the use of the elementary formula $T_0 = 2\pi(l/g)^{1/2}$.
(c) Find the approximate amount of third-harmonic content in the oscillation of the pendulum.

3.20 Verify Equations 3.9.9 and 3.9.10 in the text.

3.21 Show that the Fourier series for a periodic square wave is

$$f(t) = \frac{4}{\pi}\left[\sin(\omega t) + \frac{1}{3}\sin(3\omega t) + \frac{1}{5}\sin(5\omega t) + \cdots\right]$$

where

$$f(t) = +1 \qquad \text{for } 0 < \omega t < \pi,\ 2\pi < \omega t < 3\pi,\ \text{and so on}$$
$$f(t) = -1 \qquad \text{for } \pi < \omega t < 2\pi,\ 3\pi < \omega t < 4\pi,\ \text{and so on}$$

3.22 Use the above result to find the steady-state motion of a damped harmonic oscillator that is driven by a periodic square-wave force of amplitude F_0. In particular, find the relative amplitudes of the first three terms, A_1, A_3, and A_5 of the response function $x(t)$ in the case that the third harmonic 3ω of the driving frequency coincides with the frequency ω_0 of the undamped oscillator. Let the quality factor $Q = 100$.

3.23 (a) Derive the first-order differential equation, dy/dx, describing the phase-space trajectory of the simple harmonic oscillator.
(b) Solve the equation, proving that the trajectory is an ellipse.

3.24 Let a particle of unit mass be subject to a force $x - x^3$ where x is its displacement from the coordinate origin.
(a) Find the equilibrium points, and tell whether they are stable or unstable.
(b) Calculate the total energy of the particle, and show that it is a conserved quantity.
(c) Calculate the trajectories of the particle in phase space.

3.25 A simple pendulum whose length $l = 9.8$ m satisfies the equation

$$\ddot{\theta} + \sin\theta = 0$$

(a) If Θ_0 is the amplitude of oscillation, show that its period T is given by

$$T = 4\int_0^{\pi/2} \frac{d\phi}{(1 - \alpha \sin^2\phi)^{1/2}} \qquad \text{where } \alpha = \sin^2\frac{1}{2}\Theta_0$$

(b) Expand the integrand in powers of α, integrate term by term, and find the period T as a power series in α. Keep terms up to and including $O(\alpha^2)$.
(c) Expand α in a power series of Θ_0, insert the result into the power series found in (b), and find the period T as a power series in Θ_0. Keep terms up to and including $O(\Theta_0^2)$.

Computer Problems

C 3.1   The exact equation of motion for a simple pendulum of length L (see Example 3.2.2) is given by

$$\ddot{\theta} + \omega_0^2 \sin\theta = 0$$

where $\omega_i^2 = g/L$. Find $\theta(t)$ by numerically integrating this equation of motion. Let $L =$ 1.00 m. Let the initial conditions be $\theta_0 = \pi/2$ rad and $\dot{\theta}_0 = 0$ rad/s.

(a) Plot $\theta(t)$ from $t = 0$ to 4 s. Also, plot the solution obtained by using the small-angle approximation ($\sin\theta \approx \theta$) on the same graph.

(b) Repeat (a) for $\theta_0 = 3.10$ rad.

(c) Plot the period of the pendulum as a function of the amplitude θ_0 from 0 to 3.10 rad. At what amplitude does the period deviate by more than 2% from $2\pi\sqrt{L/g}$?

C 3.2   Assume that the damping force for the damped harmonic oscillator is proportional to the square of its velocity; that is, it is given by $-c_2\dot{x}|\dot{x}|$. The equation of motion for such an oscillator is thus

$$\ddot{x} + 2\gamma\dot{x}|\dot{x}| + \omega_0^2 x = 0$$

where $\gamma = c_2/2m$ and $\omega_0^2 = k/m$. Find $x(t)$ by numerically integrating the above equation of motion. Let $\gamma = 0.20$ m^{-1} and $\omega_0 = 2.00$ rad/s. Let the initial conditions be $x(0) = 1.00$ m and $\dot{x}(0) = 0$ m/s.

(a) Plot $x(t)$ from $t = 0$ to 20 s. Also, on the same graph, plot the solution for the damped harmonic oscillator where the damping force is linearly proportional to the velocity; that is, it is given by $-c_1\dot{x}$. Again, let $\gamma = c_1/2m = 0.20$ s^{-1} and $\omega_0 = 2.00$ rad/s.

(b) For the case of linear damping, plot the log of the absolute value of the successive extrema versus their time of occurrence. Find the slope of this plot, and use it to estimate γ. (This method works well for the case of weak damping.)

(c) Find the value of γ that results in critical damping for the linear case. Plot this solution from $t = 0$ to 5 s. Can you find a well-defined value of γ that results in critical damping for the quadratic case? If not, what value of γ is required to limit the first negative excursion of the oscillator to less than 2% of the initial amplitude?

C 3.3   The equation of motion for the van der Pol oscillator is

$$\ddot{x} - \gamma(A^2 - x^2)\dot{x} + \omega_0^2 x = 0$$

Let $A = 1$ and $\omega_0 = 1$. Solve this equation numerically, and make a phase-space plot of its motion. Let the motion evolve for 10 periods (1 period $= 2\pi/\omega_0$). Assume the following conditions.

(a) $\gamma = 0.05$, $(x_0, \dot{x}_0) = (-1.5, 0)$.

(b) $\gamma = 0.05$, $(x_0, \dot{x}_0) = (0.5, 0)$.

(c, d) Repeat (a) and (b) with $\gamma = 0.5$. Does the motion exhibit a limit cycle? Describe it.

C 3.4   The driven van der Pol oscillator is described by the equation of motion

$$\ddot{x} - \gamma(1 - x^2)\dot{x} + x = \alpha \cos\omega t$$

where α is the amplitude of the driving force and ω is the driving frequency. Let $x = x$, $y = \dot{x}$, and $z = \omega t$. Solve the equation numerically in terms of these variables. Let the oscillator start at $(x_0, y_0, z_0) = (0,0,0)$. Let the motion evolve for 100 drive cycles (1 drive period $= 2\pi/\omega$). (1) Make a phase-space plot of its motion. (2) Make a plot of its position versus number of drive cycles. (3) Make a three-dimensional phase-space plot of the first 10 drive cycles. Assume the following conditions.

(a) $\alpha = 0.1$, $\gamma = 0.05$, $\omega = 1$.

(b) $\alpha = 5$, $\gamma = 5$, $\omega = 2.466$. Which state is periodic? Chaotic?

C 3.5 Consider a simple harmonic oscillator resting on a roller belt, as shown in Figure C.3.5. Assume that the frictional force exerted on the block by the roller belt depends on the *slip velocity*, $\dot{x} - u$, where u is the speed of the belt, and that it is given by

$$
\begin{array}{ll}
\beta v & \dot{x} - u > v \\
\beta(\dot{x} - u) & |\dot{x} - u| \leq v \\
-\beta v & \dot{x} - u < v
\end{array}
$$

In other words, the force is constant when the slip velocity is outside the limits given by the constant v and is proportional to the slip velocity when it is within those limits.

(a) Write down the equation of motion for this oscillator. Solve it numerically, and make a phase-space plot for the following conditions: $k = 1$, $m = 1$, $\beta = 5$, $v = 0.2$, $u = 0.1$. Assume the following starting conditions.

(b) $(x_0, \dot{x}_0) = (0,0)$.

(c) $(x_0, \dot{x}_0) = (2,0)$.

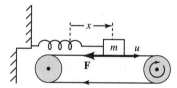

Figure C3.5

4

General Motion of a Particle in Three Dimensions

Sir Isaac Newton, and his followers, have also a very odd opinion concerning the work of God. According to their doctrine, God almighty needs to wind up his watch from time to time; otherwise it would cease to move. He had not, it seems, sufficient foresight to make it a perpetual motion.

Nay, the machine of God's making, is so imperfect, according to these gentlemen, that He is obliged to clean it now and then by an extraordinary concourse, and even to mend it, as a clockmaker mends his work; who must consequently be so much the more unskillful a workman, as He is often obliged to mend his work and set it right. According to my opinion, the same force and vigour [energy] remains always in the world, and only passes from one part to another, agreeable to the laws of nature, and the beautiful pre-established order—

Gottfried Wilhelm Leibniz—*Letter to Caroline, Princess of Wales*, 1715; The Leibniz-Clarke Correspondence, Manchester, Manchester Univ. Press, 1956

4.1 | Introduction: General Principles

We now examine the general case of the motion of a particle in three dimensions. The vector form of the equation of motion for such a particle is

$$\mathbf{F} = \frac{d\mathbf{p}}{dt} \qquad (4.1.1)$$

in which $\mathbf{p} = m\mathbf{v}$ is the linear momentum of the particle. This vector equation is equivalent to three scalar equations in Cartesian coordinates.

$$F_x = m\ddot{x}$$
$$F_y = m\ddot{y}$$
$$F_z = m\ddot{z}$$

(4.1.2)

The three force components may be explicit or implicit functions of the coordinates, their time and spatial derivatives, and possibly time itself. There is no general method for obtaining an analytic solution to the above equations of motion. In problems of even the mildest complexity, we might have to resort to the use of applied numerical techniques; however, there are many problems that can be solved using relatively simple analytical methods. It may be true that such problems are sometimes overly simplistic in their representation of reality. However, they ultimately serve as the basis of models of real physical systems, and so it is well worth the effort that we take here to develop the analytical skills necessary to solve such idealistic problems. Even these may prove capable of taxing our analytic ability.

It is rare that one knows the explicit way in which $\mathbf{F}$ depends on time; therefore, we do not worry about this situation but instead focus on the more normal situation in which $\mathbf{F}$ is known as an explicit function of spatial coordinates and their derivatives. The simplest situation is one in which $\mathbf{F}$ is known to be a function of spatial coordinates only. We devote most of our effort to solving such problems. There are many only slightly more complex situations, in which $\mathbf{F}$ is a known function of coordinate derivatives as well. Such cases include projectile motion with air resistance and the motion of a charged particle in a static electromagnetic field. We will solve problems such as these, too. Finally, $\mathbf{F}$ may be an implicit function of time, as in situations where the coordinate and coordinate derivative dependency is nonstatic. A prime example of such a situation involves the motion of a charged particle in a time-varying electromagnetic field. We will not solve problems such as these. For now, we begin our study of three-dimensional motion with a development of several powerful analytical techniques that can be applied when $\mathbf{F}$ is a known function of $\mathbf{r}$ and/or $\dot{\mathbf{r}}$.

The Work Principle

Work done on a particle causes it to gain or lose kinetic energy. The work concept was introduced in Chapter 2 for the case of motion of a particle in one dimension. We would like to generalize the results obtained there to the case of three-dimensional motion. To do so, we first take the dot product of both sides of Equation 4.1.1 with the velocity $\mathbf{v}$:

$$\mathbf{F} \cdot \mathbf{v} = \frac{d\mathbf{p}}{dt} \cdot \mathbf{v} = \frac{d(m\mathbf{v})}{dt} \cdot \mathbf{v}$$

(4.1.3)

Because $d(\mathbf{v} \cdot \mathbf{v})/dt = 2\mathbf{v} \cdot \dot{\mathbf{v}}$, and assuming that the mass is constant, independent of the velocity of the particle, we may write Equation 4.1.3 as

$$\mathbf{F} \cdot \mathbf{v} = \frac{d}{dt}\left(\tfrac{1}{2}m\mathbf{v} \cdot \mathbf{v}\right) = \frac{dT}{dt}$$

(4.1.4)

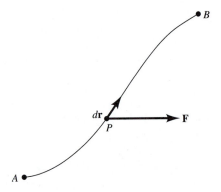

Figure 4.1.1   The work done by a force **F** is the line integral $\int_A^B \mathbf{F} \cdot d\mathbf{r}$.

in which T is the kinetic energy, $mv^2/2$. Because $\mathbf{v} = d\mathbf{r}/dt$, we can rewrite Equation 4.1.4 and then integrate the result to obtain

$$\mathbf{F} \cdot \frac{d\mathbf{r}}{dt} = \frac{dT}{dt} \tag{4.1.5a}$$

$$\therefore \int \mathbf{F} \cdot d\mathbf{r} = \int dT = T_f - T_i = \Delta T \tag{4.1.5b}$$

The left-hand side of this equation is a *line integral*, or the integral of $F_r\, dr$, the component of **F** parallel to the particle's displacement vector $d\mathbf{r}$. The integral is carried out along the trajectory of the particle from some initial point in space A to some final point B. This situation is pictured in Figure 4.1.1. The line integral represents the work done on the particle by the force **F** as the particle moves along its trajectory from A to B. The right-hand side of the equation is the net change in the kinetic energy of the particle. **F** is the net sum of all vector forces acting on the particle; hence, the equation states that the work done on a particle by the net force acting on it, in moving from one position in space to another, is equal to the difference in the kinetic energy of the particle at those two positions.

Conservative Forces and Force Fields

In Chapter 2 we introduced the concept of potential energy. We stated there that if the force acting on a particle were *conservative*, it could be derived as the derivative of a scalar potential energy function, $F_x = -dV(x)/dx$. This condition led us to the notion that the work done by such a force in moving a particle from point A to point B along the x-axis was $\int F_x dx = -\Delta V = V(A) - V(B)$, or equal to minus the change in the potential energy of the particle. Thus, we no longer required a detailed knowledge of the motion of the particle from A to B to calculate the work done on it by a conservative force. We needed to know only that it started at point A and ended up at point B. The work done depended only upon the potential energy function evaluated at the endpoints of the motion. Moreover, because the work done was also equal to the change in kinetic energy of the particle, $\Delta T = T(B) - T(A)$, we were able to establish a general conservation of total energy principle, namely, $E_{tot} = V(A) + T(A) = V(B) + T(B) = $ constant throughout the motion of the particle.

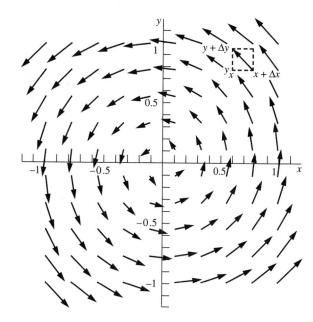

Figure 4.1.2 A nonconservative force field whose force components are $F_x = -by$ and $F_y = +bx$.

This principle was based on the condition that the force acting on the particle was conservative. Indeed, the very name implies that something is being conserved as the particle moves under the action of such a force. We would like to generalize this concept for a particle moving in three dimensions, and, more importantly, we would like to define just what is meant by the word *conservative*. Clearly, we would like to have some prescription that tells us whether or not a particular force is conservative and, thus, whether or not a potential energy function exists for the particle. Then we could invoke the powerful conservation of energy principle in solving the motion of a particle.

In searching for such a prescription, we first describe an example of a nonconservative force that, in fact, is a well-defined function of position but cannot be derived from a potential energy function. This should give us a hint of the critical characteristic that a force must have if it is to be conservative. Consider the two-dimensional force field depicted in Figure 4.1.2. The term *force field* simply means that if a small test particle[1] were to be placed at any point (x_1, y_1) on the xy plane, it would experience a force **F**. Thus, we can think of the xy plane as permeated, or "mapped out," with the potential for generating a force.

This situation can be mathematically described by assigning a vector **F** to every point in the xy plane. The field is, therefore, a vector field, represented by the function $\mathbf{F}(x, y)$. Its components are $F_x = -by$ and $F_y = +bx$, where b is some constant. The arrows

---

[1] A test particle is one whose mass is small enough that its presence does not alter its environment. Conceptually, we might imagine it placed at some point in space to serve as a "test probe" for the suspected presence of forces. The forces are "sensed" by observing any resultant acceleration of the test particle. We further imagine that its presence does not disturb the sources of those forces.

in the figure represent the vector $\mathbf{F} = -\mathbf{i}by + \mathbf{j}bx$ evaluated at each point on which the center of the arrow is located. You can see by looking at the figure that there seems to be a general counterclockwise "circulation" of the force vectors around the origin. The magnitude of the vectors increases with increasing distance from the origin. If we were to turn a small test particle loose in such a "field," the particle would tend to circulate counterclockwise, gaining kinetic energy all the while.

This situation, at first glance, does not appear to be so unusual. After all, when you drop a ball in a gravitational force field, it falls and gains kinetic energy, with an accompanying loss of an equal amount of potential energy. The question here is, can we even define a potential energy function for this circulating particle such that it would lose an amount of "potential energy" equal to the kinetic energy it gained, thus preserving its overall energy, as it travels from one point to another? That is not the case here. If we were to calculate the work done on this particle in tracing out some path that came back on itself (such as the rectangular path indicated by the dashed line in Figure 4.1.2), we would obtain a nonzero result! In traversing such a loop over and over again, the particle would continue to gain kinetic energy equal to the nonzero value of work done per loop. But if the particle could be assigned a potential energy dependent only upon its (x, y) position, then its change in potential energy upon traversing the closed loop would be zero. It should be clear that there is no way in which we could assign a unique value of potential energy for this particle at any particular point on the xy plane. Any value assigned would depend on the previous history of the particle. For example, how many loops has the particle already made before arriving at its current position?

We can further expose the nonuniqueness of any proposed potential energy function by examining the work done on the particle as it travels between two points A and B but along two different paths. First, we let the particle move from (x, y) to $(x + \Delta x, y + \Delta y)$ by traveling in the $+x$ direction to $(x + \Delta x, y)$ and then in the $+y$ direction to $(x + \Delta x, y + \Delta y)$. Then we let the particle travel first along the $+y$ direction from (x,y) to $(x,y + \Delta y)$ and then along the $+x$ direction to $(x + \Delta x, y + \Delta y)$. We see that a different amount of work is done depending upon which path we let the particle take. If this is true, then the work done cannot be set equal to the difference between the values of some scalar potential energy function evaluated at the two endpoints of the motion, because such a difference would give a unique, *path-independent* result. The difference in work done along these two paths is equal to $2b\Delta x\Delta y$ (see Equation 4.1.6). This difference is just equal to the value of the closed-loop work integral; therefore, the statement that the work done in going from one point to another in this force field is path-dependent is equivalent to the statement that the closed-loop work integral is nonzero. The particular force field represented in Figure 4.1.2 demands that we know the complete history of the particle to calculate the work done and, therefore, its kinetic energy gain. The potential energy concept, from which the force could presumably be derived, is rendered meaningless in this particular context.

The only way in which we could assign a unique value to the potential energy would be if the closed-loop work integral vanished. In such cases, the work done along a path from A to B would be path-independent and would equal both the potential energy loss and the kinetic energy gain. The total energy of the particle would be a constant, independent of its location in such a force field! We, therefore, must find the constraint that a particular force must obey if its closed-loop work integral is to vanish.

To find the desired constraint, let us calculate the work done in taking a test particle counterclockwise around the rectangular loop of area $\Delta x \Delta y$ from the point (x,y) and back again, as indicated in Figure 4.1.2. We get the following result:

$$W = \oint \mathbf{F} \cdot d\mathbf{r}$$

$$= \int_x^{x+\Delta x} F_x(y)\,dx + \int_y^{y+\Delta y} F_y(x + \Delta x)\,dy$$

$$+ \int_{x+\Delta x}^{x} F_x(y + \Delta y)\,dx + \int_{y+\Delta y}^{y} F_y(x)\,dy$$

$$= \int_y^{y+\Delta y} (F_y(x + \Delta x) - F_y(x))\,dy \tag{4.1.6}$$

$$+ \int_x^{x+\Delta x} (F_x(y) - F_x(y + \Delta y))\,dx$$

$$= (b(x + \Delta x) - bx)\,\Delta y + (b(y + \Delta y) - by)\,\Delta x$$

$$= 2b\Delta x\,\Delta y$$

The work done is nonzero and is proportional to the area of the loop, $\Delta A = \Delta x \cdot \Delta y$, which was chosen in an arbitrary fashion. If we divide the work done by the area of the loop and take limits as $\Delta A \to 0$, we obtain the value $2b$. The result is dependent on the precise nature of this particular nonconservative force field.

If we reverse the direction of one of the force components—say, let $F_x = +by$ (thus "destroying" the circulation of the force field but everywhere preserving its magnitude)—then the work done per unit area in traversing the closed loop vanishes. The resulting force field is conservative and is shown in Figure 4.1.3. Clearly, the value of the closed-loop

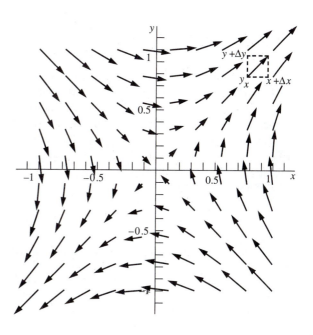

Figure 4.1.3 A conservative force field whose components are $F_x = by$ and $F_y = bx$.

integral depends upon the precise way in which the vector force $\mathbf{F}$ changes its direction as well as its magnitude as we move around on the xy plane.

There is obviously some sort of constraint that $\mathbf{F}$ must obey if the closed-loop work integral is to vanish. We can derive this condition of constraint by evaluating the forces at $x + \Delta x$ and $y + \Delta y$ using a Taylor expansion and then inserting the resultant expansion into the closed-loop work integral of Equation 4.1.6. The result follows:

$$F_x(y + \Delta y) = F_x(y) + \frac{\partial F_x}{\partial y}\Delta y$$

$$F_y(x + \Delta x) = F_y(x) + \frac{\partial F_y}{\partial x}\Delta x \tag{4.1.7}$$

$$\oint \mathbf{F} \cdot d\mathbf{r} = \int_y^{y+\Delta y}\left(\frac{\partial F_y}{\partial x}\Delta x\right)dy - \int_x^{x+\Delta x}\left(\frac{\partial F_x}{\partial y}\Delta y\right)dx$$

$$= \left(\frac{\partial F_y}{\partial x} - \frac{\partial F_x}{\partial y}\right)\Delta x\Delta y = 2b\Delta x\Delta y \tag{4.1.8}$$

This last equation contains the term $(\partial F_y/\partial x - \partial F_x/\partial y)$, whose zero or nonzero value represents the test we are looking for. If this term were identically equal to zero instead of $2b$, then the closed-loop work integral would vanish, which would ensure the existence of a potential energy function from which the force could be derived.

This condition is a rather simplified version of a very general mathematical theorem called *Stokes' theorem*.[2] It is written as

$$\oint \mathbf{F} \cdot d\mathbf{r} = \int_s \text{curl } \mathbf{F} \cdot \hat{\mathbf{n}}\, da$$

$$\text{curl } \mathbf{F} = \mathbf{i}\left(\frac{\partial F_z}{\partial y} - \frac{\partial F_y}{\partial z}\right) + \mathbf{j}\left(\frac{\partial F_x}{\partial z} - \frac{\partial F_z}{\partial x}\right) + \mathbf{k}\left(\frac{\partial F_y}{\partial x} - \frac{\partial F_x}{\partial y}\right) \tag{4.1.9}$$

The theorem states that the closed-loop line integral of *any* vector function $\mathbf{F}$ is equal to curl $\mathbf{F} \cdot \mathbf{n}\, da$ integrated over a surface S surrounded by the closed loop. The vector $\mathbf{n}$ is a unit vector normal to the surface-area integration element da. Its direction is that of the advance of a right-hand screw turned in the same rotational sense as the direction of traversal around the closed loop. In Figure 4.1.2, $\mathbf{n}$ would be directed out of the paper. The surface would be the rectangular area enclosed by the dashed rectangular loop. Thus, a vanishing curl $\mathbf{F}$ ensures that the line integral of $\mathbf{F}$ around a closed path is zero and, thus, that $\mathbf{F}$ is a conservative force.

---

[2] See any advanced calculus textbook (e.g., S. I. Grossman and W. R. Derrick, *Advanced Engineering Mathematics*, Harper Collins, New York, 1988) or any advanced electricity and magnetism textbook (e.g., J. R. Reitz, F. J. Milford, and R. W. Christy, *Foundations of Electromagnetic Theory*, Addison-Wesley, New York, 1992).

4.2 | The Potential Energy Function in Three-Dimensional Motion: The Del Operator

Assume that we have a test particle subject to some force whose curl vanishes. Then all the components of curl $\mathbf{F}$ in Equation 4.1.9 vanish. We can make certain that the curl vanishes if we derive $\mathbf{F}$ from a potential energy function $V(x,y,z)$ according to

$$F_x = -\frac{\partial V}{\partial x} \qquad F_y = -\frac{\partial V}{\partial y} \qquad F_z = -\frac{\partial V}{\partial z} \qquad (4.2.1)$$

For example, the z component of curl $\mathbf{F}$ becomes

$$\frac{\partial F_x}{\partial y} = -\frac{\partial^2 V}{\partial y \partial x} \qquad \frac{\partial F_y}{\partial x} = -\frac{\partial^2 V}{\partial x \partial y} = -\frac{\partial^2 V}{\partial y \partial x} \qquad \therefore \; \frac{\partial F_y}{\partial x} - \frac{\partial F_x}{\partial y} = 0 \quad (4.2.2)$$

This last step follows if we assume that V is everywhere continuous and differentiable. We reach the same conclusion for the other components of curl $\mathbf{F}$. One might wonder whether there are other reasons why curl $\mathbf{F}$ might vanish, besides its being derivable from a potential energy function. However, curl $\mathbf{F} = 0$ is a necessary and sufficient condition for the existence of $V(x, y, z)$ such that Equation 4.2.1 holds.[3]

We can now express a conservative force $\mathbf{F}$ vectorially as

$$\mathbf{F} = -\mathbf{i}\frac{\partial V}{\partial x} - \mathbf{j}\frac{\partial V}{\partial y} - \mathbf{k}\frac{\partial V}{\partial z} \qquad (4.2.3)$$

This equation can be written more succinctly as

$$\mathbf{F} = -\boldsymbol{\nabla}V \qquad (4.2.4)$$

where we have introduced the vector operator *del*:

$$\boldsymbol{\nabla} = \mathbf{i}\frac{\partial}{\partial x} + \mathbf{j}\frac{\partial}{\partial y} + \mathbf{k}\frac{\partial}{\partial z} \qquad (4.2.5)$$

The expression $\boldsymbol{\nabla}V$ is also called the *gradient of* V and is sometimes written grad V. Mathematically, the gradient of a function is a vector that represents the maximum spatial derivative of the function in direction and magnitude. Physically, the negative gradient of the potential energy function gives the direction and magnitude of the force that acts on a particle located in a field created by other particles. The meaning of the negative sign is that the particle is urged to move in the direction of *decreasing* potential energy rather than in the opposite direction. This is illustrated in Figure 4.2.1. Here the potential energy function is plotted out in the form of contour lines representing the curves of constant potential energy. The force at any point is always normal to the equipotential curve or surface passing through the point in question.

---

[3] See, for example, S. I. Grossman, op cit. Also, Feng presents an interesting discussion of conservancy criteria when the force field contains singularities in *Amer. J. Phys.* 37, 616 (1969).

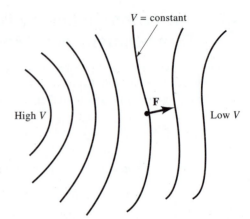

Figure 4.2.1   A force field represented by equipotential contour curves.

We can express curl **F** using the del operator. Look at the components of curl **F** in Equation 4.1.9. They are the components of the vector $\nabla \times \mathbf{F}$. Thus, $\nabla \times \mathbf{F} = \text{curl } \mathbf{F}$. The condition that a force be conservative can be written compactly as

$$\nabla \times \mathbf{F} = \mathbf{i}\left(\frac{\partial F_z}{\partial y} - \frac{\partial F_y}{\partial z}\right) + \mathbf{j}\left(\frac{\partial F_x}{\partial z} - \frac{\partial F_z}{\partial x}\right) + \mathbf{k}\left(\frac{\partial F_y}{\partial x} - \frac{\partial F_x}{\partial y}\right) = 0 \qquad (4.2.6)$$

Furthermore, if $\nabla \times \mathbf{F} = 0$, then **F** can be derived from a scalar function V by the operation $\mathbf{F} = -\nabla V$, since $\nabla \times \nabla V \equiv 0$, or the curl of any gradient is identically 0.

We are now able to generalize the conservation of energy principle to three dimensions. The work done by a conservative force in moving a particle from point A to point B can be written as

$$\int_A^B \mathbf{F} \cdot d\mathbf{r} = -\int_A^B \nabla V(\mathbf{r}) \cdot d\mathbf{r} = -\int_{A_x}^{B_x} \frac{\partial V}{\partial x}\, dx - \int_{A_y}^{B_y} \frac{\partial V}{\partial y}\, dy - \int_{A_z}^{B_z} \frac{\partial V}{\partial z}\, dz$$
$$= -\int_A^B dV(\mathbf{r}) = -\Delta V = V(A) - V(B) \qquad (4.2.7)$$

The last step illustrates the fact that $\nabla V \cdot d\mathbf{r}$ is an *exact* differential equal to dV. The work done by any net force is always equal to the change in kinetic energy, so

$$\int_A^B \mathbf{F} \cdot d\mathbf{r} = \Delta T = -\Delta V$$
$$\therefore \Delta(T + V) = 0 \qquad (4.2.8)$$
$$\therefore T(A) + V(A) = T(B) + V(B) = E = \text{constant}$$

and we have arrived at our desired law of conservation of total energy.

If $\mathbf{F}'$ is a nonconservative force, it cannot be set equal to $-\nabla V$. The work increment $\mathbf{F}' \cdot d\mathbf{r}$ is not an exact differential and cannot be equated to $-dV$. In those cases where both conservative forces **F** and nonconservative forces $\mathbf{F}'$ are present, the total work increment is $(\mathbf{F} + \mathbf{F}') \cdot d\mathbf{r} = -dV + \mathbf{F}' \cdot d\mathbf{r} = dT$, and the generalized form of the work energy theorem becomes

$$\int_A^B \mathbf{F}' \cdot d\mathbf{r} = \Delta(T + V) = \Delta E \qquad (4.2.9)$$

The total energy E does not remain a constant throughout the motion of the particle but increases or decreases depending upon the nature of the nonconservative force $\mathbf{F}'$. In the case of dissipative forces such as friction and air resistance, the direction of $\mathbf{F}'$ is always *opposite* the motion; hence, $\mathbf{F}' \cdot d\mathbf{r}$ is negative, and the total energy of the particle decreases as it moves through space.

EXAMPLE 4.2.1

Given the two-dimensional potential energy function

$$V(\mathbf{r}) = V_0 - \frac{1}{2} k\delta^2 e^{-r^2/\delta^2}$$

where $\mathbf{r} = \mathbf{i}x + \mathbf{j}y$ and V_0, k, and δ are constants, find the force function.

Solution:

We first write the potential energy function as a function of x and y,

$$V(x,y) = V_0 - \frac{1}{2} k\delta^2 e^{-(x^2+y^2)/\delta^2}$$

and then apply the gradient operator:

$$\begin{aligned}
\mathbf{F} = -\nabla V &= -\left(\mathbf{i}\frac{\partial}{\partial x} + \mathbf{j}\frac{\partial}{\partial y} \right) V(x,y) \\
&= -k(\mathbf{i}x + \mathbf{j}y)e^{-(x^2+y^2)/\delta^2} \\
&= -k\mathbf{r}e^{-r^2/\delta^2}
\end{aligned}$$

Notice that the constant V_0 does not appear in the force function; its value is arbitrary. It simply raises or lowers the value of the potential energy function by a constant everywhere on the x, y plane and, thus, has no effect on the resulting force function.

We have plotted the potential energy function in Figure 4.2.2(a) and the resulting force function in Figure 4.2.2(b). The constants were taken to be $V_0 = 1$, $\delta^2 = 1/3$, and $k = 6$. The "hole" in the potential energy surface reaches greatest depth at the origin, which is obviously the location of a source of attraction. The concentric circles around the center of the hole are *equipotentials*—lines of constant potential energy. The radial lines are lines of steepest descent that depict the gradient of the potential energy surface. The slope of a radial line at any point on the plane is proportional to the force that a particle would experience there. The force field in Figure 4.2.2(b) shows the force vectors pointing towards the origin. They weaken both far from and near to the origin, where the slope of the potential energy function approaches zero.

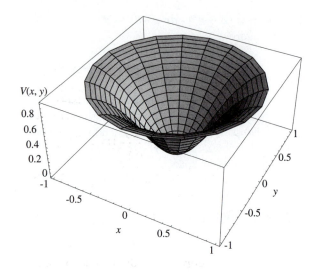

Figure 4.2.2a   The potential energy function $V(x, y) = V_0 - \frac{1}{2}k\delta^2 e^{-(x^2+y^2)/\delta^2}$.

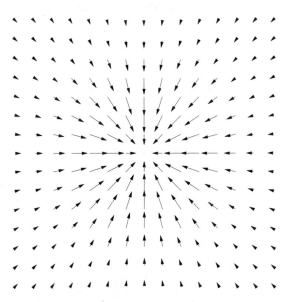

Figure 4.2.2b   Force field gradient of potential energy function in Figure 4.2.2(a); $\mathbf{F} = -\Delta V - k(\mathbf{i}x + \mathbf{j}y)e^{-(x^2+y^2)/\delta^2}$.

EXAMPLE 4.2.2

Suppose a particle of mass m is moving in the above force field, and at time $t = 0$ the particle passes through the origin with speed v_0. What will the speed of the particle be at some small distance away from the origin given by $\mathbf{r} = \mathbf{e}_r\Delta$, where $\Delta \ll \delta$?

Solution:

The force is conservative, because a potential energy function exists. Thus, the total energy $E = T + V = $ constant,

$$E = \frac{1}{2}mv^2 + V(\mathbf{r}) = \frac{1}{2}mv_0^2 + V(0)$$

and solving for v, we obtain

$$v^2 = v_0^2 + \frac{2}{m}[V(0) - V(\mathbf{r})]$$

$$= v_0^2 + \frac{2}{m}\left[\left(V_0 - \frac{1}{2}k\delta^2\right) - \left(V_0 - \frac{1}{2}k\delta^2 e^{-\Delta^2/\delta^2}\right)\right]$$

$$= v_0^2 - \frac{k\delta^2}{m}[1 - e^{-\Delta^2/\delta^2}]$$

$$\approx v_0^2 - \frac{k\delta^2}{m}[1 - (1 - \Delta^2/\delta^2)]$$

$$= v_0^2 - \frac{k}{m}\Delta^2$$

The potential energy is a quadratic function of the displacement Δ from the origin for small displacements, so this solution reduces to the conservation of energy for the simple harmonic oscillator

EXAMPLE 4.2.3

Is the force field $\mathbf{F} = \mathbf{i}xy + \mathbf{j}xz + \mathbf{k}yz$ conservative? The curl of $\mathbf{F}$ is

$$\nabla \times \mathbf{F} = \begin{vmatrix} \mathbf{i} & \mathbf{j} & \mathbf{k} \\ \partial/\partial x & \partial/\partial y & \partial/\partial z \\ xy & xz & yz \end{vmatrix} = \mathbf{i}(z - x) + \mathbf{j}0 + \mathbf{k}(z - x)$$

The final expression is not zero for all values of the coordinates; hence, the field is *not* conservative.

EXAMPLE 4.2.4

For what values of the constants a, b, and c is the force $\mathbf{F} = \mathbf{i}(ax + by^2) + \mathbf{j}cxy$ conservative? Taking the curl, we have

$$\nabla \times \mathbf{F} = \begin{vmatrix} \mathbf{i} & \mathbf{j} & \mathbf{k} \\ \partial/\partial x & \partial/\partial y & \partial/\partial z \\ ax + by^2 & cxy & 0 \end{vmatrix} = \mathbf{k}(c - 2b)y$$

This shows that the force is conservative, provided $c = 2b$. The value of a is immaterial.

EXAMPLE 4.2.5

Show that the inverse-square law of force in three dimensions $\mathbf{F} = (-k/r^2)\mathbf{e}_r$ is conservative by the use of the curl. Use spherical coordinates. The curl is given in Appendix F as

$$\nabla \times \mathbf{F} = \frac{1}{r^2 \sin\theta} \begin{vmatrix} \mathbf{e}_r & \mathbf{e}_\theta r & \mathbf{e}_\phi r \sin\theta \\ \dfrac{\partial}{\partial r} & \dfrac{\partial}{\partial \theta} & \dfrac{\partial}{\partial \phi} \\ F_r & rF_\theta & rF_\phi \sin\theta \end{vmatrix}$$

We have $F_r = -k/r^2$, $F_\theta = 0$, $F_\phi = 0$. The curl then reduces to

$$\nabla \times \mathbf{F} = \frac{\mathbf{e}_\theta}{r \sin\theta} \frac{\partial}{\partial \phi} \left(\frac{-k}{r^2} \right) - \frac{\mathbf{e}_\phi}{r} \frac{\partial}{\partial \theta} \left(\frac{-k}{r^2} \right) = 0$$

which, of course, vanishes because both partial derivatives are zero. Thus, the force in question is conservative.

4.3 | Forces of the Separable Type: Projectile Motion

A Cartesian coordinate system can be frequently chosen such that the components of a force field involve the respective coordinates alone, that is,

$$\mathbf{F} = \mathbf{i}F_x(x) + \mathbf{j}F_y(y) + \mathbf{k}F_z(z) \tag{4.3.1}$$

Forces of this type are *separable*. The curl of such a force is identically zero:

$$\nabla \times \mathbf{F} = \begin{vmatrix} \mathbf{i} & \mathbf{j} & \mathbf{k} \\ \partial/\partial x & \partial/\partial y & \partial/\partial z \\ F_x(x) & F_y(y) & F_z(z) \end{vmatrix} = 0 \tag{4.3.2}$$

The x component is $\partial F_z(z)/\partial y - \partial F_y(y)/\partial z$ and a similar expression holds for the other components; therefore, the field is conservative because each partial derivative is of the mixed type and vanishes identically, because the coordinates x, y, and z are independent variables. The integration of the differential equations of motion is then very simple because each component equation is of the type $m\ddot{x} = F_x(x)$. In this case the equations can be solved by the methods described under rectilinear motion in Chapter 2.

In the event that the force components involve the time and the time derivatives of the respective coordinates, then it is no longer true that the force is necessarily conservative. Nevertheless, if the force is separable, then the component equations of motion are of the form $m\ddot{x} = F_x(x, \dot{x}, t)$ and may be solved by the methods used in Chapter 2. Some examples of separable forces, both conservative and nonconservative, are discussed here and in the sections to follow.

Motion of a Projectile in a Uniform Gravitational Field

While a professor at Padua, Italy, during the years 1602–1608, Galileo spent much of his time projecting balls horizontally into space by rolling them down an inclined plane at the bottom of which he had attached a curved deflector. He hoped to demonstrate that the horizontal motion of objects would persist in the absence of frictional forces. If this were true, then the horizontal motion of heavy projectiles should not be affected much by air resistance and should occur at a constant speed. Galileo had already demonstrated that balls rolling down inclined planes attained a speed that was proportional to their time of roll, and so he could vary the speed of a horizontally projected ball in a controlled way. He observed that the horizontal distance traveled by a projectile increased in direct proportion to its speed of projection from the plane, thus, experimentally demonstrating his conviction. During these investigations, he was stunned to find that the paths these projectiles followed were parabolas! In 1609, already knowing the answer (affirming as gospel what every modern, problem-solving student of physics knows from experience), Galileo was able to prove mathematically that the parabolic trajectory of projectiles was a natural consequence of horizontal motion that was unaccelerated—and an independent vertical motion that was. Indeed, he understood the consequences of this motion as well. Before finally publishing his work in 1638 in *Discourse of Two New Sciences,* he wrote the following in a letter to one of his many scientific correspondents, Giovanni Baliani:

> . . . I treat also of the motion of projectiles, demonstrating various properties, among which is the proof that the projectile thrown by the projector, as would be the ball shot by firing artillery, makes its maximum flight and falls at the greatest distance when the piece is elevated at half a right angle, that is at 45°; and moreover, that other shots made at greater or less elevation come out equal when the piece is elevated an equal number of degrees above and below the said 45°.[4]

Not unlike the funding situation that science and technology finds itself in today, fundamental problems in the fledgling science of Galileo's time, which piqued the interest of the interested few, stood a good chance of being addressed if they related in some way to the military enterprise. Indeed, solving the motion of a projectile is one of the most famous problems in classical mechanics, and it is no accident that Galileo made the discovery partially supported by funds ultimately derived from wealthy patrons attempting to gain some military advantage over their enemies.

In 1597, Galileo had entered into a 10-year collaboration with a toolmaker, Marc' antonio Mazzoleni. In Galileo's day, the use of cannons to pound away at castle walls was more art than science. The Marquis del Monte in Florence and General del Monte in Padua, with whom Galileo had worked earlier, wondered if it were possible to devise a lightweight military "compass" that could be used to gauge the distance and height of a target,

---

[4] See for example, S. Drake and J. MacLachlan, Galileo's Discovery of the Parabolic Trajectory, *Scienti. Amer.* 232, 102–110, (March, 1975). Also see S. Drake, *Galileo at Work,* Mineola, NY, Dover, 1978.

to measure the angle of elevation of the cannon and to track the path of its projectile. Galileo solved the problem and developed the military compass, which his toolmaker produced in quantity in his workshop. There was a ready market for these devices, and they sold well. However, Galileo gained most of the support that enabled him to carry out his own investigation of motion by instructing students in the use of the compass and charging them 120 *lire* for the privilege. Though, like many professors today with which many readers of this text are likely familiar, Galileo more than resented any labor that prevented him from pursuing his own interests. "I'm always at the service of this or that person. I have to consume many hours of the day—often the best ones—in the service of others." Fortunately, he found enough time to carry out his experiments with rolling balls, which led to his discovery of the parabolic trajectory and ultimately helped lead Newton to the discovery of the classical laws of motion.

In 1611, Galileo informed Antonio de'Medici of his work on projectiles, which no doubt the powerful de'Medici family of Florence put to good use . . . and no doubt, went a long way towards helping Galileo secure their undying gratitude and unending patronage.

So with undying gratitude to Galileo and his successor, Newton, here we take only a few minutes—and not years—to solve the projectile problem.

No Air Resistance

For simplicity, we first consider the case of a projectile moving with no air resistance. Only one force, gravity, acts on the projectile, and, consistent with Galileo's observations as we shall see, it affects only its vertical motion. Choosing the z-axis to be vertical, we have the following equation of motion:

$$m\frac{d^2\mathbf{r}}{dt^2} = -\mathbf{k}\,mg \tag{4.3.3}$$

In the case of projectiles that don't rise too high or travel too far, we can take the acceleration of gravity, g, to be constant. Then the force function is conservative and of the separable type, because it is a special case of Equation 4.3.1. v_0 is the initial speed of the projectile, and the origin of the coordinate system is its initial position. Furthermore, there is no loss of generality if we orient the coordinate system so that the x-axis lies along the projection of the initial velocity onto the xy horizontal plane. Because there are no horizontally directed forces acting on the projectile, the motion occurs solely in the xz vertical plane. Thus, the position of the projectile at any time is (see Figure 4.3.1)

$$\mathbf{r} = \mathbf{i}x + \mathbf{k}z \tag{4.3.4}$$

The speed of the projectile can be calculated as a function of its height, z, using the energy equation (Equation 4.2.8)

$$\tfrac{1}{2}m(\dot{x}^2 + \dot{z}^2) + mgz = \tfrac{1}{2}mv_0^2 \tag{4.3.5a}$$

or equivalently,

$$v^2 = v_0^2 - 2gz \tag{4.3.5b}$$

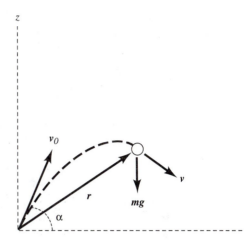

Figure 4.3.1 The parabolic path of a projectile.

We can calculate the velocity of the projectile at any instant of time by integrating Equation 4.3.3

$$\mathbf{v} = \frac{d\mathbf{r}}{dt} = -\mathbf{k}gt + \mathbf{v}_0 \tag{4.3.6a}$$

The constant of integration is the initial velocity $\mathbf{v}_0$. In terms of unit vectors, the velocity is

$$\mathbf{v} = \mathbf{i}v_0\cos\alpha + \mathbf{k}(v_0\sin\alpha - gt) \tag{4.3.6b}$$

Integrating once more yields the position vector

$$\mathbf{r} = -\mathbf{k}\frac{1}{2}gt^2 + \mathbf{v}_0 t + \mathbf{r}_0 \tag{4.3.7a}$$

The constant of integration is the initial position of the projectile, $\mathbf{r}_0$, which is equal to zero; therefore, in terms of unit vectors, Equation 4.3.7a becomes

$$\mathbf{r} = \mathbf{i}(v_0\cos\alpha)t + \mathbf{k}\left((v_0\sin\alpha)t - \frac{1}{2}gt^2\right) \tag{4.3.7b}$$

In terms of components, the position of the projectile at any instant of time is

$$\begin{aligned}
x &= \dot{x}_0 t = (v_0\cos\alpha)t \\
y &= \dot{y}_0 t \equiv 0 \\
z &= \dot{z}_0 t - \frac{1}{2}gt^2 = (v_0\sin\alpha)t - \frac{1}{2}gt^2
\end{aligned} \tag{4.3.7c}$$

$\dot{x}_0 = v_0\cos\alpha$, $\dot{y}_0 = 0$, and $\dot{z}_0 = v_0\sin\alpha$ are the components of the initial velocity $\mathbf{v}_0$.

We can now show, as Galileo did in 1609, that the path of the projectile is a parabola. We find $z(x)$ by using the first of Equations 4.3.7c to solve for t as a function of x and then substitute the resulting expression in the third of Equations 4.3.7c

$$t = \frac{x}{v_0 \cos \alpha} \tag{4.3.8}$$

$$z = (\tan \alpha)x - \left(\frac{g}{2v_0^2 \cos^2 \alpha}\right)x^2 \tag{4.3.9}$$

Equation 4.3.9 is the equation of a parabola and is shown in Figure 4.3.1.

Like Galileo, we calculate several properties of projectile motion: (1) the maximum height, z_{max}, of the projectile, (2) the time, t_{max}, it takes to reach maximum height, (3) the time of flight, T, of the projectile, and (4) the range, R, and maximum range, R_{max}, of the projectile.

- First, we calculate the maximum height obtained by the projectile by using Equation 4.3.5b and noting that at maximum height the vertical component of the velocity of the projectile is zero so that its velocity is in the horizontal direction and equal to the constant horizontal component, $v_0 \cos \alpha$. Thus

$$v_0^2 \cos^2 \alpha = v_0^2 - 2gz_{max} \tag{4.3.10}$$

We solve this to obtain

$$z_{max} = \frac{v_0^2 \sin^2 \alpha}{2g} \tag{4.3.11}$$

- The time it takes to reach maximum height can be obtained from Equation 4.3.6b where we again make use of the fact that at maximum height, the vertical component of the velocity vanishes, so

$$v_0 \sin \alpha - gt_{max} = 0$$

or

$$t_{max} = \frac{v_0 \sin \alpha}{g} \tag{4.3.12}$$

- We can obtain the total time of flight T of the projectile by setting $z = 0$ in the last of Equations 4.3.7c, which yields

$$T = \frac{2v_0 \sin \alpha}{g} \tag{4.3.13}$$

This is twice the time it takes the projectile to reach maximum height. This indicates that the upward flight of the projectile to the apex of its trajectory is symmetrical to its downward flight away from it.

- Finally, we calculate the range of the projectile by substituting the total time of flight, T, into the first of Equations 4.3.7c, obtaining

$$R = x = \frac{v_0^2 \sin^2 2\alpha}{g} \qquad (4.3.14)$$

R has its maximum value $R_{max} = v_0^2/g$ at $\alpha = 45°$.

Linear Air Resistance

We now consider the motion of a projectile subject to the force of air resistance. In this case, the motion does not conserve total energy, which continually diminishes during the flight of the projectile. To solve the problem analytically, we assume that the resisting force varies linearly with the velocity. To simplify the resulting equation of motions, we take the constant of proportionality to be $m\gamma$ where m is the mass of the projectile. The equation of motion is then

$$m\frac{d^2\mathbf{r}}{dt^2} = -m\gamma\mathbf{v} - \mathbf{k}mg \qquad (4.3.15)$$

Upon canceling m's, the equation simplifies to

$$\frac{d^2\mathbf{r}}{dt^2} = -\gamma\mathbf{v} - \mathbf{k}g \qquad (4.3.16)$$

Before integrating, we write Equation 4.3.16 in component form

$$\ddot{x} = -\gamma\dot{x}$$
$$\ddot{y} = -\gamma\dot{y} \qquad (4.3.17)$$
$$\ddot{z} = -\gamma\dot{z} - g$$

We see that the equations are separated; therefore, each can be solved individually by the methods of Chapter 2. Using the results from Example 2.4.1, we can write down the solutions immediately, noting that here $\gamma = c_1/m$, c_1 being the linear drag coefficient. The results are

$$\dot{x} = \dot{x}_0 e^{-\gamma t}$$
$$\dot{y} = \dot{y}_0 e^{-\gamma t} \qquad (4.3.18)$$
$$\dot{z} = \dot{z}_0 e^{-\gamma t} - \frac{g}{\gamma}(1 - e^{-\gamma t})$$

for the velocity components. As before, we orient the coordinate system such that the x-axis lies along the projection of the initial velocity onto the xy horizontal plane. Then $\dot{y} = \dot{y}_0 = 0$

and the motion is confined to the xz vertical plane. Integrating once more, we obtain the position coordinates

$$x = \frac{\dot{x}_0}{\gamma}(1 - e^{-\gamma t})$$

$$z = \left(\frac{\dot{z}_0}{\gamma} + \frac{g}{\gamma^2}\right)(1 - e^{-\gamma t}) - \frac{g}{\gamma}t$$

(4.3.19)

We have taken the initial position of the projectile to be zero, the origin of the coordinate system. This solution can be written vectorially as

$$\mathbf{r} = \left(\frac{\mathbf{v}_0}{\gamma} + \frac{\mathbf{k}g}{\gamma^2}\right)(1 - e^{-\gamma t}) - \mathbf{k}\frac{gt}{\gamma}$$

(4.3.20)

which can be verified by differentiation.

Contrary to the case of zero air resistance the path of the projectile is not a parabola, but rather a curve that lies below the corresponding parabolic trajectory. This is illustrated in Figure 4.3.2. Inspection of the x equation shows that, for large t, the value of x approaches the limiting value

$$x \rightarrow \frac{\dot{x}_0}{\gamma}$$

(4.3.21)

This means that the complete trajectory of the projectile, if it did not hit anything, would have a vertical asymptote as shown in Figure 4.3.2.

In the actual motion of a projectile through the atmosphere, the law of resistance is by no means linear; it is a very complicated function of the velocity. An accurate calculation of the trajectory can be done by means of numerical integration methods. (See the reference cited in Example 2.4.3.)

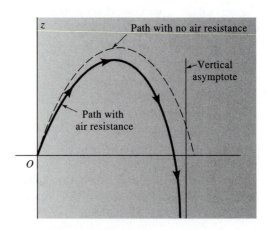

Figure 4.3.2   Comparison of the paths of a projectile with and without air resistance.

Horizontal Range

The horizontal range of a projectile with linear air drag is found by setting $z = 0$ in the second of Equations 4.3.19 and then eliminating t among the two equations. From the first of Equations 4.3.19, we have $1 - \gamma x/\dot{x}_0 = e^{-\gamma t}$, so $t = -\gamma^{-1} \ln(1 - \gamma x/\dot{x}_0)$. Thus, the horizontal range x_{max} is given by the implicit expression

$$\left(\frac{\dot{z}_0}{\gamma} + \frac{g}{\gamma^2}\right)\frac{\gamma x_{max}}{\dot{x}_0} + \frac{g}{\gamma^2}\ln\left(1 - \frac{\gamma x_{max}}{\dot{x}_0}\right) = 0 \tag{4.3.22}$$

This is a transcendental equation and must be solved by some approximation method to find x_h. We can expand the logarithmic term by use of the series

$$\ln(1 - u) = -u - \frac{u^2}{2} - \frac{u^3}{3} - \cdots \tag{4.3.23}$$

which is valid for $|u| < 1$. With $u = \gamma x_{max}/\dot{x}_0$, it is left as a problem to show that this leads to the following expression for the horizontal range:

$$x_{max} = \frac{2\dot{x}_0\dot{z}_0}{g} - \frac{8\dot{x}_0\dot{z}_0^2}{3g^2}\gamma + \cdots \tag{4.3.24a}$$

If the projectile is fired at angle of elevation α with initial speed v_0, then $\dot{x}_0 = v_0 \cos\alpha$, $\dot{z}_0 = v_0 \sin\alpha$, and $2\dot{x}_0\dot{z}_0 = 2v_0^2 \sin\alpha \cos\alpha = v_0^2 \sin 2\alpha$. An equivalent expression is then

$$x_{max} = \frac{v_0^2 \sin 2\alpha}{g} - \frac{4v_0^3 \sin 2\alpha \sin\alpha}{3g^2}\gamma + \cdots \tag{4.3.24b}$$

The first term on the right is the range in the absence of air resistance. The remainder is the decrease due to air resistance.

EXAMPLE 4.3.1

Horizontal Range of a Golf Ball

For objects of baseball or golf-ball size traveling at normal speeds, the air drag is more nearly quadratic in v, rather than linear, as pointed out in Section 2.4. However, the approximate expression found above can be used to find the range for flat trajectories by "linearizing" the force function given by Equation 2.4.3, which may be written in three dimensions as

$$\mathbf{F}(\mathbf{v}) = -\mathbf{v}(c_1 + c_2 |\mathbf{v}|)$$

To linearize it, we set $|\mathbf{v}|$ equal to the initial speed v_0, and so the constant γ is given by

$$\gamma = \frac{c_1 + c_2 v_0}{m}$$

(A better approximation would be to take the average speed, but that is not a given quantity.) Although this method exaggerates the effect of air drag, it allows a quick ballpark estimate to be found easily.

For a golf ball of diameter $D = 0.042$ m and mass $m = 0.046$ kg, we find that c_1 is negligible and so

$$\gamma = \frac{c_2 v_0}{m} = \frac{0.22 D^2 v_0}{m}$$

$$= \frac{0.22(0.042)^2 v_0}{0.046} = 0.0084 v_0$$

numerically, where v_0 is in ms^{-1}. For a chip shot with, say, $v_0 = 20$ ms^{-1}, we find $\gamma = 0.0084 \times 20 = 0.17$ s^{-1}. The horizontal range is then, for $\alpha = 30°$,

$$x_{max} = \frac{(20)^2 \sin 60°}{9.8} \text{ m} - \frac{4(20)^3 \sin 60° \sin 30° \times 0.17}{3(9.8)^2} \text{ m}$$

$$= 35.3 \text{ m} - 8.2 \text{ m} = 27.1 \text{ m}$$

Our estimate, thus, gives a reduction of about one-fourth due to air drag on the ball.

EXAMPLE 4.3.2

A "Tape Measure" Home Run

Here we calculate what is required of a baseball player to hit a *tape measure home run*, or one that travels a distance in excess of 500 feet. In Section 2.4, we mentioned that the force of air drag on a baseball is essentially proportional to the square of its speed that is, $\mathbf{F}_D(v) = -c_2 |v| \mathbf{v}$. The actual air drag force on a baseball is more complicated than that. For example, the "constant" of proportionality c_2 varies somewhat with the speed of the baseball, and the air drag depends, among other things, on its spin and the way its cover is stitched on. We assume, however, for our purposes here that the above equation describes the situation adequately enough with the caveat that we take $c_2 = 0.15$ instead of the value 0.22 that we used previously. This value "normalizes" the air drag factor of a baseball traveling at speeds near 100 mph to that used by Robert Adair in *The Physics of Baseball.*[4]

Trajectories of bodies subject to an air drag force that depends upon the square of its speed cannot not be calculated analytically, so we use *Mathematica*, a computer software tool (see Appendix I), to find a numerical solution for the trajectory of a baseball in fight. Our goal is to find the minimum velocity and optimum angle of launch that a baseball batter must achieve to propel a baseball to maximum range. The situation we analyze concerns the longest home run ever hit in a regular-season, major league baseball game according to the *Guinness Book of Sports Records,* namely, a ball struck by

[4] R. K. Adair, *The Physics of Baseball*, 2nd ed., New York, Harper Collins.

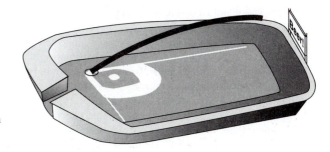

Figure 4.3.3a Trajectory of Mickey Mantle's home run on April 17, 1953, in Griffith Stadium, Washington, D.C.

Figure 4.3.3b Trajectory of Mantle's home run as seen from the batter's perspective. (Mantle, a switch hitter, was actually batting right-handed against the left-handed Stobbs. This photo, showing him batting left-handed, is for the sake of illustration only.)

Mickey Mantle in 1953 that is claimed to have traveled 565 feet over the left field bleachers in old Griffith Stadium in Washington, D.C. The following is an account of that historic home run,[5] which one of your authors (GLC) was privileged to see while watching the baseball game as a bright-eyed young boy from those very left field bleachers for which he paid an entrance fee of 25¢ (oh, how times have changed).

> The Yankees were playing the Senators at Griffith Stadium in Washington, D.C. (The Washington Senators baseball club and Griffith Stadium no longer exist.) The stadium was a little sandbox of a ballpark but, as Mickey Mantle said, "It wasn't that easy to hit a home run there. There was a 90-foot wall in centerfield and there always seemed to be a breeze blowing in."
>
> Lefty Chuck Stobbs was on the mound. A light wind was blowing out from home plate for a change. It was two years to the day since Mickey's first major league game. Mickey stepped up to the plate. Stobbs fired a fast ball just below the letters, right where the Mick liked them, and he connected full-on with it. The ball took off toward the 391-foot sign in left-centerfield. It soared past the fence, over the bleachers and was headed out of the park when it ricocheted off a beer sign on the auxiliary football scoreboard (see Figures 4.3.3a and b). Although, slightly impeded, it continued its flight over neighboring

---

[5]This account of Mantle's Guinness Book of Sports Records home run can be found at the website, http://www.themick.com/10homers.html.

Fifth Street and landed in the backyard of 434 Oakdale Street, several houses up the block.

Billy Martin was on third when Mickey connected and, as a joke, he pretended to tag up like it was just a long fly ball. Mickey didn't notice Billy's shenanigans ("I used to keep my head down as I rounded the bases after a home run. I didn't want to show up the pitcher. I figured he felt bad enough already") and almost ran into Billy! If not for third base coach, Frank Crosetti, he would have. Had Mickey touched Billy he would have automatically been declared out and would have been credited with only a double.

Meanwhile, up in the press box, Yankees PR director, Red Patterson, cried out, "That one's got to be measured!" He raced out of the park and around to the far side of the park where he found 10-year-old Donald Dunaway with the ball. Dunaway showed Red the ball's impact in the yard and Red paced off the distance to the outside wall of Griffith Stadium. Contrary to popular myth, he did not use a tape measure, although he and Mickey were photographed together with a giant tape measure shortly after the historic blast. Using the dimensions of the park, its walls, and the distance he paced off, Patterson calculated the ball traveled 565 feet. However, sports writer Joe Trimble, when adding together the distances, failed to account for the three-foot width of the wall and came up with the 562-foot figure often cited. However, 565 feet is the correct number.

This was the first ball to ever go over Griffith Stadium's leftfield bleachers. Most believe the ball would have gone even further had it not hit the scoreboard (see Figure 4.3.3b). At any rate, it became one of the most famous home runs ever. It was headline news in a number of newspapers and a major story across the country. From that date forward, long home runs were referred to as "tape measure home runs."

So, did Mickey Mantle really hit a 565 foot home run, and, if so, at what angle did he strike the ball and what initial velocity did he impart to it? The equation of motion of a baseball subject to quadratic air drag is

$$m\ddot{\mathbf{r}} = -c_2 \left| v \right| \mathbf{v} - mg\mathbf{k}$$

This separates into two component equations

$$m\ddot{x} = -c_2 \left| v \right| \dot{x}$$
$$m\ddot{z} = -c_2 \left| v \right| \dot{z} - mg$$

Letting $\gamma = c_2 / m$, we obtain

$$\ddot{x} = -\gamma (\dot{x}^2 + \dot{z}^2)^{1/2} \dot{x}$$
$$\ddot{z} = -\gamma (\dot{x}^2 + \dot{z}^2)^{1/2} \dot{z} - g$$

Understandably, the game of baseball being the great American pastime, the weight (5.125 oz) and diameter (2.86 in) of the baseball are given in English units. In metric units, they are $m = 0.145$ kg and $D = 0.0728$ m respectively, so

$$\gamma = \frac{c_2}{m} = \frac{0.15\,D^2}{m} = \frac{0.15(0.0728)^2}{0.145} \text{ meters}^{-1} = 0.0055 \text{ meters}^{-1}$$

The numerical solution to these second-order, coupled nonlinear differential equations can be generated by using the discussion of Mathematica given in Appendix I. Here we

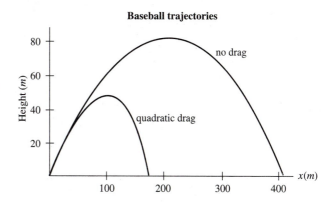

Figure 4.3.4 The calculated range of a baseball with quadratic air drag and without air drag. The range of the baseball is 172.2 m (565 ft) for an initial speed of 143.2 mph and elevation angle of 39 degrees.

simply outline the solution process, which involves an iterative procedure.

- First, we make reasonable guesses for the initial velocity (v_0) and angle (θ_0) of the baseball and then solve the coupled differential equations using these values.
- Plot the trajectory and find the x-axis intercept (the range)
- Hold v_0 fixed, and repeat the above using different values of θ_0 until we find the value of θ_0 that yields the maximum range
- Hold θ_0 fixed at the value found above (that yields the maximum range), and repeat the procedure again, but varying v_0 until we find the value that yields the required range of Mickey's tape measure home run, 565 feet (172.2 m)

The resultant trajectory is shown in Figure 4.3.4 along with the parameters that generated that trajectory. For comparison, we also show the trajectory of a similarly struck baseball in the absence of air resistance. We find that Mickey had to strike the ball at an elevation angle of $\theta_0 = 39°$ with an initial velocity of $v_0 = 143.2$ mph. Are these values reasonable? We would guess that the initial angle ought to be a bit less than the 45° one finds for the case of no air resistance. With resistance, a smaller launch angle (rather than one greater than 45°) corresponds to less time spent in flight during which air resistance can effectively act. What about the initial speed? Chuck Stobbs threw a baseball not much faster than 90 mph. Mantle could swing a bat such that its speed when striking the ball was approximately 90 mph. The coefficient of restitution (see Chapter 7) of baseballs is such that the resultant velocity imparted to the batted ball would be about 130 mph, so the value we've estimated is somewhat high but not outrageously so. If the ball Mantle hit in Griffith Stadium was assisted by a moderate tailwind, his Herculean swat seems possible. Wouldn't it have been spectacular to have seen Mantle hit one like that—in a vacuum?

4.4 | The Harmonic Oscillator in Two and Three Dimensions

Consider the motion of a particle subject to a linear restoring force that is always directed toward a fixed point, the origin of our coordinate system. Such a force can be represented by the expression

$$\mathbf{F} = -k\mathbf{r} \qquad (4.4.1)$$

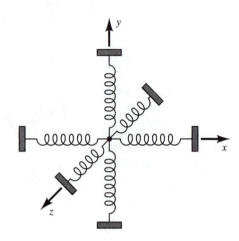

Figure 4.4.1   A model of a three-dimensional harmonic oscillator.

Accordingly, the differential equation of motion is simply expressed as

$$m\frac{d^2\mathbf{r}}{dt^2} = -k\mathbf{r} \tag{4.4.2}$$

The situation can be represented approximately by a particle attached to a set of elastic springs as shown in Figure 4.4.1. This is the three-dimensional generalization of the linear oscillator studied earlier. Equation 4.4.2 is the differential equation of the *linear isotropic oscillator.*

The Two-Dimensional Isotropic Oscillator

In the case of motion in a single plane, Equation 4.4.2 is equivalent to the two compo-nent equations

$$
\begin{aligned}
m\ddot{x} &= -kx \\
m\ddot{y} &= -ky
\end{aligned}
\tag{4.4.3}
$$

These are separated, and we can immediately write down the solutions in the form

$$x = A\,\cos\,(\omega t + \alpha) \qquad y = B\,\cos\,(\omega t + \beta) \tag{4.4.4}$$

in which

$$\omega = \left(\frac{k}{m}\right)^{1/2} \tag{4.4.5}$$

The constants of integration A, B, α, and β are determined from the initial conditions in any given case.

To find the equation of the path, we eliminate the time t between the two equations. To do this, let us write the second equation in the form

$$y = B\,\cos\,(\omega t + \alpha + \Delta) \tag{4.4.6}$$

where

$$\Delta = \beta - \alpha \tag{4.4.7}$$

Then

$$y = B[\cos(\omega t + \alpha)\cos\Delta - \sin(\omega t + \alpha)\sin\Delta] \tag{4.4.8}$$

Combining the above with the first of Equations 4.4.4, we then have

$$\frac{y}{B} = \frac{x}{A}\cos\Delta - \left(1 - \frac{x^2}{A^2}\right)^{1/2}\sin\Delta \tag{4.4.9}$$

and upon transposing and squaring terms, we obtain

$$\frac{x^2}{A^2} - xy\frac{2\cos\Delta}{AB} + \frac{y^2}{B^2} = \sin^2\Delta \tag{4.4.10}$$

which is a quadratic equation in x and y. Now the general quadratic

$$ax^2 + bxy + cy^2 + dx + ey = f \tag{4.4.11}$$

represents an ellipse, a parabola, or a hyperbola, depending on whether the discriminant

$$b^2 - 4ac \tag{4.4.12}$$

is negative, zero, or positive, respectively. In our case the discriminant is equal to $-(2\sin\Delta/AB)^2$, which is negative, so the path is an ellipse as shown in Figure 4.4.2.

In particular, if the phase difference Δ is equal to $\pi/2$, then the equation of the path reduces to the equation

$$\frac{x^2}{A^2} + \frac{y^2}{B^2} = 1 \tag{4.4.13}$$

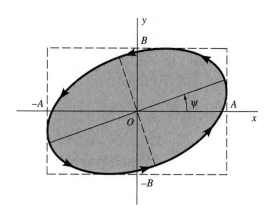

Figure 4.4.2 The elliptical path of a two-dimensional isotropic oscillator.

which is the equation of an ellipse whose axes coincide with the coordinate axes. On the other hand, if the phase difference is 0 or π, then the equation of the path reduces to that of a straight line, namely,

$$y = \pm \frac{B}{A} x \tag{4.4.14}$$

The positive sign is taken if $\Delta = 0$, and the negative sign, if $\Delta = \pi$. In the general case it is possible to show that the axis of the elliptical path is inclined to the x-axis by the angle ψ, where

$$\tan 2\psi = \frac{2AB \cos \Delta}{A^2 - B^2} \tag{4.4.15}$$

The derivation is left as an exercise.

The Three-Dimensional Isotropic Harmonic Oscillator

In the case of three-dimensional motion, the differential equation of motion is equivalent to the three equations

$$m\ddot{x} = -kx \qquad m\ddot{y} = -ky \qquad m\ddot{z} = -kz \tag{4.4.16}$$

which are separated. Hence, the solutions may be written in the form of Equations 4.4.4, or, alternatively, we may write

$$\begin{aligned} x &= A_1 \sin \omega t + B_1 \cos \omega t \\ y &= A_2 \sin \omega t + B_2 \cos \omega t \\ z &= A_3 \sin \omega t + B_3 \cos \omega t \end{aligned} \tag{4.4.17a}$$

The six constants of integration are determined from the initial position and velocity of the particle. Now Equations 4.4.16 can be expressed vectorially as

$$\mathbf{r} = \mathbf{A} \sin \omega t + \mathbf{B} \cos \omega t \tag{4.4.17b}$$

in which the components of $\mathbf{A}$ are A_1, A_2, and A_3, and similarly for $\mathbf{B}$. It is clear that the motion takes place entirely in a single plane, which is common to the two constant vectors $\mathbf{A}$ and $\mathbf{B}$, and that the path of the particle in that plane is an ellipse, as in the two-dimensional case. Hence, the analysis concerning the shape of the elliptical path under the two-dimensional case also applies to the three-dimensional case.

Nonisotropic Oscillator

The previous discussion considered the motion of the isotropic oscillator, wherein the restoring force is independent of the direction of the displacement. If the magnitudes of the components of the restoring force depend on the direction of the displacement, we have the case of the *nonisotropic oscillator*. For a suitable choice of axes, the differential equations for the nonisotropic case can be written

$$m\ddot{x} = -k_1 x$$
$$m\ddot{y} = -k_2 y \qquad (4.4.18)$$
$$m\ddot{z} = -k_3 z$$

Here we have a case of *three* different frequencies of oscillation, $\omega_1 = \sqrt{k_1/m}$, $\omega_2 = \sqrt{k_2/m}$, and $\omega_3 = \sqrt{k_3/m}$, and the motion is given by the solutions

$$x = A\cos(\omega_1 t + \alpha)$$
$$y = B\cos(\omega_2 t + \beta) \qquad (4.4.19)$$
$$z = C\cos(\omega_3 t + \gamma)$$

Again, the six constants of integration in the above equations are determined from the initial conditions. The resulting oscillation of the particle lies entirely within a rectangular box (whose sides are $2A$, $2B$, and $2C$) centered on the origin. In the event that ω_1, ω_2, and ω_3 are commensurate—that is, if

$$\frac{\omega_1}{n_1} = \frac{\omega_2}{n_2} = \frac{\omega_3}{n_3} \qquad (4.4.20)$$

where n_1, n_2, and n_3 are integers—the path, called a *Lissajous* figure, is closed, because after a time $2\pi n_1/\omega_1 = 2\pi n_2/\omega_2 = 2\pi n_3/\omega_3$ the particle returns to its initial position and the motion is repeated. (In Equation 4.4.20 we assume that any common integral factor is canceled out.) On the other hand, if the ω's are *not* commensurate, the path is not closed. In this case the path may be said to completely fill the rectangular box mentioned above, at least in the sense that if we wait long enough, the particle comes arbitrarily close to any given point.

The net restoring force exerted on a given atom in a solid crystalline substance is approximately linear in the displacement in many cases. The resulting frequencies of oscillation usually lie in the infrared region of the spectrum: 10^{12} to 10^{14} vibrations per second.

Energy Considerations

In the preceding chapter we showed that the potential energy function of the one-dimensional harmonic oscillator is quadratic in the displacement, $V(x) = \frac{1}{2}kx^2$. For the general three-dimensional case, it is easy to verify that

$$V(x,y,z) = \frac{1}{2}k_1 x^2 + \frac{1}{2}k_2 y^2 + \frac{1}{2}k_3 z^2 \qquad (4.4.21)$$

because $F_x = -\partial V/\partial x = -k_1 x$, and similarly for F_y and F_z. If $k_1 = k_2 = k_3 = k$, we have the isotropic case, and

$$V(x,y,z) = \frac{1}{2}k(x^2 + y^2 + z^2) = \frac{1}{2}kr^2 \qquad (4.4.22)$$

The total energy in the isotropic case is then given by the simple expression

$$\tfrac{1}{2}mv^2 + \tfrac{1}{2}kr^2 = E \qquad (4.4.23)$$

which is similar to that of the one-dimensional case discussed in the previous chapter.

EXAMPLE 4.4.1

A particle of mass m moves in two dimensions under the following potential energy function:

$$V(\mathbf{r}) = \tfrac{1}{2}k(x^2 + 4y^2)$$

Find the resulting motion, given the initial condition at $t = 0$: $x = a$, $y = 0$, $\dot{x} = 0$, $\dot{y} = v_0$.

Solution:

This is a nonisotropic oscillator potential. The force function is

$$\mathbf{F} = -\nabla V = -\mathbf{i}\,kx - \mathbf{j}4ky = m\ddot{\mathbf{r}}$$

The component differential equations of motion are then

$$m\ddot{x} + kx = 0 \qquad m\ddot{y} + 4ky = 0$$

The x-motion has angular frequency $\omega = (k/m)^{1/2}$, while the y-motion has angular frequency just twice that, namely, $\omega_y = (4k/m)^{1/2} = 2\omega$. We shall write the general solution in the form

$$x = A_1 \cos\omega t + B_1 \sin\omega t$$
$$y = A_2 \cos 2\omega t + B_2 \sin 2\omega$$

To use the initial condition we must first differentiate with respect to t to find the general expression for the velocity components:

$$\dot{x} = -A_1\omega \sin\omega t + B_1\omega \cos\omega t$$
$$\dot{y} = -2A_2\omega \sin 2\omega t + 2B_2\omega \cos 2\omega t$$

Thus, at $t = 0$, we see that the above equations for the components of position and velocity reduce to

$$a = A_1 \qquad 0 = A_2 \qquad 0 = B_1\omega \qquad v_0 = 2B_2\omega$$

These equations give directly the values of the amplitude coefficients, $A_1 = a$, $A_2 = B_1 = 0$, and $B_2 = v_0/2\omega$, so the final equations for the motion are

$$x = a \cos\omega t$$

$$y = \frac{v_0}{2\omega} \sin 2\omega t$$

The path is a Lissajous figure having the shape of a figure-eight as shown in Figure 4.4.3.

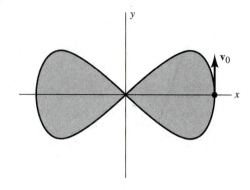

Figure 4.4.3 A Lissajous figure.

4.5 | Motion of Charged Particles in Electric and Magnetic Fields

When an electrically charged particle is in the vicinity of other electric charges, it experiences a force. This force **F** is said to be caused by the electric field **E**, which arises from these other charges. We write

$$\mathbf{F} = q\mathbf{E} \tag{4.5.1}$$

where q is the electric charge carried by the particle in question.[6] The equation of motion of the particle is then

$$m\frac{d^2\mathbf{r}}{dt^2} = q\mathbf{E} \tag{4.5.2a}$$

or, in component form,

$$
\begin{aligned}
m\ddot{x} &= qE_x \\
m\ddot{y} &= qE_y \\
m\ddot{z} &= qE_z
\end{aligned} \tag{4.5.2b}
$$

The field components are, in general, functions of the position coordinates x, y, and z. In the case of time-varying fields (that is, if the charges producing **E** are moving), the components also involve t.

Let us consider a simple case, namely, that of a uniform constant electric field. We can choose one of the axes—say, the z-axis—to be in the direction of the field. Then $E_x = E_y = 0$, and $E = E_z$. The differential equations of motion of a particle of charge q moving in this field are then

$$\ddot{x} = 0 \qquad \ddot{y} = 0 \qquad \ddot{z} = \frac{qE}{m} = \text{constant} \tag{4.5.3}$$

---

[6] In SI units, F is in newtons, q in coulombs, and E in volts per meter.

These are of exactly the same form as those for a projectile in a uniform gravitational field. The path is, therefore, a parabola, if $\dot{x}$ and $\dot{y}$ are not both zero initially. Otherwise, the path is a straight line, as with a body falling vertically.

Textbooks dealing with electromagnetic theory[7] show that

$$\nabla \times \mathbf{E} = 0 \qquad (4.5.4)$$

if $\mathbf{E}$ is due to static charges. This means that motion in such a field is conservative, and that there exists a potential function Φ such that $\mathbf{E} = -\nabla\Phi$. The potential energy of a particle of charge q in such a field is then $q\Phi$, and the total energy is constant and is equal to $\frac{1}{2}mv^2 + q\Phi$.

In the presence of a static magnetic field $\mathbf{B}$ (called the magnetic induction), the force acting on a moving particle is conveniently expressed by means of the cross product, namely,

$$\mathbf{F} = q(\mathbf{v} \times \mathbf{B}) \qquad (4.5.5)$$

where $\mathbf{v}$ is the velocity and q is the charge.[8] The differential equation of motion of a particle moving in a purely magnetic field is then

$$m\frac{d^2\mathbf{r}}{dt^2} = q(\mathbf{v} \times \mathbf{B}) \qquad (4.5.6)$$

Equation 4.5.6 states that the acceleration of the particle is always at right angles to the direction of motion. This means that the tangential component of the acceleration ($\dot{v}$) is zero, and so the particle moves with constant speed. This is true even if $\mathbf{B}$ is a varying function of the position $\mathbf{r}$, as long as it does not vary with time.

EXAMPLE 4.5.1

Let us examine the motion of a charged particle in a uniform constant magnetic field. Suppose we choose the z-axis to be in the direction of the field; that is, we write

$$\mathbf{B} = \mathbf{k}B$$

The differential equation of motion now reads

$$m\frac{d^2\mathbf{r}}{dt^2} = q(\mathbf{v} \times \mathbf{k}B) = qB\begin{vmatrix} \mathbf{i} & \mathbf{j} & \mathbf{k} \\ \dot{x} & \dot{y} & \dot{z} \\ 0 & 0 & 1 \end{vmatrix}$$

$$m(\mathbf{i}\ddot{x} + \mathbf{j}\ddot{y} + \mathbf{k}\ddot{z}) = qB(\mathbf{i}\dot{y} - \mathbf{j}\dot{x})$$

---

[7] For example, Reitz, Milford, and Christy, op cit.

[8] Equation 4.5.5 is valid for SI units: F is in newtons, q in coulombs, v in meters per second, and B in webers per square meter.

Equating components, we have

$$m\ddot{x} = qB\dot{y}$$
$$m\ddot{y} = -qB\dot{x} \qquad (4.5.7)$$
$$\ddot{z} = 0$$

Here, for the first time we meet a set of differential equations of motion that are *not* of the separated type. The solution is relatively simple, however, for we can integrate at once with respect to t, to obtain

$$m\dot{x} = qBy + c_1$$
$$m\dot{y} = -qBx + c_2$$
$$\dot{z} = \text{constant} = \dot{z}_0$$

or

$$\dot{x} = \omega y + C_1 \qquad \dot{y} = -\omega x + C_2 \qquad \dot{z} = \dot{z}_0 \qquad (4.5.8)$$

where we have used the abbreviation $\omega = qB/m$. The c's are constants of integration, and $C_1 = c_1/m$, $C_2 = c_2/m$. Upon inserting the expression for $\dot{y}$ from the second part of Equation 4.5.8 into the first part of Equation 4.5.7, we obtain the following separated equation for x:

$$\ddot{x} + \omega^2 x = \omega^2 a \qquad (4.5.9)$$

where $a = C_2/\omega$. The solution is

$$x = a + A\cos(\omega t + \theta_0) \qquad (4.5.10)$$

where A and θ_0 are constants of integration. Now, if we differentiate with respect to t, we have

$$\dot{x} = -A\omega\sin(\omega t + \theta_0) \qquad (4.5.11)$$

The above expression for $\dot{x}$ may be substituted for the left-hand side of the first of Equations 4.5.8 and the resulting equation solved for y. The result is

$$y = b - A\sin(\omega t + \theta_0) \qquad (4.5.12)$$

where $b = -C_1/\omega$. To find the form of the path of motion, we eliminate t between Equation 4.5.10 and Equation 4.5.12 to get

$$(x - a)^2 + (y - b)^2 = A^2 \qquad (4.5.13)$$

Thus, the projection of the path of motion on the xy plane is a circle of radius A centered at the point (a, b). Because, from the third of Equations 4.5.8, the speed in the z direction is constant, we conclude that the path is a *helix*. The axis of the winding path is in the direction of the magnetic field, as shown in Figure 4.5.1. From Equation 4.5.12 we have

$$\dot{y} = -A\omega\cos(\omega t + \theta_0) \qquad (4.5.14)$$

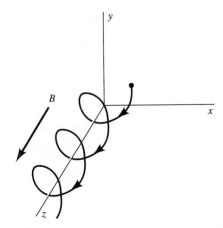

Figure 4.5.1   The helical path of a particle moving in a magnetic field.

Upon eliminating t between Equation 4.5.11 and Equation 4.5.14, we find

$$\dot{x}^2 + \dot{y}^2 = A^2\omega^2 = A^2\left(\frac{qB}{m}\right)^2 \tag{4.5.15}$$

Letting $v_1 = (\dot{x}^2 + \dot{y}^2)^{1/2}$, we see that the radius A of the helix is given by

$$A = \frac{v_1}{\omega} = v_1 \frac{m}{qB} \tag{4.5.16}$$

If there is no component of the velocity in the z direction, the path is a circle of radius A. It is evident that A is directly proportional to the speed v_1 and that the angular frequency ω of motion in the circular path is independent of the speed. The angular frequency ω is known as the cyclotron frequency. The cyclotron, invented by Ernest Lawrence, depends for its operation on the fact that ω is independent of the speed of the charged particle.

4.6 | Constrained Motion of a Particle

When a moving particle is restricted geometrically in the sense that it must stay on a certain definite surface or curve, the motion is said to be *constrained*. A piece of ice sliding around a bowl and a bead sliding on a wire are examples of constrained motion. The constraint may be complete, as with the bead, or it may be one-sided, as with the ice in the bowl. Constraints may be fixed, or they may be moving. In this chapter we study only fixed constraints.

The Energy Equation for Smooth Constraints

The total force acting on a particle moving under constraint can be expressed as the vector sum of the net external force $\mathbf{F}$ and the force of constraint $\mathbf{R}$. The latter force is the reaction of the constraining agent upon the particle. The equation of motion may, therefore, be written

$$m\frac{d\mathbf{v}}{dt} = \mathbf{F} + \mathbf{R} \tag{4.6.1}$$

If we take the dot product with the velocity $\mathbf{v}$, we have

$$m\frac{d\mathbf{v}}{dt}\cdot\mathbf{v} = \mathbf{F}\cdot\mathbf{v}+\mathbf{R}\cdot\mathbf{v} \tag{4.6.2}$$

Now in the case of a *smooth* constraint—for example, a frictionless surface—the reaction $\mathbf{R}$ is normal to the surface or curve while the velocity $\mathbf{v}$ is tangent to the surface. Hence, $\mathbf{R}$ is perpendicular to $\mathbf{v}$, and the dot product $\mathbf{R}\cdot\mathbf{v}$ vanishes. Equation 4.6.2 then reduces to

$$\frac{d}{dt}\left(\tfrac{1}{2}m\mathbf{v}\cdot\mathbf{v}\right) = \mathbf{F}\cdot\mathbf{v} \tag{4.6.3}$$

Consequently, if $\mathbf{F}$ is conservative, we can integrate as in Section 4.2, and we find that, even though the particle is constrained to move along the surface or curve, its total energy remains constant, namely,

$$\tfrac{1}{2}mv^2 + V(x,y,z) = \text{constant} = E \tag{4.6.4}$$

We might, of course, have expected this to be the case for frictionless constraints.

EXAMPLE 4.6.1

A particle is placed on top of a smooth sphere of radius a. If the particle is slightly disturbed, at what point will it leave the sphere?

Solution:

The forces acting on the particle are the downward force of gravity and the reaction $\mathbf{R}$ of the spherical surface. The equation of motion is

$$m\frac{d\mathbf{v}}{dt} = m\mathbf{g}+\mathbf{R}$$

Let us choose coordinate axes as shown in Figure 4.6.1. The potential energy is then mgz, and the energy equation reads

$$\tfrac{1}{2}mv^2 + mgz = E$$

From the initial conditions ($v = 0$ for $z = a$) we have $E = mga$, so, as the particle slides down, its speed is given by the equation

$$v^2 = 2g(a-z)$$

Now, if we take radial components of the equation of motion, we can write the force equation as

$$-\frac{mv^2}{a} = -mg\,\cos\theta + R = -mg\frac{z}{a} + R$$

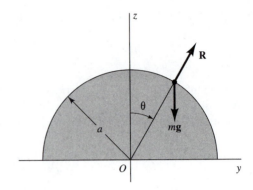

Figure 4.6.1 A particle sliding on a smooth sphere.

Hence,

$$R = mg\frac{z}{a} - \frac{mv^2}{a} = mg\frac{z}{a} - \frac{m}{a}2g(a-z)$$

$$= \frac{mg}{a}(3z - 2a)$$

Thus, R vanishes when $z = \frac{2}{3}a$, at which point the particle leaves the sphere. This may be argued from the fact that the sign of R changes from positive to negative there.

EXAMPLE 4.6.2

Constrained Motion on a Cycloid

Consider a particle sliding under gravity in a smooth cycloidal trough, Figure 4.6.2, represented by the parametric equations

$$x = A(2\phi + \sin 2\phi)$$
$$z = A(1 - \cos 2\phi)$$

where ϕ is the parameter. Now the energy equation for the motion, assuming no y-motion, is

$$E = \frac{m}{2}v^2 + V(z) = \frac{m}{2}(\dot{x}^2 + \dot{z}^2) + mgz$$

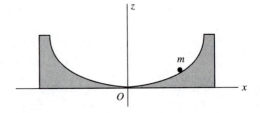

Figure 4.6.2 A particle sliding in a smooth cycloidal trough.

Because $\dot{x} = 2A\dot{\phi}(1 + \cos 2\phi)$ and $\dot{z} = 2A\dot{\phi} \sin 2\phi$, we find the following expression for the energy in terms of ϕ:

$$E = 4mA^2\dot{\phi}^2(1 + \cos 2\phi) + mgA(1 - \cos 2\phi)$$

or, by use of the identities $1 + \cos 2\phi = 2 \cos^2 \phi$ and $1 - \cos 2\phi = 2 \sin^2 \phi$,

$$E = 8mA^2\dot{\phi}^2 \cos^2 \phi + 2mgA \sin^2 \phi$$

Let us introduce the variable s defined by $s = 4A \sin \phi$. The energy equation can then be written

$$E = \frac{m}{2}\dot{s}^2 + \frac{1}{2}\left(\frac{mg}{4A}\right)s^2$$

This is just the energy equation for harmonic motion in the single variable s. Thus, the particle undergoes periodic motion whose frequency is independent of the amplitude of oscillation, unlike the simple pendulum for which the frequency depends on the amplitude. The periodic motion in the present case is said to be *isochronous*. (The linear harmonic oscillator under Hooke's law is, of course, isochronous.)

The Dutch physicist and mathematician Christiaan Huygens discovered the above fact in connection with attempts to improve the accuracy of pendulum clocks. He also discovered the theory of evolutes and found that the evolute of a cycloid is also a cycloid. Hence, by providing cycloidal "cheeks" for a pendulum, the motion of the bob must follow a cycloidal path, and the period is, thus, independent of the amplitude. Though ingenious, the invention never found extensive practical use.

Problems

4.1 Find the force for each of the following potential energy functions:
(a) $V = cxyz + C$
(b) $V = \alpha x^2 + \beta y^2 + \gamma z^2 + C$
(c) $V = ce^{-(\alpha x + \beta y + \gamma z)}$
(d) $V = cr^n$ in spherical coordinates

4.2 By finding the curl, determine which of the following forces are conservative:
(a) $\mathbf{F} = \mathbf{i}x + \mathbf{j}y + \mathbf{k}z$
(b) $\mathbf{F} = \mathbf{i}y - \mathbf{j}x + \mathbf{k}z^2$
(c) $\mathbf{F} = \mathbf{i}y + \mathbf{j}x + \mathbf{k}z^3$
(d) $\mathbf{F} = -kr^{-n}\mathbf{e}_r$ in spherical coordinates

4.3 Find the value of the constant c such that each of the following forces is conservative:
(a) $\mathbf{F} = \mathbf{i}xy + \mathbf{j}cx^2 + \mathbf{k}z^3$
(b) $\mathbf{F} = \mathbf{i}(z/y) + c\mathbf{j}(xz/y^2) + \mathbf{k}(x/y)$

4.4 A particle of mass m moving in three dimensions under the potential energy function $V(x, y, z) = \alpha x + \beta y^2 + \gamma z^3$ has speed v_0 when it passes through the origin.
(a) What will its speed be if and when it passes through the point $(1, 1, 1)$?
(b) If the point $(1, 1, 1)$ is a turning point in the motion $(v = 0)$, what is v_0?
(c) What are the component differential equations of motion of the particle?
(*Note*: It is *not* necessary to solve the differential equations of motion in this problem.)

4.5   Consider the two force functions
(a) $\mathbf{F} = \mathbf{i}x + \mathbf{j}y$
(b) $\mathbf{F} = \mathbf{i}y - \mathbf{j}x$
Verify that (a) is conservative and that (b) is nonconservative by showing that the integral $\int \mathbf{F} \cdot d\mathbf{r}$ is independent of the path of integration for (a), but not for (b), by taking two paths in which the starting point is the origin (0, 0), and the endpoint is (1, 1). For one path take the line $x = y$. For the other path take the x-axis out to the point (1, 0) and then the line $x = 1$ up to the point (1, 1).

4.6   Show that the variation of gravity with height can be accounted for approximately by the following potential energy function:

$$V = mgz\left(1 - \frac{z}{r_e}\right)$$

in which r_e is the radius of the Earth. Find the force given by the above potential function. From this find the component differential equations of motion of a projectile under such a force. If the vertical component of the initial velocity is v_{0z}, how high does the projectile go? (Compare with Example 2.3.2.)

4.7   Particles of mud are thrown from the rim of a rolling wheel. If the forward speed of the wheel is v_0, and the radius of the wheel is b, show that the greatest height above the ground that the mud can go is

$$b + \frac{v_0^2}{2g} + \frac{gb^2}{2v_0^2}$$

At what point on the rolling wheel does this mud leave?
(*Note:* It is necessary to assume that $v_0^2 \geq bg$.)

4.8   A gun is located at the bottom of a hill of constant slope ϕ. Show that the range of the gun measured up the slope of the hill is

$$\frac{2v_0^2 \cos\alpha \, \sin(\alpha - \phi)}{g \cos^2\phi}$$

where α is the angle of elevation of the gun, and that the maximum value of the slope range is

$$\frac{v_0^2}{g(1 + \sin\phi)}$$

4.9   A cannon that is capable of firing a shell at speed V_0 is mounted on a vertical tower of height h that overlooks a level plain below.
(a) Show that the elevation angle α at which the cannon must be set to achieve maximum range is given by the expression

$$\csc^2\alpha = 2\left(1 + \frac{gh}{V_0^2}\right)$$

(b) What is the maximum range R of the cannon?

4.10   A movable cannon is positioned somewhere on the level plain below the cannon mounted on the tower of Problem 4.9. How close must it be positioned from the tower to fire a shell that can hit the cannon in that tower? Assume the two cannons have identical muzzle velocities V_0.

4.11   While playing in Yankee, Stadium, Mickey Mantle hits a baseball that attains a maximum height of 69 ft and strikes the ground 328 ft away from home plate unless it is caught by an outfielder. Assume that the outfielder can catch the ball sometime before it strikes the ground—only if it is less than 9.8 ft above the ground. Assume that Mantle hit the ball when it was 3.28 ft above the ground, and assume no air resistance. Within what horizontal distance can the fielder catch the ball?

4.12   A baseball pitcher can throw a ball more easily horizontally than vertically. Assume that the pitcher's throwing speed varies with elevation angle approximately as $v_0 \cos \frac{1}{2}\theta_0$ m/s, where θ_0 is the initial elevation angle and v_0 is the initial velocity when the ball is thrown horizontally. Find the angle θ_0 at which the ball must be thrown to achieve maximum **(a)** height and **(b)** range.
Find the values of the maximum **(c)** height and **(d)** range. Assume no air resistance and let $v_0 = 25$ m/s.

4.13   A gun can fire an artillery shell with a speed V_0 in any direction. Show that a shell can strike any target within the surface given by

$$g^2 r^2 = V_0^4 - 2gV_0^2 z$$

where z is the height of the target and r is its horizontal distance from the gun. Assume no air resistance.

4.14   Write down the component form of the differential equations of motion of a projectile if the air resistance is proportional to the square of the speed. Are the equations separated? Show that the x component of the velocity is given by

$$\dot{x} = \dot{x}_0 e^{-\gamma s}$$

where s is the distance the projectile has traveled along the path of motion, and $\gamma = c_2/m$.

4.15   Fill in the steps leading to Equations 4.3.24a and b, giving the horizontal range of a projectile that is subject to linear air drag.

4.16   The initial conditions for a two-dimensional isotropic oscillator are as follows: $t = 0$, $x = A$, $y = 4A$, $\dot{x} = 0$, $\dot{y} = 3\omega A$ where ω is the angular frequency. Find x and y as functions of t. Show that the motion takes place entirely within a rectangle of dimensions $2A$ and $10A$. Find the inclination ψ of the elliptical path relative to the x-axis. Make a sketch of the path.

4.17   A small lead ball of mass m is suspended by means of six light springs as shown in Figure 4.4.1. The stiffness constants are in the ratio 1:4:9, so that the potential energy function can be expressed as

$$V = \frac{k}{2}(x^2 + 4y^2 + 9z^2)$$

At time $t = 0$ the ball receives a push in the (1, 1, 1) direction that imparts to it a speed v_0 at the origin. If $k = \pi^2 m$, numerically find x, y, and z as functions of the time t. Does the ball ever retrace its path? If so, for what value of t does it first return to the origin with the same velocity that it had at $t = 0$?

4.18   Complete the derivation of Equation 4.4.15.

4.19   An atom is situated in a simple cubic crystal lattice. If the potential energy of interaction between any two atoms is of the form $cr^{-\alpha}$, where c and α are constants and r is the distance between the two atoms, show that the total energy of interaction of a given atom with its six nearest neighbors is approximately that of the three-dimensional harmonic oscillator potential

$$V \simeq A + B(x^2 + y^2 + z^2)$$

where A and B are constants.

[*Note:* Assume that the six neighboring atoms are fixed and are located at the points $(\pm d, 0, 0), (0, \pm d, 0), (0, 0, \pm d)$, and that the displacement (x, y, z) of the given atom from the equilibrium position $(0, 0, 0)$ is small compared to d. Then $V = \Sigma\, cr_i^{-\alpha}$ where

$$r_1 = [(d - x)^2 + y^2 + z^2]^{1/2},$$

with similar expressions for $r_2, r_3, \ldots, r_6$. See the approximation formulas in Appendix D.]

4.20   An electron moves in a force field due to a uniform electric field $\mathbf{E}$ and a uniform magnetic field $\mathbf{B}$ that is at right angles to $\mathbf{E}$. Let $\mathbf{E} = \mathbf{j}E$ and $\mathbf{B} = \mathbf{k}B$. Take the initial position of the electron at the origin with initial velocity $\mathbf{v}_0 = \mathbf{i}v_0$ in the x direction. Find the resulting motion of the particle. Show that the path of motion is a cycloid:

$$x = a \sin \omega t + bt$$
$$y = a(1 - \cos \omega t)$$
$$z = 0$$

Cycloidal motion of electrons is used in an electronic tube called a magnetron to produce the microwaves in a microwave oven.

4.21   A particle is placed on a smooth sphere of radius b at a distance $b/2$ above the central plane. As the particle slides down the side of the sphere, at what point will it leave?

4.22   A bead slides on a smooth rigid wire bent into the form of a circular loop of radius b. If the plane of the loop is vertical, and if the bead starts from rest at a point that is level with the center of the loop, find the speed of the bead at the bottom and the reaction of the wire on the bead at that point.

4.23   Show that the period of the particle sliding in the cycloidal trough of Example 4.6.2 is $4\pi(A/g)^{1/2}$.

Computer Problems

C 4.1   A bomber plane, about to drop a bomb, suffers a malfunction of its targeting computer. The pilot notes that there is a strong horizontal wind, so she decides to release the bomb anyway, directly over the visually sighted target, as the plane flies over it directly into the wind. She calculates the required *ground speed* of the aircraft for her flying altitude of 50,000 feet and realizes that there is no problem flying her craft at that speed. She is perfectly confident that the wind speed will offset the plane's speed and blow the bomb "backwards" onto the intended target. She adjusts the ground speed of the aircraft accordingly and informs the bombardier to release the bomb at the precise instant that the target appears directly below the plane in the crosshairs of his visual targeting device. Assume that the wind is blowing horizontally throughout the entire space below the plane with a speed of 60 mph, and that the air density does not vary with altitude. The bomb has a mass of 100 kg. Assume that it is spherical in shape with a radius of 0.2 m.

(a) Calculate the required *ground speed* of the plane if the bomb is to strike the target.

(b) Plot the trajectory of the bomb. Explain why the "trailing side" of the trajectory is linear.

(c) How precisely must the pilot control the speed of the plane if the bomb is to strike within 100 m of the target?

C 4.2 Assume that a projectile subject to a linear air resistance drag is fired with an initial speed v_0 equal to its terminal speed v_t at an elevation angle θ_0.

(a) Find parametric solutions $x(t)$ and $z(t)$ for the trajectory of the projectile in terms of the above parameters. Convert the solution to parametric equations of dimensionless variables $X(s)$, $Z(s)$, and s, where $X = (g/v_t^2)x$, $Z = (g/v_t^2)z$, and $s = (g/v_t)t$.

(b) Solve numerically the dimensionless parametric equations obtained above to find the angle θ_0 at which the projectile must be fired to achieve maximum range.

(c) Plot the trajectory of the missile corresponding to its maximum range, along with the trajectory that would occur under these same firing conditions but in the absence of air resistance. Use the above dimensionless parameters as plotting variables.

(d) Using the mass and dimensions of a baseball given in Example 4.3.2, calculate (i) the terminal velocity of the baseball for linear air drag and (ii) its maximum range when launched at this initial velocity and optimum elevation angle.

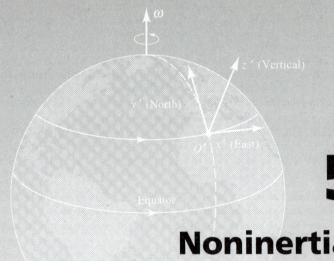

5

Noninertial Reference Systems

"I was sitting in a chair at the patent office in Bern, when all of a sudden a thought occurred to me: If a person falls freely, he will not feel his own weight. I was startled. This simple thought made a deep impression on me. It impelled me toward a theory of gravitation."

— Albert Einstein, *"The Happiest Thought of My Life"*; see A. Pais, *Inward Bound*, New York, Oxford Univ. Press, 1986

5.1 | Accelerated Coordinate Systems and Inertial Forces

In describing the motion of a particle, it is frequently convenient, and sometimes necessary, to employ a coordinate system that is not inertial. For example, a coordinate system fixed to the Earth is the most convenient one to describe the motion of a projectile, even though the Earth is accelerating and rotating.

We shall first consider the case of a coordinate system that undergoes pure translation. In Figure 5.1.1 $Oxyz$ are the primary coordinate axes (assumed fixed), and $O'x'y'z'$ are the moving axes. In the case of pure translation, the respective axes Ox and $O'x'$, and so on, remain parallel. The position vector of a particle P is denoted by $\mathbf{r}$ in the fixed system and by $\mathbf{r}'$ in the moving system. The displacement OO' of the moving origin is denoted by $\mathbf{R}_0$. Thus, from the triangle $OO'P$, we have

$$\mathbf{r} = \mathbf{R}_0 + \mathbf{r}' \tag{5.1.1}$$

Taking the first and second time derivatives gives

$$\mathbf{v} = \mathbf{V}_0 + \mathbf{v}' \tag{5.1.2}$$

$$\mathbf{a} = \mathbf{A}_0 + \mathbf{a}' \tag{5.1.3}$$

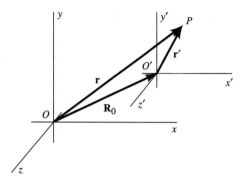

Figure 5.1.1 Relationship between the position vectors for two coordinate systems undergoing pure translation relative to each other.

in which $\mathbf{V}_0$ and $\mathbf{A}_0$ are, respectively, the velocity and acceleration of the moving system, and $\mathbf{v}'$ and $\mathbf{a}'$ are the velocity and acceleration of the particle *in* the moving system.

In particular, if the moving system is not accelerating, so that $\mathbf{A}_0 = 0$, then

$$\mathbf{a} = \mathbf{a}'$$

so the acceleration is the same in either system. Consequently, if the primary system is inertial, Newton's second law $\mathbf{F} = m\mathbf{a}$ becomes $\mathbf{F} = m\mathbf{a}'$ in the moving system; that is, the moving system is also an inertial system (provided it is not rotating). Thus, as far as Newtonian mechanics is concerned, we cannot specify a unique coordinate system; if Newton's laws hold in one system, they are also valid in any other system moving with uniform velocity relative to the first.

On the other hand if the moving system is accelerating, then Newton's second law becomes

$$\mathbf{F} = m\mathbf{A}_0 + m\mathbf{a}' \qquad (5.1.4a)$$

or

$$\mathbf{F} - m\mathbf{A}_0 = m\mathbf{a}' \qquad (5.1.4b)$$

for the equation of motion in the accelerating system. If we wish, we can write Equation 5.1.4b in the form

$$\mathbf{F}' = m\mathbf{a}' \qquad (5.1.5)$$

in which $\mathbf{F}' = \mathbf{F} + (-m\mathbf{A}_0)$. That is, an acceleration $\mathbf{A}_0$ of the reference system can be taken into account by adding an *inertial term* $-m\mathbf{A}_0$ to the force $\mathbf{F}$ and equating the result to the product of mass and acceleration in the moving system. Inertial terms in the equations of motion are sometimes called *inertial forces,* or *fictitious forces*. Such "forces" are not due to interactions with other bodies, rather, they stem from the acceleration of the reference system. Whether or not one wishes to call them forces is purely a matter of taste. In any case, inertial terms are present if a noninertial coordinate system is used to describe the motion of a particle.

EXAMPLE 5.1.1

A block of wood rests on a rough horizontal table. If the table is accelerated in a horizontal direction, under what conditions will the block slip?

Solution:

Let μ_s be the coefficient of static friction between the block and the table top. Then the force of friction $\mathbf{F}$ has a maximum value of $\mu_s mg$, where m is the mass of the block. The condition for slipping is that the inertial force $-m\mathbf{A}_0$ exceeds the frictional force, where $\mathbf{A}_0$ is the acceleration of the table. Hence, the condition for slipping is

$$|-m\mathbf{A}_0| > \mu_s m\mathbf{g}$$

or

$$A_0 > \mu_s g$$

EXAMPLE 5.1.2

A pendulum is suspended from the ceiling of a railroad car, as shown in Figure 5.1.2a. Assume that the car is accelerating uniformly toward the right ($+x$ direction). A noninertial observer, the boy inside the car, sees the pendulum hanging at an angle θ, left of vertical. He believes it hangs this way because of the existence of an inertial force $\mathbf{F}'_x$, which acts on all objects in his accelerated frame of reference (Figure 5.1.2b). An inertial observer, the girl outside the car, sees the same thing. She knows, however, that there is no real force $\mathbf{F}'_x$ acting on the pendulum. She knows that it hangs this way because a net force in the horizontal direction is required to accelerate it at the rate $\mathbf{A}_0$ that she observes (Figure 5.1.2c). Calculate the acceleration $\mathbf{A}_0$ of the car from the inertial observer's point of view. Show that, according to the noninertial observer, $\mathbf{F}'_x = -m\mathbf{A}_0$ is the force that causes the pendulum to hang at the angle θ.

Solution:

The inertial observer writes down Newton's second law for the hanging pendulum as

$$\sum \mathbf{F}_i = m\mathbf{a}$$

$$T \sin\theta = mA_0 \qquad T \cos\theta - mg = 0$$

$$\therefore A_0 = g \tan\theta$$

She concludes that the suspended pendulum hangs at the angle θ because the railroad car is accelerating in the horizontal direction and a horizontal force is needed to make it accelerate. This force is the x-component of the tension in the string. The acceleration of the car is proportional to the tangent of the angle of deflection. The pendulum, thus, serves as a linear accelerometer.

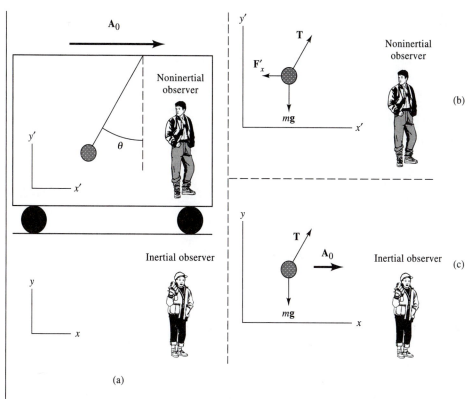

Figure 5.1.2   (a) Pendulum suspended in an accelerating railroad car as seen by (b) the noninertial observer and (c) the inertial observer.

On the other hand the noninertial observer, unaware of the outside world (assume the railroad track is perfectly smooth—no vibration—and that the railroad car has no windows or other sensory clues for another reference point), observes that the pendulum just hangs there, tilted to the left of vertical. He concludes that

$$\sum \mathbf{F}_i' = m\mathbf{a}' = 0$$

$$T\sin\theta - F_x' = 0 \qquad T\cos\theta - mg = 0$$

$$\therefore F_x' = mg \tan\theta$$

All the forces acting on the pendulum are in balance, and the pendulum hangs left of vertical due to the force $\mathbf{F}_x' (=-m\mathbf{A}_0)$. In fact if this observer were to do some more experiments in the railroad car, such as drop balls or stones or whatever, he would see that they would also be deflected to the left of vertical. He would soon discover that the amount of the deflection would be independent of their mass. In other words he would conclude that there was a force, quite like a gravitational one (to be discussed in Chapter 6), pushing things to the left of the car with an acceleration $\mathbf{A}_0$ as well as the force pulling them down with an acceleration $\mathbf{g}$.

EXAMPLE 5.1.3

Two astronauts are standing in a spaceship accelerating upward with an acceleration $\mathbf{A}_0$ as shown in Figure 5.1.3. Let the magnitude of $\mathbf{A}_0$ equal g. Astronaut #1 throws a ball directly toward astronaut #2, who is 10 m away on the other side of the ship. What must be the initial speed of the ball if it is to reach astronaut #2 before striking the floor? Assume astronaut #1 releases the ball at a height $h = 2$ m above the floor of the ship. Solve the problem from the perspective of both (a) a noninertial observer (inside the ship) and (b) an inertial observer (outside the ship).

Solution:

(a) The noninertial observer believes that a force $-m\mathbf{A}_0$ acts upon all objects in the ship. Thus, in the noninertial (x', y') frame of reference, we conclude that the trajectory of the ball is a parabola, that is,

$$x'(t) = \dot{x}_0' t \qquad\qquad y'(t) = y_0' - \tfrac{1}{2} A_0 t^2$$

$$\therefore y'(x') = y_0' - \frac{1}{2} A_0 \left(\frac{x'}{\dot{x}_0'} \right)^2$$

Setting $y'(x')$ equal to zero when $x' = 10$ m and solving for $\dot{x}_0'$ yields

$$\dot{x}_0' = \left(\frac{A_0}{2y_0'} \right)^{1/2} x'$$

$$= \left(\frac{9.8 \text{ ms}^{-2}}{4 \text{ m}} \right)^{1/2} (10 \text{ m}) = 15.6 \text{ ms}^{-1}$$

(b) The inertial observer sees the picture a little differently. It appears to him that the ball travels at constant velocity in a straight line after it is released and that the floor of the spaceship accelerates upward to intercept the ball. A plot of the vertical position of the ball and the floor of the spaceship is shown schematically in Figure 5.1.4. Both the ball and the rocket have the same initial upward speed $\dot{y}_0$ at the moment the ball is released by astronaut #1.

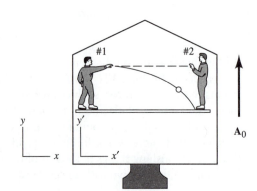

Figure 5.1.3   Two astronauts throwing a ball in a spaceship accelerating at $|\mathbf{A}_0| = |\mathbf{g}|$.

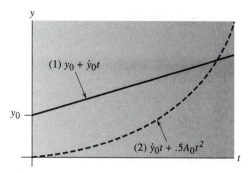

Figure 5.1.4 Vertical position of (1) a ball thrown in an accelerating rocket and (2) the floor of the rocket as seen by an inertial observer.

The vertical positions of the ball and the floor coincide at a time t that depends on the initial height of the ball

$$y_0 + \dot{y}_0 t = \dot{y}_0 t + \frac{1}{2} A_0 t^2$$

$$y_0 = \frac{1}{2} A_0 t^2$$

During this time t, the ball has traveled a horizontal distance x, where

$$x = \dot{x}_0 t \qquad \text{or} \qquad t = \frac{x}{\dot{x}_0}$$

Inserting this time into the relation for y_0 above yields the required initial horizontal speed of the ball

$$y_0 = \frac{1}{2} A_0 \left(\frac{x}{\dot{x}_0} \right)^2$$

$$\dot{x}_0 = \left(\frac{A_0}{2 y_0} \right)^{1/2} x$$

Thus, each observer calculates the same value for the initial horizontal velocity, as well they should.

The analysis seems less complex from the perspective of the noninertial observer. In fact the noninertial observer would physically experience the inertial force $-m\mathbf{A}_0$. It would seem every bit as real as the gravitational force we experience here on Earth. Our astronaut might even invent the concept of gravity to "explain" the dynamics of moving objects observed in the spaceship.

5.2 | Rotating Coordinate Systems

In the previous section, we showed how velocities, accelerations, and forces transform between an inertial frame of reference and a noninertial one that is accelerating at a constant rate. In this section and the following one, we show how these quantities transform between an inertial frame and a noninertial one that is rotating as well.

We start our discussion with the case of a primed coordinate system rotating with respect to an unprimed, fixed, inertial one. The axes of the coordinate systems have a common origin (see Figure 5.2.1). At any given instant the rotation of the primed system takes place about some specific axis of rotation, whose direction is designated by a unit vector, $\mathbf{n}$. The instantaneous angular speed of the rotation is designated by ω. The product, $\omega\mathbf{n}$, is the *angular velocity* of the rotating system

$$\boldsymbol{\omega} = \omega\mathbf{n} \tag{5.2.1}$$

The sense direction of the angular velocity vector is given by the right-hand rule (see Figure 5.2.1), as in the definition of the cross product.

The position of any point P in space can be designated by the vector $\mathbf{r}$ in the fixed, unprimed system and by the vector $\mathbf{r}'$ in the rotating, primed system (see Figure 5.2.2).

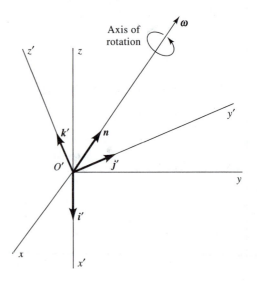

Figure 5.2.1   The angular velocity vector of a rotating coordinate system.

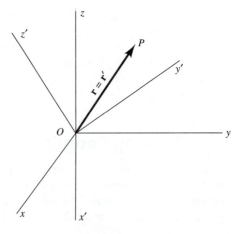

Figure 5.2.2   Rotating coordinate system (primed system).

Because the coordinate axes of the two systems have the same origin, these vectors are equal, that is,

$$\mathbf{r} = \mathbf{i}x + \mathbf{j}y + \mathbf{k}z = \mathbf{r}' = \mathbf{i}'x' + \mathbf{j}'y' + \mathbf{k}'z' \qquad (5.2.2)$$

When we differentiate with respect to time to find the velocity, we must keep in mind the fact that the unit vectors $\mathbf{i}'$, $\mathbf{j}'$, and $\mathbf{k}'$ in the rotating system are *not* constant, whereas the primary unit vectors $\mathbf{i}$, $\mathbf{j}$, and $\mathbf{k}$ are. Thus, we can write

$$\mathbf{i}\frac{dx}{dt} + \mathbf{j}\frac{dy}{dt} + \mathbf{k}\frac{dz}{dt} = \mathbf{i}'\frac{dx'}{dt} + \mathbf{j}'\frac{dy'}{dt} + \mathbf{k}'\frac{dz'}{dt} + x'\frac{d\mathbf{i}'}{dt} + y'\frac{d\mathbf{j}'}{dt} + z'\frac{d\mathbf{k}'}{dt} \qquad (5.2.3)$$

The three terms on the left-hand side of the preceding equation clearly give the velocity vector $\mathbf{v}$ in the fixed system, and the first three terms on the right are the components of the velocity *in* the rotating system, which we shall call $\mathbf{v}'$, so the equation may be written

$$\mathbf{v} = \mathbf{v}' + x'\frac{d\mathbf{i}'}{dt} + y'\frac{d\mathbf{j}'}{dt} + z'\frac{d\mathbf{k}'}{dt} \qquad (5.2.4)$$

The last three terms on the right represent the velocity due to rotation of the primed coordinate system. We must now determine how the time derivatives of the basis vectors are related to the rotation.

To find the time derivatives $d\mathbf{i}'/dt$, $d\mathbf{j}'/dt$, and $d\mathbf{k}'/dt$, consider Figure 5.2.3. Here is shown the change $\Delta\mathbf{i}'$ in the unit vector $\mathbf{i}'$ due to a small rotation $\Delta\theta$ about the axis of rotation. (The vectors $\mathbf{j}'$ and $\mathbf{k}'$ are omitted for clarity.) From the figure we see that the magnitude of $\Delta\mathbf{i}'$ is given by the approximate relation

$$|\Delta\mathbf{i}'| \approx (|\mathbf{i}'|\sin\phi)\Delta\theta = (\sin\phi)\Delta\theta$$

where ϕ is the angle between $\mathbf{i}'$ and $\boldsymbol{\omega}$. Let Δt be the time interval for this change. Then we can write

$$\left|\frac{d\mathbf{i}'}{dt}\right| = \lim_{\Delta t \to 0}\left|\frac{\Delta\mathbf{i}'}{\Delta t}\right| = \sin\phi\frac{d\theta}{dt} = (\sin\phi)\omega \qquad (5.2.5)$$

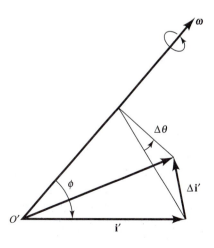

Figure 5.2.3   Change in the unit vector $\mathbf{i}'$ produced by a small rotation $\Delta\theta$.

Now the direction of $\Delta \mathbf{i}'$ is perpendicular to *both* $\boldsymbol{\omega}$ and $\mathbf{i}'$; consequently, from the definition of the cross product, we can write Equation 5.2.5 in vector form

$$\frac{d\mathbf{i}'}{dt} = \boldsymbol{\omega} \times \mathbf{i}' \qquad (5.2.6)$$

Similarly, we find $d\mathbf{j}'/dt = \boldsymbol{\omega} \times \mathbf{j}'$, and $d\mathbf{k}'/dt = \boldsymbol{\omega} \times \mathbf{k}'$.

We now apply the preceding result to the last three terms in Equation 5.2.4 as follows:

$$x'\frac{d\mathbf{i}'}{dt} + y'\frac{d\mathbf{j}'}{dt} + z'\frac{d\mathbf{k}'}{dt} = x'(\boldsymbol{\omega} \times \mathbf{i}') + y'(\boldsymbol{\omega} \times \mathbf{j}') + z'(\boldsymbol{\omega} \times \mathbf{k}')$$

$$= \boldsymbol{\omega} \times (\mathbf{i}'x' + \mathbf{j}'y' + \mathbf{k}'z') \qquad (5.2.7)$$

$$= \boldsymbol{\omega} \times \mathbf{r}'$$

This is the velocity of P due to rotation of the primed coordinate system. Accordingly, Equation 5.2.4 can be shortened to read

$$\mathbf{v} = \mathbf{v}' + \boldsymbol{\omega} \times \mathbf{r}' \qquad (5.2.8)$$

or, more explicitly

$$\left(\frac{d\mathbf{r}}{dt}\right)_{fixed} = \left(\frac{d\mathbf{r}'}{dt}\right)_{rot} + \boldsymbol{\omega} \times \mathbf{r}' = \left[\left(\frac{d}{dt}\right)_{rot} + \boldsymbol{\omega} \times\right]\mathbf{r}' \qquad (5.2.9)$$

that is, the operation of differentiating the position vector with respect to time in the fixed system is equivalent to the operation of taking the time derivative in the rotating system plus the operation $\boldsymbol{\omega} \times$. A little reflection shows that the same applies to *any* vector $\mathbf{Q}$, that is,

$$\left(\frac{d\mathbf{Q}}{dt}\right)_{fixed} = \left(\frac{d\mathbf{Q}}{dt}\right)_{rot} + \boldsymbol{\omega} \times \mathbf{Q} \qquad (5.2.10a)$$

In particular, if that vector is the velocity, then we have

$$\left(\frac{d\mathbf{v}}{dt}\right)_{fixed} = \left(\frac{d\mathbf{v}}{dt}\right)_{rot} + \boldsymbol{\omega} \times \mathbf{v} \qquad (5.2.10b)$$

But $\mathbf{v} = \mathbf{v}' + \boldsymbol{\omega} \times \mathbf{r}'$, so

$$\left(\frac{d\mathbf{v}}{dt}\right)_{fixed} = \left(\frac{d}{dt}\right)_{rot}(\mathbf{v}' + \boldsymbol{\omega} \times \mathbf{r}') + \boldsymbol{\omega} \times (\mathbf{v}' + \boldsymbol{\omega} \times \mathbf{r}')$$

$$= \left(\frac{d\mathbf{v}'}{dt}\right)_{rot} + \left[\frac{d(\boldsymbol{\omega} \times \mathbf{r}')}{dt}\right]_{rot} + \boldsymbol{\omega} \times \mathbf{v}' + \boldsymbol{\omega} \times (\boldsymbol{\omega} \times \mathbf{r}') \qquad (5.2.11)$$

$$= \left(\frac{d\mathbf{v}'}{dt}\right)_{rot} + \left(\frac{d\boldsymbol{\omega}}{dt}\right)_{rot} \times \mathbf{r}' + \boldsymbol{\omega} \times \left(\frac{d\mathbf{r}'}{dt}\right)_{rot}$$

$$+ \boldsymbol{\omega} \times \mathbf{v}' + \boldsymbol{\omega} \times (\boldsymbol{\omega} \times \mathbf{r}')$$

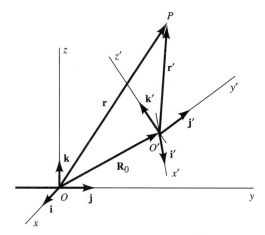

Figure 5.2.4 Geometry for the general case of translation and rotation of the moving coordinate system (primed system).

Now concerning the term involving the time derivative of $\boldsymbol{\omega}$, we have $(d\boldsymbol{\omega}/dt)_{fixed} = (d\boldsymbol{\omega}/dt)_{rot} + \boldsymbol{\omega} \times \boldsymbol{\omega}$. But the cross product of any vector with itself vanishes, so $(d\boldsymbol{\omega}/dt)_{fixed} = (d\boldsymbol{\omega}/dt)_{rot} = \dot{\boldsymbol{\omega}}$. Because $\mathbf{v}' = (d\mathbf{r}'/dt)_{rot}$ and $\mathbf{a}' = (d\mathbf{v}'/dt)_{rot}$, we can express the final result as follows:

$$\mathbf{a} = \mathbf{a}' + \dot{\boldsymbol{\omega}} \times \mathbf{r}' + 2\boldsymbol{\omega} \times \mathbf{v}' + \boldsymbol{\omega} \times (\boldsymbol{\omega} \times \mathbf{r}') \tag{5.2.12}$$

giving the acceleration in the fixed system in terms of the position, velocity, and acceleration in the rotating system.

In the general case in which the primed system is undergoing *both* translation and rotation (Figure 5.2.4), we must add the velocity of translation $\mathbf{V}_0$ to the right-hand side of Equation 5.2.8 and the acceleration $\mathbf{A}_0$ of the moving system to the right-hand side of Equation 5.2.12. This gives the general equations for transforming from a fixed system to a moving and rotating system:

$$\mathbf{v} = \mathbf{v}' + \boldsymbol{\omega} \times \mathbf{r}' + \mathbf{V}_0 \tag{5.2.13}$$

$$\mathbf{a} = \mathbf{a}' + \dot{\boldsymbol{\omega}} \times \mathbf{r}' + 2\boldsymbol{\omega} \times \mathbf{v}' + \boldsymbol{\omega} \times (\boldsymbol{\omega} \times \mathbf{r}') + \mathbf{A}_0 \tag{5.2.14}$$

The term $2\boldsymbol{\omega} \times \mathbf{v}'$ is known as the *Coriolis acceleration,* and the term $\boldsymbol{\omega} \times (\boldsymbol{\omega} \times \mathbf{r}')$ is called the *centripetal acceleration.* The Coriolis acceleration appears whenever a particle moves in a rotating coordinate system (except when the velocity $\mathbf{v}'$ is parallel to the axis of rotation), and the centripetal acceleration is the result of the particle being carried around a circular path in the rotating system. The centripetal acceleration is always directed toward the axis of rotation and is perpendicular to the axis as shown in Figure 5.2.5. The term $\dot{\boldsymbol{\omega}} \times \mathbf{r}'$ is called the *transverse acceleration,* because it is perpendicular to the position vector $\mathbf{r}'$. It appears as a result of any angular acceleration of the rotating system, that is, if the angular velocity vector is changing in either magnitude or direction, or both.

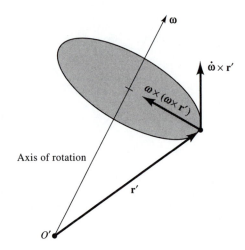

Figure 5.2.5 Illustrating the centripetal acceleration.

EXAMPLE 5.2.1

A wheel of radius b rolls along the ground with constant forward speed V_0. Find the acceleration, relative to the ground, of any point on the rim.

Solution:

Let us choose a coordinate system fixed to the rotating wheel, and let the moving origin be at the center with the x'-axis passing through the point in question, as shown in Figure 5.2.6. Then we have

$$\mathbf{r}' = \mathbf{i}'b \qquad \mathbf{a}' = \ddot{\mathbf{r}}' = 0 \qquad \mathbf{v}' = \dot{\mathbf{r}}' = 0$$

The angular velocity vector is given by

$$\boldsymbol{\omega} = \mathbf{k}'\omega = \mathbf{k}'\frac{V_0}{b}$$

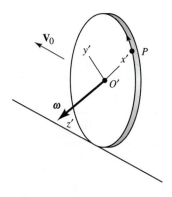

Figure 5.2.6 Rotating coordinates fixed to a rolling wheel.

for the choice of coordinates shown; therefore, all terms in the expression for acceleration vanish except the centripetal term:

$$\mathbf{a} = \boldsymbol{\omega} \times (\boldsymbol{\omega} \times \mathbf{r}') = \mathbf{k}' \boldsymbol{\omega} \times (\mathbf{k}' \boldsymbol{\omega} \times \mathbf{i}' b)$$

$$= \frac{V_0^2}{b} \mathbf{k}' \times (\mathbf{k}' \times \mathbf{i}')$$

$$= \frac{V_0^2}{b} \mathbf{k}' \times \mathbf{j}'$$

$$= \frac{V_0^2}{b} (-\mathbf{i}')$$

Thus, $\mathbf{a}$ is of magnitude V_0^2/b and is always directed toward the center of the rolling wheel.

EXAMPLE 5.2.2

A bicycle travels with constant speed around a track of radius ρ. What is the acceleration of the highest point on one of its wheels? Let V_0 denote the speed of the bicycle and b the radius of the wheel.

Solution:

We choose a coordinate system with origin at the center of the wheel and with the x'-axis horizontal pointing toward the center of curvature C of the track. Rather than have the moving coordinate system rotate with the wheel, we choose a system in which the z'-axis remains vertical as shown in Figure 5.2.7. Thus, the $O'x'y'z'$ system rotates

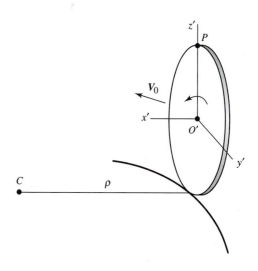

Figure 5.2.7 Wheel rolling on a curved track. The z'-axis remains vertical as the wheel turns.

with angular velocity $\boldsymbol{\omega}$, which can be expressed as

$$\boldsymbol{\omega} = \mathbf{k}'\frac{V_0}{\rho}$$

and the acceleration of the moving origin $\mathbf{A}_0$ is given by

$$\mathbf{A}_0 = \mathbf{i}'\frac{V_0^2}{\rho}$$

Because each point on the wheel is moving in a circle of radius b with respect to the moving origin, the acceleration in the $O'x'y'z'$ system of any point on the wheel is directed toward O' and has magnitude V_0^2/b. Thus, in the moving system we have

$$\ddot{\mathbf{r}}' = -\mathbf{k}'\frac{V_0^2}{b}$$

for the point at the top of the wheel. Also, the velocity of this point in the moving system is given by

$$\mathbf{v}' = -\mathbf{j}'V_0$$

so the Coriolis acceleration is

$$2\boldsymbol{\omega} \times \mathbf{v}' = 2\left(\frac{V_0}{\rho}\mathbf{k}'\right) \times (-\mathbf{j}'V_0) = 2\frac{V_0^2}{\rho}\mathbf{i}'$$

Because the angular velocity $\boldsymbol{\omega}$ is constant, the transverse acceleration is zero. The centripetal acceleration is also zero because

$$\boldsymbol{\omega} \times (\boldsymbol{\omega} \times \mathbf{r}') = \frac{V_0^2}{\rho^2}\mathbf{k}' \times (\mathbf{k}' \times b\mathbf{k}') = 0$$

Thus, the net acceleration, relative to the ground, of the highest point on the wheel is

$$\mathbf{a} = 3\frac{V_0^2}{\rho}\mathbf{i}' - \frac{V_0^2}{b}\mathbf{k}'$$

5.3 | Dynamics of a Particle in a Rotating Coordinate System

The fundamental equation of motion of a particle in an inertial frame of reference is

$$\mathbf{F} = m\mathbf{a} \tag{5.3.1}$$

where $\mathbf{F}$ is the vector sum of all real, physical forces acting on the particle. In view of Equation 5.2.14, we can write the equation of motion in a noninertial frame of reference as

$$\mathbf{F} - m\mathbf{A}_0 - 2m\boldsymbol{\omega} \times \mathbf{v}' - m\dot{\boldsymbol{\omega}} \times \mathbf{r}' - m\boldsymbol{\omega} \times (\boldsymbol{\omega} \times \mathbf{r}') = m\mathbf{a}' \tag{5.3.2}$$

All the terms from Equation 5.2.14, except $\mathbf{a}'$, have been multiplied by m and transposed to show them as inertial forces added to the real, physical forces $\mathbf{F}$. The $\mathbf{a}'$ term has been multiplied by m also, but kept on the right-hand side. Thus, Equation 5.3.2 represents the dynamical equation of motion of a particle in a noninertial frame of reference subjected to both real, physical forces as well as those inertial forces that appear as a result of the acceleration of the noninertial frame of reference. The inertial forces have names corresponding to their respective accelerations, discussed in Section 5.2. The *Coriolis force* is

$$\mathbf{F}'_{Cor} = -2m\boldsymbol{\omega} \times \mathbf{v}' \tag{5.3.3}$$

The *transverse force* is

$$\mathbf{F}'_{trans} = -m\dot{\boldsymbol{\omega}} \times \mathbf{r}' \tag{5.3.4}$$

The *centrifugal force* is

$$\mathbf{F}'_{centrif} = -m\boldsymbol{\omega} \times (\boldsymbol{\omega} \times \mathbf{r}') \tag{5.3.5}$$

The remaining inertial force $-m\mathbf{A}_0$ appears whenever the (x', y', z') coordinate system is undergoing a translational acceleration, as discussed in Section 5.1.

A noninertial observer in an accelerated frame of reference who denotes the acceleration of a particle by the vector $\mathbf{a}'$ is forced to include any or all of these inertial forces along with the real forces to calculate the correct motion of the particle. In other words, such an observer writes the fundamental equation of motion as

$$\mathbf{F}' = m\mathbf{a}'$$

in which the sum of the vector forces $\mathbf{F}'$ acting on the particle is given by

$$\mathbf{F}' = \mathbf{F}_{physical} + \mathbf{F}'_{Cor} + \mathbf{F}'_{trans} + \mathbf{F}'_{centrif} - m\mathbf{A}_0$$

We have emphasized the real, physical nature of the force term $\mathbf{F}$ in Equation 5.3.2 by appending the subscript *physical* to it here. $\mathbf{F}$ (or $\mathbf{F}_{physical}$) forces are the only forces that a noninertial observer claims are actually acting upon the particle. The inclusion of the remaining four inertial terms depends critically on the exact status of the noninertial frame of reference being used to describe the motion of the particle. They arise because of the inertial property of the matter whose motion is under investigation, rather than from the presence or action of any surrounding matter.

The Coriolis force is particularly interesting. It is present only if a particle is *moving* in a rotating coordinate system. Its direction is always perpendicular to the velocity vector of the particle in the moving system. The Coriolis force thus seems to deflect a moving particle at right angles to its direction of motion. (The Coriolis force has been rather fancifully called "the merry-go-round force." Try walking radially inward or outward on a moving merry-go-round to experience its effect.) This force is important in computing the trajectory of a projectile. Coriolis effects are responsible for the circulation of air around high- or low-pressure systems on Earth's surface. In the case of a high-pressure area,[1] as air spills down from the high, it flows outward and away, deflecting toward the right as it moves into the surrounding low, setting up a clockwise circulation pattern. In the Southern Hemisphere the reverse is true.

---

[1] A high-pressure system is essentially a bump in Earth's atmosphere where more air is stacked up above some region on Earth's surface than it is for surrounding regions.

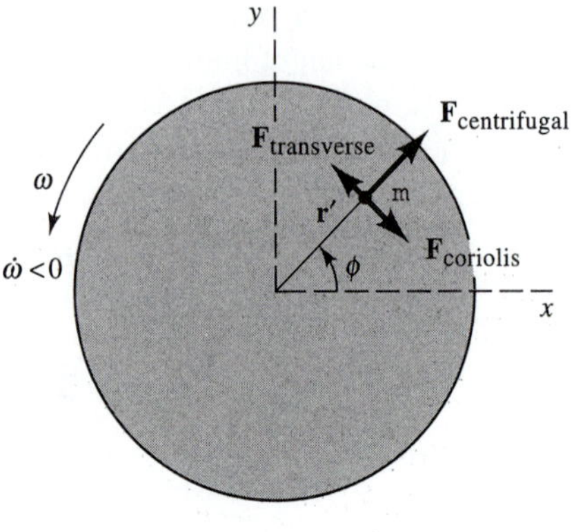

Figure 5.3.1   Inertial forces acting on a mass m moving radially outward on a platform rotating with angular velocity $\boldsymbol{\omega}$ and angular acceleration $\dot{\boldsymbol{\omega}} < 0$. The xy-axes are fixed. The direction of $\boldsymbol{\omega}$ is out of the paper.

The transverse force is present only if there is an angular acceleration (or deceleration) of the rotating coordinate system. This force is always perpendicular to the radius vector $\mathbf{r}'$ in the rotating coordinate system.

The centrifugal force is the familiar one that arises from rotation about an axis. It is directed outward away from the axis of rotation and is perpendicular to that axis. These three inertial forces are illustrated in Figure 5.3.1 for the case of a mass m moving radially outward on a rotating platform, whose rate of rotation is decreasing ($\dot{\omega} < 0$). The z-axis is the axis of rotation, directed out of the paper. That is also the direction of the angular velocity vector $\boldsymbol{\omega}$. Because $\mathbf{r}'$, the radius vector denoting the position of m in the rotating system, is perpendicular to $\boldsymbol{\omega}$, the magnitude of the centrifugal force is $mr'\omega^2$. In general if the angle between $\boldsymbol{\omega}$ and $\mathbf{r}'$ is θ, then the magnitude of the centripetal force is $mr'\omega^2 \sin\theta$ where $r' \sin\theta$ is the shortest distance from the mass to the axis of rotation.

EXAMPLE 5.3.1

A bug crawls outward with a constant speed v' along the spoke of a wheel that is rotating with constant angular velocity $\boldsymbol{\omega}$ about a vertical axis. Find all the apparent forces acting on the bug (see Figure 5.3.2).

Solution:

First, let us choose a coordinate system fixed on the wheel, and let the x'-axis point along the spoke in question. Then we have

$$\dot{\mathbf{r}}' = \mathbf{i}'x' = \mathbf{i}'v'$$

$$\ddot{\mathbf{r}}' = 0$$

for the velocity and acceleration of the bug as described in the rotating system. If we choose the z'-axis to be vertical, then

$$\boldsymbol{\omega} = \mathbf{k}'\omega$$

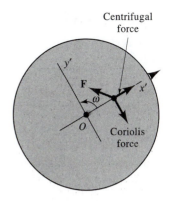

Figure 5.3.2   Forces on an insect crawling outward along a radial line on a rotating wheel.

The various forces are then given by the following:

$$-2m\boldsymbol{\omega} \times \dot{\mathbf{r}}' = -2m\omega v'(\mathbf{k}' \times \mathbf{i}') = -2m\omega v'\mathbf{j}' \qquad \textit{Coriolis force}$$

$$-m\dot{\boldsymbol{\omega}} \times \mathbf{r}' = 0 \qquad (\omega = \text{constant}) \qquad \textit{transverse force}$$

$$-m\boldsymbol{\omega} \times (\boldsymbol{\omega} \times \mathbf{r}') = -m\omega^2[\mathbf{k}' \times (\mathbf{k}' \times \mathbf{i}'x')] \qquad \textit{centrifugal force}$$

$$= -m\omega^2(\mathbf{k}' \times \mathbf{j}'x')$$

$$= m\omega^2 x'\mathbf{i}'$$

Thus, Equation 5.3.2 reads

$$\mathbf{F} - 2m\omega v'\mathbf{j}' + m\omega^2 x'\mathbf{i}' = 0$$

Here **F** is the real force exerted on the bug by the spoke. The forces are shown in Figure 5.3.2.

EXAMPLE 5.3.2

In Example 5.3.1, find how far the bug can crawl before it starts to slip, given the coefficient of static friction μ_s between the bug and the spoke.

Solution:

Because the force of friction **F** has a maximum value of $\mu_s mg$, slipping starts when

$$|\mathbf{F}| = \mu_s mg$$

or

$$[(2m\omega v')^2 + (m\omega^2 x')^2]^{1/2} = \mu_s mg$$

On solving for x', we find

$$x' = \frac{[\mu_s^2 g^2 - 4\omega^2(v')^2]^{1/2}}{\omega^2}$$

for the distance the bug can crawl before slipping.

EXAMPLE 5.3.3

A smooth rod of length l rotates in a plane with a constant angular velocity $\boldsymbol{\omega}$ about an axis fixed at the end of the rod and perpendicular to the plane of rotation. A bead of mass m is initially positioned at the stationary end of the rod and given a slight push such that its initial speed directed down the rod is $\epsilon = \omega l$ (see Figure 5.3.3). Calculate how long it takes for the bead to reach the other end of the rod.

Solution:

The best way to solve this problem is to examine it from the perspective of an (x', y') frame of reference rotating with the rod. If we let the x'-axis lie along the rod, then the problem is one-dimensional along that direction. The only real force acting on the bead is $\mathbf{F}$, the reaction force that the rod exerts on the bead. It points perpendicular to the rod, along the y'-direction as shown in Figure 5.3.3. $\mathbf{F}$ has no x'-component because there is no friction. Thus, applying Equation 5.3.2 to the bead in this rotating frame, we obtain

$$F\mathbf{j}' - 2m\omega\mathbf{k}' \times \dot{x}'\mathbf{i}' - m\omega\mathbf{k}' \times (\omega\mathbf{k}' \times x'\mathbf{i}') = m\ddot{x}'\mathbf{i}'$$

$$F\mathbf{j} - 2m\omega\dot{x}'\mathbf{j}' + m\omega^2 x'\mathbf{i}' = m\ddot{x}'\mathbf{i}'$$

The first inertial force in the preceding equation is the Coriolis force. It appears in the expression because of the bead's velocity $\dot{x}'\,\mathbf{i}'$ along the x'-axis in the rotating frame. Note that it balances out the reaction force $\mathbf{F}$ that the rod exerts on the bead. The second inertial force is the centrifugal force, $m\omega^2 x'$. From the bead's perspective, this force shoves it down the rod. These ideas are embodied in the two scalar equivalents of the above vector equation

$$F = 2m\omega\dot{x}' \qquad m\omega^2 x' = m\ddot{x}'$$

Solving the second equation above yields $x'(t)$, the position of the bead along the rod as a function of time

$$x'(t) = Ae^{\omega t} + Be^{-\omega t}$$

$$\dot{x}'(t) = \omega Ae^{\omega t} - \omega Be^{-\omega t}$$

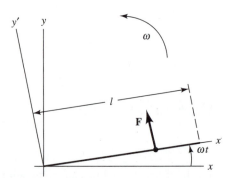

Figure 5.3.3   Bead sliding along a smooth rod rotating at constant angular velocity ω about an axis fixed at one end.

The boundary conditions, $x'(t=0) = 0$ and $\dot{x}'(t=0) = \epsilon$, allow us to determine the constants A and B

$$x'(0) = 0 = A + B \qquad \dot{x}'(0) = \epsilon = \omega(A - B)$$

$$A = -B = \frac{\epsilon}{2\omega}$$

which lead to the explicit solution

$$x'(t) = \frac{\epsilon}{2\omega}(e^{\omega t} - e^{-\omega t})$$

$$= \frac{\epsilon}{\omega} \sinh \omega t$$

The bead flies off the end of the rod at time T, where

$$x'(T) = \frac{\epsilon}{\omega} \sinh \omega T = l$$

$$T = \frac{1}{\omega} \sinh^{-1}\left(\frac{\omega l}{\epsilon}\right)$$

Because the initial speed of the bead is $\epsilon = \omega l$, the preceding equation becomes

$$T = \frac{1}{\omega} \sinh^{-1}(1) = \frac{0.88}{\omega}$$

5.4 | Effects of Earth's Rotation

Let us apply the theory developed in the foregoing sections to a coordinate system that is moving with the Earth. Because the angular speed of Earth's rotation is 2π radians per day, or about 7.27×10^{-5} radians per second, we might expect the effects of such rotation to be relatively small. Nevertheless, it is the spin of the Earth that produces the equatorial bulge; the equatorial radius is some 13 miles greater than the polar radius.

Static Effects: The Plumb Line

Let us consider the case of a plumb bob that is normally used to define the direction of the local "vertical" on the surface of the Earth. We discover that the plumb bob hangs perpendicular to the local surface (discounting bumps and surface irregularities). Because of the Earth's rotation, however, it does not point toward the center of the Earth unless it is suspended somewhere along the equator or just above one of the poles. Let us describe the motion of the plumb bob in a local frame of reference whose origin is at the position of the bob. Our frame of reference is attached to the surface of the Earth. It is undergoing translation as well as rotation. The translation of the frame takes place along a circle whose radius is $\rho = r_e \cos \lambda$, where r_e is the radius of the Earth and λ is the geocentric latitude of the plumb bob (see Figure 5.4.1).

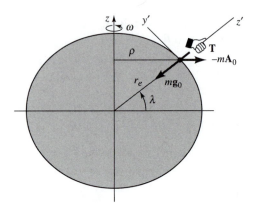

Figure 5.4.1 Gravitational force $m\mathbf{g}_0$, inertial force $-m\mathbf{A}_0$, and tension $\mathbf{T}$ acting on a plumb bob hanging just above the surface of the Earth at latitude λ.

Its rate of rotation is $\boldsymbol{\omega}$, the same as that of the Earth about its axis. Let us now examine the terms that make up Equation 5.3.2. The acceleration of the bob $\mathbf{a}'$ is zero; the bob is at rest in the local frame of reference. The centrifugal force on the bob relative to our local frame is zero because $\mathbf{r}'$ is zero; the origin of the local coordinate system is centered on the bob. The transverse force is zero because $\dot{\boldsymbol{\omega}} = 0$; the rotation of the Earth is constant. The Coriolis force is zero because $\mathbf{v}'$, the velocity of the plumb bob, is zero; the plumb bob is at rest in the local frame. The only surviving terms in Equation 5.3.2 are the real forces $\mathbf{F}$ and the inertial term $-m\mathbf{A}_0$, which arises because the local frame of reference is accelerating. Thus,

$$\mathbf{F} - m\mathbf{A}_0 = 0 \tag{5.4.1}$$

The rotation of the Earth causes the acceleration of the local frame. In fact, the situation under investigation here is entirely analogous to that of Example 5.1.2—the linear accelerometer. There, the pendulum bob did not hang vertically because it experienced an inertial force directed opposite to the acceleration of the railroad car. The case here is almost completely identical. The bob does not hang on a line pointing toward the center of the Earth because the inertial force $-m\mathbf{A}_0$ throws it outward, away from Earth's axis of rotation. This force, like the one of Example 5.1.2, is also directed opposite to the acceleration of the local frame of reference. It arises from the centripetal acceleration of the local frame toward Earth's axis. The magnitude of this force is $m\omega^2 r_e \cos \lambda$. It is a maximum when $\lambda = 0$ at the Earth's equator and a minimum at either pole when $\lambda = \pm 90°$. It is instructive to compare the value of the acceleration portion of this term, $A_0 = \omega^2 r_e \cos \lambda$, to g, the acceleration due to gravity. At the equator, it is $3.4 \times 10^{-3} g$ or less than 1% of g.

$\mathbf{F}$ is the vector sum of all real, physical forces acting on the plumb bob. All forces, including the inertial force $-m\mathbf{A}_0$, are shown in the vector diagram of Figure 5.4.2a. The tension $\mathbf{T}$ in the string balances out the real gravitational force $m\mathbf{g}_0$ and the inertial force $-m\mathbf{A}_0$. In other words

$$(\mathbf{T} + m\mathbf{g}_0) - m\mathbf{A}_0 = 0 \tag{5.4.2}$$

Now, when we hang a plumb bob, we normally think that the tension $\mathbf{T}$ balances out the local force of gravity, which we call $m\mathbf{g}$. We can see from Equation 5.4.2 and Figure 5.4.2b

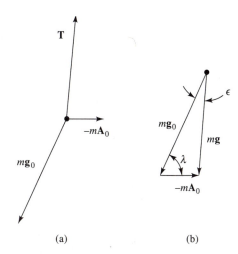

Figure 5.4.2 (a) Forces acting on a plumb
bob at latitude λ. (b) Forces defining the
weight of the plumb bob, $m\mathbf{g}$.

that $m\mathbf{g}$ is actually the vector sum of the real gravitational force $m\mathbf{g}_0$ and the inertial
force $-m\mathbf{A}_0$. Thus,

$$m\mathbf{g}_0 - m\mathbf{g} - m\mathbf{A}_0 = 0 \qquad \therefore \ \mathbf{g} = \mathbf{g}_0 - \mathbf{A}_0 \tag{5.4.3}$$

As can be seen from Figure 5.4.2b, the local acceleration $\mathbf{g}$ due to gravity contains a term
$\mathbf{A}_0$ due to the rotation of the Earth. The force $m\mathbf{g}_0$ is the true force of gravity and is
directed toward the center of the Earth. The inertial reaction $-m\mathbf{A}_0$, directed away from
Earth's axis, causes the direction of the plumb line to deviate by a small angle ϵ away from
the direction toward Earth's center. The plumb line direction defines the local direction
of the vector $\mathbf{g}$. The shape of the Earth is also defined by the direction of $\mathbf{g}$. Hence, the
plumb line is always perpendicular to Earth's surface, which is not shaped like a true
sphere but is flattened at the poles and bulged outward at the equator as depicted in
Figure 5.4.1.

We can easily calculate the value of the angle ϵ. It is a function of the geocentric latitude of the plumb bob. Applying the law of sines to Figure 5.4.2b, we have

$$\frac{\sin \epsilon}{m\omega^2 r_e \cos \lambda} = \frac{\sin \lambda}{mg} \tag{5.4.4a}$$

or, because ϵ is small

$$\sin \epsilon \approx \epsilon = \frac{\omega^2 r_e}{g} \cos \lambda \, \sin \lambda = \frac{\omega^2 r_e}{2g} \sin 2\lambda \tag{5.4.4b}$$

Thus, ϵ vanishes at the equator ($\lambda = 0$) and the poles ($\lambda = \pm 90°$) as we have already surmised. The maximum deviation of the direction of the plumb line from the center of the
Earth occurs at $\lambda = 45°$ where

$$\epsilon_{max} = \frac{\omega^2 r_e}{2g} \approx 1.7 \times 10^{-3} \ \text{radian} \approx 0.1° \tag{5.4.4c}$$

In this analysis, we have assumed that the real gravitational force $m\mathbf{g}_0$ is constant and directed toward the center of the Earth. This is not valid, because the Earth is not a true sphere. Its cross section is approximately elliptical as we indicated in Figure 5.4.1; therefore, $\mathbf{g}_0$ varies with latitude. Moreover, local mineral deposits, mountains, and so on, affect the value of $\mathbf{g}_0$. Clearly, calculating the shape of the Earth (essentially, the angle ϵ as a function of λ) is difficult. A more accurate solution can only be obtained numerically. The corrections to the preceding analysis are small.

Dynamic Effects: Motion of a Projectile

The equation of motion for a projectile near the Earth's surface (Equation 5.3.2) can be written

$$m\ddot{\mathbf{r}}' = \mathbf{F} + m\mathbf{g}_0 - m\mathbf{A}_0 - 2m\boldsymbol{\omega} \times \dot{\mathbf{r}}' - m\boldsymbol{\omega} \times (\boldsymbol{\omega} \times \mathbf{r}') \tag{5.4.5}$$

where $\mathbf{F}$ represents any applied forces other than gravity. From the static case considered above, however, the combination $m\mathbf{g}_0 - m\mathbf{A}_0$ is called $m\mathbf{g}$; hence, we can write the equation of motion as

$$m\ddot{\mathbf{r}}' = \mathbf{F} + m\mathbf{g} - 2m\boldsymbol{\omega} \times \dot{\mathbf{r}}' - m\boldsymbol{\omega} \times (\boldsymbol{\omega} \times \mathbf{r}') \tag{5.4.6}$$

Let us consider the motion of a projectile. If we ignore air resistance, then $\mathbf{F} = 0$. Furthermore, the term $-m\boldsymbol{\omega} \times (\boldsymbol{\omega} \times \mathbf{r}')$ is very small compared with the other terms, so we can ignore it. The equation of motion then reduces to

$$m\ddot{\mathbf{r}}' = m\mathbf{g} - 2m\boldsymbol{\omega} \times \dot{\mathbf{r}}' \tag{5.4.7}$$

in which the last term is the Coriolis force.

To solve the preceding equation we choose the directions of the coordinate axes $O'x'y'z'$ such that the z'-axis is vertical (in the direction of the plumb line), the x'-axis is to the east, and the y'-axis points north (Figure 5.4.3). With this choice of axes, we have

$$\mathbf{g} = -\mathbf{k}'g \tag{5.4.8}$$

The components of $\boldsymbol{\omega}$ in the primed system are

$$\omega_{x'} = 0 \qquad \omega_{y'} = \omega \cos \lambda \qquad \omega_{z'} = \omega \sin \lambda \tag{5.4.9}$$

The cross product is, therefore, given by

$$\boldsymbol{\omega} \times \dot{\mathbf{r}}' = \begin{vmatrix} \mathbf{i}' & \mathbf{j}' & \mathbf{k}' \\ \omega_{x'} & \omega_{y'} & \omega_{z'} \\ \dot{x}' & \dot{y}' & \dot{z}' \end{vmatrix} \tag{5.4.10}$$

$$= \mathbf{i}'(\omega \dot{z}' \cos \lambda - \omega \dot{y}' \sin \lambda) + \mathbf{j}'(\omega \dot{x}' \sin \lambda) + \mathbf{k}'(-\omega \dot{x}' \cos \lambda)$$

Using the results for $\boldsymbol{\omega} \times \dot{\mathbf{r}}'$ in Equation 5.4.10 and canceling the m's and equating components, we find

$$\ddot{x}' = -2\omega(\dot{z}' \cos \lambda - \dot{y}' \sin \lambda) \tag{5.4.11a}$$

$$\ddot{y}' = -2\omega(\dot{x}' \sin \lambda) \tag{5.4.11b}$$

$$\ddot{z}' = -g + 2\omega \dot{x}' \cos \lambda \tag{5.4.11c}$$

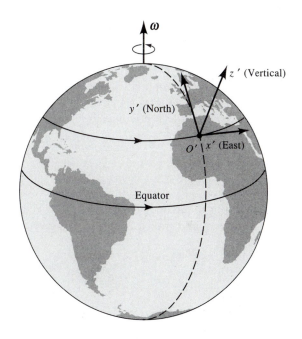

Figure 5.4.3 Coordinate axes for analyzing projectile motion.

for the component differential equations of motion. These equations are not of the separated type, but we can integrate once with respect to t to obtain

$$\dot{x}' = -2\omega(z' \cos\lambda - y' \sin\lambda) + \dot{x}'_0 \qquad (5.4.12a)$$

$$\dot{y}' = -2\omega x' \sin\lambda + \dot{y}'_0 \qquad (5.4.12b)$$

$$\dot{z}' = -gt + 2\omega x' \cos\lambda + \dot{z}'_0 \qquad (5.4.12c)$$

The constants of integration $\dot{x}'_0, \dot{y}'_0$, and $\dot{z}'_0$ are the initial components of the velocity. The values of $\dot{y}'$ and $\dot{z}'$ from Equations 5.4.12b and c may be substituted into Equation 5.4.11a. The result is

$$\ddot{x}' = 2\omega gt \cos\lambda - 2\omega(\dot{z}'_0 \cos\lambda - \dot{y}'_0 \sin\lambda) \qquad (5.4.13)$$

where terms involving ω^2 have been ignored. We now integrate again to get

$$\dot{x}' = \omega gt^2 \cos\lambda - 2\omega t(\dot{z}'_0 \cos\lambda - \dot{y}'_0 \sin\lambda) + \dot{x}'_0 \qquad (5.4.14)$$

and finally, by a third integration, we find x' as a function of t:

$$x'(t) = \tfrac{1}{3}\omega gt^3 \cos\lambda - \omega t^2(\dot{z}'_0 \cos\lambda - \dot{y}'_0 \sin\lambda) + \dot{x}'_0 t + x'_0 \qquad (5.4.15a)$$

The preceding expression for x' may be inserted into Equations 5.4.12b and c. The resulting equations, when integrated, yield

$$y'(t) = \dot{y}'_0 t - \omega\dot{x}'_0 t^2 \sin\lambda + y'_0 \qquad (5.4.15b)$$

$$z'(t) = -\tfrac{1}{2}gt^2 + \dot{z}'_0 t + \omega\dot{x}'_0 t^2 \cos\lambda + z'_0 \qquad (5.4.15c)$$

where, again, terms of order ω^2 have been ignored.

In Equations 5.4.15a–c, the terms involving ω express the effect of Earth's rotation on the motion of a projectile in a coordinate system fixed to the Earth.

EXAMPLE 5.4.1

Falling Body

Suppose a body is dropped from rest at a height h above the ground. Then at time $t = 0$ we have $\dot{x}_0' = \dot{y}_0' = \dot{z}_0' = 0$, and we set $x_0' = y_0' = 0, z_0' = h$ for the initial position. Equations 5.4.15a–c then reduce to

$$x'(t) = \tfrac{1}{3}\omega g t^3 \cos\lambda$$
$$y'(t) = 0$$
$$z'(t) = -\tfrac{1}{2}g t^2 + h$$

Thus, as it falls, the body drifts to the east. When it hits the ground ($z' = 0$), we see that $t^2 = 2h/g$, and the eastward drift is given by the corresponding value of $x'(t)$, namely,

$$x_h' = \tfrac{1}{3}\omega\left(\frac{8h^3}{g}\right)^{1/2}\cos\lambda$$

For a height of, say, 100 m at a latitude of $45°$, the drift is

$$\tfrac{1}{3}(7.27\times10^{-5}\ \mathrm{s}^{-1})(8\times100^3\ \mathrm{m}^3/9.8\ \mathrm{m\cdot s}^{-2})^{1/2}\ \cos45° = 1.55\times10^{-2}\ \mathrm{m} = 1.55\ \mathrm{cm}$$

Because Earth turns to the east, common sense would seem to say that the body should drift westward. Can the reader think of an explanation?

EXAMPLE 5.4.2

Deflection of a Rifle Bullet

Consider a projectile that is fired with high initial speed v_0 in a nearly horizontal direction, and suppose this direction is east. Then $\dot{x}_0' = v_0$ and $\dot{y}_0' = \dot{z}_0' = 0$. If we take the origin to be the point from which the projectile is fired, then $x_0' = y_0' = z_0' = 0$ at time $t = 0$. Equation 5.4.15b then gives

$$y'(t) = -\omega v_0 t^2 \sin\lambda$$

which says that the projectile veers to the south or to the right in the Northern Hemisphere ($\lambda > 0$) and to the left in the Southern Hemisphere ($\lambda < 0$). If H is the horizontal range of the projectile, then we know that $H \approx v_0 t_1$, where t_1 is the time of flight. The transverse deflection is then found by setting $t = t_1 = H/v_0$ in the above expression for $y'(t)$. The result is

$$\Delta \approx \frac{\omega H^2}{v_0}\,|\sin\lambda|$$

for the magnitude of the deflection. This is the same for *any* direction in which the projectile is initially aimed, provided the trajectory is flat. This follows from the fact that the magnitude of the horizontal component of the Coriolis force on a body traveling parallel to the ground is independent of the direction of motion. (See Problem 5.12.) Because the deflection is proportional to the square of the horizontal range, it becomes of considerable importance in long-range gunnery.

*5.5 | Motion of a Projectile in a Rotating Cylinder

Here is one final example concerning the dynamics of projectiles in rotating frames of reference. The example is rather involved and makes use of applied numerical techniques. We hope its inclusion gives you a better appreciation for the connection between the geometry of straight-line, force-free trajectories seen in an inertial frame of reference and the resulting curved geometry seen in a noninertial rotating frame of reference. The inertial forces that appear in a noninertial frame lead to a curved trajectory that may be calculated from the perspective of an inertial frame solely on the basis of geometrical considerations. This must be the case if the validity of Newton's laws of motion is to be preserved in noninertial frames of reference. Such a realization, although completely obvious with hindsight, should not be trivialized. It was ultimately just this sort of realization that led Einstein to formulate his general theory of relativity.

EXAMPLE 5.5.1

In several popular science fiction novels[2] spacecraft capable of supporting entire populations have been envisioned as large, rotating toroids or cylinders. Consider a cylinder of radius $R = 1000$ km and, for our purposes here, infinite length. Let it rotate about its axis with an angular velocity of $\omega = 0.18°/s$. It completes one revolution every 2000 s. This rotation rate leads to an apparent centrifugal acceleration for objects on the interior surface of $\omega^2 R$ equal to 1 g. Imagine several warring factions living on the interior of the cylinder. Let them fire projectiles at each other.

(a) Show that when projectiles are fired at low speeds ($v \ll \omega R$) and low "altitudes" at nearby points (say, $\Delta r' \le R/10$), the equations of motion governing the resulting trajectories are identical to those of a similarly limited projectile on the surface of the Earth.

(b) Find the general equations of motion for a projectile of unlimited speed and range using cylindrical coordinates rotating with the cylinder.

(c) Find the trajectory h versus ϕ' of a projectile fired vertically upward with a velocity $v' = \omega R$ in this noninertial frame of reference. $h = R - r'$ is the altitude of the projectile and ϕ' is its angular position in azimuth relative to the launch point. Calculate the angle Φ where it lands relative to the launch point. Also, calculate the maximum height H reached by the projectile.

---

[2] For example, *Rendezvous with Rama* and *Rama II* by Arthur C. Clarke (Bantam Books) or *Titan* by John Varley (Berkeley Books).

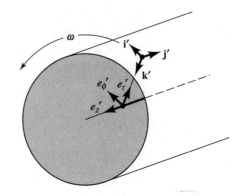

Figure 5.5.1 Coordinates denoted by unit vectors $\mathbf{i}'$, $\mathbf{j}'$, $\mathbf{k}'$ on the interior surface of a rotating cylinder. Unit vectors $\mathbf{e}_r$, $\mathbf{e}_\phi$, $\mathbf{e}_{z'}$ denote cylindrical coordinates. Each set is embedded in and rotates with the cylinder.

(d) Finally, calculate h versus ϕ' solely from the geometrical basis that an inertial observer would employ to predict what the noninertial observer would see. Show that this result agrees with that of part (c), calculated from the perspective of the noninertial observer. In particular, show that Φ and H agree.

Solution:

(a) Because we first consider short, low-lying trajectories, we choose Cartesian coordinates (x', y', z') denoted by the unit vectors $\mathbf{i}'$, $\mathbf{j}'$, $\mathbf{k}'$ attached to and rotating with the cylinder shown in Figure 5.5.1.

The coordinate system is centered on the launch point. Because no real force is acting on the projectile after it is launched, the mass m common to all remaining terms in Equation 5.3.2 can be stripped and the equation then written in terms of accelerations only

$$-\mathbf{A}_0 - 2\boldsymbol{\omega} \times \mathbf{v}' - \boldsymbol{\omega} \times (\boldsymbol{\omega} \times \mathbf{r}') = \mathbf{a}' \qquad (5.5.1)$$

The transverse acceleration is zero because the cylinder rotates at a constant rate. The first term on the left is the acceleration of the coordinate system origin. It is given by

$$\mathbf{A}_0 = \omega^2 R \mathbf{k}' \qquad (5.5.2)$$

The second term is the Coriolis acceleration and is given by

$$\begin{aligned} \mathbf{a}_{Cor} &= 2\boldsymbol{\omega} \times \mathbf{v}' = 2(-\mathbf{j}'\omega) \times (\mathbf{i}'\dot{x}' + \mathbf{j}'\dot{y}' + \mathbf{k}'\dot{z}') \\ &= 2\omega\dot{x}'\mathbf{k}' - 2\omega\dot{z}'\mathbf{i}' \end{aligned} \qquad (5.5.3)$$

The third term is the centrifugal acceleration given by

$$\begin{aligned} \mathbf{a}_{centrif} &= -\mathbf{j}'\omega \times [(-\mathbf{j}'\omega) \times \mathbf{r}'] \\ &= \mathbf{j}'\omega \times [(\mathbf{j}'\omega) \times (\mathbf{i}'x' + \mathbf{j}'y' + \mathbf{k}'z')] \\ &= \mathbf{j}'\omega \times (-\mathbf{k}'\omega x' + \mathbf{i}'\omega z') \\ &= -\mathbf{i}'\omega^2 x' - \mathbf{k}'\omega^2 z' \end{aligned} \qquad (5.5.4)$$

After gathering all appropriate terms, the x'-, y'-, and z'-components of the resultant acceleration become

$$\ddot{x}' = 2\omega\dot{z}' + \omega^2 x'$$
$$\ddot{y}' = 0 \qquad\qquad\qquad (5.5.5)$$
$$\ddot{z}' = -2\omega\dot{x}' + \omega^2 z' - \omega^2 R$$

If projectiles are limited in both speed and range such that

$$|\dot{x}'| \sim |\dot{z}'| \ll \omega R \qquad\qquad |x'| \sim |z'| \ll R \qquad\qquad (5.5.6)$$

and recalling that the rotation rate of the cylinder has been adjusted to $\omega^2 R = g$, the above acceleration components reduce to

$$\ddot{x}' \approx 0 \qquad\qquad \ddot{y}' = 0 \qquad\qquad \ddot{z}' \approx -g \qquad\qquad (5.5.7)$$

which are equivalent to the equations of motion for a projectile of limited speed and range on the surface of the Earth.

(b) In this case no limit is placed on projectile velocity or range. We describe the motion using cylindrical coordinates (r', ϕ', z') attached to and rotating with the cylinder as indicated in Figure 5.5.1. r' denotes the radial position of the projectile measured from the central axis of the cylinder; ϕ' denotes its azimuthal position and is measured from the radius vector directed outward to the launch point; z' represents its position along the cylinder ($z' = 0$ corresponds to the z'-position of the launch point). The overall position, velocity, and acceleration of the projectile in cylindrical coordinates are given by Equations 1.12.1–1.12.3. We can use these relations to evaluate all the acceleration terms in Equation 5.5.1. The term $\mathbf{A}_0$ is zero, because the rotating coordinate system is centered on the axis of rotation. The Coriolis acceleration is

$$2\boldsymbol{\omega} \times \boldsymbol{v}' = 2\omega \boldsymbol{e}_{z'} \times (\dot{r}'\boldsymbol{e}_{r'} + r'\dot{\phi}'\boldsymbol{e}_{\phi'} + \dot{z}'\boldsymbol{e}_{z'})$$
$$= 2\omega\dot{r}'(\boldsymbol{e}_{z'} \times \boldsymbol{e}_{r'}) + 2\omega r'\dot{\phi}'(\boldsymbol{e}_{z'} \times \boldsymbol{e}_{\phi'}) \qquad\qquad (5.5.8)$$
$$= 2\omega\dot{r}'\boldsymbol{e}_{\phi'} - 2\omega r'\dot{\phi}'\boldsymbol{e}_{r'}$$

The centrifugal acceleration is

$$\boldsymbol{\omega} \times (\boldsymbol{\omega} \times \boldsymbol{r}') = \omega^2 \boldsymbol{e}_{z'} \times [\boldsymbol{e}_{z'} \times (r'\boldsymbol{e}_{r'} + z'\boldsymbol{e}_{z'})]$$
$$= \omega^2 \boldsymbol{e}_{z'} \times r'\boldsymbol{e}_{\phi'} \qquad\qquad (5.5.9)$$
$$= -\omega^2 r'\boldsymbol{e}_{r'}$$

We can now rewrite Equation 5.5.1 in terms of components by gathering together all the previous corresponding elements and equating them to those in Equation 1.12.3

$$\ddot{r}' - r'\dot{\phi}'^2 = 2\omega r'\dot{\phi}' + \omega^2 r'$$
$$2\dot{r}'\dot{\phi}' + r'\ddot{\phi}' = -2\omega\dot{r}' \qquad\qquad (5.5.10)$$
$$\ddot{z}' = 0$$

In what follows we ignore the z'-equation of motion because it contains no nonzero acceleration terms and simply gives rise to a "drift" along the axis of the cylinder of any trajectory seen in the $r'\phi'$ plane. Finally, we rewrite the radial and azimuthal equations in such a way that we can more readily see the dependency of the acceleration upon velocities and positions

$$\ddot{r}' = 2\omega r' \dot{\phi}' + (\omega^2 + \dot{\phi}'^2)r' \tag{5.5.11a}$$

$$\ddot{\phi}' = -\frac{2\dot{r}'}{r'}(\omega + \dot{\phi}') \tag{5.5.11b}$$

(c) Before solving these equations of motion for a projectile fired vertically upward (from the viewpoint of a cylinder dweller), we investigate the situation from the point of view of an inertial observer located outside the rotating cylinder. The rotational speed of the cylinder is ωR. If the projectile is fired vertically upward with a speed ωR from the point of view of the noninertial observer, the inertial observer sees the projectile launched with a speed $v = \sqrt{2}\,\omega R$ at 45° with respect to the vertical. Furthermore, according to this observer, no real forces act on the projectile. Travel appears to be in a straight line. Its flight path is a chord of a quadrant. This situation is depicted in Figure 5.5.2.

As can be seen in Figure 5.5.2, by the time the projectile reaches a point in its trajectory denoted by the vector $\mathbf{r}'$, the cylinder has rotated such that the launch point a has moved to the position labeled b. Therefore, the inertial observer concludes that the noninertial observer thinks that the projectile has moved through the angle ϕ' and attained an altitude of $R - r'$. When the projectile lands, the noninertial observer finds that the projectile has moved through a total angle of $\Phi = \pi/2 - \omega T$, where T is the total time of flight. But $T = L/v = \sqrt{2}\,R/(\sqrt{2}\,\omega R) = 1/\omega$, or $\omega T = 1$ radian. Hence, the apparent deflection angle should be $\Phi = \pi/2 - 1$ radians, or about 32.7°. The maximum height reached by the projectile occurs midway through its trajectory when $\omega t + \phi' = \pi/4$ radians. At this point $r' = R/\sqrt{2}$ or $H = R - R/\sqrt{2} = 290$ km. At least, this is what the inertial observer believes the noninertial observer would see. Let us see what the noninertial observer does see according to Newton's laws of motion.

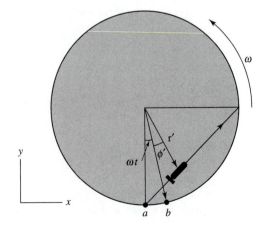

Figure 5.5.2 Trajectory of a projectile launched inside a rotating cylinder at 45° with respect to the "vertical" from the point of view of an inertial observer.

We have used *Mathematica* to solve the differential equations of motion (Equation 5.5.11a and b) numerically as in Example 4.3.2 and the result is shown in Figure 5.5.3.

It can be seen that the projectile is indeed launched vertically upward according to the rotating observer. But the existence of the centrifugal and Coriolis inertial forces causes the projectile to accelerate back toward the surface and toward the east, in the direction of the angular rotation of the cylinder. Note that the rotating observer concludes that the vertically launched projectile has been pushed sideways by the Coriolis force such that it lands 32.7° to the east of the launch point. The centrifugal force has limited its altitude to a maximum value of 290 km. Each value is in complete agreement with the conclusion of the noninertial observer. Clearly, an intelligent military, aware of the dynamical equations of motion governing projectile trajectories on this cylindrical world, could launch all their missiles vertically upward and hit any point around the cylinder by merely adjusting launch velocities. (Positions located up or down the cylindrical axis could be hit by tilting the launcher in that direction and firing the projectile at the required initial and thereafter constant axial velocity $\dot{z}_0$.)

(d) The inertial observer calculates the trajectory seen by the rotating observer in the following way: first, look at Figure 5.5.4. It is a blow-up of the geometry illustrated in Figure 5.5.2.

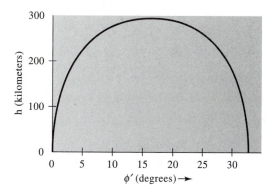

Figure 5.5.3 Trajectory of a projectile fired vertically upward (toward the central axis) from the interior surface of a large cylinder rotating with an angular velocity ω, such that $\omega^2 R = g$.

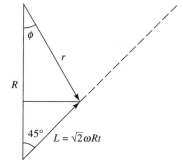

Figure 5.5.4 Geometry used to calculate the trajectory seen by a rotating observer according to an inertial observer.

ϕ is the azimuthal angle of the projectile as measured in the fixed, inertial frame. The azimuthal angle in the noninertial frame is $\phi' = \phi - \omega t$ (see Figure 5.5.2). As can be seen from the geometry of Figure 5.5.4, we can calculate the functional dependency of ϕ upon time

$$\tan \phi(t) = \frac{L(t) \sin 45°}{R - L(t) \cos 45°} = \frac{L(t)}{\sqrt{2} R - L(t)}$$

$$= \frac{\sqrt{2} \, \omega R t}{\sqrt{2} R - \sqrt{2} \, \omega R t} = \frac{\omega t}{1 - \omega t}$$

(5.5.12)

The projectile appears to be deflected toward the east by the angle ϕ' as a function of time given by

$$\phi'(t) = \phi(t) - \omega t = \tan^{-1}\left(\frac{\omega t}{1 - \omega t}\right) - \omega t \qquad (5.5.13)$$

The dependency of r' on time is given by

$$\begin{aligned} r'^2(t) &= [L(t) \sin 45°]^2 + [R - L(t) \cos 45°]^2 \\ &= L(t)^2 + R^2 - \sqrt{2} L(t)R \\ &= 2(\omega R t)^2 + R^2 - \sqrt{2}\,(\sqrt{2}\,\omega R t)R \\ &= R^2 [1 - 2\omega t(1 - \omega t)] \end{aligned}$$

(5.5.14)

$$\therefore r'(t) = R[1 - 2\omega t(1 - \omega t)]^{1/2}$$

These final two parametric equations $r'(t)$ and $\phi'(t)$ describe a trajectory that the inertial observer predicts the noninertial observer should see. If we let time evolve and then plot $h = R - r'$ versus ϕ', we obtain exactly the same trajectory shown in Figure 5.5.3. That trajectory was calculated by the noninertial observer who used Newton's dynamical equations of motion in the rotating frame of reference. Thus, we see the equivalence between the curved geometry of straight lines seen from the perspective of an accelerated frame of reference and the existence of inertial forces that produce that geometry in the accelerated frame.

5.6 | The Foucault Pendulum

In this section we study the effect of Earth's rotation on the motion of a pendulum that is free to swing in any direction, the so-called *spherical pendulum*. As shown in Figure 5.6.1, the applied force acting on the pendulum bob is the vector sum of the weight $m\mathbf{g}$ and the tension $\mathbf{S}$ in the cord. The differential equation of motion is then

$$m\ddot{\mathbf{r}}' = m\mathbf{g} + \mathbf{S} - 2m\boldsymbol{\omega} \times \dot{\mathbf{r}}' \qquad (5.6.1)$$

Here we ignored the term $-m\boldsymbol{\omega} \times (\boldsymbol{\omega} \times \mathbf{r}')$. It is vanishingly small in this context. Previously, we worked out the components of the cross product $\boldsymbol{\omega} \times \mathbf{r}'$ (see Equation 5.4.10). Now the x'- and y'-components of the tension can be found simply by noting that the

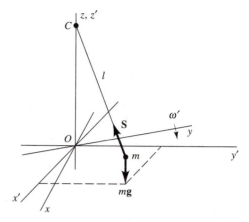

Figure 5.6.1   The Foucault pendulum.

direction cosines of the vector **S** are $-x'/l$, $-y'/l$, and $-(l-z')/l$, respectively. Consequently $S_x = -x'S/l$, $S_y = -y'S/l$, and the corresponding components of the differential equation of motion (5.6.1) are

$$m\ddot{x}' = \frac{-x'}{l}S - 2m\omega(\dot{z}' \cos \lambda - \dot{y}' \sin \lambda) \tag{5.6.2a}$$

$$m\ddot{y}' = \frac{-y'}{l}S - 2m\omega\dot{x}' \sin \lambda \tag{5.6.2b}$$

We are interested in the case in which the amplitude of oscillation of the pendulum is small so that the magnitude of the tension S is very nearly constant and equal to mg. Also, we shall ignore $\dot{z}'$ compared with $\dot{y}'$ in Equation 5.6.2a. The $x'y'$ motion is then governed by the following differential equations:

$$\ddot{x}' = -\frac{g}{l}x' + 2\omega'\dot{y}' \tag{5.6.3a}$$

$$\ddot{y}' = -\frac{g}{l}y' - 2\omega'\dot{x}' \tag{5.6.3b}$$

in which we have introduced the quantity $\omega' = \omega \sin \lambda = \omega_{z'}$, which is the local vertical component of Earth's angular velocity.

Again we are confronted with a set of differential equations of motion that are not in separated form. A heuristic method of solving the equations is to transform to a new coordinate system $Oxyz$ that rotates relative to the primed system in such a way as to cancel the vertical component of Earth's rotation, namely, with angular rate $-\omega'$ about the vertical axis as shown in Figure 5.4.3. Thus, the unprimed system has no rotation about the vertical axis. The equations of transformation are

$$x' = x \cos\omega't + y \sin\omega't \tag{5.6.4a}$$

$$y' = -x \sin\omega't + y \cos\omega't \tag{5.6.4b}$$

On substituting the expressions for the primed quantities and their derivatives from the preceding equations into Equations 5.6.3a and b, the following result is obtained, after collecting terms and dropping terms involving ω'^2,

$$\left(\ddot{x}+\frac{g}{l}x\right)\cos\omega't+\left(\ddot{y}+\frac{g}{l}y\right)\sin\omega't=0 \qquad (5.6.5)$$

and an identical equation, except that the sine and cosine are reversed. Clearly, the preceding equation is satisfied if the coefficients of the sine and cosine terms both vanish, namely,

$$\ddot{x}+\frac{g}{l}x=0 \qquad (5.6.6a)$$

$$\ddot{y}+\frac{g}{l}y=0 \qquad (5.6.6b)$$

These are the differential equations of the two-dimensional harmonic oscillator discussed previously in Section 4.4. Thus, the path, projected on the xy plane, is an ellipse with *fixed* orientation in the unprimed system. In the primed system the path is an ellipse that undergoes a steady precession with angular speed $\omega' = \omega \sin \lambda$.

In addition to this type of precession, there is another *natural* precession of the spherical pendulum, which is ordinarily much larger than the rotational precession under discussion. However, if the pendulum is carefully started by drawing it aside with a thread and letting it start from rest by burning the thread, the natural precession is rendered negligibly small.[3]

The rotational precession is clockwise in the Northern Hemisphere and counterclockwise in the Southern. The period is $2\pi/\omega' = 2\pi/(\omega \sin \lambda) = 24/\sin \lambda$ h. Thus, at a latitude of 45°, the period is $(24/0.707)$ h $= 33.94$ h. The result was first demonstrated by the French physicist Jean Foucault in Paris in the year 1851. The Foucault pendulum has come to be a traditional display in major planetariums throughout the world.

Problems

5.1   A 120-lb person stands on a bathroom spring scale while riding in an elevator. If the elevator has (a) upward and (b) downward acceleration of $g/4$, what is the weight indicated on the scale in each case?

5.2   An ultracentrifuge has a rotational speed of 500 rps. (a) Find the centrifugal force on a 1-μg particle in the sample chamber if the particle is 5 cm from the rotational axis. (b) Express the result as the ratio of the centrifugal force to the weight of the particle.

5.3   A plumb line is held steady while being carried along in a moving train. If the mass of the plumb bob is m, find the tension in the cord and the deflection from the local vertical if the train is accelerating forward with constant acceleration $g/10$. (Ignore any effects of Earth's rotation.)

5.4   If, in Problem 5.3, the plumb line is not held steady but oscillates as a simple pendulum, find the period of oscillation for small amplitude.

5.5   A hauling truck is traveling on a level road. The driver suddenly applies the brakes, causing the truck to decelerate by an amount $g/2$. This causes a box in the rear of the truck to slide forward. If the coefficient of sliding friction between the box and the truckbed is $\frac{1}{3}$, find the acceleration of the box relative to (a) the truck and (b) the road.

---

[3]The natural precession will be discussed briefly in Chapter 10.

5.6 The position of a particle in a fixed inertial frame of reference is given by the vector

$$\mathbf{r} = \mathbf{i}(x_0 + R \cos \Omega t) + \mathbf{j} R \sin \Omega t$$

where x_0, R, and Ω are constants.

(a) Show that the particle moves in a circle with constant speed.

(b) Find two coupled, first-order differential equations of motion that relate the components of position, x' and y', and the components of velocity, $\dot{x}'$ and $\dot{y}'$, of the particle relative to a frame of reference rotating with an angular velocity $\boldsymbol{\omega} = \mathbf{k}\omega$.

(c) Letting the fixed and rotating frames of reference coincide at times $t = 0$ and letting $u' = x' + iy'$, find $u'(t)$, assuming that $\Omega \neq -\omega$. (*Note:* $i = \sqrt{-1}$.)

5.7 A new asteroid is discovered in a circular orbit or radius $4^{1/3}$ AU about the Sun.[4] Its period of revolution is precisely 2 years. Assume that at $t = 0$ it is at its distance of closest approach to the Earth.

(a) Find its coordinates $[x(t), y(t)]$ in a frame of reference fixed to Earth but whose axes remain fixed in orientation relative to the Sun. Assume that the x-axis points toward the asteroid at $t = 0$.

(b) Calculate the velocity of the asteroid, relative to Earth, at $t = 0$.

(c) Find the x- and y-components of the acceleration of the asteroid in this frame of reference. Integrate them twice to show that the resulting positions as a function of time agree with part (a).

(d) Plot the trajectory of the asteroid as seen from the rotating frame of reference attached to Earth for its 2-year orbital period. (*Hint: Use Mathematica's* ParametricPlot *graphing tool.*)

5.8 A cockroach crawls with constant speed in a circular path of radius b on a phonograph turntable rotating with constant angular speed ω. The circular path is concentric with the center of the turntable. If the mass of the insect is m and the coefficient of static friction with the surface of the turntable is μ_s, how fast, relative to the turntable, can the cockroach crawl before it starts to slip if it goes (a) in the direction of rotation and (b) opposite to the direction of rotation?

5.9 In the problem of the bicycle wheel rounding a curve, Example 5.2.2, what is the acceleration relative to the ground of the point at the very front of the wheel?

5.10 If the bead on the rotating rod of Example 5.3.3 is initially released from rest (relative to the rod) at its midpoint calculate (a) the displacement of the bead along the rod as a function of time; (b) the time; and (c) the velocity (relaive to the rod) when the bead leaves the end of the rod.

5.11 On the salt flats at Bonneville, Utah (latitude = 41°N) the British auto racer John Cobb in 1947 became the first man to travel at a speed of 400 mph on land. If he was headed due north at this speed, find the ratio of the magnitude of the Coriolis force on the racing car to the weight of the car. What is the direction of the Coriolis force?

5.12 A particle moves in a horizontal plane on the surface of the Earth. Show that the magnitude of the horizontal component of the Coriolis force is independent of the direction of the motion of the particle.

5.13 If a pebble were dropped down an elevator shaft of the Empire State Building ($h = 1250$ ft, latitude = 41°N), find the deflection of the pebble due to Coriolis force. Assume no air resistance.

---

[4]The radius of the Earth's nearly circular orbit is 1 AU, or 1 astronomical unit. Both the Earth and the asteroid revolve counterclockwise in a common plane about the Sun as seen from the north pole star, Polaris.

5.14   In Yankee Stadium, New York, a batter hits a baseball a distance of 200 ft in a fairly flat trajectory. Is the amount of deflection due to the Coriolis force alone of much importance? (Let the angle of elevation be 15°.) Assume no air resistance.

5.15   Show that the third derivative with respect to time of the position vector (jerk) of a particle moving in a rotating coordinate system in terms of appropriate derivatives in the rotating system is given by

$$\dddot{\mathbf{r}} = \dddot{\mathbf{r}}' + 3\dot{\boldsymbol{\omega}} \times \dot{\mathbf{r}}' + 3\boldsymbol{\omega} \times \ddot{\mathbf{r}}' + \ddot{\boldsymbol{\omega}} \times \mathbf{r}' + 3\boldsymbol{\omega} \times (\boldsymbol{\omega} \times \dot{\mathbf{r}}')$$

$$+ \dot{\boldsymbol{\omega}} \times (\boldsymbol{\omega} \times \mathbf{r}') + 2\boldsymbol{\omega} \times (\dot{\boldsymbol{\omega}} \times \mathbf{r}') - \omega^2 (\boldsymbol{\omega} \times \mathbf{r}')$$

5.16   A bullet is fired straight up with initial speed v_0'. Assuming g is constant and ignoring air resistance, show that the bullet will hit the ground west of the initial point of upward motion by an amount $4\omega v_0'^3 \cos \lambda / 3g^2$, where λ is the latitude and ω is Earth's angular velocity.

5.17   If the bullet in Problem 5.16 is fired due east at an elevation angle α from a point on Earth whose latitude is $+\lambda$, show that it will strike the Earth with a lateral deflection given by $4\omega v_0'^3 \sin \lambda \sin^2 \alpha \cos \alpha / g^2$.

5.18   A satellite travels around the Earth in a circular orbit of radius R. The angular speed of a satellite varies inversely with its distance from Earth according to $\omega^2 = k/R^3$, where k is a constant. Observers in the satellite see an object moving nearby, also presumably in orbit about Earth. To describe its motion, they use a coordinate system fixed to the satellite with x-axis pointing away from Earth and y-axis pointing in the direction in which the satellite is moving. Show that the equations of motion for the nearby object with respect to the observers' frame of reference are given approximately by

$$\ddot{x} - 2\omega\dot{y} - 3\omega^2 x = 0$$

$$\ddot{y} + 2\omega\dot{x} = 0$$

(see Example 2.3.2 for the force of gravity that Earth exerts on an object at a distance r from it and ignore the gravitational effect of the satellite on the object).

5.19   The force on a charged particle in an electric field **E** and a magnetic field **B** is given by

$$\mathbf{F} = q(\mathbf{E} + \mathbf{v} \times \mathbf{B})$$

in an inertial system, where q is the charge and **v** is the velocity of the particle in the inertial system. Show that the differential equation of motion referred to a rotating coordinate system with angular velocity $\boldsymbol{\omega} = -(q/2m)\mathbf{B}$ is, for small **B**

$$m\ddot{\mathbf{r}}' = q\mathbf{E}$$

that is, the term involving **B** is eliminated. This result is known as *Larmor's theorem*.

5.20   Complete the steps leading to Equation 5.6.5 for the differential equation of motion of the Foucault pendulum.

5.21   The latitude of Mexico City is approximately 19°N. What is the period of precession of a Foucault pendulum there?

5.22   Work Example 5.2.2 using a coordinate system that is fixed to the bicycle wheel and rotates with it, as in Example 5.2.1.

Computer Problems

C 5.1   (a) Solve parts (c) and (d) of Example 5.5.1. Plot the trajectory h versus ϕ' as seen from each observer's (inertial and noninertial) point of view, as explained in the text. Your graphs should be identical to those in Figure 5.5.1. (b) Repeat part (a) when the missile is fired with an initial velocity of $\mathbf{v}' = (2\omega R/\pi)\mathbf{e}_{r'} - \omega R\mathbf{e}_{\phi'}$ [see Figure 5.5.1]. In this case, what is the maximum altitude H attained by the missile and the angle Φ where it lands relative to the launch point?

C 5.2   A small mass is free to move on a frictionless horizontal surface. Let the horizontal surface be a circle of radius $R = 1$ m, and let it rotate counterclockwise about a vertical axis with a constant angular velocity $\omega = 1$ rad/s. Let the coordinates of the mass be described by a rectangular (x, y) coordinate system centered on the axis of the rotating system and rotating with it. (a) Find the equations of motion of the mass in terms of the rotating (x, y) coordinate system. (b) If the initial position of the mass is $(-R, 0)$, what initial y-component of the velocity (relative to the rotating frame) is necessary if the mass is to be projected across a diameter of the circular surface, from the perspective of a fixed, inertial observer? (c) Find an expression in terms of integers $n = 1, 2, 3 \ldots$ for the initial x-component of the velocities that results in the mass traversing a diameter of the circular surface in a fixed inertial frame of reference *and landing at the same point $(-R, 0)$ from which it was projected in the rotating frame* (let $n = 1$ represent the fastest initial velocity). (d) Plot these trajectories from the perspective of the rotating frame of reference for the five largest initial x-components of the velocities ($n = 1 \ldots 5$). (e) Describe the resultant trajectory as seen from the rotating frame of reference as the x-component of the velocity approaches zero ($n \to \infty$).

C 5.3   Find the equations of motion for the asteroid in Problem 5.7, as seen from Earth frame of reference described in that problem, and solve them numerically using *Mathematica*. As starting conditions, assume that the asteroid is in direct opposition to the Sun relative to Earth at $t = 0$, and use initial velocities appropriate to circular orbits about the Sun to calculate the initial velocity of the asteroid relative to Earth. Plot the trajectory of the satellite for one orbital period.

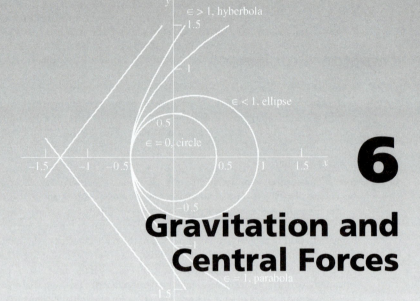

6

Gravitation and Central Forces

"We have explained the phenomena of the heavens and of our sea by the power of gravity, but have not yet assigned the cause of this power . . . I have not been able to discover the cause of those properties of gravity from phenomena, and I frame no hypotheses;—"

—Sir Isaac Newton, *The Principia*, 1687; Florian Cajori's translation, Berkeley, Univ. of Calif. Press, 1966

"Gravity must be a scholastic occult quality or the effect of a miracle."

—Gottfried Wilhelm Leibniz; See *Let Newton Be!*, by J. Fauvel, R. Flood, M. Shorthand, and R. Wilson, Oxford Univ. Press, 1988

6.1 | Introduction

Throughout the year ancient peoples observed the five visible planets slowly move through the fixed constellations of the zodiac in a fairly regular fashion. But occasionally, at times that occurred with astonishing predictability, they mysteriously halted their slow forward progression, suddenly reversing direction for as long as a few weeks before again resuming their steady march through the sky. This apparent quirk of planetary behavior is called *retrograde motion*. Unmasking its origin would consume the intellectual energy of ancient astronomers for centuries to come. Indeed, horribly complicated concoctions from minds shackled by philosophical dogma and fuzzy notions of physics, such as the cycles and epicycles of Ptolemy (125 C.E.) and others of like-minded mentality, would serve as models of physical reality for more than 2000 years. Ultimately, Nicolaus Copernicus (1473–1543) demonstrated that retrograde motion was nothing other than a simple consequence of the relative motion between Earth and the other planets each moving in a heliocentric orbit. Nonetheless, even Copernicus could not purge

218

himself of the Ptolemaic epicycles, constrained by the dogma of uniform circular motion and the requirement of obtaining agreement between the observed and predicted irregularities of planetary motion.

It was not until Johannes Kepler (1571–1630) turned loose his potent intellect on the problem of solving the orbit of Mars, an endeavor that was to occupy him intensely for 20 years, that for the first time in history, scientists glimpsed the precise mathematical nature of the heavenly motions. Kepler painstakingly constructed a concise set of three mathematical laws that accurately described the orbits of the planets around the Sun. These three laws of planetary motion were soon seen by Newton as nothing other than simple consequences of the interplay of a law of universal gravitation with three fundamental laws of mechanics that Newton had developed mostly from Galileo's investigations of the motions of terrestrial objects. Thus, Newton was to incorporate the physical workings of all the heavenly bodies within a framework of natural law that resided on Earth. The world would never be seen in quite the same way again.

Newton's Law of Universal Gravitation

Newton formally announced the law of universal gravitation in the *Principia*, published in 1687. He actually worked out much of the theory at his family home in Woolsthorpe, England, as early as 1665–1666, during a six-month hiatus from Cambridge University, which was closed while a plague ravaged most of London.

The law can be stated as follows:

Every particle in the universe attracts every other particle with a force whose magnitude is proportional to the product of the masses of the two particles and inversely proportional to the square of the distance between them. The direction of the force lies along the straight line connecting the two particles.

We can express the law vectorially by the equation

$$\mathbf{F}_{ij} = G \frac{m_i m_j}{r_{ij}^2} \left(\frac{\mathbf{r}_{ij}}{r_{ij}} \right) \tag{6.1.1}$$

where $\mathbf{F}_{ij}$ is the force on particle i of mass m_i exerted by particle j of mass m_j. The vector $\mathbf{r}_{ij}$ is the directed line segment running from particle i to particle j, as shown in Figure 6.1.1. The law of action and reaction requires that $\mathbf{F}_{ij} = -\mathbf{F}_{ji}$. The constant of proportionality G

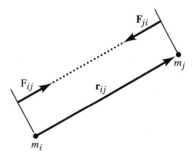

Figure 6.1.1   Action and reaction in Newton's law of gravity.

is known as the *universal constant of gravitation*. Its value is determined in the laboratory by carefully measuring the force between two bodies of known mass. The internationally accepted value at present is, in SI units,

$$G = (6.67259 \pm 0.00085) \times 10^{-11} \text{ Nm}^2 \text{ kg}^{-2}$$

All our present knowledge of the masses of astronomical bodies, including Earth, is based on the value of this fundamental constant.[1]

This law is an example of a general class of forces termed *central;* that is, forces whose lines of action either emanate from or terminate on a single point or center. Furthermore, if the magnitude of the force, as is the case with gravitation, is independent of any direction, the force is isotropic. The behavior of such a force may be visualized in the following way: Imagine being confined to a hypothetical, spherical surface centered about a massive particle that serves as a source of gravity. When walking around that surface, one would discover that the force of attraction would always be directed toward the center, and the magnitude of this force would be independent of position on the spherical surface. Nothing about this force could be used to determine position on the sphere.

The main purpose of this chapter is to study the motion of a particle subject to a central, isotropic force with particular emphasis on the force of gravity. In carrying out this study, we follow Newton's original line of inquiry, which led to the formulation of his universal law of gravitation. In so doing, we hope to engender an appreciation for the tremendous depth of Newton's intellectual achievement.

Gravitation: An Inverse-Square Law?

While home at Woolsthorpe in 1665, Newton took up the studies that were to occupy him for the rest of his life: mathematics, mechanics, optics, and gravitation. Perhaps the most classic image we have of Newton depicts him sitting under an apple tree and being struck by a falling apple. This visual image is meant to convey the notion of Newton pondering the nature of gravity, most probably wondering whether or not the force that caused the apple to fall could be the same one that held the Moon in its orbit about the Earth.

Galileo, who very nearly postulated the law of inertia in its Newtonian form, inexplicably failed to apply it correctly to the motion of heavenly objects. He missed the most fundamental point of circular motion, namely, that objects moving in circles are accelerated inward and, therefore, require a resultant force in that direction. By Newton's time, a number of natural philosophers had come to the conclusion that some sort of force was required, not to accelerate a planet or satellite inward toward its parent body, but to "maintain it in its orbit." In 1665, the Italian astronomer Giovanni Borelli had presented a theory of the motion of the Galilean Moons of Jupiter in which he stated that the centrifugal

---

[1] G is the least accurately known of all the basic physical constants. This stems from the fact that the gravitational force between two bodies of laboratory size is extremely small. For a review of the current situation regarding the determination of G, see an article by J. Maddox, *Nature,* **30,** 723 (1984). Also, H. de Boer, "Experiments Relating to the Newtonian Gravitational Constant," in B. N. Taylor and W. D. Phillips, eds., *Precision Measurements and Fundamental Constants* (Natl. Bur. Stand. U.S., Spec. Publ., 617, 1984).

force of a Moon's orbital motion was exactly in equilibrium with the attractive force of Jupiter.[2]

Newton was the first to realize that Earth's Moon was not "balanced in its orbit" but was undergoing a centripetal acceleration toward Earth that had to be caused by a centripetal force. Newton surmised that this force was the same one that attracted all Earth-bound objects toward its surface. This had to be the case, because the kinematical behavior of the Moon was no different from that of any object falling toward Earth. The falling Moon never hits Earth because the Moon has such a large tangential velocity that, as it falls a given distance, it moves far enough sideways that Earth's surface has curved away by that same distance. No one at the time even remotely suspected that the centripetal acceleration of the Moon and the gravitational acceleration of an apple falling on the surface of Earth had a common origin.

Newton demonstrated that if a falling apple could also be given a large enough horizontal velocity, its motion would be identical to that of the orbiting Moon (the apple's orbit would just be closer to Earth), thus making the argument for a common origin of an attractive gravitational force even more convincing. Newton further reasoned that the centripetal acceleration of an apple put in orbit about Earth just above its surface would be identical to its gravitational free-fall acceleration. (Imagine an apple shot horizontally out of a powerful cannon. Let there be no air resistance. If the initial horizontal speed of the projected apple were adjusted just right, the apple would never hit Earth because Earth's surface would fall away at the same rate that the apple would fall toward it, just as is the case for the orbiting Moon. In other words, the apple would be in orbit and its centripetal acceleration would exactly equal the g of an apple falling from rest.) Thus, with this single brilliant mental leap, Newton was about to uncover the first and one of the most beautiful of all unifying principles in physics, the law of universal gravitation.

The critical question for Newton was figuring out just how this attractive force depended on distance away from Earth's center. Newton knew that the strength of Earth's attractive force was proportional to the acceleration of falling objects at whatever distance from Earth they happened to be. The Moon's acceleration toward Earth is $a = v^2/r$, where v is the speed of the Moon and r is the radius of its circular orbit. (This is equal to the local value of g.) Newton deduced, with the aid of Kepler's third law (the square of the orbital period τ^2 is proportional to the cube of the distance from the center of the orbit r^3), that this acceleration should vary as $1/r^2$. For example, if the Moon were 4 times farther away from Earth than it actually is, then by Kepler's third law its period of revolution would be 8 times longer, and its orbital speed 2 times slower; consequently, its centripetal acceleration would be 16 times less than it is—or weaker as the inverse square of the distance.

---

[2] Recall from Chapter 5 that centrifugal force is an inertial force exerted on an object in a rotating frame of reference. In the context here, it arises from the centripetal acceleration of a Galilean moon traveling in essentially a circular orbit around Jupiter. To most pre-Newtonian thinkers, the centrifugal force acting on planets or satellites was a real one. Many of their arguments centered on the nature of the force required to "balance out" the centrifugal force. They completely missed the point that from the perspective of an inertial observer, the satellite was undergoing centripetal acceleration inward. They were thus arguing from the perspective of a noninertial observer, although none of them had such a precise understanding regarding the distinction.

Newton thus hypothesized that the local value of g for all falling objects and, hence, the attractive force of gravity, should vary accordingly. To confirm this hypothesis Newton had to calculate the centripetal acceleration of the Moon, compare it to the acceleration g of a falling apple, and see if the ratio were equal to that of the inverse square of their respective distances from the center of the Earth. The Moon's distance is 60 Earth radii. The force of Earth's gravity must, therefore, weaken by a factor of 3600. The rate of fall of an apple must be 3600 times larger than that of the Moon or, put another way, the distance an apple falls in 1 s should equal the distance the Moon falls toward Earth in 1 min, the distance of fall being proportional to time squared. Unfortunately, Newton made a mistake in carrying out this calculation. He assumed that an angle of 1° subtended an arc length of 60 miles on the surface of Earth. He got this from a sailor's manual, the only book at hand. (This distance is, in fact, 60 nautical miles, or 69 English miles.) Setting this equal to 60 English miles of 5280 ft each, however, he computed the Moon's distance of fall in 1 s to be 0.0036 ft, or 13 ft in 1 min. Through Galileo's experiments with falling bodies, repeated later with more accuracy, an apple (or any other body) had been measured to fall about 15 ft in 1 s on Earth. The values are very close, differing by about 1 part in 8, but such a difference was great enough that Newton abandoned his brilliant idea! Later, he was to use the correct values, get it exactly right, and, thus, demonstrate an inverse-square law for the law of gravity.

Proportional to Mass?

Newton also concluded that the force of gravity acting on any object must be proportional to its mass (as opposed to, say, mass squared or something else). This conclusion is derivable from his second law of motion and Galileo's finding that the rate of fall of all objects is independent of their weight and composition. For example, let the force of gravity of Earth acting on some object of inertial mass m be proportional to that mass. Then, according to Newton's second law of motion, $F_{grav} = k \cdot m/r^2 = m \cdot a = m \cdot g$. Thus, $g = k/r^2$. The masses cancel out in this dynamical equation, and the acceleration g depends only on some constant k (which, in some way, must depend on the mass of the Earth but, obviously, is the same for all bodies attracted to the Earth) and the distance r to the center of the Earth. So all bodies fall with the same acceleration regardless of their mass or composition. The gravitational force must be directly proportional to the inertial mass, or this precise cancellation would not occur. Then all falling bodies would exhibit mass-dependent accelerations, contrary to all experiments designed to test such a hypothesis. In fact the equivalence of inertial and gravitational mass of all objects is one of the cornerstones of Einstein's general theory of relativity. For Newton this equivalence remained a mystery to his death.

Product of Masses, Universality?

Newton also realized that if the force of gravity were to obey his third law of motion and if the force of gravity were proportional to the mass of the object being attracted, then it must also be proportional to the mass of the attracting object. Such a requirement leads us inevitably to the conclusion that the law of gravity must, therefore, be "universal";

that is, every object in the universe must attract (albeit very weakly, in most cases) every other object in the universe. Let us see how this comes about. Imagine two masses m_1 and m_2 separated by a distance r. The forces of attraction on 1 by 2 and on 2 by 1 are $F_{12} = k_2 m_1 / r^2$ and $F_{21} = k_1 m_2 / r^2$, where k_1 and k_2 are "constants" that, as we are forced to conclude, must depend on the mass of the attracting object. According to Newton's third law, these forces have to be equal in magnitude (and opposite in direction); therefore, $k_2 m_1 = k_1 m_2$ or $k_2 / k_1 = m_2 / m_1$. To ensure the equality of this ratio, the strength of the attraction of gravity must be proportional to the mass of the attractive body, that is, $k_i = Gm_i$. Thus, the force of gravity between two particles is a central, isotropic law of force possessing a wonderful symmetry: particle 1 attracts particle 2 and particle 2 attracts particle 1 with a magnitude and direction, obeying Newton's third law, proportional to the product of each of their masses and varying inversely as the square of their distance of separation. This conclusion was the work of true genius!

6.2 | Gravitational Force between a Uniform Sphere and a Particle

Newton did not publish the *Principia* until 1687. There was one particular problem that bothered him and made him reluctant to publish. We quickly glossed over that problem in our preceding discussion. Newton derived the inverse-square law by assuming that the relevant distance of separation between two objects, such as Earth and Moon, is the distance between their respective geometrical centers. This does not seem to be unreasonable for spherical objects like the Sun and the planets, or Earth and Moon, whose distances of separation are large compared with their radii. But what about Earth and the apple? If you or I were the apple, looking around at all the stuff in Earth attracting us, we would see lines of gravitational force tugging on us from directions all over the place. There is stuff to the east and stuff to the west whose directions of pull differ by 180°. Who is to say that when we properly add up all the force vectors, due to all this attractive stuff, we get a resultant vector that points to the center of Earth and whose strength depends on the mass of Earth and inversely on the distance to its center squared, as though all Earth's mass were completely concentrated at its center?

Yet, this is the way it works out. It is a tricky problem in calculus that requires a vector sum of infinitessimal contributions over an infinite number of mass elements that lead to a finite result. At that time no one knew calculus because Newton had just invented it, probably to solve this very problem! He was understandably reluctant to publish such a proof, couched in a framework of nonexistent mathematics. Because everyone knows calculus in our present age of enlightenment, we will go ahead and use it to solve the problem, proving that, for any uniform spherical body or any spherically symmetric distribution of matter, the gravitational force exerted by it on any external particle can be calculated by simply assuming that the entire mass of the distribution acts as though concentrated at its geometric center. Only an inverse-square force law works this way.

Consider first a thin uniform shell of mass M and radius R. Let r be the distance from the center O to a test particle P of mass m (Fig. 6.2.1). We assume that $r > R$. We shall divide the shell into circular rings of width $R\,\Delta\theta$ where, as shown in the figure, the angle

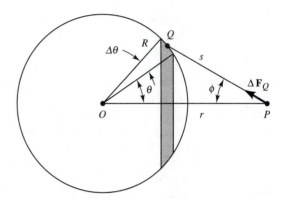

Figure 6.2.1   Coordinates for calculating the gravitational field of a spherical shell.

POQ is denoted by θ, Q being a point on the ring. The circumference of our representative ring element is, therefore, $2\pi R \sin\theta$, and its mass ΔM is given by

$$\Delta M \approx \rho 2\pi R^2 \sin\theta\, \Delta\theta \qquad (6.2.1)$$

where ρ is the mass per unit area of the shell.

Now the gravitational force exerted on P by a small subelement Q of the ring (which we regard as a particle) is in the direction PQ. Let us resolve this force $\Delta\mathbf{F}_Q$ into two components, one component along PO, of magnitude $\Delta F_Q \cos\phi$, the other perpendicular to PO, of magnitude $\Delta F_Q \sin\phi$. Here ϕ is the angle OPQ, as shown in Figure 6.2.1. From symmetry we can easily see that the vector sum of all of the perpendicular components exerted on P by the whole ring vanishes. The force $\Delta\mathbf{F}$ exerted by the entire ring is, therefore, in the direction PO, and its magnitude ΔF is obtained by summing the components $\Delta F_Q \cos\phi$. The result is

$$\Delta F = G\frac{m\,\Delta M}{s^2}\cos\phi = G\frac{m2\pi\rho R^2 \sin\theta\,\cos\phi}{s^2}\Delta\theta \qquad (6.2.2)$$

where s is the distance PQ (the distance from the particle P to the ring) as shown. The magnitude of the force exerted on P by the whole shell is then obtained by taking the limit of $\Delta\theta$ and integrating

$$F = Gm2\pi\rho R^2 \int_0^\pi \frac{\sin\theta\,\cos\phi\,d\theta}{s^2} \qquad (6.2.3)$$

The integral is most easily evaluated by expressing the integrand in terms of s. From the triangle OPQ we have, from the law of cosines,

$$r^2 + R^2 - 2rR\cos\theta = s^2 \qquad (6.2.4)$$

Differentiating, because both R and r are constant, we have,

$$rR\sin\theta\,d\theta = s\,ds \qquad (6.2.5)$$

Also, in the same triangle OPQ, we can write

$$\cos\phi = \frac{s^2 + r^2 - R^2}{2rs} \qquad (6.2.6)$$

On performing the substitutions given by the preceding two equations, and changing the limits of integration from $[0, \pi]$ to $[r - R, r + R]$ we obtain

$$
\begin{aligned}
F &= Gm2\pi\rho R^2 \int_{r-R}^{r+R} \frac{s^2 + r^2 - R^2}{2Rr^2 s^2} \, ds \\
&= \frac{GmM}{4Rr^2} \int_{r-R}^{r+R} \left(1 + \frac{r^2 - R^2}{s^2}\right) ds \\
&= \frac{GmM}{r^2}
\end{aligned}
\tag{6.2.7}
$$

where $M = 4\pi\rho R^2$ is the mass of the shell. We can then write vectorially

$$
\mathbf{F} = -G\frac{Mm}{r^2}\mathbf{e}_r
\tag{6.2.8}
$$

where $\mathbf{e}_r$ is the unit radial vector from the origin O. The preceding result means that a uniform spherical shell of matter attracts an external particle as if the whole mass of the shell were concentrated at its center. This is true for every concentric spherical portion of a solid uniform sphere. *A uniform spherical body, therefore, attracts an external particle as if the entire mass of the sphere were located at the center.* The same is true also for a nonuniform sphere provided the density depends only on the radial distance r.

The gravitational force on a particle located *inside* a uniform spherical shell is zero. The proof is left as an exercise (see Problem 6.2).

6.3 | Kepler's Laws of Planetary Motion

Kepler's laws of planetary motion were a landmark in the history of physics. They played a crucial role in Newton's development of the law of gravitation. Kepler deduced these laws from a detailed analysis of planetary motions, primarily the motion of Mars, the closest outer planet and one whose orbit is, unlike that of Venus, highly elliptical. Mars had been most accurately observed, and its positions on the celestial sphere dutifully recorded by Kepler's irascible but brilliant patron, Tycho de Brahe (1546–1601). Kepler even used some sightings that had been made by the early Greek astronomer Hipparchus (190–125 B.C.E.). Kepler's three laws are:

I. Law of Ellipses (1609)
   The orbit of each planet is an ellipse, with the Sun located at one of its foci.

II. Law of Equal Areas (1609)
   A line drawn between the Sun and the planet sweeps out equal areas in equal times as the planet orbits the Sun.

III. Harmonic Law (1618)
   The square of the sidereal period of a planet (the time it takes a planet to complete one revolution about the Sun relative to the stars) is directly proportional to the cube of the semimajor axis of the planet's orbit.

The derivation of these laws from Newton's theories of gravitation and mechanics was one of the most stupendous achievements in the annals of science. A number of Newton's colleagues who were prominent members of the British Royal Society were convinced that the Sun exerted a force of gravitation on the planets, that the strength of that force must diminish by the square of the distance between the Sun and the planet, and that this fact could be used to explain Kepler's laws. (Kepler's second law is, however, a statement that the angular momentum of a planet in orbit is conserved, a consequence only of the central nature of the gravitational force, not its inverse-square feature.) The trouble was, as noted by Edmond Halley (1656–1742) over lunch with Robert Hooke (1635–1703) and Christopher Wren (1632–1723) in January of 1684, that no one could make the connection mathematically. Part of the problem was that no one, except the silent Newton, could show that the gravitational forces of spherical bodies could be treated as though they emanated from and terminated on their geometric centers. Hooke brashly stated that he could prove the fact that the planets traveled in elliptical orbits but had not told anyone how to do it so that they might, in attempting a solution themselves, appreciate the magnitude of the problem. Wren offered a prize of 40 shillings—in those days the price of an expensive book—to the one who could produce such a proof within two months. Neither Hooke, nor anyone else, won the prize!

In August of 1684, while visiting Cambridge, Halley stopped in to see Newton and asked him what would be the shape of the planets' orbits if they were subject to an inverse-square attractive force by the Sun? Newton replied, without hesitation, "An ellipse!" Halley wanted to know how Newton knew this, and Newton said that he had calculated it years ago. Halley was stunned. They looked through thousands of Newton's papers but could not find the calculation. Newton told Halley that he would redo it and send it to him.

Newton had actually done the calculation five years earlier, in 1679, stimulated in part by Robert Hooke, the aforementioned claimant to the inverse-square law, who had written Newton with questions about the trajectory of objects falling toward a gravitationally attractive body. Unfortunately, there was a mistake in the calculation of Newton's written reply to Hooke. Hooke, with glee, pointed out the mistake, and the angry Newton, concentrating on the problem with renewed vigor, apparently straightened things out. These subsequent calculations, however, also contained a mistake, which is perhaps why Newton failed to find them when queried by Halley. At any rate, Newton furiously attacked the problem again and within three months sent Halley a paper in which he correctly derived all of Kepler's laws from an inverse-square law of gravitation and the laws of mechanics. Thus was the *Principia* born. In the sections that follow we, too, derive Kepler's laws from Newton's fundamental principles.

6.4 | Kepler's Second Law: Equal Areas

Conservation of Angular Momentum

Kepler's second law is nothing other than the statement that the angular momentum of a planet about the Sun is a conserved quantity. To show this, we first define angular momentum and then show that its conservation is a general consequence of the central nature of the gravitational force.

The *angular momentum* of a particle located a vector distance $\mathbf{r}$ from a given origin and moving with momentum $\mathbf{p}$ is defined to be the quantity $\mathbf{L} = \mathbf{r} \times \mathbf{p}$. The time derivative of this quantity is

$$\frac{d\mathbf{L}}{dt} = \frac{d(\mathbf{r} \times \mathbf{p})}{dt} = \mathbf{v} \times \mathbf{p} + \mathbf{r} \times \frac{d\mathbf{p}}{dt} \tag{6.4.1}$$

but

$$\mathbf{v} \times \mathbf{p} = \mathbf{v} \times m\mathbf{v} = m\mathbf{v} \times \mathbf{v} = 0 \tag{6.4.2}$$

Thus,

$$\mathbf{r} \times \mathbf{F} = \mathbf{r} \times \frac{d\mathbf{p}}{dt} = \frac{d\mathbf{L}}{dt} \tag{6.4.3}$$

where we have used Newton's second law, $\mathbf{F} = d\mathbf{p}/dt$.

The cross product $\mathbf{N} = \mathbf{r} \times \mathbf{F}$ is the moment of force, or torque, on the particle about the origin of the coordinate system. If $\mathbf{r}$ and $\mathbf{F}$ are collinear, this cross product vanishes and so does $\mathbf{L}$. The angular momentum $\mathbf{L}$, in such cases, is a constant of the motion. This is quite obviously the case for a particle (or a planet) subject to a *central force* $\mathbf{F}$, that is, one that either emanates or terminates from a single point and whose line of action lies along the radius vector $\mathbf{r}$.

Furthermore, because the vectors $\mathbf{r}$ and $\mathbf{v}$ define an "instantaneous" plane within which the particle moves, and because the angular momentum vector $\mathbf{L}$ is normal to this plane and is constant in both magnitude and direction, the orientation of this plane is fixed in space. Thus, the problem of motion of a particle in a central field is really a two-dimensional problem and can be treated that way without any loss of generality.

Angular Momentum and Areal Velocity of a Particle Moving in a Central Field

As previously mentioned, Kepler's second law, the constancy of the *areal velocity*, $\dot{A}$, of a planet about the Sun, depends only upon the central nature of the gravitational force and not upon how the strength of the force varies with radial distance from the Sun. Here we show that this law is equivalent to the more general result that the angular momentum of any particle moving in a central field of force is conserved, as shown in the preceding section.

To do so, we first calculate the magnitude of the angular momentum of a particle moving in a central field. We use polar coordinates to describe the motion. The velocity of the particle is

$$\mathbf{v} = \mathbf{e}_r \dot{r} + \mathbf{e}_\theta r \dot{\theta} \tag{6.4.4}$$

where $\mathbf{e}_r$ is the unit radial vector and $\mathbf{e}_\theta$ is the unit transverse vector. (see Figure 6.4.1(a)).

The magnitude of the angular momentum is

$$L = |\mathbf{r} \times m\mathbf{v}| = |r\mathbf{e}_r \times m(\dot{r}\mathbf{e}_r + r\dot{\theta}\mathbf{e}_\theta)| \tag{6.4.5}$$

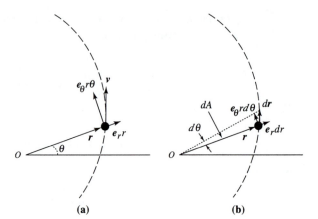

Figure 6.4.1   (a) Angular
momentum $L = |\mathbf{r} \times m\mathbf{v}|$ of a
particle moving in a central
field. (b) Area $dA = \frac{1}{2}|\mathbf{r} \times d\mathbf{r}|$
swept out by the radius vector $\mathbf{r}$
of the particle as it moves in a
central field.

Because $|\mathbf{e}_r \times \mathbf{e}_r| = 0$ and $|\mathbf{e}_r \times \mathbf{e}_\theta| = 1$, we find

$$L = mr^2\dot{\theta} = \text{constant} \tag{6.4.6}$$

for *any particle moving in a central field of force* including a planet moving in the gravitational field of the Sun.

Now, we calculate the "areal velocity," $\dot{A}$, of the particle. Figure 6.4.1(b) shows the triangular area, dA, swept out by the radius vector $\mathbf{r}$ as a planet moves a vector distance $d\mathbf{r}$ in a time dt along its trajectory relative to the origin of the central field. The area of this small triangle is

$$dA = \tfrac{1}{2}|\mathbf{r} \times d\mathbf{r}| = \tfrac{1}{2}|r\mathbf{e}_r \times (\mathbf{e}_r dr + \mathbf{e}_\theta r d\theta)| = \tfrac{1}{2}r(rd\theta) \tag{6.4.7}$$

(**Note:** any increment of motion along the radial direction, $\mathbf{e}_r$, does not add to, or subtract from, the area dA—nor does it contribute anything to the angular momentum of the particle about the center of force.)

Thus, the areal velocity, or the rate at which "area is swept out" by the radius vector pointing to the moving particle is

$$\frac{dA}{dt} = \dot{A} = \tfrac{1}{2}r^2\dot{\theta} = \frac{L}{2m} = \text{constant} \tag{6.4.8}$$

An equivalent way to see this relation is to note that, because $d\mathbf{r} = \mathbf{v}dt$, Equation 6.4.7 can be written as

$$dA = \tfrac{1}{2}|\mathbf{r} \times d\mathbf{r}| = \tfrac{1}{2}|\mathbf{r} \times \mathbf{v}dt| = \frac{L}{2m}dt \tag{6.4.9}$$

which also reduces to Equation 6.4.8.

Thus, the areal velocity, $\dot{A}$, of a particle moving in a central field is directly proportional to its angular momentum and, therefore, is also a constant of the motion, exactly as Kepler discovered for planets moving in the central gravitational field of the Sun.

EXAMPLE 6.4.1

Let a particle be subject to an attractive central force of the form $f(r)$ where r is the distance between the particle and the center of the force. Find $f(r)$ if all circular orbits are to have identical areal velocities, $\dot{A}$.

Solution:

Because the orbits are circular, the acceleration, $\ddot{\mathbf{r}}$, has no transverse component and is entirely in the radial direction. In polar coordinates, it is given by Equation 1.11.10

$$a_r = \ddot{r} - r\dot{\theta}^2 = -r\dot{\theta}^2$$

because $\ddot{r} = 0$. Thus,

$$-mr\dot{\theta}^2 = f(r)$$

Because the areal velocity is the same for all circular orbits, then the angular momentum of the particle, $L = mr^2\dot{\theta}$, must be also. Multiplying and dividing the above expression by the factor, r^3, yields the relation

$$-\frac{mr^4\dot{\theta}^2}{r^3} = -\frac{L^2}{mr^3} = f(r)$$

or in terms of the areal velocity, $\dot{A} = L/2m$

$$f(r) = -\frac{4m\dot{A}^2}{r^3}$$

Therefore, the attractive force for which all circular orbits have identical areal velocities (and angular momenta) is the *inverse r-cube*.

6.5 | Kepler's First Law: The Law of Ellipses

To prove Kepler's first law, we develop a general differential equation for the orbit of a particle in any central, isotropic field of force. Then we solve the orbital equation for the specific case of an inverse-square law of force.

First we express Newton's differential equations of motion using two-dimensional polar coordinates instead of three, remembering from our previous discussion that no loss of generality is incurred because the motion is confined to a plane. The equation of motion in polar coordinates is

$$m\ddot{\mathbf{r}} = f(r)\mathbf{e}_r \tag{6.5.1}$$

where $f(r)$ is the central, isotropic force that acts on the particle of mass m. It is a function only of the scalar distance r to the force center (hence, it is isotropic), and its direction is along the radius vector (hence, it is central). As shown in Equations 1.11.9 and 1.11.10, the radial component of $\ddot{\mathbf{r}}$ is $\ddot{r} - r\dot{\theta}^2$ and the transverse component is $2\dot{r}\dot{\theta} + r\ddot{\theta}$. Thus, the component differential equations of motion are

$$m(\ddot{r} - r\dot{\theta}^2) = f(r) \tag{6.5.2a}$$

$$m(2\dot{r}\dot{\theta} + r\ddot{\theta}) = 0 \tag{6.5.2b}$$

From the latter equation it follows that (see Equation 1.11.11)

$$\frac{d}{dt}(r^2\dot{\theta}) = 0 \tag{6.5.3}$$

or

$$r^2\dot{\theta} = \text{constant} = l \tag{6.5.4}$$

From Equation 6.4.6 we see that

$$|l| = \frac{L}{m} = |\mathbf{r} \times \mathbf{v}| \tag{6.5.5}$$

Thus, l is the angular momentum per unit mass. Its constancy is simply a restatement of a fact that we already know, namely, that the angular momentum of a particle is constant when it is moving under the action of a central force.

Given a certain radial force function $f(r)$, we could, in theory, solve the pair of differential equations (Equations 6.5.2a and b) to obtain r and θ as functions of t. Often one is interested only in the path in space (the *orbit*) without regard to the time t. To find the equation of the orbit, we use the variable u defined by

$$r = \frac{1}{u} \tag{6.5.6}$$

Then

$$\dot{r} = -\frac{1}{u^2}\dot{u} = -\frac{1}{u^2}\dot{\theta}\frac{du}{d\theta} = -l\frac{du}{d\theta} \tag{6.5.7}$$

The last step follows from the fact that

$$\dot{\theta} = lu^2 \tag{6.5.8}$$

according to Equations 6.5.4 and 6.5.6.

Differentiating a second time, we obtain

$$\ddot{r} = -l\frac{d}{dt}\frac{du}{d\theta} = -l\frac{d\theta}{dt}\frac{d}{d\theta}\frac{du}{d\theta} = -l\dot{\theta}\frac{d^2u}{d\theta^2} = -l^2u^2\frac{d^2u}{d\theta^2} \tag{6.5.9}$$

Substituting the values found for r, $\dot{\theta}$, and $\ddot{r}$ into Equation 6.5.2a, we obtain

$$m\left[-l^2u^2\frac{d^2u}{d\theta^2} - \frac{1}{u}(l^2u^4)\right] = f(u^{-1}) \tag{6.5.10a}$$

which reduces to

$$\frac{d^2u}{d\theta^2} + u = -\frac{1}{ml^2u^2}f(u^{-1}) \tag{6.5.10b}$$

Equation 6.5.10b is the *differential equation of the orbit* of a particle moving under a central force. The solution gives u (hence, r) as a function of θ. Conversely, if one is given the polar equation of the orbit, namely, $r = r(\theta) = u^{-1}$, then the force function can be found by differentiating to get $d^2u/d\theta^2$ and inserting this into the differential equation.

EXAMPLE 6.5.1

A particle in a central field moves in the spiral orbit

$$r = c\theta^2$$

Determine the force function.

Solution:

We have

$$u = \frac{1}{c\theta^2}$$

and

$$\frac{du}{d\theta} = \frac{-2}{c}\theta^{-3} \qquad \frac{d^2u}{d\theta^2} = \frac{6}{c}\theta^{-4} = 6cu^2$$

Then from Equation 6.5.10b

$$6cu^2 + u = -\frac{1}{ml^2u^2}f(u^{-1})$$

Hence,

$$f(u^{-1}) = -ml^2(6cu^4 + u^3)$$

and

$$f(r) = -ml^2\left(\frac{6c}{r^4} + \frac{1}{r^3}\right)$$

Thus, the force is a combination of an inverse cube and inverse–fourth power law.

EXAMPLE 6.5.2

In Example 6.5.1 determine how the angle θ varies with time.

Solution:

Here we use the fact that $l = r^2\dot{\theta}$ is constant. Thus,

$$\dot{\theta} = lu^2 = l\frac{1}{c^2\theta^4}$$

or

$$\theta^4 d\theta = \frac{l}{c^2} dt$$

and so, by integrating, we find

$$\frac{\theta^5}{5} = lc^{-2}t$$

where the constant of integration is taken to be zero, so that $\theta = 0$ at $t = 0$. Then we can write

$$\theta = \alpha t^{1/5}$$

where $\alpha = \text{constant} = (5lc^{-2})^{1/5}$.

Inverse-Square Law

We can now solve Equation 6.5.10b for the orbit of a particle subject to the force of gravity. In this case

$$f(r) = -\frac{k}{r^2} \tag{6.5.11}$$

where the constant $k = GMm$. In this chapter we always assume that $M \gg m$ and remains fixed in space. The small mass m is the one whose orbit we calculate. (A modification in our treatment is required when $M \approx m$, or at least not much greater than m. We present such a treatment in Chapter 7.) The equation of the orbit (Equation 6.5.10b) then becomes

$$\frac{d^2u}{d\theta^2} + u = \frac{k}{ml^2} \tag{6.5.12}$$

Equation 6.5.12 has the same form as the one that describes the simple harmonic oscillator, but with an additive constant. The general solution is

$$u = A \cos(\theta - \theta_0) + \frac{k}{ml^2} \tag{6.5.13}$$

or

$$r = \frac{1}{k/ml^2 + A \cos(\theta - \theta_0)} \tag{6.5.14}$$

The constants of integration, A and θ_0, are determined from initial conditions or from the values of the position and velocity of the particle at some particular instant of time. The value of θ_0, however, can always be adjusted by a simple rotation of the coordinate system used to measure the polar angle of the particle. Consistent with convention, we set $\theta_0 = 0$, which corresponds to a direction toward the point of the particle's closest approach

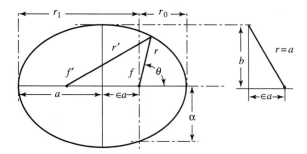

Figure 6.5.1   The ellipse
    f, f' The two foci of the ellipse
    a Semimajor axis
    b Semiminor axis: $b = (1 - \epsilon^2)^{1/2} a$
    ϵ Eccentricity: each focus displaced from center by ϵa
    α Latus rectum: Distance of focus from point on the ellipse
          perpendicular to major axis: $\alpha = (1 - \epsilon^2)\, a$
    r_0 Distance from the focus to the pericenter: $r_0 = (1 - \epsilon)a$
    r_1 Distance from the focus to the apocenter: $r_1 = (1 + \epsilon)a$

to the origin. We can then rewrite Equation 6.5.14 as

$$r = \frac{ml^2/k}{1 + (Aml^2/k)\cos\theta} \tag{6.5.15}$$

This equation describes an ellipse (see Figure 6.5.1) with the origin at one of its foci, in the case where the motion of the particle is bound.

An ellipse is defined to be the locus of points whose total distance from two *foci*, f and f', is a constant, that is,

$$r + r' = constant = 2a \tag{6.5.16}$$

where a is the semimajor axis of the ellipse, and the two foci are offset from its center, each by an amount ϵa. ϵ is called the *eccentricity* of the ellipse. We can show that Equation 6.5.15 is equivalent to this fundamental definition of the ellipse by first finding a relation between r and r' using the Pythagorean theorem (see Figure 6.5.1).

$$r'^2 = r^2 \sin^2\theta + (2\epsilon a + r\cos\theta)^2$$
$$= r^2 + 4\epsilon a(\epsilon a + r\cos\theta) \tag{6.5.17}$$

and then substituting the fundamental definition, $r' = 2a - r$, into it to obtain

$$r = \frac{a(1 - \epsilon^2)}{1 + \epsilon\cos\theta} \tag{6.5.18a}$$

As can be seen from Figure 6.5.1, at $\theta = \pi/2$, $r = a(1 - \epsilon^2) = \alpha$ the *latus rectum* of the ellipse. Hence, Equation 6.5.18a can be put into the form

$$r = \frac{\alpha}{1 + \epsilon\cos\theta} \tag{6.5.18b}$$

Thus, it is equivalent to Equation 6.5.15 with

$$\alpha = \frac{ml^2}{k} \tag{6.5.19}$$

and

$$\epsilon = \frac{Aml^2}{k} \tag{6.5.20}$$

Even though Equations 6.5.18a and b were derived for an ellipse, they are more general than that: they actually describe any *conic section* and all possible orbits other than elliptical around the gravitational source.

A conic section is formed by the intersection of a plane and a cone (see Figure 6.5.2a–d). The angle of tilt between the plane and the axis of the cone determines the resulting section. This angle is related to the eccentricity ϵ in Equations 6.5.18a and b. When $0 < \epsilon < 1$, Equations 6.5.18a and b describe an ellipse. It is formed when the angle between the plane and the axis of the cone is less than $\pi/2$ but greater than β, where β is the generating angle of the cone (see Figure 6.5.2b). The expression for a circle, $r = a$, is retrieved when $\epsilon = 0$, and it is formed when the plane is perpendicular to the cone's axis (see Figure 6.5.2a). As $\epsilon \to 1$, $a \to \infty$, but the product $\alpha = a(1 - \epsilon^2)$ remains finite and when $\epsilon = 1$ Equation 6.5.18a and b then describe a parabola. The angle between the plane and the axis of the cone is then equal to β (see Figure 6.5.2c). Finally, when $\epsilon > 1$, Equations 6.5.18a and b describe a hyperbola, and the angle of tilt lies between 0 and β (see Figure 6.5.2d). The different conic sections as seen by an observer positioned perpendicular to the plane in each of Figures 6.5.2a–d are shown in Figure 6.5.2e. They correspond to different possible orbits of the particle.

In reference to the elliptical orbits of the planets around the Sun (see Figure 6.5.1), r_0, the *pericenter* of the orbit, is called the *perihelion*, or distance of closest approach to the Sun; r_1, the *apocenter* of the orbit, is called the *aphelion*, or the distance at which the planet is farthest from the Sun. The corresponding distances for the orbit of the Moon around the Earth—and for the orbits of the Earth's artificial satellites—are called the *perigee* and *apogee*, respectively. From Equation 6.5.18b, it can be seen that these are the values of r at $\theta = 0$ and $\theta = \pi$, respectively:

$$r_0 = \frac{\alpha}{1+\epsilon} \tag{6.5.21a}$$

$$r_1 = \frac{\alpha}{1-\epsilon} \tag{6.5.21b}$$

The orbital eccentricities of the planets are quite small. (See Table 6.5.1.) For example, in the case of Earth's orbit $\epsilon = 0.017$, $r_0 = 91{,}000{,}000$ mi, and $r_1 = 95{,}000{,}000$ mi. On the other hand, the comets generally have large orbital eccentricities (highly elongated orbits). Halley's Comet, for instance, has an orbital eccentricity of 0.967 with a perihelion distance of only 55,000,000 mi, while at aphelion it is beyond the orbit of Neptune. Many comets (the nonrecurring type) have parabolic or hyperbolic orbits.

The energy of the object is the primary factor that determines whether or not its orbit is an open (parabola, hyperbola) or closed (circle, ellipse) conic section. "High"-energy

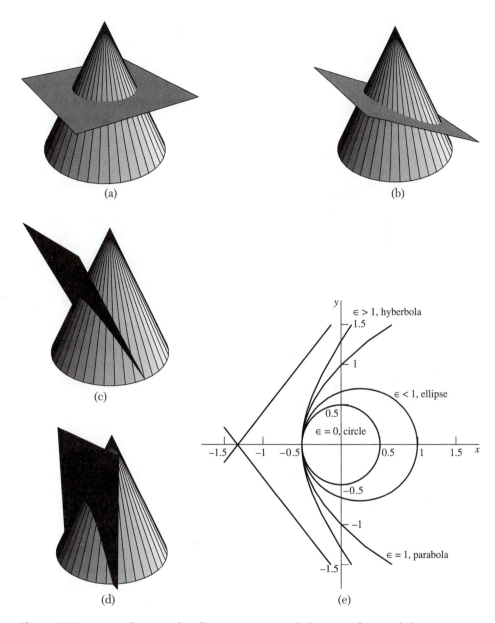

Figure 6.5.2   (a) Circle: $\epsilon = 0$. (b) Ellipse: $\epsilon < 1$. (c) Parabola: $\epsilon = 1$. (d) Hyperbola: $\epsilon > 1$.
(e) The family of conic sections.

objects follow open, unbound orbits, and "low"-energy objects follow closed, bound ones. We treat this subject in greater detail in Section 6.10. Using the language of a noninertial observer, perfectly circular orbits correspond to a situation when the gravitational and centrifugal forces of a planet are exactly balanced. It should surprise you to see that the orbits of the planets are nearly circular.

It is very difficult to envision just how such a situation might arise from initial conditions. If a planet is hurtling around the Sun a little too fast, the centrifugal force slightly outweighs the gravitational force, and the planet moves away from the Sun a little bit. In doing so, it slows down until the force of gravity begins to overwhelm the centrifugal force. The planet then falls back in a little bit closer to the Sun, picking up speed along the way. The centrifugal force builds up to a point where it again outweighs the force of gravity and the process repeats itself. Thus, elliptical orbits can be seen as the result of a continuing tug of war between the slightly unbalanced gravitational and centrifugal forces that inevitably occurs whenever the tangential velocity of the planet is not adjusted just so. These forces must grow and shrink in such a way that the stability of the orbit is ensured. The criterion for stability is discussed in Section 6.13.

One way in which these two forces could be perfectly balanced all the way around the orbit would be if the planet started off just right; that is, very special initial conditions would have to have been set up in the beginning, so to speak. It is difficult to imagine how any natural process could have established such nearly perfect prerequisites. Hence, if planets are bound to the Sun at all, one would think that they would be most likely to travel in elliptical orbits just like Kepler said, unless something happened during the course of solar system evolution that brought the planets into circular orbits. We leave it to the student to think about just what sort of thing might do this.

EXAMPLE 6.5.3

Calculate the speed of a satellite in circular orbit about Earth.

Solution:

In the case of circular motion, the orbital radius is given by $r_c = a = \alpha = ml^2/k$ because the eccentricity is zero (Equations 6.5.18a and 6.5.19). In Earth's gravitational field, the force constant is $k = GM_e m$ in which M_e is the mass of Earth and m is the mass of the satellite. The angular momentum of the satellite per unit mass $l = v_c r_c$, where v_c is the speed of the satellite. Thus,

$$r_c = \frac{m(v_c r_c)^2}{GM_e m}$$

$$\therefore v_c^2 = \frac{GM_e}{r_c}$$

As shown in Example 2.3.2, the product GM_e can be found by noting that the force of gravity at Earth's surface is $mg = GM_e/R_e^2$ or $GM_e = gR_e^2$, where R_e is the radius of Earth. Thus, the speed of a satellite in circular orbit is

$$v_c = \left(\frac{gR_e^2}{r_c} \right)^{1/2}$$

For satellites in low-lying orbits close to Earth's surface, $r_c \approx R_e$, so the speed is $v_c \approx (gR_e)^{1/2} = (9.8 \text{ ms}^{-1} \times 6.4 \times 10^6 \text{ m})^{1/2} = 7920$ m/s, or about 8 km/s.

EXAMPLE 6.5.4

The most energy-efficient way to send a spacecraft to the Moon is to boost its speed while it is in circular orbit about the Earth such that its new orbit is an ellipse. The boost point is the perigee of the ellipse, and the point of arrival at the Moon is the apogee (see Figure 6.5.3). Calculate the percentage increase in speed required to achieve such an orbit. Assume that the spacecraft is initially in a low-lying circular orbit about Earth. The distance between Earth and the Moon is approximately $60R_e$, where R_e is the radius of Earth.

Solution:

The radius and speed of the craft in its initial circular orbit was calculated in Example 6.5.3. That radius is the perigee distance of the new orbit, $r_0 = R_e$. Let v_0 be the velocity required at perigee to send the craft to an apogee at $r_1 = 60R_e$. Because the eccentricity of the initial circular orbit is zero, we have (Equations 6.5.19, 6.5.21a)

$$r_0 = \frac{\alpha_c}{\epsilon + 1} = \alpha_c = \frac{ml_c^2}{k}$$

But the angular momentum per unit mass for the circular orbit, l_c (Equation 6.5.4), is a constant and can be set equal to

$$l_c = r^2\dot{\theta} = r_0^2\dot{\theta}_0 = r_0 v_c$$

On substituting this into the preceding expression, we obtain

$$r_0 = \frac{k}{mv_c^2}$$

After the speed boost from v_c to v_0 at perigee, from Equation 6.5.21a we obtain an elliptical orbit of eccentricity ϵ given by

$$\epsilon = \frac{\alpha}{r_0} - 1 = \frac{ml^2}{kr_0} - 1 = \frac{mv_0^2 r_0}{k} - 1$$

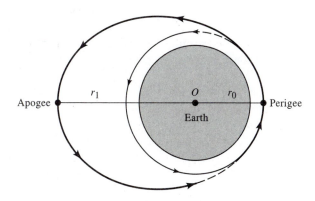

Apogee r_1 O r_0 Perigee

Earth

Figure 6.5.3 Spacecraft changing from a circular to an elliptical orbit.

where we made use of the fact that the new angular momentum per unit mass is $l = v_0 r_0$. Inserting the previous expression for r_0 into the preceding one gives us the ratio of the speeds required to achieve the new eccentric orbit

$$\left(\frac{v_0}{v_c}\right)^2 = \epsilon + 1$$

We can find the eccentricity in terms of the distances of perigee and apogee from the geometry of an ellipse

$$r_1 = (1 + \epsilon)a = (1 + \epsilon)\frac{(r_1 + r_0)}{2}$$

$$\therefore (1 + \epsilon) = \left(\frac{v_0}{v_c}\right)^2 = \frac{2r_1}{r_1 + r_0}$$

Putting in numbers for the required orbit, we obtain

$$\frac{v_0}{v_c} = \sqrt{\frac{2r_1}{r_1 + r_0}} = \sqrt{\frac{120R_e}{61R_e}} = 1.40$$

Thus, a 40% boost up to a speed of about 11.2 km/s is required.

6.6 | Kepler's Third Law: The Harmonic Law

Why is Kepler's third law, relating orbital period to distance from the Sun, called the harmonic law? Kepler's work, more than that of any of the other great scientists who were trying to unlock the mysteries of planetary motion, is a wonderful illustration of the profound effect that intense hunger for knowledge and personal belief have on the growth of science. Kepler held the conviction that the world, which had treated him so harshly at times, nonetheless, was fundamentally a beautiful place. Kepler believed in the Pythagorean doctrine of celestial harmony. The world was a tumultuous place, and the planets were discordant only because humankind had not yet learned how to hear the true harmony of the worlds. In his work *Harmonice Mundi* (*The Harmony of the World*), Kepler, like the Pythagoreans almost 2000 years earlier, tried to connect the planetary motions with all fields of abstraction and harmony: geometrical figures, numbers, and musical harmonies. In this attempt he failed. But in the midst of all this work, indicative of his yearnings and strivings, we find his final precious jewel, always cited as Kepler's third law, the harmonic law. It was this law that gave us the sheets to the music of the spheres.

We show how the third law can be derived from Newton's laws of motion and the inverse-square law of gravity. Starting with Equation 6.4.7b, Kepler's second law

$$\dot{A} = \frac{L}{2m} \tag{6.6.1}$$

we can relate the area of the orbit to its period and angular momentum per unit mass, $l = L/m$, by integrating the areal velocity over the entire orbital period

$$\int_0^\tau \dot{A}\,dt = A = \frac{l}{2}\tau$$

$$\tau = \frac{2A}{l}$$

(6.6.2)

The industrious student can easily prove that the area of an ellipse is πab. Thus, we have (see Figure 6.5.1)

$$\tau = \frac{2\pi ab}{l} = \frac{2\pi a^2 \sqrt{1-\epsilon^2}}{l}$$

(6.6.3)

or

$$\tau^2 = \frac{4\pi^2 a^4}{l^2}(1-\epsilon^2)$$

$$= \frac{4\pi^2 a^4}{l^2}\frac{\alpha}{a} = 4\pi^2 a^3 \frac{\alpha}{l^2}$$

(6.6.4)

On inserting the relation $\alpha = ml^2/k$ (Equation 6.5.19) and the force constant $k = GM_\odot m$, appropriate for planetary motion about the Sun (whose mass is $M_\odot$), into Equation 6.6.4, we arrive at Kepler's third law:

$$\tau^2 = \frac{4\pi^2}{GM_\odot}a^3$$

(6.6.5)

—the square of a planet's orbital period is proportional to the cube of its "distance" from the Sun. The relevant "distance" is the semimajor axis a of the elliptical orbit. In the case of a circular orbit, this distance reduces to the radius.

The constant $4\pi^2/GM_\odot$ is the same for all objects in orbit about the Sun, regardless of their mass.[3] If distances are measured in astronomical units (1 AU $= 1.50 \cdot 10^8$ km) and periods are expressed in Earth years, then $4\pi^2/GM_\odot = 1$. Kepler's third law then takes the very simple form $\tau^2 = a^3$. Listed in Table 6.6.1 are the periods and their squares, the semimajor axes and their cubes, along with the eccentricities of all the planets.

(**Note:** Most of the planets have nearly circular orbits, with the exception of Pluto, Mercury, and Mars.)

---

[3] Actually the Sun and a planet orbit their common center of mass (if only a two-body problem is being considered). A more accurate treatment of the orbital motion is carried out in Section 7.3 where it is shown that a more correct value for this "constant" is given by $4\pi^2/G(M_\odot + m)$, where m is the mass of the orbiting planet. The correction is very small.

TABLE 6.6.1	Planetary Data				
	Period		**Semimajor Cube**		**Eccentricity**
Planet	$\tau(yr)$	**Square** $\tau^2(yr^2)$	**Axis** $a(AU)$	$a^3(AU^3)$	ϵ
Mercury	0.241	**0.0581**	0.387	**0.0580**	0.206
Venus	0.615	**0.378**	0.723	**0.378**	0.007
Earth	1.000	**1.000**	1.000	**1.000**	0.017
Mars	1.881	**3.538**	1.524	**3.540**	0.093
Jupiter	11.86	**140.7**	5.203	**140.8**	0.048
Saturn	29.46	**867.9**	9.539	**868.0**	0.056
Uranus	84.01	**7058.**	19.18	**7056.**	0.047
Neptune	164.8	**27160.**	30.06	**27160.**	0.009
Pluto	247.7	**61360.**	39.440	**61350.**	0.249

EXAMPLE 6.6.1

Find the period of a comet whose semimajor axis is 4 AU.

Solution:

With τ measured in years and a in astronomical units, we have

$$\tau = 4^{3/2} \text{ years} = 8 \text{ years}$$

About 20 comets in the solar system have periods like this, whose aphelia lie close to Jupiter's orbit. They are known as Jupiter's family of comets. They do not include Halley's Comet.

EXAMPLE 6.6.2

The altitude of a near circular, low earth orbit (LEO) satellite is about 200 miles. (a) Calculate the period of this satellite.

Solution:

For circular orbits, we have

$$\frac{GM_E m}{R^2} = m\frac{v^2}{R} = m\frac{4\pi^2 R^2/\tau^2}{R}$$

Solving for τ

$$\tau^2 = \frac{4\pi^2}{GM_E} R^3$$

which—no surprise—is Kepler's third law for objects in orbit about the Earth. Let $R = R_E + h$ where h is the altitude of the satellite above the Earth's surface. Then

$$\tau^2 = \frac{4\pi^2}{GM_E} R_E^3 \left(1 + \frac{h}{R_E}\right)^3$$

But $GM_E/R_E^2 = g$, so we have

$$\tau = 2\pi \sqrt{\frac{R_E}{g}} \left(1 + \frac{h}{R_E}\right)^{3/2} \approx 2\pi \sqrt{\frac{R_E}{g}} \left(1 + \frac{3h}{2R_E}\right)$$

Putting in numbers $R_E = 6371$ km, $h = 322$ km, we get $\tau \approx 90.8$ min ≈ 1.51 hr.

There is another way to do this if you realize that Kepler's third law, being a derivative of Newton's laws of motion and his law of gravitation, applies to any set of bodies in orbit about another. The Moon orbits the Earth once every 27.3 days[4] at a radius of 60.3 R_E. Thus, scaling Kepler's third law to these values (1 month = 27.3 days and 1 lunar unit (LU) = 60.3 R_E), we have

$$\tau^2 \text{ (months)} = R^3 \text{ (LU)}$$

Thus, for our LEO satellite $R = \dfrac{6693}{6371} R_E = 1.051 R_E = \dfrac{1.051 R_E}{60.3\, R_E/LU} = 0.01743\, LU$

$$\tau \text{ (months)} = R^{3/2}[LU] = (0.01743)^{3/2} \text{ months} = 0.002301 \text{ months} \equiv 1.51 \text{ hr}$$

(b) A geosynchronous satellite orbits the Earth in its equatorial plane with a period of 24 hr. Thus, it seems to hover above a fixed point on the ground (which is why you can point your TV satellite receiver dish towards a fixed direction in the sky). What is the radius of its orbit?

Solution:

Using Kepler's third law again, we get

$$R_{geo} = \tau^{2/3} = \left(\frac{1}{27.3}\right)^{2/3} = 0.110\, LU \equiv 6.65 R_E \approx 42{,}400 \text{ km}$$

Universality of Gravitation

A tremendous triumph of Newtonian physics ushered in the 19th century—Urbain Jean Leverrier's (1811–1877) discovery of Neptune. It signified a turning point in the history of science, when the newly emerging methodology, long embroiled in a struggle with biblical ideas, began to dominate world concepts. The episode started when Alexis Bouvard, a farmer's boy from the Alps, came to Paris to study science and there perceived irregularities in the motion of Uranus that could not be accounted for by the attraction

---

[4]This is the *sidereal month*, or the time it takes the Moon to complete one orbit of 360 degrees relative to the stars. The month that we are all familiar with is the *synodic month* of 29.5 days, which is the time it takes for the Moon to go through all of its phases.

of the other known planets. In the years to follow, as the irregularities in the motion of Uranus mounted up, the opinion became fairly widespread among astronomers that there had to be an unknown planet disturbing the motion of Uranus.

In 1842–1843 John Couch Adams (1819–1892), a gifted student at Cambridge, began work on this problem, and by September of 1845 he presented Sir George Airy (1801–1892), the Astronomer Royal, and James Challis, the director of the Cambridge Observatory, the likely coordinates of the offending, but then unknown, planet. It seemed impossible to them that a mere student, armed only with paper and pencil, could take observations of Uranus, invoke the known laws of physics, and predict the existence and precise location of an undiscovered planet. Besides, Airy had grave doubts concerning the validity of the inverse square law of gravity. In fact, he believed that the law of gravity fell off faster than inverse square at great distances. Thus, somewhat understandably, Airy was reluctant to place much credence in Adams' work, and so the two great men chose to ignore him, sealing forever their fate as the astronomers who failed to discover Neptune.

It was about this time that Leverrier began work on the problem. By 1846, he had calculated the orbit of the unknown planet and made a precise prediction of its position on the celestial sphere. Airy and Challis saw that Leverrier's result miraculously agreed with the prediction of Adams. Challis immediately initiated a search for the unknown planet in the suspicious sector of the sky, but owing to Cambridge's lack of detailed star maps in that area, the search was laboriously painstaking and the data reduction problem prodigious. Had Challis proceeded with vigor and tenacity, he most assuredly would have found Neptune, for it was there on his photographic plates. Unfortunately, he dragged his feet. By this time, however, an impatient Leverrier had written Johann Galle (1812–1910), astronomer at the Berlin Observatory, asking him to use their large refractor to examine the stars in the suspect area to see if one showed a disc, a sure signature of a planet. A short time before the arrival of Leverrier's letter containing this request, the Berlin Observatory had received a detailed star map of this sector of the sky from the Berlin Academy. On receipt of Leverrier's letter, September 23, 1846, the map was compared with an image of the sky taken that night, and the planet was identified as a foreign star of eighth magnitude, not seen on the star map. It was named Neptune. Newtonian physics had triumphed in a way never seen before—laws of physics had been used to make a verifiable prediction to the world at large, an unexpected demonstration of the power of science.

Since that time, celestial objects observed at increasingly remote distances continue to exhibit behavior consistent with the laws of Newtonian physics (ignoring those special cases involving large gravitational fields or very extreme distances, each requiring treatment with general relativity). The behavior of binary star systems within our galaxy serves as a classic example. Such stars are bound together gravitationally, and their orbital dynamics are well described by Newtonian mechanics. We discuss them in the next chapter. So strongly do we believe in the universality of gravitation and the laws of physics, that apparent violations by celestial objects, as in the case of Uranus and the subsequent discovery of Neptune, are usually greeted by searches for unseen disturbances. Rarely do we instead demand the overthrow of the laws of physics. (Although, astonishingly enough, two famous examples discussed subsequently in this chapter had precisely this effect and helped revolutionize physics.)

The more likely scenario, ferreting out the unseen disturbance, is currently in progress in many areas of contemporary astronomical research, as illustrated by the

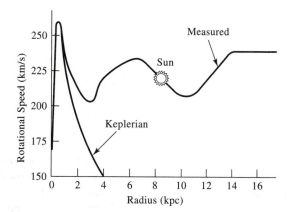

Figure 6.6.1 Galactic rotation curve. The Sun's speed is about 220 km/s and its distance from the galactic center is about 8.5 kpc ($\approx$28,000 light years).

search for dark matter in the universe. One of the reasons we think that an enormous amount of unseen matter fills the universe (perhaps 10 times as much as is visible) can be gleaned from the dynamics of spiral galaxies—thin disc-shaped, rotating aggregates of as many as 100 billion stars. At first sight the rotation curve for spiral galaxies (a plot of the rotational velocity of the stars as a function of their radial distance from the galactic center) seems to violate Kepler's laws. An example of such a curve is shown in Figure 6.6.1. Most of the luminous matter of a spiral galaxy is contained in its central nucleus whose extent is on the order of several thousand light years in radius. The rest of the luminous matter is in the spiral arms that extend in radius out to a distance of about 50,000 light years. The whole thing slowly rotates about its center of gravity, exactly as one would expect for a self-gravitating conglomerate of stars, gas, dust, and so on. The surprising thing about the rotation curve in Figure 6.6.1 is that it is apparently non-Keplerian.

We can illustrate what we mean by this with a simple example. Assume that the entire galactic mass is concentrated within a nucleus of radius R and that stars fill the nucleus at uniform density. This is an oversimplification, but a calculation based on such a model should serve as a guide for what we might expect a rotation curve to look like. The rotational velocity of stars at some radius $r < R$ within the nucleus is determined only by the amount of mass M within the radius r. Stars external to r have no effect. Because the density of stars within the total nuclear radius R is constant, we calculate M as

$$M = \tfrac{4}{3}\pi\rho r^3 \tag{6.6.6}$$

where

$$\rho = \frac{M_{gal}}{\left(\tfrac{4}{3}\right)\pi R^3} \tag{6.6.7}$$

and from Newton's second law

$$\frac{GMm}{r^2} = \frac{mv^2}{r} \tag{6.6.8}$$

for the gravitational force exerted on a star of mass m at a distance r from the center of the nucleus by the mass M interior to that distance r. Solving for v, we get

$$v = \sqrt{GM_{gal}/R^3}\ r \tag{6.6.9}$$

or, the rotational velocity of stars at $r < R$ is proportional to r. For stars in the spiral arms at distances $r > R$, we obtain

$$\frac{GM_{gal}m}{r^2} = \frac{mv^2}{r} \tag{6.6.10}$$

thus,

$$v = \sqrt{\frac{GM_{gal}}{r}} \tag{6.6.11}$$

or, the rotational velocity of stars at $r > R$ is proportional to $1/\sqrt{r}$. This is what we mean by Keplerian rotation. It is the way the velocities of planets depend on their distance from the Sun. We show such a curve in Figure 6.6.1, where we have assumed that the entire mass of the galaxy is uniformly distributed in a sphere whose radius is 1 kpc [1 parsec (pc) = 3.26 light years (ly)].

Let us examine the measured rotation curve. Initially, it climbs rapidly from zero at the galactic center to about 250 km/s at 1 kpc, more or less as expected, but the astonishing thing is that the curve does not fall off in the expected Keplerian manner. It stays more or less flat all the way out to the edges of the spiral arms (the zero on the vertical axis has been suppressed so the curve is flatter than it appears to be). The conclusion is inescapable. As we move away from the galactic center, we must "pick up" more and more matter within any given radius, which causes even the most remote objects in the galaxy to orbit at velocities that exceed those expected for a highly centralized matter distribution. Because most of the luminosity of a galaxy comes from its nucleus, we conclude that dark, unseen matter must permeate spiral galaxies all the way out to their very edges and beyond. (In fact it should be a simple matter to deduce the radial distribution of dark matter required to generate this flat rotation curve.) Of course, Newton's laws could be wrong, but we think not. It looks like a case of Neptune revisited.

6.7 | Potential Energy in a Gravitational Field: Gravitational Potential

In Example 2.3.2 we showed that the inverse-square law of force leads to an inverse first power law for the potential energy function. In this section we derive this same relationship in a more physical way.

Let us consider the work W required to move a test particle of mass m along some prescribed path in the gravitational field of another particle of mass M.

We place the particle of mass M at the origin of our coordinate system, as shown in Figure 6.7.1a. Because the force $\mathbf{F}$ on the test particle is given by $\mathbf{F} = -(GMm/r^2)\mathbf{e}_r$, then

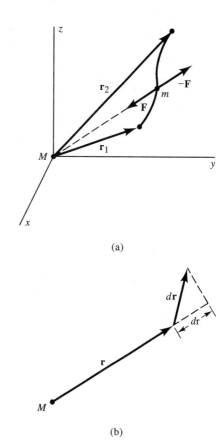

(a)

Figure 6.7.1 Diagram for finding the work required to move a test particle in a gravitational field.

(b)

to overcome this force an external force $-\mathbf{F}$ must be applied. The work dW done in moving the test particle through a displacement $d\mathbf{r}$ is, thus, given by

$$dW = -\mathbf{F} \cdot d\mathbf{r} = \frac{GMm}{r^2}\mathbf{e}_r \cdot d\mathbf{r} \tag{6.7.1}$$

Now we can resolve $d\mathbf{r}$ into two components: $\mathbf{e}_r\, dr$ parallel to $\mathbf{e}_r$ (the radial component) and the other at right angles to $\mathbf{e}_r$ (Figure 6.7.1b). Clearly,

$$\mathbf{e}_r \cdot d\mathbf{r} = dr \tag{6.7.2}$$

and so W is given by

$$W = GMm \int_{r_1}^{r_2} \frac{dr}{r^2} = -GMm\left(\frac{1}{r_2} - \frac{1}{r_1}\right) \tag{6.7.3}$$

where r_1 and r_2 are the radial distances of the particle at the beginning and end of the path. Thus, the work is independent of the particular path taken; it depends only on the endpoints. This verifies a fact we already knew, namely that a central force described by an inverse-square law is conservative.

We can, thus, define the potential energy of a test particle of mass m at a given point in the gravitational field of another particle of mass M as the work done in moving the test particle from some arbitrary reference position r_1 to the position r_2. We take the reference position to be $r_1 = \infty$. This assignment is usually a convenient one, because the gravitational force between two particles vanishes when they are separated by ∞. Thus, putting $r_1 = \infty$ and $r_2 = r$ in Equation 6.7.3, we have

$$V(r) = GMm \int_{\infty}^{r} \frac{dr}{r^2} = -\frac{GMm}{r} \tag{6.7.4}$$

Like the gravitational force, the gravitational potential energy of two particles separated by ∞ also vanishes. Note that for finite separations, the gravitational potential energy is negative.

Both the gravitational force and potential energy between two particles involve the concept of action at a distance. Newton himself was never able to explain or describe the mechanism by which such a force worked. We do not attempt to either, but we would like to introduce the concept of field in such a way that forces and potential energies can be thought of as being generated not by actions at a distance but by local actions of matter with an existing field. To do this, we introduce the quantity Φ, called the *gravitational potential*

$$\Phi = \lim_{m \to 0} \left(\frac{V}{m} \right) \tag{6.7.5}$$

In essence Φ is the gravitational potential energy per unit mass that a very small test particle would have in the presence of other surrounding masses. We take the limit as $m \to 0$ to ensure that the presence of the test particle does not affect the distribution of the other matter and change the thing we are trying to define. Clearly, the potential should depend only on the magnitude of the other masses and their positions in space, not those of the particle we are using to test for the presence of gravitation. We can think of the potential as a scalar function of spatial coordinates, $\Phi(x, y, z)$, or a field, set up by all the other surrounding masses. We test for its presence by placing the test mass m at any point (x, y, z). The potential energy of that test particle is then given by

$$V(x, y, z) = m\Phi(x, y, z) \tag{6.7.6}$$

We can think of this potential energy as being generated by the local interaction of the mass m and the field Φ that is present at the point (x, y, z).

The gravitational potential at a distance r from a particle of mass M is

$$\Phi = -\frac{GM}{r} \tag{6.7.7}$$

If we have a number of particles $M_1, M_2, \ldots, M_i, \ldots$ located at positions $\mathbf{r}_1, \mathbf{r}_2, \ldots, \mathbf{r}_i, \ldots$, then the gravitational potential at the point $\mathbf{r}(x, y, z)$ is the sum of the gravitational potentials of all the particles, that is,

$$\Phi(x, y, z) = \sum \Phi_i = -G \sum \frac{M_i}{s_i} \tag{6.7.8}$$

in which s_i is the distance of the field point $\mathbf{r}(x, y, z)$ from the position $\mathbf{r}_i(x_i, y_i, z_i)$ of the ith particle

$$s_i = |\mathbf{r} - \mathbf{r}_i| \tag{6.7.9}$$

We define a vector field $\mathbf{g}$, called the *gravitational field intensity*, in a way that is completely analogous to the preceding definition of the gravitational potential scalar field

$$\mathbf{g} = \lim_{m \to 0} \left(\frac{\mathbf{F}}{m} \right) \tag{6.7.10}$$

Thus, the gravitational field intensity is the gravitational force per unit mass acting on a test particle of mass m positioned at the point (x, y, z). Clearly, if the test particle experiences a gravitational force given by

$$\mathbf{F} = m\mathbf{g} \tag{6.7.11}$$

then we know that other nearby masses are responsible for the presence of the local field intensity $\mathbf{g}$.[5]

The relationship between field intensity and the potential is the same as that between the force $\mathbf{F}$ and the potential energy V, namely

$$\mathbf{g} = -\nabla\Phi \tag{6.7.12a}$$

$$\mathbf{F} = -\nabla V \tag{6.7.12b}$$

The gravitational field intensity can be calculated by first finding the potential function from Equation 6.7.8 and then calculating the gradient. This method is usually simpler than the method of calculating the field directly from the inverse-square law. The reason is that the potential energy is a scalar sum, whereas the field intensity is given by a vector sum. The situation is analogous to the theory of electrostatic fields. In fact one can apply any of the corresponding results from electrostatics to find gravitational fields and potentials with the proviso, of course, that there are no negative masses analogous to negative charge.

EXAMPLE 6.7.1

Potential of a Uniform Spherical Shell

As an example, let us find the potential function for a uniform spherical shell.

Solution:

By using the same notation as that of Figure 6.2.1, we have

$$\Phi = -G \int \frac{dM}{s} = -G \int \frac{2\pi\rho R^2 \sin\theta \, d\theta}{s}$$

---

[5] $\mathbf{g}$ is the local acceleration of a mass m due to gravity. On the surface of Earth, its value is 9.8 m/s^2 and is primarily due to the mass of the Earth.

From the relation between s and θ that we used in Equation 6.2.5, we find that the preceding equation may be simplified to read

$$\Phi = -G\frac{2\pi\rho R^2}{rR}\int_{r-R}^{r+R} ds = -\frac{GM}{r} \tag{6.7.13}$$

where M is the mass of the shell. This is the same potential function as that of a single particle of mass M located at O. Hence, the gravitational field outside the shell is the same as if the entire mass were concentrated at the center. It is left as a problem to show that, with an appropriate change of the integral and its limits, the potential inside the shell is constant and, hence, that the field there is zero.

EXAMPLE 6.7.2

Potential and Field of a Thin Ring

We now wish to find the potential function and the gravitational field intensity in the plane of a thin circular ring.

Solution:

Let the ring be of radius R and mass M. Then, for an exterior point lying in the plane of the ring, Figure 6.7.2, we have

$$\Phi = -G\int\frac{dM}{s} = -G\int_0^{2\pi}\frac{\mu R\,d\theta}{s}$$

where μ is the linear mass density of the ring. To evaluate the integral, we first express s as a function of θ using the law of cosines

$$s^2 = R^2 + r^2 - 2Rr\cos\theta$$

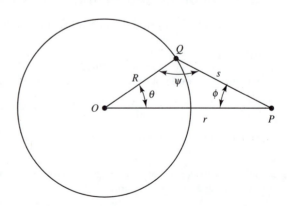

Figure 6.7.2   Coordinates for calculating the gravitational field of a ring.

The integral becomes

$$\Phi = -2R\mu G \int_0^\pi \frac{d\theta}{(r^2 + R^2 - 2Rr \cos \theta)^{1/2}}$$

$$= -\frac{2R\mu G}{r} \int_0^\pi \frac{d\theta}{[1 + (R^2/r^2) - 2(R/r) \cos \theta]^{1/2}}$$

First, let us use the so-called far field approximation $r > R$ and expand the integrand in a power series of $x (= R/r)$, making certain to keep all terms of order x^2.

$$\Phi = -2x\mu G \int_0^\pi \left[\left(1 - \frac{1}{2}x^2 + x \cos \theta \right) + \frac{3}{8}(x^2 - 2x \cos \theta)^2 + \cdots \right] d\theta$$

$$= -2x\mu G \int_0^\pi \left(1 - \frac{1}{2}x^2 + x \cos \theta + \frac{3}{2}x^2 \cos^2 \theta - \frac{3}{2}x^3 \cos \theta + \frac{3}{8}x^4 + \cdots \right) d\theta$$

Now, dropping all terms of order x^3 or higher and noting that the term containing $\cos \theta$ has zero integral over a half cycle, we obtain

$$\Phi = -2x\mu G \left(\pi + \pi \frac{x^2}{4} + \cdots \right)$$

$$= \frac{-2\pi R\mu G}{r} \left(1 + \frac{R^2}{4r^2} + \cdots \right)$$

$$= -\frac{GM}{r} \left(1 + \frac{R^2}{4r^2} + \cdots \right)$$

The field intensity at a distance r from the center of the ring is in the radial direction (because Φ is not a function of θ) and is given by

$$\mathbf{g} = -\frac{\partial \Phi}{\partial r} \mathbf{e}_r = -\frac{GM}{r^2} \left[1 + \frac{3}{4} \left(\frac{R}{r} \right)^2 + \cdots \right] \mathbf{e}_r$$

The field is not given by an inverse-square law. If $r \gg R$, the term in parentheses approaches unity, and the field intensity approaches the inverse-square field of a single particle of mass M. This is true for a finite-sized body of any shape; that is, for distances large compared with the linear dimensions of the body, the field intensity approaches that of a single particle of mass M.

The potential for a point near the center of the ring can be found by invoking the near field, or $r < R$, approximation. The solution proceeds more or less as before, but in this case we expand the preceding integrand in powers of r/R to obtain

$$\Phi = -\frac{GM}{R} \left(1 + \frac{r^2}{4R^2} + \cdots \right)$$

$\mathbf{g}$ can again be found by differentiation

$$\mathbf{g} = \left(\frac{GM}{2R^3} r \right) \mathbf{e}_r + \cdots$$

Thus, a ring of matter exerts an approximately linear *repulsive* force, directed away from the center, on a particle located somewhere near the center of the ring. It is easy to see that this must be so. Imagine that you are a small mass at the center of such a ring of radius R with a field of view both in front of you and behind you that subtends some definite angle. If you move slowly a distance r away from the center, the matter you see attracting you in the forward direction diminishes by a factor of r, whereas the matter you see attracting you from behind grows by r. But because the force of gravity from any material element falls as $1/r^2$, the force exerted on you by the forward mass and the backward mass varies as $1/r$, and, thus, the force difference between the two of them is $[1/(R - r) - 1/(R + r)]$ or proportional to r for $r < R$. The gravitational force of the ring repels objects from its center.

6.8 | Potential Energy in a General Central Field

We showed previously that a central field of the inverse-square type is conservative. Let us now consider the question of whether or not any (isotropic) central field of force is conservative. A general isotropic central field can be expressed in the following way:

$$\mathbf{F} = f(r)\mathbf{e}_r \tag{6.8.1}$$

in which $\mathbf{e}_r$ is the unit radial vector. To apply the test for conservativeness, we calculate the curl of $\mathbf{F}$. It is convenient here to employ spherical coordinates for which the curl is given in Appendix F. We find

$$\nabla \times \mathbf{F} = \frac{1}{r^2 \sin\theta} \begin{vmatrix} \mathbf{e}_r & \mathbf{e}_\theta r & \mathbf{e}_\phi r \sin\theta \\ \dfrac{\partial}{\partial r} & \dfrac{\partial}{\partial \theta} & \dfrac{\partial}{\partial \phi} \\ F_r & rF_\theta & rF_\phi \sin\theta \end{vmatrix} \tag{6.8.2}$$

For our central force $F_r = f(r)$, $F_\theta = 0$, and $F_\phi = 0$. The curl then reduces to

$$\nabla \times \mathbf{F} = \frac{\mathbf{e}_\theta}{r \sin\theta} \frac{\partial f}{\partial \phi} - \frac{\mathbf{e}_\phi}{r} \frac{\partial f}{\partial \theta} = 0 \tag{6.8.3}$$

The two partial derivatives both vanish because $f(r)$ does not depend on the angular coordinates ϕ and θ. Thus, the curl vanishes, and so the general central field defined by Equation 6.8.1 is conservative. We recall that the same test was applied to the inverse-square field in Example 4.2.5.

We can now define a potential energy function

$$V(r) = -\int_{r_{ref}}^{r} \mathbf{F} \cdot d\mathbf{r} = -\int_{r_{ref}}^{r} f(r)dr \tag{6.8.4}$$

where the lower limit r_{ref} is the reference value of r at which the potential energy is defined to be zero. For inverse-power type forces, r_{ref} is often taken to be at infinity. This allows us to calculate the potential energy function, given the force function. Conversely, if we know the potential energy function, we have

$$f(r) = -\frac{dV(r)}{dr} \tag{6.8.5}$$

giving the force function for a central field.

6.9 | Energy Equation of an Orbit in a Central Field

The square of the speed is given in polar coordinates from Equation 1.11.7

$$\mathbf{v} \cdot \mathbf{v} = v^2 = \dot{r}^2 + r^2\dot{\theta}^2 \tag{6.9.1}$$

Because a central force is conservative, the total energy $T + V$ is constant and is given by

$$\tfrac{1}{2}m(\dot{r}^2 + r^2\dot{\theta}^2) + V(r) = E = \text{constant} \tag{6.9.2}$$

We can also write Equation 6.9.2 in terms of the variable $u = 1/r$. From Equations 6.5.7 and 6.5.8 we obtain

$$\tfrac{1}{2}ml^2\left[\left(\frac{du}{d\theta}\right)^2 + u^2\right] + V(u^{-1}) = E \tag{6.9.3}$$

The preceding equation is called *the energy equation of the orbit.*

EXAMPLE 6.9.1

In Example 6.5.1 we had for the spiral orbit $r = c\theta^2$

$$\frac{du}{d\theta} = \frac{-2}{c}\theta^{-3} = -2c^{1/2}u^{3/2}$$

so the energy equation of the orbit is

$$\tfrac{1}{2}ml^2(4cu^3 + u^2) + V = E$$

Thus,

$$V(r) = E - \tfrac{1}{2}ml^2\left(\frac{4c}{r^3} + \frac{1}{r^2}\right)$$

This readily gives the force function of Example 6.5.1, because $f(r) = -dV/dr$.

6.10 | Orbital Energies in an Inverse-Square Field

The potential energy function for an inverse-square force field is

$$V(r) = -\frac{k}{r} = -ku \tag{6.10.1}$$

so the energy equation of the orbit (Equation 6.9.3) becomes

$$\tfrac{1}{2}ml^2\left[\left(\frac{du}{d\theta}\right)^2 + u^2\right] - ku = E \tag{6.10.2}$$

Solving for $du/d\theta$, we first get

$$\left(\frac{du}{d\theta}\right)^2 + u^2 = \frac{2E}{ml^2} + \frac{2ku}{ml^2} \tag{6.10.3a}$$

and then

$$\frac{du}{d\theta} = \sqrt{\frac{2E}{ml^2} + \frac{2ku}{ml^2} - u^2} \tag{6.10.3b}$$

Separating variables yields

$$d\theta = \frac{du}{\sqrt{\dfrac{2E}{ml^2} + \dfrac{2ku}{ml^2} - u^2}} \tag{6.10.3c}$$

We introduce three constants, a, b, and c

$$a = -1 \qquad b = \frac{2k}{ml^2} \qquad c = \frac{2E}{ml^2} \tag{6.10.4}$$

to write Equation 6.10.3c in a standard form to carry out its integration

$$\theta - \theta_0 = \int \frac{du}{\sqrt{au^2 + bu + c}} = \frac{1}{\sqrt{-a}} \cos^{-1}\left(-\frac{b + 2au}{\sqrt{b^2 - 4ac}}\right) \tag{6.10.5}$$

where θ_0 is a constant of integration. Rewriting Equation 6.10.5 first gives us

$$-\frac{b + 2au}{\sqrt{b^2 - 4ac}} = \cos\left[\sqrt{-a}(\theta - \theta_0)\right] \tag{6.10.6a}$$

and then solving for u

$$u = \frac{\sqrt{b^2 - 4ac}}{-2a} \cos\left[\sqrt{-a}(\theta - \theta_0)\right] + \frac{b}{-2a} \tag{6.10.6b}$$

Now we replace u with $1/r$ and insert the values for the constants $a, b,$ and c from Equation 6.10.4 into Equation 6.10.6b.

$$\frac{1}{r} = \frac{1}{2}\sqrt{\frac{4k^2}{m^2 l^4} + \frac{8E}{ml^2}} \cos(\theta - \theta_0) + \frac{k}{ml^2} \tag{6.10.7a}$$

and factoring out the quantity k/ml^2 yields

$$\frac{1}{r} = \frac{k}{ml^2}\left[\sqrt{1 + \frac{2Eml^2}{k^2}} \cos(\theta - \theta_0) + 1\right] \tag{6.10.7b}$$

which, upon simplifying, yields the polar equation of the orbit analogous to Equation 6.5.18a,

$$r = \frac{ml^2/k}{1 + \sqrt{1 + 2Eml^2/k^2} \cos(\theta - \theta_0)} \tag{6.10.7c}$$

If, as before, we set $\theta_0 = 0$, which defines the direction toward the orbital pericenter to be the reference direction for measuring polar angles, and we compare Equation 6.10.7c with 6.5.18b, we see that again it represents a conic section whose eccentricity is

$$\epsilon = \sqrt{1 + \frac{2E}{k}\frac{ml^2}{k}} \tag{6.10.8}$$

From Equations 6.5.18a, b and 6.5.19, $\alpha = ml^2/k = (1 - \epsilon^2)a$ and on inserting these relations into Equation 6.10.8, we see that

$$-\frac{2E}{k} = \frac{1-\epsilon^2}{\alpha} = \frac{1}{a} \tag{6.10.9}$$

or

$$E = -\frac{k}{2a} \tag{6.10.10}$$

Thus, the total energy of the particle determines the semimajor axis of its orbit. We now see from Equation 6.10.8, as stated in Section 6.5, that the total energy E of the particle completely determines the particular conic section that describes the orbit:

$$E < 0 \qquad \epsilon < 1 \qquad \textit{closed orbits (ellipse or circle)}$$
$$E = 0 \qquad \epsilon = 1 \qquad \textit{parabolic orbit}$$
$$E > 0 \qquad \epsilon > 1 \qquad \textit{hyperbolic orbit}$$

Because $E = T + V$ and is constant, the closed orbits are those for which $T < |V|$, and the open orbits are those for which $T \geq |V|$.

In the Sun's gravitational field the force constant $k = GM_\odot m$, where $M_\odot$ is the mass of the Sun and m is the mass of the body. The total energy is then

$$\frac{mv^2}{2} - \frac{GM_\odot m}{r} = E = \text{constant} \tag{6.10.11}$$

so the orbit is an ellipse, a parabola, or a hyperbola depending on whether v^2 is less than, equal to, or greater than the quantity $2GM_\odot/r$, respectively.

EXAMPLE 6.10.1

A comet is observed to have a speed v when it is a distance r from the Sun, and its direction of motion makes an angle ϕ with the radius vector from the Sun, Figure 6.10.1. Find the eccentricity of the comet's orbit.

Solution:

To use the formula for the eccentricity (Equation 6.10.8), we need the square of the angular momentum constant l. It is given by

$$l^2 = |\mathbf{r} \times \mathbf{v}|^2 = (rv \sin \phi)^2$$

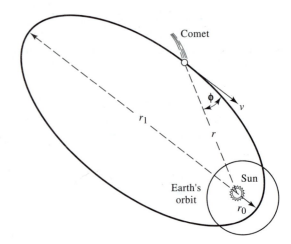

Figure 6.10.1   Orbit of a comet.

The eccentricity, therefore, has the value

$$\epsilon = \left[1 + \left(v^2 - \frac{2GM_\odot}{r}\right)\left(\frac{rv \sin \phi}{GM_\odot}\right)^2\right]^{1/2}$$

Note that the mass m of the comet cancels out. Now the product $GM_\odot$ can be expressed in terms of Earth's speed v_e and orbital radius a_e (assuming a circular orbit), namely

$$GM_\odot = a_e v_e^2$$

The preceding expression for the eccentricity then becomes

$$\epsilon = \left[1 + \left(V^2 - \frac{2}{R}\right)(RV \sin \phi)^2\right]^{1/2}$$

where we have introduced the *dimensionless ratios*

$$V = \frac{v}{v_e} \qquad R = \frac{r}{a_e}$$

which simplify the computation of ϵ.

As a numerical example, let v be one-half the Earth's speed, let r be four times Earth–Sun distance, and $\phi = 30°$. Then $V = 0.5$ and $R = 4$, so the eccentricity is

$$\epsilon = [1 + (0.25 - 0.5)(4 \times 0.5 \times 0.5)^2]^{1/2} = (0.75)^{1/2} = 0.866$$

For an ellipse the quantity $(1 - \epsilon^2)^{-1/2}$ is equal to the ratio of the major (long) axis to the minor (short) axis. For the orbit of the comet in this example this ratio is $(1 - 0.75)^{-1/2} = 2$, or 2:1, as shown in Figure 6.10.1.

EXAMPLE 6.10.2

When a spacecraft is placed into geosynchronous orbit (Example 6.6.2), it is first launched, along with a propulsion stage, into a near circular, low earth orbit (LEO) using an appropriate booster rocket. Then the propulsion stage is fired and the spacecraft is

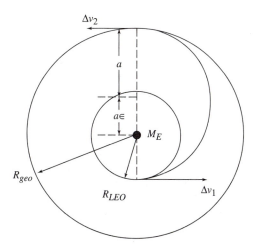

Figure 6.10.2   Boosting a satellite
from low earth orbit (LEO) to a
geosynchronous (geo) orbit.

transferred to an elliptic orbit designed to take it to geosynchronous altitude at orbital
apogee (see Figure 6.10.2). At apogee, the propulsion stage is fired again to take it out
of its elliptical orbit and put it into a circular, geocentric (geo) orbit. Thus, two velocity
boosts are required of the propulsion stage: (a) Δv_1, to move the satellite from its circular LEO into the elliptical transfer orbit and (b) Δv_2, to circularize the orbit of the satellite at the geosynchronous altitude. Calculate the required velocity boosts, Δv_1 and Δv_2.

(a) Solution:

We essentially solved this problem in Example 6.5.4. We do it here in a slightly different way. First, we note that the radii of the two circular orbits and the semimajor
axis of the transfer elliptical orbit (see Figure 6.10.2) are related

$$R_{LEO} + R_{geo} = 2a$$

From the figure we see that $R_{LEO} = a(1 - \epsilon)$ and $R_{geo} = a(1 + \epsilon)$ are the perigee and
apogee distances of the transfer orbit.

Now, we use the energy equation (6.10.11) to calculate the velocity at perigee,
v_p, of the spacecraft, after boost to the transfer elliptical orbit. The energy of the elliptical orbit is

$$E = -\frac{GM_E m}{2a} = \frac{1}{2}mv_p^2 - \frac{GM_E m}{a(1 - \epsilon)}$$

Solving for v_p gives

$$v_p^2 = \frac{GM_E}{a}\left(\frac{1+\epsilon}{1-\epsilon}\right)$$

Substituting for a, $1 + \epsilon$ and $1 - \epsilon$ into the above gives

$$v_a^2 = \frac{2GM_E}{R_{LEO} + R_{geo}}\left(\frac{R_{geo}}{R_{LEO}}\right)$$

Now, we can calculate the velocity of the satellite in *circular* LEO from the condition

$$\frac{m v_{LEO}^2}{R_{LEO}} = \frac{GM_E m}{R_{LEO}^2}$$

or

$$v_{LEO}^2 = \frac{GM_E}{R_{LEO}}$$

After a little algebra, we find that the required velocity boost is

$$\Delta v_1 = v_p - v_{LEO} = \sqrt{\frac{GM_E}{R_{LEO}}} \left[\sqrt{\frac{2R_{geo}}{R_{LEO} + R_{geo}}} - 1 \right]$$

Remembering that $g = GM_E / R_E^2$ we have

$$\Delta v_1 = R_E \sqrt{\frac{g}{R_{LEO}}} \left[\sqrt{\frac{2R_{geo}}{R_{LEO} + R_{geo}}} - 1 \right]$$

Putting in numbers: $R_E = 6371$ km, $R_{LEO} = 6693$ km, $R_{geo} = 42,400$ km we get

$$\Delta v_1 = 8,600 \text{ km/hr}$$

(b) Solution:

The energy of the spacecraft at apogee is

$$E = -\frac{GM_E m}{2a} = \frac{1}{2} m v_a^2 - \frac{GM_E m}{a(1 + \epsilon)}$$

Solving for the velocity at apogee, v_a

$$v_a^2 = \frac{GM_E}{a} \left(\frac{1 - \epsilon}{1 + \epsilon} \right)$$

Substituting for a, $1 + \epsilon$ and $1 - \epsilon$ into the above gives

$$v_a^2 = \frac{2GM_E}{R_{LEO} + R_{geo}} \left(\frac{R_{LEO}}{R_{geo}} \right)$$

As before, the condition for a *circular* orbit at this radius is

$$\frac{m v_{geo}^2}{R_{geo}} = \frac{GM_E m}{R_{geo}^2}$$

Thus,

$$v_{geo}^2 = \frac{GM_E}{R_{geo}}$$

$$\Delta v_2 = v_{geo} - v_a = \sqrt{\frac{GM_E}{R_{geo}}} \left[1 - \sqrt{\frac{2R_{LEO}}{R_{LEO} + R_{geo}}} \right] = R_E \sqrt{\frac{g}{R_{geo}}} \left[1 - \sqrt{\frac{2R_{LEO}}{R_{LEO} + R_{geo}}} \right]$$

Putting in numbers, we get

$$\Delta v_2 = 5269 \text{ km/hr}$$

Note, the total boost, $\Delta v_1 + \Delta v_2 = 8{,}600 \text{ km/hr} + 5269 \text{ km/hr} = 13{,}869 \text{ km/hr}$, required of the spacecraft propulsion system to place it into geo orbit is almost 50% of the boost required by the launcher to place it into LEO!

6.11 | Limits of the Radial Motion: Effective Potential

We have seen that the angular momentum of a particle moving in any isotropic central field is a constant of the motion, as expressed by Equations 6.5.4 and 6.5.5 defining l. This fact allows us to write the general energy equation (Equation 6.9.2) in the following form:

$$\frac{m}{2}\left(\dot{r}^2 + \frac{l^2}{r^2} \right) + V(r) = E \tag{6.11.1a}$$

or

$$\frac{m}{2}\dot{r}^2 + U(r) = E \tag{6.11.1b}$$

in which

$$U(r) = \frac{ml^2}{2r^2} + V(r) \tag{6.11.1c}$$

The function $U(r)$ defined here is called the *effective potential*. The term $ml^2/2r^2$ is called the *centrifugal potential*. Looking at Equation 6.11.1b we see that, as far as the radial motion is concerned, the particle behaves in exactly the same way as a particle of mass m moving in one-dimensional motion under a potential energy function $U(r)$. As in Section 3.3 in which we discussed harmonic motion, the limits of the radial motion (turning points) are given by setting $\dot{r} = 0$ in Equation 6.11.1b. These limits are, therefore, the roots of the equation

$$U(r) - E = 0 \tag{6.11.2a}$$

or

$$\frac{ml^2}{2r^2} + V(r) - E = 0 \tag{6.11.2b}$$

Furthermore, the *allowed* values of r are those for which $U(r) \le E$, because $\dot{r}^2$ is necessarily positive or zero.

Thus, it is possible to determine the range of the radial motion without knowing anything about the orbit. A plot of $U(r)$ is shown in Figure 6.11.1. Also shown are the radial limits r_0 and r_1 for a particular value of the total energy E. The graph is drawn for the inverse-square law, namely,

$$U(r) = \frac{ml^2}{2r^2} - \frac{k}{r} \tag{6.11.3}$$

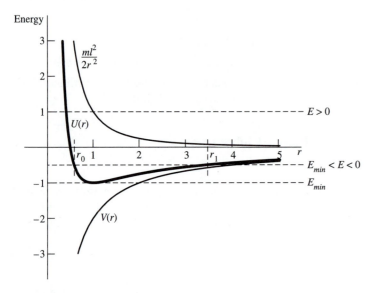

Figure 6.11.1   The effective potential for the inverse-square law of force and limits of the radial motion.

In this case Equation 6.11.2a, on rearranging terms, becomes

$$-2Er^2 - 2kr + ml^2 = 0 \qquad (6.11.4)$$

which is a quadratic equation in r. The two roots are

$$r_{1,0} = \frac{k \pm (k^2 + 2Eml^2)^{1/2}}{-2E} \qquad (6.11.5)$$

giving the maximum (upper sign) and minimum (lower sign) values of the radial distance r under the inverse-square law of force.

When $E < 0$, the orbits are bound, the two roots are both positive, and the resulting orbit is an ellipse in which r_0 and r_1 are the pericenter and apocenter, respectively. When the energy is equal to its minimum possible value

$$E_{min} = -\frac{k^2}{2ml^2} \qquad (6.11.6)$$

Equation 6.11.5 then has a single root given by

$$r_0 = -\frac{k}{2E_{min}} \qquad (6.11.7)$$

and the orbit is a circle. Note, that this result can also be obtained from Equation 6.10.10 ($a = -k/2E$) because $r_0 = a$ in the case of a circular orbit. When $E \geq 0$, Equation 6.11.5 has only a single, positive real root corresponding to a parabola ($E = 0$) or a hyperbola ($E > 0$).

Because the effective potential of the particle is axially symmetric, its shape in two dimensions can be formed by rotating the curve in Figure 6.11.1 about the vertical

(a)

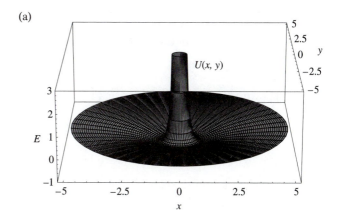

(b)

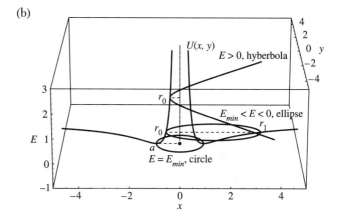

Figure 6.11.2 (a) The effective potential for the inverse-square law of force in two dimensions. (b) The relationship between total energy E, the effective potential, and the resulting orbits.

(Energy) axis. The resulting two-dimensional shape is shown in Figure 6.11.2a. The orbit of the particle can be visualized as taking place in that figure, constrained to a horizontal plane of constant energy E. If the energy of the particle is $E < 0$, the intersection of the plane and the effective potential surface forms an inner circle of radius r_0 that marks a central, impenetrable centrifugal barrier and an outer circle of radius r_1, that marks the farthest point of escape from the center of attraction. The motion of the particle is bound to a region between these two limits. If the energy of the particle is equal to its minimum possible value $E = E_{min}$, these two circles converge to a single one that traces out the minimum of the effective potential surface. The particle is constrained to that circle. If the energy of the particle is $E \geq 0$, the intersection forms only a single circle of radius r_0 around the centrifugal barrier and the particle is only inhibited from passing into that region; otherwise, it is free to escape to infinity along either a parabolic or hyperbolic trajectory.

Figure 6.11.2b suppresses the display of the effective potential *mesh* in Figure 6.11.2a (except for the two radial curves in the $\pm x$ direction) and shows the circular, elliptical, and hyperbolic orbits that result when the energy (in appropriately scaled units) is equal to -1, $-\frac{1}{2}$, and $+1$ units, respectively. Note that, in all cases, the particle is constrained to move in a region of its plane of constant energy, in which the value of its energy exceeds the value of the effective potential at that point.

EXAMPLE 6.11.1

Find the semimajor axis of the orbit of the comet of Example 6.10.1.

Solution:

Equation 6.10.10 gives directly

$$a = \frac{k}{-2E} = \frac{GM_\odot m}{-2\left(\dfrac{mv^2}{2} - \dfrac{GM_\odot m}{r}\right)}$$

where m is the mass of the comet. Clearly, m again cancels out. Also, as stated previously, $GM_\odot = a_e v_e^2$. So the final result is the simple expression

$$a = \frac{a_e}{(2/R) - V^2}$$

where R and V are as defined in Example 6.10.1.

For the previous numerical values, $R = 4$ and $V = 0.5$, we find $a = a_e/[0.5 - (0.5)^2] = 4a_e$.

Examples 6.10.1 and 6.10.2 bring out an important fact, namely, that the orbital parameters are independent of the mass of a body. Given the same initial position, speed, and direction of motion, a grain of sand, a coasting spaceship, or a comet would all have identical orbits, provided that no other bodies came near enough to have an effect on the motion of the body. (We also assume, of course, that the mass of the body in question is small compared with the Sun's mass.)

6.12 | Nearly Circular Orbits in Central Fields: Stability

A circular orbit is possible under any attractive central force, but not all central forces result in *stable* circular orbits. We wish to investigate the following question: If a particle traveling in a circular orbit suffers a slight disturbance, does the ensuing orbit remain close to the original circular path? To answer the query, we refer to the radial differential equation of motion (Equation 6.5.2a). Because $\dot{\theta} = l/r^2$, we can write the radial equation as follows:

$$m\ddot{r} = \frac{ml^2}{r^3} + f(r) \tag{6.12.1}$$

[This is the same as the differential equation for one-dimensional motion under the effective potential $U(r) = (ml^2/2r^2) + V(r)$, so that $m\ddot{r} = -dU(r)/dr = (ml^2/r^3) - dV(r)/dr$.]

Now for a circular orbit, r is constant, and $\ddot{r} = 0$. Thus, calling a the radius of the circular orbit, we have

$$-\frac{ml^2}{a^3} = f(a) \tag{6.12.2}$$

for the force at $r = a$. It is convenient to express the radial motion in terms of the variable x defined by

$$x = r - a \qquad (6.12.3)$$

The differential equation for radial motion then becomes

$$m\ddot{x} = ml^2(x+a)^{-3} + f(x+a) \qquad (6.12.4)$$

Expanding the two terms involving $x + a$ as power series in x, we obtain

$$m\ddot{x} = ml^2 a^{-3}\left(1 - 3\frac{x}{a} + \cdots\right) + [f(a) + f'(a)x + \cdots] \qquad (6.12.5)$$

Equation 6.12.5, by virtue of the relation shown in Equation 6.12.2, reduces to

$$m\ddot{x} + \left[\frac{-3}{a}f(a) - f'(a)\right]x = 0 \qquad (6.12.6)$$

if we ignore terms involving x^2 and higher powers of x. Now, if the coefficient of x (the quantity in brackets) in Equation 6.12.6 is positive, then the equation is the same as that of the simple harmonic oscillator. In this case the particle, if perturbed, oscillates harmonically about the circle $r = a$, so the circular orbit is a stable one. On the other hand, if the coefficient of x is negative, the motion is nonoscillatory, and the result is that x eventually increases exponentially with time; the orbit is unstable. (If the coefficient of x is zero, then higher terms in the expansion must be included to determine the stability.) Hence, we can state that a circular orbit of radius a is stable if the force function $f(r)$ satisfies the inequality

$$f(a) + \frac{a}{3}f'(a) < 0 \qquad (6.12.7)$$

For example, if the radial force function is a power law, namely,

$$f(r) = -cr^n \qquad (6.12.8)$$

then the condition for stability reads

$$-ca^n - \frac{a}{3}cna^{n-1} < 0 \qquad (6.12.9)$$

which reduces to

$$n > -3 \qquad (6.12.10)$$

Thus, the inverse-square law ($n = -2$) gives stable circular orbits, as does the law of direct distance ($n = 1$). The latter case is that of the two-dimensional isotropic harmonic oscillator. For the inverse–fourth power ($n = -4$) circular orbits are unstable. It can be shown that circular orbits are also unstable for the inverse-cube law of force ($n = -3$). To show this it is necessary to include terms of higher power than 1 in the radial equation. (See Problem 6.26.)

6.13 | Apsides and Apsidal Angles for Nearly Circular Orbits

An *apsis*, or *apse*, is a point in an orbit at which the radius vector assumes an extreme value (maximum or minimum). The perihelion and aphelion points are the apsides of planetary orbits. The angle swept out by the radius vector between two consecutive apsides is called the *apsidal angle*. Thus, the apsidal angle is π for elliptic orbits under the inverse-square law of force.

In the case of motion in a nearly circular oribt, we have seen that r oscillates about the circle $r = a$ (if the orbit is stable). From Equation 6.12.6 it follows that the period τ_r of this oscillation is given by

$$\tau_r = 2\pi \left[\frac{m}{-(3/a)f(a) - f'(a)} \right]^{1/2} \tag{6.13.1}$$

The apsidal angle in this case is just the amount by which the polar angle θ increases during the time that r oscillates from a minimum value to the succeeding maximum value. This time is clearly $\tau_r/2$. Now $\dot{\theta} = l/r^2$; therefore, $\dot{\theta}$ remains approximately constant, and we can write

$$\dot{\theta} \approx \frac{l}{a^2} = \left[-\frac{f(a)}{ma} \right]^{1/2} \tag{6.13.2}$$

The last step in Equation 6.13.2 follows from Equation 6.12.2; hence, the apsidal angle is given by

$$\psi = \tfrac{1}{2}\tau_r \dot{\theta} = \pi \left[3 + a\frac{f'(a)}{f(a)} \right]^{-1/2} \tag{6.13.3}$$

Thus, for the power law of force $f(r) = -cr^n$, we obtain

$$\psi = \pi(3 + n)^{-1/2} \tag{6.13.4}$$

The apsidal angle is independent of the size of the orbit in this case. The orbit is *reentrant*, or repetitive, in the case of the inverse-square law ($n = -2$) for which $\psi = \pi$ and in the case of the linear law ($n = 1$) for which $\psi = \pi/2$. If, however, say $n = 2$, then $\psi = \pi/\sqrt{5}$, which is an irrational multiple of π, and so the motion does not repeat itself.

If the law of force departs slightly from the inverse-square law, then the apsides either advances or regresses steadily, depending on whether the apsidal angle is slightly greater or slightly less than π. (See Figure 6.13.1.)

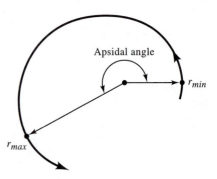

Figure 6.13.1   Illustrating the apsidal angle.

EXAMPLE 6.13.1

Let us assume that the gravitational force field acting on the planet Mercury takes the form

$$f(r) = -\frac{k}{r^2} + \epsilon r$$

where ϵ is very small. The first term is the gravitational field due to the Sun, and the second term is a repulsive perturbation due to a surrounding ring of matter. We assume this matter distribution as a simple model to represent the gravitational effects of all the other planets, primarily Jupiter. The perturbation is linear for points near the Sun and in the plane of the surrounding ring, as previously explained in Example 6.7.2. The apsidal angle, from Equation 6.13.3, is

$$\Psi = \pi \left(3 + a\frac{2ka^{-3} + \epsilon}{-ka^{-2} + \epsilon a} \right)^{-1/2}$$

$$= \pi \left(\frac{1 - 4k^{-1}\epsilon a^3}{1 - k^{-1}\epsilon a^3} \right)^{-1/2} = \pi \left(1 - \frac{\epsilon}{k}a^3 \right)^{1/2} \left(1 - 4\frac{\epsilon}{k}a^3 \right)^{-1/2}$$

$$\approx \pi \left(1 - \frac{1}{2}\frac{\epsilon}{k}a^3 \right) \left(1 + 2\frac{\epsilon}{k}a^3 \right)$$

$$\approx \pi \left(1 + \frac{3}{2}\frac{\epsilon}{k}a^3 \right)$$

In the last step, we used the binomial expansion theorem to expand the terms in brackets in powers of ϵ/k and kept only the first-order term. The apsidal angle advances if ϵ is positive and regresses if it is negative.

By 1877, Urbain Leverrier, using perturbation methods, had succeeded in calculating the gravitational effects of all the known planets on one another's orbit. Depending on the planet, the apsidal angles were found to advance or regress in good agreement with theory with the sole exception of the planet Mercury. Observations of Mercury's solar transits since 1631 indicated an advance of the perihelion of its orbit by 565″ of arc per century. According to Leverrier, it should advance only 527″ per century, a discrepancy of 38″. Simon Newcomb (1835–1909), chief of the office for the American Nautical Almanac, improved Leverrier's calculations, and by the beginning of the 20th century, the accepted values for the advance of Mercury's perihelion per century were 575″ and 534″, respectively, or a discrepancy of 41″ ± 2″ of arc. Leverrier himself had decided that the discrepancy was real and that it could be accounted for by an as yet unseen planet with a diameter of about 1000 miles circling the Sun within Mercury's orbit at a distance of about 0.2 AU. (You can easily extend the preceding example to show that an interior planet would lead to an advance in the perihelion of Mercury's orbit by the factor δ/ka^2.) Leverrier called the unseen planet Vulcan. No such planet was found.

Another possible explanation was put forward by Asaph Hall (1829–1907), the discoverer of the satellites of Mars in 1877. He proposed that the exponent in Newton's law of gravitation might not be exactly 2, that instead, it might be 2.0000001612 and that this would do the trick. Einstein was to comment that the discrepancy in Mercury's orbit "could be explained by means of classic mechanics only on the assumption of hypotheses which have little probability and which were devised solely for this purpose." The discrepancy, of course, was nicely explained by Einstein himself in a paper presented to the Berlin Academy in 1915. The paper was based on Einstein's calculations of general relativity even before he had fully completed the theory. Thus, here we have the highly remarkable event of a discrepancy between observation and existing theory leading to the confirmation of an entirely new superceding theory.

If the Sun were oblate (football-shaped) enough, its gravitational field would depart slightly from an inverse-square law, and the perihelion of Mercury's orbit would advance. Measurements to date have failed to validate this hypothesis as a possible explanation. Similar effects, however, have been observed in the case of artificial satellites in orbit about Earth. Not only does the perihelion of a satellite's orbit advance, but the plane of the orbit precesses if the satellite is not in Earth's equatorial plane. Detailed analysis of these orbits shows that Earth is basically "pear-shaped and somewhat lumpy."

6.14 | Motion in an Inverse-Square Repulsive Field: Scattering of Alpha Particles

Ironically one of the crowning achievements of Newtonian mechanics contained its own seeds of destruction. In 1911, Ernest Rutherford (1871–1937), attempting to solve the problem of the scattering of alpha particles by thin metal foils, went for help back to the very source of classical mechanics, the *Principia* of Sir Isaac Newton. Paradoxically, in the process of finding a solution to the problem based on classical mechanics, the idea of the nuclear atom was born, an idea that would forever remain incomprehensible within the confines of the classic paradigm. A complete, self-consistent theory of the nuclear atom would emerge only when many of the notions of Newtonian mechanics were given up and replaced by the novel and astounding concepts of quantum mechanics. It is not that Newtonian mechanics was "wrong"; its concepts, which worked so well time and again when applied to the macroscopic world of falling balls and orbiting planets, simply broke down when applied to the microscopic world of atoms and nuclei. Indeed, the architects of the laws of quantum physics constructed them in such a way that the results of calculations based on the new laws agreed with those of Newtonian mechanics when applied to problems in the macroscopic world. The domain of Newtonian physics would be seen to be merely limited, rather than "wrong," and its practitioners from that time on would now have to be aware of these limits.

In the early 1900s the atom was thought to be a sort of distributed blob of positive charge within which were embedded the negatively charged electrons discovered in 1897 by J. J. Thomson (1856–1940). The model was first suggested by Lord Kelvin (1824–1907) in 1902 but mathematically refined a year later by Thomson. Thomson developed the model with emphasis on the mechanical and electrical stability of the system. In his honor it became known as the *Thomson atom*.

In 1907, Rutherford accepted a position at the University of Manchester where he encountered Hans Geiger (1882–1945), a bright young German experimental physicist, who was about to embark on an experimental program designed to test the validity of the Thomson atom. His idea was to direct a beam of the recently discovered alpha particles emitted from radioactive atoms toward thin metal foils. A detailed analysis of the way they scattered should provide information on the structure of the atom. With the help of Ernest Marsden, a young undergraduate, Geiger would carry out these investigations over several years. Things behaved more or less as expected, except there were many more large-angle scatterings than could be accounted for by the Thomson model. In fact, some of the alpha particles scattered completely backward at angles of 180°. When Rutherford heard of this, he was dumbfounded. It was as though an onrushing freight train had been hurled backward on striking a chicken sitting in the middle of the track.

In searching for a model that would lead to such a large force being exerted on a fast-moving projectile, Rutherford envisioned a comet swinging around the Sun and coming back out again, just like the alpha particles scattered at large angles. This suggested the idea of a hyperbolic orbit for a positively charged alpha particle attracted by a negatively charged nucleus. Of course, Rutherford realized that the only important thing in the dynamics of the problem was the inverse-square nature of the law, which, as we have seen, leads to conic sections as solutions for the orbit. Whether the force is attractive or repulsive is completely irrelevant. Rutherford then remembered a theorem about conics from geometry that related the eccentricity of the hyperbola to the angle between its asymptotes. Using this relation, along with conservation of angular momentum and energy, he obtained a complete solution to the alpha particle-scattering problem, which agreed well with the data of Geiger and Marsden. Thus, the current model of the nuclear atom was born.

We solve this problem next but be aware that an identical solution could be obtained for an attractive force. The solution says nothing about the sign of the nuclear charge. The sign becomes obvious from other arguments.

Consider a particle of charge q and mass m (the incident high-speed particle) passing near a heavy particle of charge Q (the nucleus, assumed fixed). The incident particle is repelled with a force given by Coulomb's law:

$$f(r) = \frac{Qq}{r^2} \tag{6.14.1}$$

where the position of Q is taken to be the origin. (We shall use cgs electrostatic units for Q and q. Then r is in centimeters, and the force is in dynes.) The differential equation of the orbit (Equation 6.5.12) then takes the form

$$\frac{d^2u}{d\theta^2} + u = -\frac{Qq}{ml^2} \tag{6.14.2}$$

and so the equation of the orbit is

$$u^{-1} = r = \frac{1}{A\,\cos(\theta - \theta_0) - Qq/ml^2} \tag{6.14.3}$$

We can also write the equation of the orbit in the form given by Equation 6.10.7c, namely,

$$r = \frac{ml^2Q^{-1}q^{-1}}{-1+(1+2Eml^2Q^{-2}q^{-2})^{1/2}\,\cos(\theta-\theta_0)} \tag{6.14.4}$$

because $k = -Qq$. The orbit is a hyperbola. This may be seen from the physical fact that the energy E is always greater than zero in a repulsive field of force. (In our case $E = \frac{1}{2}mv^2 + Qq/r$.) Hence, the eccentricity ϵ, the coefficient of $\cos(\theta-\theta_0)$, is greater than unity, which means that the orbit must be hyperbolic.

The incident particle approaches along one asymptote and recedes along the other, as shown in Figure 6.14.1. We have chosen the direction of the polar axis such that the initial position of the particle is $\theta = 0$, $r = \infty$. It is clear from either of the two equations of the orbit that r assumes its minimum value when $\cos(\theta - \theta_0) = 1$, that is, when $\theta = \theta_0$. Because $r = \infty$ when $\theta = 0$, then r is also infinite when $\theta = 2\theta_0$. Hence, the angle between the two asymptotes of the hyperbolic path is $2\theta_0$, and the angle θ_s through which the incident particle is deflected is given by

$$\theta_s = \pi - 2\theta_0 \tag{6.14.5}$$

Furthermore, in Equation 6.14.4 the denominator vanishes at $\theta = 0$ and $\theta = 2\theta_0$. Thus,

$$-1 + (1 + 2Eml^2Q^{-2}q^{-2})^{1/2}\cos\theta_0 = 0 \tag{6.14.6}$$

from which we readily find

$$\tan\theta_0 = (2Em)^{1/2}lQ^{-1}q^{-1} = \cot\frac{\theta_s}{2} \tag{6.14.7}$$

The last step follows from the angle relationship given above.

In applying Equation 6.14.7 to scattering problems, the constant l is usually expressed in terms of another quantity b called the *impact parameter*. The impact parameter is the

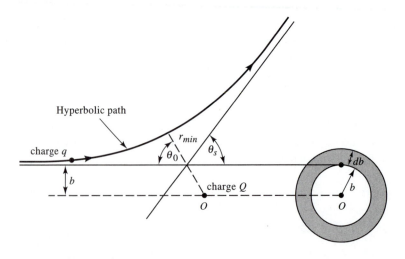

Figure 6.14.1   Hyperbolic path (orbit) of a charged particle moving in the inverse-square repulsive force field of another charged particle.

perpendicular distance from the origin (scattering center) to the initial line of motion of the particle, as shown in Figure 6.14.1. Thus

$$|l| = |\mathbf{r} \times \mathbf{v}| = bv_0 \tag{6.14.8}$$

where v_0 is the initial speed of the particle. We know also that the energy E is constant and is equal to the initial kinetic energy $\frac{1}{2}mv_0^2$, because the initial potential energy is zero ($r = \infty$). Accordingly, we can write the scattering formula (Equation 6.14.7) in the form

$$\cot\frac{\theta_s}{2} = \frac{bmv_0^2}{Qq} = \frac{2bE}{Qq} \tag{6.14.9}$$

giving the relationship between the scattering angle and the impact parameter.

In a typical scattering experiment a beam of particles is projected at a target, such as a thin foil. The nuclei of the target atoms are the scattering centers. The fraction of incident particles that are deflected through a given angle θ_s can be expressed in terms of a *differential scattering cross section* $\sigma(\theta_s)$ defined by the equation

$$\frac{dN}{N} = n\sigma(\theta_s)\,d\Omega \tag{6.14.10}$$

Here dN is the number of incident particles scattered through an angle between θ_s and $\theta_s + d\theta_s$, N is the total number of incident particles, n is the number of scattering centers per unit area of the target foil, and $d\Omega$ is the element of solid angle corresponding to the increment $d\theta_s$. Thus, $d\Omega = 2\pi \sin\theta_s\,d\theta_s$.

Now an incident particle approaching a scattering center has an impact parameter lying between b and $b + db$ if the projection of its path lies in a ring of inner radius b and outer radius $b + db$ (see Figure 6.14.1). The area of this ring is $2\pi b\,db$. The total number of such particles must correspond to the number scattered through a given angle, that is

$$dN = Nn\sigma(\theta_s)2\pi \sin\theta_s\,d\theta_s = Nn2\pi b\,db \tag{6.14.11}$$

Thus,

$$\sigma(\theta_s) = \frac{b}{\sin\theta_s}\left|\frac{db}{d\theta_s}\right| \tag{6.14.12}$$

To find the scattering cross section for charged particles, we differentiate with respect to θ_s in Equation 6.14.9:

$$\frac{1}{2\sin^2\left(\dfrac{\theta_s}{2}\right)} = \frac{2E}{Qq}\left|\frac{db}{d\theta_s}\right| \tag{6.14.13}$$

(The absolute value sign is inserted because the derivative is negative.) By eliminating b and $|db/d\theta_s|$ among Equations 6.14.9, .12, and .13 and using the identity

$$\sin\theta_s = 2\sin(\theta_s/2)\cos(\theta_s/2),$$

we find the following result:

$$\sigma(\theta_s) = \frac{Q^2q^2}{16E^2}\frac{1}{\sin^4(\theta_s/2)} \tag{6.14.14}$$

This is the famous Rutherford scattering formula. It shows that the differential cross section varies as the inverse fourth power of $\sin(\theta_s/2)$. Its experimental verification in the first part of this century marked one of the early milestones of nuclear physics.

EXAMPLE 6.14.1

An alpha particle emitted by radium ($E = 5$ million eV $= 5 \times 10^6 \times 1.6 \times 10^{-12}$ erg) suffers a deflection of 90° on passing near a gold nucleus. What is the value of the impact parameter?

Solution:

For alpha particles $q = 2e$, and for gold $Q = 79e$, where e is the elementary charge. (The charge carried by a single electron is $-e$.) In egs units $e = 4.8 \times 10^{-10}$ esu. Thus, from Equation 6.14.9

$$b = \frac{Qq}{2E} \cot 45° = \frac{2 \times 79 \times (4.8)^2 \times 10^{-20} \text{ cm}}{2 \times 5 \times 1.6 \times 10^{-6}}$$

$$= 2.1 \times 10^{-12} \text{ cm}$$

EXAMPLE 6.14.2

Calculate the distance of closest approach of the alpha particle in Example 6.14.1.

Solution:

The distance of closest approach is given by the equation of the orbit (Equation 6.14.4) for $\theta = \theta_0$; thus,

$$r_{min} = \frac{ml^2 Q^{-1} q^{-1}}{-1 + (1 + 2Eml^2 Q^{-2} q^{-2})^{1/2}}$$

On using Equation 6.14.9 and a little algebra, the preceding equation can be written

$$r_{min} = \frac{b \cot(\theta_s/2)}{-1 + [1 + \cot^2(\theta_s/2)]^{1/2}} = \frac{b \cos(\theta_s/2)}{1 - \sin(\theta_s/2)}$$

Thus, for $\theta_s = 90°$, we find $r_{min} = 2.41\, b = 5.1 \times 10^{-12}$ cm.

Notice that the expressions for r_{min} become indeterminate when $l = b = 0$. In this case the particle is aimed directly at the nucleus. It approaches the nucleus along a straight line, and, being continually repelled by the Coulomb force, its speed is reduced to zero when it reaches a certain point r_{min}, from which point it returns along the same straight line. The angle of deflection is 180°. The value of r_{min} in this case is found by using the fact that the energy E is constant. At the turning point the potential energy is Qq/r_{min}, and the kinetic energy is zero. Hence, $E = \frac{1}{2} m v_0^2 = Qq/r_{min}$, and

$$r_{min} = \frac{Qq}{E}$$

For radium alpha particles and gold nuclei we find $r_{min} \approx 10^{-12}$ cm when the angle of deflection is 180°. The fact that such deflections are actually observed shows that the order of magnitude of the radius of the nucleus is at least as small as 10^{-12} cm.

Problems

6.1   Find the gravitational attraction between two solid lead spheres of 1 kg mass each if the spheres are almost in contact. Express the answer as a fraction of the weight of either sphere. (The density of lead is 11.35 g/cm^3.)

6.2   Show that the gravitational force on a test particle inside a thin uniform spherical shell is zero
(a) By finding the force directly
(b) By showing that the gravitational potential is constant

6.3   Assuming Earth to be a uniform solid sphere, show that if a straight hole were drilled from pole to pole, a particle dropped into the hole would execute simple harmonic motion. Show also that the period of this oscillation depends only on the density of Earth and is independent of the size. What is the period in hours? ($R_{earth} = 6.4 \times 10^6$ m.)

6.4   Show that the motion is simple harmonic with the same period as the previous problem for a particle sliding in a straight, smooth tube passing obliquely through Earth. (Ignore any effects of rotation.)

6.5   Assuming a circular orbit, show that Kepler's third law follows directly from Newton's second law and his law of gravity: $GMm/r^2 = mv^2/r$.

6.6   (a) Show that the radius for a circular orbit of a synchronous (24-h) Earth satellite is about 6.6 Earth radii.
(b) The distance to the Moon is about 60.3 Earth radii. From this calculate the length of the sidereal month (period of the Moon's orbital revolution).

6.7   Show that the orbital period for an Earth satellite in a circular orbit just above Earth's surface is the same as the period of oscillation of the particle dropped into a hole drilled through Earth (see Problem 6.3).

6.8   Calculate Earth's velocity of approach toward the Sun, when it is at an extremum of the latus rectum through the Sun. Take the eccentricity of Earth's orbit to be $\frac{1}{60}$ and its semimajor axis to be 93,000,000 miles (see Figure 6.5.1).

6.9   If the solar system were embedded in a uniform dust cloud of density ρ, show that the law of force on a planet a distance r from the center of the Sun would be given by

$$F(r) = -\frac{GMm}{r^2} - \left(\tfrac{4}{3}\right)\pi \rho m Gr$$

6.10   A particle moving in a central field describes the spiral orbit $r = r_0 e^{k\theta}$. Show that the force law is inverse cube and that θ varies logarithmically with t.

6.11   A particle moves in an inverse-cube field of force. Show that, in addition to the exponential spiral orbit of Problem 6.10, two other types of orbit are possible and give their equations.

6.12   The orbit of a particle moving in a central field is a circle passing through the origin, namely $r = r_0 \cos\theta$. Show that the force law is inverse–fifth power.

6.13 A particle moves in a spiral orbit given by $r = a\theta$. If θ increases linearly with t, is the force a central field? If not, determine how θ must vary with t for a central force.

6.14 A particle of unit mass is projected with a velocity v_0 at right angles to the radius vector at a distance a from the origin of a center of attractive force, given by

$$f(r) = -k\left(\frac{4}{r^3} + \frac{a^2}{r^5}\right)$$

If $v_0^2 = 9k/2a^2$,
(a) Find the polar equation of the resulting orbit.
(b) How long does it take the particle to travel through an angle $3\pi/2$? Where is the particle at that time?
(c) What is the velocity of the particle at that time?

6.15 (a) In Example 6.5.4, find the fractional change in the apogee $\delta r_0/r_1$ as a function of a small fractional change in the ratio of boost speed to circular orbit speed, $\delta(v_0/v_c)/(v_0/v_c)$
(b) If the speed ratio is 1% too great, by how much would the spacecraft miss the Moon? [This problem illustrates the extreme accuracy needed to achieve a circumlunar orbit.]

6.16 Compute the period of Halley's Comet from the data given in Section 6.5. Find also the comet's speed at perihelion and aphelion.

6.17 A comet is first seen at a distance of d astronomical units from the Sun and it is traveling with a speed of q times the Earth's speed. Show that the orbit of the comet is hyperbolic, parabolic, or elliptic, depending on whether the quantity q^2d is greater than, equal to, or less than 2, respectively.

6.18 A particle moves in an elliptic orbit in an inverse-square force field. Prove that the product of the minimum and maximum speeds is equal to $(2\pi a/\tau)^2$, where a is the semimajor axis and τ is the periodic time.

6.19 At a certain point in its elliptical orbit about the Sun, a planet receives a small tangential impulse so that its velocity changes from v to $v + \delta v$. Find the resultant small changes in a, the semi-major axis.

6.20 (a) Prove that the time average of the potential energy of a planet in an elliptical orbit about the Sun is $-k/a$.
(b) Calculate the time average of the kinetic energy of the planet.

6.21 A satellite is placed into a low-lying orbit by launching it with a two-stage rocket from Cape Canaveral with speed v_0 inclined from the vertical by an elevation angle θ_0. On reaching apogee of the initial orbit, the second stage is ignited, generating a velocity boost Δv_1 that places the payload into a circular orbit (see Figure P6.21).
(a) Calculate the additional speed boost Δv_1 required of the second stage to make the final orbit circular.
(b) Calculate the altitude h of the final orbit. Ignore air resistance and the rotational motion of the Earth. The mass and radius of the Earth are $M_E = 5.98 \times 10^{24}$ kg and $R_E = 6.4 \times 10^3$ km, respectively. Let $v_0 = 6$ km/s and $\theta_0 = 30°$.

6.22 Find the apsidal angle for nearly circular orbits in a central field for which the law of force is

$$f(r) = -k\frac{e^{-br}}{r^2}$$

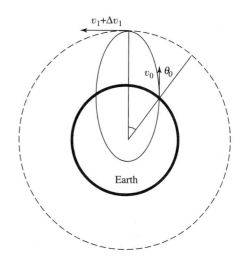

Figure P6.21 Two-stage launch to place satellite in a circular orbit.

6.23 If the solar system were embedded in a uniform dust cloud (see Problem 6.9), what would the apsidal angle of a planet be for motion in a nearly circular orbit? This was once suggested as a possible explanation for the advance of the perihelion of Mercury.

6.24 Show that the stability condition for a circular orbit of radius a is equivalent to the condition that $d^2U/dr^2 > 0$ for $r = a$, where $U(r)$ is the effective potential defined in Section 6.11.

6.25 Find the condition for which circular orbits are stable if the force function is of the form

$$f(r) = -\frac{k}{r^2} - \frac{\epsilon}{r^4}$$

6.26 (a) Show that a circular orbit of radius r is stable in Problem 6.22 if r is less than b^{-1}.
(b) Show that circular orbits are unstable in an inverse-cube force field.

6.27 A comet is going in a parabolic orbit lying in the plane of Earth's orbit. Regarding Earth's orbit as circular of radius a, show that the points where the comet intersects Earth's orbit are given by

$$\cos\theta = -1 + \frac{2p}{a}$$

where p is the perihelion distance of the comet defined at $\theta = 0$.

6.28 Use the result of Problem 6.27 to show that the time interval that the comet remains inside Earth's orbit is the fraction

$$\frac{2^{1/2}}{3\pi}\left(\frac{2p}{a}+1\right)\left(1-\frac{p}{a}\right)^{1/2}$$

of a year and that the maximum value of this time interval is $2/3\pi$ year, or 77.5 days, corresponding to $p = a/2$. Compute the time interval for Halley's Comet ($p = 0.6a$).

6.29   In advanced texts on potential theory, it is shown that the potential energy of a particle of mass m in the gravitational field of an oblate spheroid, like Earth, is approximately

$$V(r) = -\frac{k}{r}\left(1 + \frac{\epsilon}{r^2}\right)$$

where r refers to distances in the equatorial plane, $k = GMm$ as before, and $\epsilon = 2/5R\,\Delta R$, in which R is the equatorial radius and ΔR is the difference between the equatorial and polar radii. From this, find the apsidal angle for a satellite moving in a nearly circular orbit in the equatorial plane of the Earth, where $R = 4000$ miles and $\Delta R = 13$ miles.

6.30   According to the special theory of relativity, a particle moving in a central field with potential energy $V(r)$ describes the same orbit that a particle with a potential energy

$$V(r) - \frac{[E - V(r)]^2}{2m_0c^2}$$

would describe according to nonrelativistic mechanics. Here E is the total energy, m_0 is the rest mass of the particle, and c is the speed of light. From this, find the apsidal angle for motion in an inverse-square force field, $V(r) = -k/r$.

6.31   A comet is observed to have a speed v when it is a distance r from the Sun, and its direction of motion makes an angle ϕ with the radius vector from the Sun. Show that the major axis of the elliptical orbit of the comet makes an angle θ with the initial radius vector of the comet given by

$$\theta = \cot^{-1}\left(\tan\phi - \frac{2}{V^2 R}\csc 2\phi\right)$$

where $V = v/v_e$ and $R = r/a_e$ are dimensionless ratios as defined in Example 6.10.1. Use the numerical values of Example 6.10.1 to calculate a value for the angle θ.

6.32   Two spacecraft (A and B) are in circular orbit about the Earth, traveling in the same plane in the same directional, sense, Spacecraft A is in LEO and satellite B is in geo orbit, as described in Example 6.6.2. The astronauts on board spacecraft A wish to rendezvous with those on board spacecraft B. They must do so by firing their propulsion rockets when spacecraft B is in the right place in its orbit for each craft to reach the rendezvous point at apogee at the same time. (a) Calculate how long it takes spacecraft A to reach apogee and (b) how far in angular advance spacecraft B must be relative to A when A fires its propulsion rockets.

6.33   Show that the differential scattering cross section for a particle of mass m subject to central force field $f(r) = k/r^3$ is given by the expression

$$\sigma(\theta_s)\,d\Omega = 2\pi\,|b\,db| = \frac{k\pi^3}{E}\left[\frac{\pi - \theta_s}{(2\pi - \theta_s)^2\theta_s^2}\right]d\theta_s$$

where θ_s is the scattering angle and E is the energy of the particle.

Computer Problems

C 6.1   In Example 6.7.2 we calculated the gravitational potential at a point P external to a ring of matter of mass M and radius R. P was in the same plane as the ring and a distance $r > R$ from its center. Assume now that the point P is at a distance $r < R$ from the center of the ring but still in the same plane.

(a) Show that the gravitational potential acting at the point r due to the ring of mass is given by

$$\Phi = -\frac{GM}{R}\left(1 + \frac{r^2}{4R^2} + \cdots\right)$$

Let r = radius of Earth's orbit = 1.496×10^{11} m, R = radius of Jupiter's orbit = 7.784×10^{11} m, and M = mass of Jupiter = 1.90×10^{27} kg. Assume that the average gravitational potential produced by Jupiter on Earth is equivalent to that of a uniform ring of matter around the Sun whose mass is equal to that of Jupiter and whose radius from the Sun is equal to Jupiter's radius.

(b) Using this assumption and the values given in part (a), calculate a numerical value for the average gravitational potential that Jupiter exerts on Earth.

(c) Assume that we can approximate this ring of mass by a sum of N strategically deployed discrete point masses M_i, such that $NM_i = M$. As a first approximation, let $N = 2$ and $M_i = M/2$. Deploy these two masses at radii $= R$ and along a line directed between the center of the ring and the position of Earth at radius r. Calculate a numerical value for the potential at r due to these two masses. Repeat this calculation for the case $N = 4$, with the four masses being deployed at quadrants of the circle of radius R and two of them again lining up along the line connecting the center of the circle and Earth. Continue approximating the ring of matter in this fashion (successive multiplications of N by 2 and divisions of M_i by 2) and calculating the resultant potential at r. Stop the iteration when the calculated potential changes by no more than 1 part in 10^4 from the previous value. Compare your result with that obtained in part (b). How many individual masses were required to achieve this accuracy?

(d) Repeat part (c) for values of r equal to 0, 0.2, 0.4, 0.6, 0.8 times r given above. Plot the absolute values of the difference $|\Phi(r) - \Phi(0)|$ versus r, and show that this difference varies quadratically with r as predicted by the equation given in part (a).

C 6.2   Consider a satellite initially placed in a highly elliptical orbit about Earth such that at perigee it just grazes Earth's upper atmosphere as in the figure shown here. Assume that the drag experienced by the satellite can be modeled by a small impulsive force that serves to reduce its velocity by a small fraction δ as it passes through the atmosphere at perigee on each of its orbits. Thus, the position of perigee remains fixed but the distance to apogee decreases on each successive orbit. In the limit, the final orbit is a circle, given this idealized model. Assume that the eccentricity of the initial orbit is $\epsilon_0 = 0.9656$ and that the distance to perigee is $r_p = 6.6 \times 10^3$ km or at about 200 km altitude. Thus, the initial orbit takes the satellite as far away as the distance to the Moon.

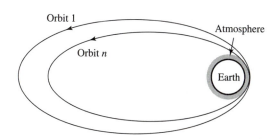

Figure C6.2   Satellite in elliptical orbit grazing Earth's atmosphere.

(a) Let $\delta = 0.01$, and find an approximate analytical solution for how many orbits n it will take until the orbit "circularizes" (we define the orbit to be "circular" when its eccentricity drops below $\epsilon_f = \frac{1}{7}$. At this point, the ratio of the semiminor to semimajor axis exceeds 0.99. Clearly, the model breaks down here because the drag becomes fairly continuous thoughout the orbit.). Express your answer in terms of ϵ_0, ϵ_f, and δ.

(b) Calculate numerically how many orbits it will take for the orbit to "circularize."

(c) How long will it take?

(d) Plot the ratio of the semiminor to semimajor axis as a function of the orbit number n.

(e) Calculate the speed of the satellite at apogee of its last orbit. Compare this speed to its speed at apogee on its initial orbit. Explain why the speed of the satellite increases at each successive apogee when it is being acted on by a resistive air drag force. (The mass and radius of Earth are given in Problem 6.21.)

7

Dynamics of Systems of Particles

"Two equal bodies which are in direct impact with each other and have equal and opposite velocities before impact, rebound with velocities that are, apart from the sign, the same." "The sum of the products of the magnitudes of each hard body, multiplied by the square of the velocities, is always the same, before and after the collision."

— Christiaan Huygens, memoir, *De Motu Corporum ex mutuo impulsu Hypothesis,* composed in Paris, 5-Jan-1669, to Oldenburg, Secretary of the Royal Society

7.1 | Introduction: Center of Mass and Linear Momentum of a System

We now expand our study of mechanics of systems of many particles (two or more). These particles may or may not move independently of one another. Special systems, called *rigid bodies,* in which the relative positions of all the particles are fixed are taken up in the next two chapters. For the present, we develop some general theorems that apply to all systems. Then we apply them to some simple systems of free particles.

Our general system consists of n particles of masses $m_1, m_2, \ldots, m_n$ whose position vectors are, respectively, $\mathbf{r}_1, \mathbf{r}_2, \ldots, \mathbf{r}_n$. We define the *center of mass* of the system as the point whose position vector $\mathbf{r}_{cm}$ (Figure 7.1.1) is given by

$$\mathbf{r}_{cm} = \frac{m_1\mathbf{r}_1 + m_2\mathbf{r}_2 + \cdots + m_n\mathbf{r}_n}{m_1 + m_2 + \cdots + m_n} = \frac{\sum_i m_i\mathbf{r}_i}{m} \tag{7.1.1}$$

where $m = \Sigma\, m_i$ is the total mass of the system. The definition in Equation 7.1.1 is equivalent to the three equations

$$x_{cm} = \frac{\sum_i m_i x_i}{m} \qquad y_{cm} = \frac{\sum_i m_i y_i}{m} \qquad z_{cm} = \frac{\sum_i m_i z_i}{m} \tag{7.1.2}$$

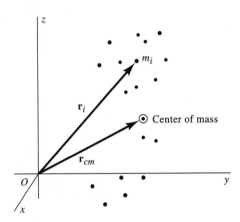

Figure 7.1.1   Center of mass of a system of particles.

We define the *linear momentum* **p** of the system as the vector sum of the linear momenta of the individual particles, namely,

$$\mathbf{p} = \sum_i \mathbf{p}_i = \sum_i m_i \mathbf{v}_i \tag{7.1.3}$$

On calculating $\dot{\mathbf{r}}_{cm} = \mathbf{v}_{cm}$ from Equation 7.1.1 and comparing with Equation 7.1.3, it follows that

$$\mathbf{p} = m\mathbf{v}_{cm} \tag{7.1.4}$$

that is, *the linear momentum of a system of particles is equal to the velocity of the center of mass multiplied by the total mass of the system.*

Suppose now that there are external forces $\mathbf{F}_1, \mathbf{F}_2, \ldots, \mathbf{F}_i, \ldots, \mathbf{F}_n$ acting on the respective particles. In addition, there may be internal forces of interaction between any two particles of the system. We denote these internal forces by $\mathbf{F}_{ij}$, meaning the force exerted on particle i by particle j, with the understanding that $\mathbf{F}_{ii} = 0$. The equation of motion of particle i is then

$$\mathbf{F}_i + \sum_{j=1}^{n} \mathbf{F}_{ij} = m_i \ddot{\mathbf{r}}_i = \dot{\mathbf{p}}_i \tag{7.1.5}$$

where $\mathbf{F}_i$ means the total external force acting on particle i. The second term in Equation 7.1.5 represents the vector sum of all the internal forces exerted on particle i by all other particles of the system. Adding Equation 7.1.5 for the n particles, we have

$$\sum_{i=1}^{n} \mathbf{F}_i + \sum_{i=1}^{n} \sum_{j=1}^{n} \mathbf{F}_{ij} = \sum_{i=1}^{n} \dot{\mathbf{p}}_i \tag{7.1.6}$$

In the double summation in Equation 7.1.6, for every force $\mathbf{F}_{ij}$ there is also a force $\mathbf{F}_{ji}$, and these two forces are equal and opposite

$$\mathbf{F}_{ij} = -\mathbf{F}_{ji} \tag{7.1.7}$$

from the law of action and reaction, Newton's third law. Consequently, the internal forces cancel in pairs, and the double sum vanishes. We can, therefore, write Equation 7.1.7 in the following way:

$$\sum_i \mathbf{F}_i = \dot{\mathbf{p}} = m\mathbf{a}_{cm} \tag{7.1.8}$$

In words: *The acceleration of the center of mass of a system of particles is the same as that of a single particle having a mass equal to the total mass of the system and acted on by the sum of the external forces.*

Consider, for example, a swarm of particles moving in a *uniform* gravitational field. Then, because $\mathbf{F}_i = m_i \mathbf{g}$ for each particle,

$$\sum_i \mathbf{F}_i = \sum m_i \mathbf{g} = m\mathbf{g} \tag{7.1.9}$$

The last step follows from the fact that $\mathbf{g}$ is constant. Hence,

$$\mathbf{a}_{cm} = \mathbf{g} \tag{7.1.10}$$

This is the same as the equation for a single particle or projectile. Thus, the center of mass of the shrapnel from an artillery shell that has burst in midair follows the same parabolic path that the shell would have taken had it not burst (until any of the pieces strikes something).

In the special case in which no external forces are acting on a system (or if $\Sigma \, \mathbf{F}_i = 0$), then $\mathbf{a}_{cm} = 0$ and $\mathbf{v}_{cm} = $ constant; thus, the linear momentum of the system remains constant:

$$\sum_i \mathbf{p}_i = \mathbf{p} = m\mathbf{v}_{cm} = \text{constant} \tag{7.1.11}$$

This is the *principle of conservation of linear momentum.* In Newtonian mechanics the constancy of the linear momentum of an isolated system is directly related to, and is in fact a consequence of, the third law. But even in those cases in which the forces between particles do not directly obey the law of action and reaction, such as the magnetic forces between moving charges, the principle of conservation of linear momentum still holds when due account is taken of the total linear momentum of the particles and the electromagnetic field.[1]

EXAMPLE 7.1.1

At some point in its trajectory a ballistic missile of mass m breaks into three fragments of mass $m/3$ each. One of the fragments continues on with an initial velocity of one-half the velocity $\mathbf{v}_0$ of the missile just before breakup. The other two pieces go off at right angles to each other with equal speeds. Find the initial speeds of the latter two fragments in terms of v_0.

Solution:

At the point of breakup, conservation of linear momentum is expressed as

$$m\mathbf{v}_{cm} = m\mathbf{v}_0 = \frac{m}{3}\mathbf{v}_1 + \frac{m}{3}\mathbf{v}_2 + \frac{m}{3}\mathbf{v}_3$$

---

[1] See, for example, P. M. Fishbane, S. Gasiorowicz, S. T. Thornton, *Physics for Scientists and Engineers.* Prentice-Hall, Englewood Cliffs, NJ, 1993.

The given conditions are: $\mathbf{v}_1 = \mathbf{v}_0/2$, $\mathbf{v}_2 \cdot \mathbf{v}_3 = 0$, and $v_2 = v_3$. From the first we get, on cancellation of the m's, $3\mathbf{v}_0 = (\mathbf{v}_0/2) + \mathbf{v}_2 + \mathbf{v}_3$, or

$$\frac{5}{2}\mathbf{v}_0 = \mathbf{v}_2 + \mathbf{v}_3$$

Taking the dot product of each side with itself, we have

$$\frac{25}{4}v_0^2 = (\mathbf{v}_2 + \mathbf{v}_3) \cdot (\mathbf{v}_2 + \mathbf{v}_3) = v_2^2 + 2\mathbf{v}_2 \cdot \mathbf{v}_3 + v_3^2 = 2v_2^2$$

Therefore,

$$v_2 = v_3 = \frac{5}{2\sqrt{2}}v_0 = 1.77v_0$$

7.2 | Angular Momentum and Kinetic Energy of a System

We previously stated that the angular momentum of a single particle is defined as the cross product $\mathbf{r} \times m\mathbf{v}$. The angular momentum $\mathbf{L}$ of a system of particles is defined accordingly, as the vector sum of the individual angular momenta, namely,

$$\mathbf{L} = \sum_{i=1}^{n} (\mathbf{r}_i \times m_i \mathbf{v}_i) \tag{7.2.1}$$

Let us calculate the time derivative of the angular momentum. Using the rule for differentiating the cross product, we find

$$\frac{d\mathbf{L}}{dt} = \sum_{i=1}^{n} (\mathbf{v}_i \times m_i \mathbf{v}_i) + \sum_{i=1}^{n} (\mathbf{r}_i \times m_i \mathbf{a}_i) \tag{7.2.2}$$

Now the first term on the right vanishes, because, $\mathbf{v}_i \times \mathbf{v}_i = 0$ and, because $m_i \mathbf{a}_i$ is equal to the total force acting on particle i, we can write

$$\frac{d\mathbf{L}}{dt} = \sum_{i=1}^{n} \left[\mathbf{r}_i \times \left(\mathbf{F}_i + \sum_{j=1}^{n} \mathbf{F}_{ij} \right) \right]$$

$$= \sum_{i=1}^{n} \mathbf{r}_i \times \mathbf{F}_i + \sum_{i=1}^{n} \sum_{j=1}^{n} \mathbf{r}_i \times \mathbf{F}_{ij} \tag{7.2.3}$$

where, as in Section 7.1, $\mathbf{F}_i$ denotes the total external force on particle i, and $\mathbf{F}_{ij}$ denotes the (internal) force exerted on particle i by any other particle j. Now the double summation on the right consists of pairs of terms of the form

$$(\mathbf{r}_i \times \mathbf{F}_{ij}) + (\mathbf{r}_j \times \mathbf{F}_{ji}) \tag{7.2.4}$$

Denoting the vector displacement of particle j relative to particle i by $\mathbf{r}_{ij}$, we see from the triangle shown in Figure 7.2.1 that

$$\mathbf{r}_{ij} = \mathbf{r}_j - \mathbf{r}_i \tag{7.2.5}$$

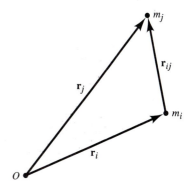

Figure 7.2.1   Definition of the vector $\mathbf{r}_{ij}$.

Therefore, because $\mathbf{F}_{ji} = -\mathbf{F}_{ij}$, expression 7.2.4 reduces to

$$-\mathbf{r}_{ij} \times \mathbf{F}_{ij} \tag{7.2.6}$$

which clearly vanishes if the internal forces are central, that is, if they act along the lines connecting pairs of particles. Hence, the double sum in Equation 7.2.3 vanishes. Now the cross product $\mathbf{r}_i \times \mathbf{F}_i$ is the moment of the external force $\mathbf{F}_i$. The sum $\Sigma\, \mathbf{r}_i \times \mathbf{F}_i$ is, therefore, the total moment of all the external forces acting on the system. If we denote the total external torque, or moment of force, by $\mathbf{N}$, Equation 7.2.3 takes the form

$$\frac{d\mathbf{L}}{dt} = \mathbf{N} \tag{7.2.7}$$

That is, *the time rate of change of the angular momentum of a system is equal to the total moment of all the external forces acting on the system.*

If a system is isolated, then $\mathbf{N} = 0$, and the angular momentum remains constant in both magnitude and direction:

$$\mathbf{L} = \sum_i \mathbf{r}_i \times m_i\mathbf{v}_i = \text{constant vector} \tag{7.2.8}$$

This is a statement of the *principle of conservation of angular momentum.* It is a generalization for a single particle in a central field. Like the constancy of linear momentum discussed in the preceding section, the angular momentum of an isolated system is also constant in the case of a system of moving charges when the angular momentum of the electromagnetic field is considered.[2]

It is sometimes convenient to express the angular momentum in terms of the motion of the center of mass. As shown in Figure 7.2.2, we can express each position vector $\mathbf{r}_i$ in the form

$$\mathbf{r}_i = \mathbf{r}_{cm} + \bar{\mathbf{r}}_i \tag{7.2.9}$$

where $\bar{\mathbf{r}}_i$ is the position of particle i relative to the center of mass. Taking the derivative with respect to t, we have

$$\mathbf{v}_i = \mathbf{v}_{cm} + \bar{\mathbf{v}}_i \tag{7.2.10}$$

---

[2] See footnote 1.

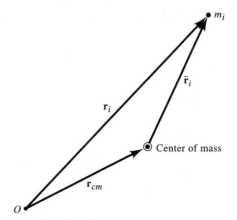

Figure 7.2.2   Definition of the vector $\bar{\mathbf{r}}_i$.

Here $\mathbf{v}_{cm}$ is the velocity of the center of mass and $\bar{\mathbf{v}}_i$ is the velocity of particle i relative to the center of mass. The expression for $\mathbf{L}$ can, therefore, be written

$$\mathbf{L} = \sum_i (\mathbf{r}_{cm} + \bar{\mathbf{r}}_i) \times m_i (\mathbf{v}_{cm} + \bar{\mathbf{v}}_i)$$

$$= \sum_i (\mathbf{r}_{cm} \times m_i \mathbf{v}_{cm}) + \sum_i (\mathbf{r}_{cm} \times m_i \bar{\mathbf{v}}_i)$$

$$+ \sum_i (\bar{\mathbf{r}}_i \times m_i \mathbf{v}_{cm}) + \sum_i (\bar{\mathbf{r}}_i \times m_i \bar{\mathbf{v}}_i)$$

$$= \mathbf{r}_{cm} \times \left(\sum_i m_i \right) \mathbf{v}_{cm} + \mathbf{r}_{cm} \times \sum_i m_i \bar{\mathbf{v}}_i \qquad (7.2.11)$$

$$+ \left(\sum_i m_i \bar{\mathbf{r}}_i \right) \times \mathbf{v}_{cm} + \sum_i (\bar{\mathbf{r}}_i \times m_i \bar{\mathbf{v}}_i)$$

Now, from Equation 7.2.9, we have

$$\sum_i m_i \bar{\mathbf{r}}_i = \sum_i m_i (\mathbf{r}_i - \mathbf{r}_{cm}) = \sum_i m_i \mathbf{r}_i - m\mathbf{r}_{cm} = 0 \qquad (7.2.12)$$

Similarly, we obtain

$$\sum_i m_i \bar{\mathbf{v}}_i = \sum_i m_i \mathbf{v}_i - m\mathbf{v}_{cm} = 0 \qquad (7.2.13)$$

by differentiation with respect to t. (These two equations merely state that the position and velocity of the center of mass, relative to the center of mass, are both zero.) Consequently, the second and third summations in the expansion of $\mathbf{L}$ vanish, and we can write

$$\mathbf{L} = \mathbf{r}_{cm} \times m\mathbf{v}_{cm} + \sum_i \bar{\mathbf{r}}_i \times m_i \bar{\mathbf{v}}_i \qquad (7.2.14)$$

expressing the angular momentum of a system in terms of an "orbital" part (motion of the center of mass) and a "spin" part (motion about the center of mass).

EXAMPLE 7.2.1

A long, thin rod of length l and mass m hangs from a pivot point about which it is free to swing in a vertical plane like a simple pendulum. Calculate the total angular momentum of the rod as a function of its instantaneous angular velocity ω. Show that the theorem represented by Equation 7.2.14 is true by comparing the angular momentum obtained using that theorem to that obtained by direct calculation.

Solution:

The rod is shown in Figure 7.2.3a. First we calculate the angular momentum $\mathbf{L}_{cm}$ of the center of mass of the rod about the pivot point. Because the velocity $\mathbf{v}_{cm}$ of the center of mass is always perpendicular to the radius vector $\mathbf{r}$ denoting its location relative to the pivot point, the sine of the angle between those two vectors is unity. Thus, the magnitude of $\mathbf{L}_{cm}$ is given by

$$L_{cm} = \frac{l}{2} p_{cm} = m \frac{l}{2} v_{cm} = m \frac{l}{2}\left(\frac{l}{2}\omega\right) = \tfrac{1}{4} m l^2 \omega$$

Figure 7.2.3b depicts the motion of the rod as seen from the perspective of its center of mass. The angular momentum dL_{rel} of two small mass elements, each of size dm symmetrically disposed about the center of mass of the rod, is given by

$$dL_{rel} = 2r\,dp = 2rv\,dm = 2r(r\omega)\lambda\,dr$$

where λ is the mass per unit length of the rod. The total relative angular momentum is obtained by integrating this expression from $r = 0$ to $r = l/2$.

$$L_{rel} = 2\lambda\omega \int_0^{l/2} r^2 dr = \tfrac{1}{12}(\lambda l)l^2 \omega = \left(\tfrac{1}{12} m l^2\right)\omega$$

We can see in the preceding equation that the angular momentum of the rod about its center of mass is directly proportional to the angular velocity ω of the rod. The constant of proportionality $ml^2/12$ is called the *moment of inertia* I_{cm} of the rod about its center of mass. Moment of inertia plays a role in rotational motion similar to that of inertial mass in translational motion as we shall see in the next chapter.

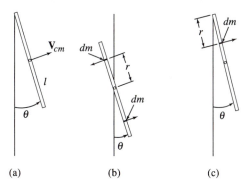

Figure 7.2.3   Rod of mass m and length l free to swing in a vertical plane about a fixed pivot.

(a)   (b)   (c)

Finally, the total angular momentum of the rod is

$$L_{tot} = L_{cm} + L_{rel} = \frac{1}{3}ml^2\omega$$

Again, the total angular momentum of the rod is directly proportional to the angular velocity of the rod. Here, though, the constant of proportionality is the moment of inertia of the rod about the pivot point at the end of the rod. This moment of inertia is larger than that about the center of mass. The reason is that more of the mass of the rod is distributed farther away from its end than from its center, thus, making it more difficult to rotate a rod about an end.

The total angular momentum can also be obtained by integrating down the rod, starting from the pivot point, to obtain the contribution from each mass element dm, as shown in Figure 7.2.3c

$$dL_{tot} = r\, dp = r(v\, dm) = r(r\omega)\lambda\, dr$$

$$L_{tot} = \lambda\omega \int_0^l r^2 dr = \frac{1}{3}ml^2\omega$$

And, indeed, the two methods yield the same result.

Kinetic Energy of a System

The total kinetic energy T of a system of particles is given by the sum of the individual energies, namely,

$$T = \sum_i \frac{1}{2}m_i v_i^2 = \sum_i \frac{1}{2}m_i(\mathbf{v}_i \cdot \mathbf{v}_i) \tag{7.2.15}$$

As before, we can express the velocities relative to the mass center giving

$$
\begin{aligned}
T &= \sum_i \frac{1}{2}m_i(\mathbf{v}_{cm} + \bar{\mathbf{v}}_i)\cdot(\mathbf{v}_{cm} + \bar{\mathbf{v}}_i) \\
&= \sum_i \frac{1}{2}m_i v_{cm}^2 + \sum_i m_i(\mathbf{v}_{cm}\cdot\bar{\mathbf{v}}_i) + \sum_i \frac{1}{2}m_i \bar{v}_i^2 \\
&= \frac{1}{2}v_{cm}^2 \sum_i m_i + \mathbf{v}_{cm}\cdot\sum_i m_i\bar{\mathbf{v}}_i + \sum_i \frac{1}{2}m_i\bar{v}_i^2
\end{aligned}
\tag{7.2.16}
$$

Because the second summation $\sum_i m_i \bar{\mathbf{v}}_i$ vanishes, we can express the kinetic energy as follows:

$$T = \frac{1}{2}mv_{cm}^2 + \sum_i \frac{1}{2}m_i\bar{v}_i^2 \tag{7.2.17}$$

The first term is the kinetic energy of translation of the whole system, and the second is the kinetic energy of motion relative to the mass center.

The separation of angular momentum and kinetic energy into a center-of-mass part and a relative-to-center-of-mass part finds important applications in atomic and molecular physics and in astrophysics. We find the preceding two theorems useful in the study of rigid bodies in the following chapters.

EXAMPLE 7.2.2

Calculate the total kinetic energy of the rod of Example 7.2.1. Use the theorem represented by Equation 7.2.17. As in Example 7.2.1, show that the total energy obtained for the rod according to this theorem is equivalent to that obtained by direct calculation.

Solution:

The translational kinetic energy of the center of mass of the rod is

$$T_{cm} = \tfrac{1}{2}m\mathbf{v}_{cm} \cdot \mathbf{v}_{cm} = \tfrac{1}{2}m\left(\frac{l}{2}\omega\right)^2 = \tfrac{1}{8}ml^2\omega^2$$

The kinetic energy of two equal mass elements dm symetrically disposed about the center of mass is

$$dT_{rel} = \tfrac{1}{2}(2dm)\mathbf{v}\cdot\mathbf{v} = \lambda\,dr(r\omega)^2 = \lambda\omega^2 r^2 dr$$

where λ, again, is the mass per unit length of the rod. The total energy relative to the center of mass can be obtained by integrating the preceding expression from $r = 0$ to $r = l/2$.

$$T_{rel} = \lambda\omega^2 \int_0^{l/2} r^2 dr = \tfrac{1}{24}\lambda\omega^2 l^3 = \tfrac{1}{2}\left(\tfrac{1}{12}ml^2\right)\omega^2 = \tfrac{1}{2}I_{cm}\omega^2$$

(**Note:** As in Example 7.2.1, the moment of inertia term I_{cm} appears as the constant of proportionality to ω^2 in the previous expression for the rotational kinetic energy of the rod about its center of mass. Again, the moment of inertia term that occurs in the expression for rotational kinetic energy can be seen to be completely analogous to the inertial mass term in an expression for the translational kinetic energy of a particle.)

The total kinetic energy of the rod is then

$$T = T_{cm} + T_{rel} = \tfrac{1}{8}ml^2\omega^2 + \tfrac{1}{24}ml^2\omega^2 = \tfrac{1}{2}\left(\tfrac{1}{3}ml^2\right)\omega^2 = \tfrac{1}{2}I\omega^2$$

where we have expressed the final result in terms of the total moment of inertia of the rod about its endpoint, exactly as in Example 7.2.1.

We leave it as an exercise for the reader to calculate the kinetic energy directly and show that it is equal to the value obtained previously. The calculation proceeds in a fashion completely analogous to that in Example 7.2.1.

7.3 | Motion of Two Interacting Bodies: The Reduced Mass

Let us consider the motion of a system consisting of two bodies, treated here as particles, that interact with each other by a central force. We assume the system is isolated, and, hence, the center of mass moves with constant velocity. For simplicity, we take the center of mass as the origin. We have then

$$m_1\bar{\mathbf{r}}_1 + m_2\bar{\mathbf{r}}_2 = 0 \tag{7.3.1}$$

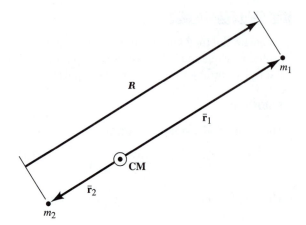

Figure 7.3.1   The relative position vector **R** for the two-body problem.

where, as shown in Figure 7.3.1, the vectors $\bar{\mathbf{r}}_1$ and $\bar{\mathbf{r}}_2$ represent the positions of the particles m_1 and m_2, respectively, relative to the center of mass. Now, if **R** is the position vector of particle 1 relative to particle 2, then

$$\mathbf{R} = \bar{\mathbf{r}}_1 - \bar{\mathbf{r}}_2 = \bar{\mathbf{r}}_1 \left(1 + \frac{m_1}{m_2} \right) \tag{7.3.2}$$

The last step follows from Equation 7.3.1.

The differential equation of motion of particle 1 relative to the center of mass is

$$m_1 \frac{d^2 \bar{\mathbf{r}}_1}{dt^2} = \mathbf{F}_1 = f(R) \frac{\mathbf{R}}{R} \tag{7.3.3}$$

in which $|f(R)|$ is the magnitude of the mutual force between the two particles. By using Equation 7.3.2, we can write

$$\mu \frac{d^2 \mathbf{R}}{dt^2} = f(R) \frac{\mathbf{R}}{R} \tag{7.3.4}$$

where

$$\mu = \frac{m_1 m_2}{m_1 + m_2} \tag{7.3.5}$$

The quantity μ is called the *reduced mass*. The new equation of motion (Equation 7.3.4) gives the motion of particle 1 relative to particle 2, and an exactly similar equation gives the motion of particle 2 relative to particle 1. This equation is precisely the same as the ordinary equation of motion of a single particle of mass μ moving in a central field of force given by $f(R)$. Thus, the fact that both particles are moving relative to the center of mass is automatically accounted for by replacing m_1 by the reduced mass μ. If the bodies are of equal mass m, then $\mu = m/2$. On the other hand, if m_2 is very much greater than m_1, so that m_1/m_2 is very small, then μ is nearly equal to m_1.

For two bodies attracting each other by gravitation

$$f(R) = -\frac{Gm_1m_2}{R^2} \tag{7.3.6}$$

In this case the equation of motion is

$$\mu\ddot{\mathbf{R}} = -\frac{Gm_1m_2}{R^2}\mathbf{e}_R \tag{7.3.7}$$

or, equivalently,

$$m_1\ddot{\mathbf{R}} = -\frac{G(m_1+m_2)m_1}{R^2}\mathbf{e}_R \tag{7.3.8}$$

where $\mathbf{e}_R = \mathbf{R}/R$ is a unit vector in the direction of $\mathbf{R}$.

In Section 6.6 we derived an equation giving the periodic time of orbital motion of a planet of mass m moving in the Sun's gravitational field, namely, $\tau = 2\pi (GM_\odot)^{-1/2}a^{3/2}$, where $M_\odot$ is the Sun's mass and a is the semimajor axis of the elliptical orbit of the planet about the Sun. In that derivation we assumed that the Sun was stationary, with the origin of our coordinate system at the center of the Sun. To account for the Sun's motion about the common center of mass, the correct equation is Equation 7.3.8 in which $m = m_1$ and $M_\odot = m_2$. The constant k, which was taken to be $GM_\odot m$ in the earlier treatment, should be replaced by $G(M_\odot + m)m$ so that the correct equation for the period is

$$\tau = 2\pi [G(M_\odot + m)]^{-1/2}a^{3/2} \tag{7.3.9a}$$

or, for any two-body system held together by gravity, the orbital period is

$$\tau = 2\pi [G(m_1 + m_2)]^{-1/2}a^{3/2} \tag{7.3.9b}$$

If m_1 and m_2 are expressed in units of the Sun's mass and a is in astronomical units (the mean distance from Earth to the Sun), then the orbital period in years is given by

$$\tau = (m_1 + m_2)^{-1/2}a^{3/2} \tag{7.3.9c}$$

For most planets in our solar system, the added mass term in the preceding expression for the period makes very little difference—Earth's mass is only 1/330,000 the Sun's mass. The most massive planet, Jupiter, has a mass of about 1/1000 the mass of the Sun, so the effect of the reduced-mass formula is to change the earlier calculation in the ratio $(1.001)^{-1/2} = 0.9995$ for the period of Jupiter's revolution about the Sun.

Binary Stars: White Dwarfs and Black Holes

About half of all the stars in the galaxy in the vicinity of the Sun are binary, or double; that is, they occur in pairs held together by their mutual gravitational attraction, with each member of the pair revolving about their common center of mass. From the preceding analysis we can infer that either member of a binary system revolves about the other in an elliptical orbit for which the orbiting period is given by Equations 7.3.9b and c, where a is the semimajor axis of the ellipse and m_1 and m_2 are the masses of the two stars. Values of a for known binary systems range from the very least (*contact binaries* in which the stars touch each other) to values so large that the period is measured in millions of years. A typical example is the brightest star in the night sky, Sirius, which consists of a very luminous star with a mass of 2.1 $M_\odot$ and a very small dim star, called a *white dwarf*, which can only be seen in large telescopes. The mass of this small companion is 1.05 $M_\odot$, but its size is roughly that of a large planet, so its density is extremely large (30,000 times the density of water). The value of a for the Sirius system is approximately 20 AU (about the distance from the Sun to the planet Uranus), and the period, as calculated from Equation 7.3.9c, should be about

$$\tau = (2.1 + 1.05)^{-1/2} (20)^{3/2} \text{ years} = 50 \text{ years}$$

which is what it is observed to be.

A binary system that is believed to harbor a black hole as one of its components is the x-ray source known as Cygnus X-1.[3] The visible component is the normal star HDE 226868. Spectroscopic observation of the optical light from this star indicates that the period and semimajor axis of the orbit are 5.6 days and about 30×10^6 km, respectively. The optically invisible companion is the source of an x-ray flux that exhibits fluctuations that vary as rapidly as a millisecond, indicating that it can be no larger than 300 km across. These observations, as well as a number of others, indicate that the mass of HDE 226868 is at least as large as 20 $M_\odot$, while that of its companion is probably as large as 16 $M_\odot$ but surely exceeds 7 $M_\odot$. It is difficult to conclude that this compact, massive object could be anything other than a *black hole*. Black holes are objects that contain so much mass within a given radius[4] that nothing, not even light, can escape their gravitational field. If black holes are located in binary systems, however, mass can "leak over" from the large companion star and form an accretion disk about the black hole. As the matter in this disk orbits the black hole, it can lose energy by frictional heating and crash down into it, ultimately heating to temperatures well in excess of tens of millions of degrees. X-rays are emitted by this hot matter before it falls completely into the hole (Figure 7.3.2). Black holes are predicted mathematically by the general theory of relativity, and unequivocal proof of their existence would constitute a milestone in astrophysics.

---

[3] A. P. Cowley, *Ann. Rev. Astron. Astrophys.* 30, 287 (1992).

[4] According to the theory of general relatively, a nonrotating, spherically symmetric body of mass m becomes a Schwarzschild black hole if it is compressed to a radius r_s, Schwarzschild radius, where

$$r_s = \frac{2Gm}{c^2}$$

in which c is the speed of light. The Earth would become a black hole if compressed to the size of a small marble; the Sun would become one if compressed to a radius of about 3 km, much smaller than the white dwarf companion of Sirius.

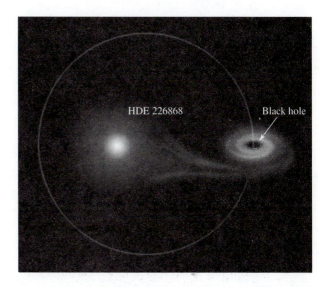

Figure 7.3.2   The Cygnus X-1 System.

EXAMPLE 7.3.1

A certain binary star system is observed to be both eclipsing and spectroscopic. This means that the system is seen from Earth with its orbital plane edge-on and that the orbital velocities v_1 and v_2 of the two stars that constitute the system can be determined from Doppler shift measurements of observed spectral lines. You don't need to understand the details of this last statement. The important point is that we know the orbital velocities. They are, in appropriate units, $v_1 = 1.257$ AU/year and $v_2 = 5.027$ AU/year. The period of revolution of each star about its center of mass is $\tau = 5$ years. (That can be ascertained from the observed frequency of eclipses.) Calculate the mass (in solar mass units $M_\odot$) of each star. Assume circular orbits.

Solution:

The radius of the orbit of each star about their common center of mass can be calculated from its velocity and period

$$r_1 = \frac{1}{2\pi} v_1 \tau = 1 \text{ AU} \qquad r_2 = \frac{1}{2\pi} v_2 \tau = 4 \text{ AU}$$

Thus, the semimajor axis a of the orbit is

$$a = r_1 + r_2 = 5 \text{ AU}$$

The sum of the masses can be obtained from Equation 7.3.9c

$$m_1 + m_2 = \frac{a^3}{\tau^2} = 5 \, M_\odot$$

The ratio of the two masses can be determined by differentiating Equation 7.3.1

$$m_1 \mathbf{v}_1 + m_2 \mathbf{v}_2 = 0 \qquad \frac{m_2}{m_1} = \left| \frac{\mathbf{v}_1}{\mathbf{v}_2} \right| = \frac{1}{4}$$

Combining these last two expressions yields the values for each mass, $m_1 = 4\,M_\odot$ and $m_2 = 1\,M_\odot$.

*7.4 | The Restricted Three-Body Problem[5]

In Chapter 6, we considered the motion of a single particle subject to a central force. The motion of a planet in the gravitational field of the Sun is well described by such a theory because the mass of the Sun is so large compared with that of a planet that its own motion can be ignored. In the previous section, we relaxed this condition and found that we could still apply the techniques of Newtonian analysis to this more general case and find an analytic solution for their motion. If we add just one more, third body, however, the problem becomes completely intractable. The general three-body problem, namely the calculation of the motion of three bodies of different masses, initial positions, and velocities, subject to the combined gravitational field of the others, confounded some of the greatest minds in the post-Newtonian era. It is not possible to solve this problem analytically because of insurmountable mathematical difficulties. Indeed, the problem is described by a system of nine second-order differential equations: three bodies moving in three dimensions. Even after a mathematical reduction accomplished by a judicious choice of coordinate system and by invoking laws of conservation to find invariants of the motion, the problem continues to defy assault by modern analytic techniques.

Fortunately, it is possible to solve a simplified case of the general problem that nonetheless describes a wide variety of phenomena. This special case is called the *restricted three-body problem.* The simplifications involved are both physical and mathematical: We assume that two of the bodies (called the *primaries*[6]) are much more massive than the third body (called the *tertiary*) and that they move in a plane—in circular orbits about their center of mass. The tertiary has a negligible mass compared with either of the primaries, moves in their orbital plane, and exerts no gravitational influence on either of them.

No physical system meets these requirements exactly. The tertiary always perturbs the orbits of the primaries. Perfectly circular orbits never occur, although most of the orbits of bodies in the solar system come very close—with the exception of comets. The orbit of the tertiary is almost never coplanar with those of the primaries, although deviations from coplanarity are often quite small. Gravitational systems with a dominant central mass exhibit a remarkable propensity for coplanarity. Again, disregarding the comets, the remaining members of the solar system exhibit a high degree of coplanarity, as do the individual systems of the large Jovian planets and their assemblage of moons.

---

[5] Our analysis of the restricted three-body problem is based on P. Hellings, *Astrophysics with a PC, An Introduction to Computational Astrophysics,* Willman-Bell, Inc., Richmond, VA (1994). Also, for an even more in-depth analysis see V. Szebehely, *Theory of Orbits,* Academic Press, New York (1967).

[6] Usually, the most massive of the pair is called the primary and the least massive is called the secondary. Here, we lump them together as the two primaries because their motion is only incidental to our main interest—the motion of the third body.

The restricted three-body problem serves as an excellent model for calculating the orbital motion of a small tertiary in the gravitational field of the other two. It is fairly easy to see two possible solutions depicting two extreme situations. One occurs when the tertiary more or less orbits the center of mass of the other two at such a remote distance that the two primaries appear to blur together as a single gravitational source. A second occurs when the tertiary is bound so closely to one of the primaries that it orbits it in Keplerian fashion, seemingly oblivious to the presence of the second primary. Both of these possibilities are realized in nature. In this section, however, we attempt to find a third, not so obvious, "stationary" solution; that is, one in which the tertiary is "held fixed" by the other two and partakes of their overall rotational motion. In other words, it remains more or less at rest relative to the two primaries; the orientation in space of the entire system rotates with a constant angular speed, but its relative configuration remains fixed in time. The great 18th-century mathematician Joseph-Louis Lagrange (1736–1813) solved this problem and showed that such orbits are possible.

Equations of Motion for the Restricted Three-Body Problem

The restricted problem is a two-dimensional one: All orbits lie within a single, fixed plane in space. The orbit of each of the two primaries is a circle with common angular velocity ω about their center of mass. We assume that the center of mass of the two primaries remains fixed in space and that the rotational sense of their orbital motion viewed from above is counterclockwise as shown in Figure 7.4.1

We designate M_1 the mass of the most massive primary, M_2 the mass of the least massive one, and m the small mass of the tertiary whose orbit we wish to calculate. We choose a coordinate system x'-y' that rotates with the two primaries and whose origin is their center of mass. We let the $+x$-axis lie along the direction toward the most massive primary M_1. The radii of the circular orbits of M_1 and M_2 are designated a and b, respectively. These distances remain fixed along the x'-axis in the rotating coordinate system.

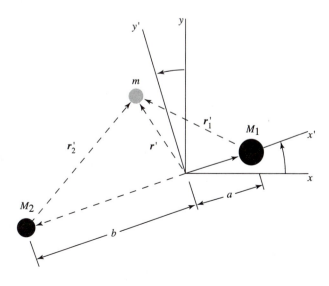

Figure 7.4.1 Coordinate system for restricted three-body problem.

Letting the coordinates of the tertiary be (x', y'), the distance between it and each of the two primaries is

$$r_1' = \sqrt{(x' - a)^2 + y'^2} \tag{7.4.1a}$$

$$r_2' = \sqrt{(x' + b)^2 + y'^2} \tag{7.4.1b}$$

The net gravitational force exerted on m (see Equation 6.1.1) is thus

$$\mathbf{F} = -m\frac{GM_1}{r_1'^2}\left(\frac{\mathbf{r}_1'}{r_1'}\right) - m\frac{GM_2}{r_2'^2}\left(\frac{\mathbf{r}_2'}{r_2'}\right) \tag{7.4.2}$$

where $\mathbf{r}_1'$ and $\mathbf{r}_2'$ are the vector positions of m with respect of M_1 and M_2. This force is the only real one that acts on m, but because we have effectively nullified the motion of the two primaries by choosing to calculate the motion in a frame of reference that rotates with them, we must include the effect of the noninertial forces that are introduced as a result of this choice.

The general equation of motion for a particle in a rotating frame of reference was given by Equation 5.3.2. Because the origin of the rotating coordinate system remains fixed in space, $\mathbf{A}_0 = 0$, and because the rate of rotation is a constant, $\dot{\boldsymbol{\omega}} = 0$ and Equation 5.3.2 takes the form

$$\mathbf{F}' = m\mathbf{a}' = \mathbf{F} - 2m\boldsymbol{\omega} \times \mathbf{v}' - m\boldsymbol{\omega} \times (\boldsymbol{\omega} \times \mathbf{r}') \tag{7.4.3}$$

Because m is common to all terms in Equation 7.4.3, we can rewrite it in terms of accelerations as

$$\mathbf{a}' = \frac{\mathbf{F}}{m} - 2\boldsymbol{\omega} \times \mathbf{v}' - \boldsymbol{\omega} \times (\boldsymbol{\omega} \times \mathbf{r}') \tag{7.4.4}$$

We are now in a position to calculate the later two noninertial accelerations in Equation 7.4.4—the Coriolis and centrifugal accelerations

$$2\boldsymbol{\omega} \times \mathbf{v}' = 2\omega\mathbf{k}' \times (\mathbf{i}'\dot{x}' + \mathbf{j}'\dot{y}') = -\mathbf{i}'2\omega\dot{y}' + \mathbf{j}'2\omega\dot{x}' \tag{7.4.5}$$

and

$$\boldsymbol{\omega} \times (\boldsymbol{\omega} \times \mathbf{r}') = \omega\mathbf{k}' \times [\omega\mathbf{k}' \times (\mathbf{i}'x' + \mathbf{j}'y')]$$
$$= -\mathbf{i}'\omega^2 x' - \mathbf{j}'\omega^2 y' \tag{7.4.6}$$

We now insert Equations 7.4.1a and b, 7.4.2, 7.4.5, and 7.4.6 into 7.4.4 to obtain the equations of motion of mass m in the x' and y' coordinates

$$\ddot{x}' = -GM_1\frac{x' - a}{[(x' - a)^2 + y'^2]^{3/2}} - GM_2\frac{x' + b}{[(x' + b)^2 + y'^2]^{3/2}}$$
$$+ \omega^2 x' + 2\omega\dot{y}' \tag{7.4.7a}$$

$$\ddot{y}' = -GM_1\frac{y'}{[(x' - a)^2 + y'^2]^{3/2}} - GM_2\frac{y'}{[(x' + b)^2 + y'^2]^{3/2}}$$
$$+ \omega^2 y' - 2\omega\dot{x}' \tag{7.4.7b}$$

The Effective Potential: The Five Lagrangian Points

Before solving Equations 7.4.7a and b, we would like to speculate about the possible solutions that we might obtain. Toward this end, we note that the first three terms in each of those equations can be expressed as the gradient of an effective potential function, $V(r')$ in polar coordinates

$$V(r') = -\frac{GM_1}{|\mathbf{r'} - \mathbf{a}|} - \frac{GM_2}{|\mathbf{r'} - \mathbf{b}|} - \frac{1}{2}\omega^2 r'^2 \qquad (7.4.8a)$$

or $V(x', y')$ in Cartesian coordinates

$$V(x', y') = -\frac{GM_1}{\sqrt{(x' - a)^2 + y'^2}} - \frac{GM_2}{\sqrt{(x' + b)^2 + y'^2}} - \frac{1}{2}\omega^2(x'^2 + y'^2) \qquad (7.4.8b)$$

The last term in Equations 7.4.7a and b is velocity-dependent and cannot be expressed as the gradient of an effective potential. Thus, we must include the Coriolis term as an additional term in any equation that derives the force from the effective potential. For example, Equation 7.4.3 becomes

$$\mathbf{F'} = -\nabla V(x', y') - 2m\boldsymbol{\omega} \times \mathbf{v} \qquad (7.4.9)$$

A considerable simplification in all further calculations may be achieved by expressing mass, length, and time in units that transform $V(x', y')$ into an invariant form that makes it applicable to all restricted three-body situations regardless of the values of their masses. First, we scale all distances to the total separation of the two primaries; that is, we let $a + b$ equal *one length unit.* This is analogous to the convention in which the *astronomical unit,* or AU, the mean distance between the Earth and the Sun, is used to express distances to the other planets in the solar system. Next, we set the factor $G(M_1 + M_2)$, equal to one *"gravitational"* mass unit. The "gravitational" masses GM_i of each body can then be expressed as fractional multiples α_i of this unit. Finally, we set the orbital period of the primaries τ equal to 2π *time units.* This implies that the angular velocity of the two primaries about their center of mass and, by association, the rate of rotation of the x'-y' frame of reference, is $\omega = 1$ *inverse time unit.* Use of these scaled units allows us to characterize the equations of motion by the single parameter α, where $0 < \alpha < 0.5$. In addition, it has the added benefit of riding our expressions of the obnoxious factor G.

In terms of α, the distance of each primary from the center of mass is then

$$\alpha = \frac{a}{a+b} \qquad \beta = \frac{b}{a+b} = 1 - \alpha \qquad (7.4.10)$$

The coordinates of the first primary are, thus, $(\alpha, 0)$ and those of the second primary are $(1 - \alpha, 0)$. Furthermore, because the origin of the coordinate system is the center of mass, from Equation 7.3.1, we have

$$M_1 a = M_2 b \qquad (7.4.11)$$

and the "gravitational" masses of each primary can then be expressed also in terms of the factor α

$$\alpha_1 = \frac{GM_1}{G(M_1 + M_2)} = \frac{b}{a+b} = 1 - \alpha \qquad (7.4.12a)$$

$$\alpha_2 = \frac{GM_2}{G(M_1 + M_2)} = \frac{b}{a+b} = \alpha \qquad (7.4.12b)$$

M_1 is the mass of the larger primary, and M_2 is the mass of the smaller one, hence, $0 < \alpha < 0.5$ and $0.5 < 1 - \alpha < 1$.

EXAMPLE 7.4.1

Using the previously discussed units, describe the general properties for the binary star system in Example 7.3.1. The mass of the Sun is $M_\odot = 1.99 \times 10^{30}$ kg. The astronomical unit is 1 AU $= 1.496 \times 10^{11}$ m.

Solution:

The masses of the two primaries: M_i	4 $M_\odot$ and 1 $M_\odot$, respectively
The parameter α:	$1/(1 + 4) = 0.2$.
The scaled masses of the two primaries α_i:	$1 - \alpha = 0.8$; $\alpha = 0.2$.
Coordinates (x'_i, y'_i) of the two primaries:	$(0.2, 0), (-0.8, 0)$
Unit of "gravitational" mass $G(M_1 + M_2)$:	6.6×10^{20} m^3/s^2
Orbital period: $\tau = 5$ years $= 2\pi$ time units	1.58×10^8 s
Unit of time: $\tau/2\pi$	2.51×10^7 s (0.796 year)
Angular speed: $\omega = 2\pi/\tau$ (=1 inverse time unit)	3.98×10^{-8} s^{-1}
Unit of length: $a + b = 5$ AU	7.48×10^{11} m

In terms of these new units, the effective potential function of Equation 7.4.8b becomes

$$V(x', y') = -\frac{1-\alpha}{\sqrt{(x'-\alpha)^2 + y'^2}} - \frac{\alpha}{\sqrt{(x'+1-\alpha)^2 + y'^2}} - \frac{x'^2 + y'^2}{2} \qquad (7.4.13)$$

A plot of the effective potential $V(x', y')$ is shown in Figure 7.4.2 for the Earth–Moon primary system, where the parameter $\alpha = 0.0121$. Plots of the effective potential of other binary systems, such as binary stars where the parameter α is rarely less than 20% or, at the other extreme, the Sun–Jupiter system where $\alpha = 0.000953875$, are qualitatively identical.

It is worth taking the time to examine this plot closely because it exhibits a number of features that give us some insight into the possible orbits of the tertiary.

- $V(x', y') \to -\infty$ at the location of the two primaries. These points are singularities. This is a consequence of the fact that each primary has been treated as though it was a point mass. We might imagine that, if a tertiary were embedded somewhere

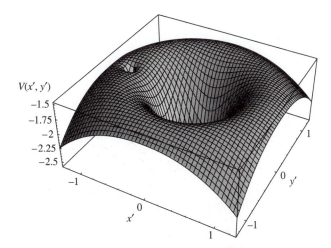

Figure 7.4.2 Effective potential $V(x', y')$ for the Earth–Moon system.

within one of those potential "holes," it might orbit that primary as though the other primary didn't even exist. As an example, consider the Sun–Jupiter system: Each primary is the source of an accouterment of "satellites"; Jupiter has its moons and the Sun has its four inner, terrestrial planets. Neither primary interferes with the attachments of the other (at least not very much). Note, though, that the angular speeds of all these "satellites" about their respective primary are much greater than the angular speed of the two primaries about their center of mass. In addition, tertiaries in such orbits are dragged along by the primary in its own orbit.

- $V(x', y') \to -\infty$ as either x' or $y' \to \infty$. This is a consequence of the rotation of the $x'y'$ coordinate system. In essence, any tertiary initially at rest with respect to this rotating system, but far from the center of mass of the two primaries, experiences a large centrifugal force that tends to move the body even farther from the origin. Eventually, such a body might find itself in a stable orbit at some remote distance $[r' > (a + b)]$ around the center of mass of the two primaries but *not at rest in the rotating frame of reference.* The angular speed of such a tertiary would be so much smaller than the angular speed of the two primaries that a stable, counterclockwise, *prograde* orbit in a fixed frame of reference would appear to be a stable, clockwise, *retrograde* orbit in the rotating system, with an angular velocity that is the negative of that of the primaries. An example of this is our nearest stellar neighbor, the three-body, α-Centauri star system, made up of two primaries, α-Centauri A and B and a tertiary, Proxima Centauri (Figure 7.4.3).
- There are five locations where $\nabla V(x', y') = 0$, or where the force on a particle at rest in the $x'y'$ frame of reference vanishes. These points are called the *Lagrangian points,* after Joseph-Louis Lagrange. They are designated L_1–L_5 in his honor. Three of these points are collinear, lying along the x'-axis. L_1 lies between the two primaries. L_2 lies on the side opposite the least massive primary, and L_3 lies on the side opposite the most massive primary. These three points are saddle points of $V(x', y')$. Along the x' direction they are local maxima, but along the y' direction they are local minima.
- The two primaries form a common base of two equilateral triangles at whose apex lie the points L_4 and L_5, which are absolute maxima of the function $V(x', y')$. As

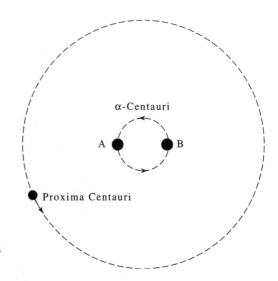

Figure 7.4.3 The α-Centauri system. The masses of the primaries are 1.1 and 0.88$M_\odot$. The mass of Proxima Centauri is 0.1$M_\odot$. A and B separated by 25 AU, and Proxima Centauri orbits the pair at a distance of 50,000 AU.

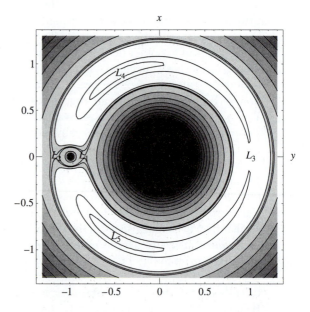

Figure 7.4.4 Contour plot of the effective potential $V(x', y')$ for the Earth–Moon system.

the primaries rotate about their center of mass, L_4 remains $60°$ ahead of the least massive primary (in the $+y'$ direction), and L_5 remains $60°$ behind it (in the $-y'$ direction). The location of these five points can be more easily visualized by examining a contour plot of the effective potential function shown in Figure 7.4.4.

- Each line in the contour plot is an equipotential, that is, a line that satisfies the condition $V(x_i', y_i') = V_i$, where V_i is a constant. Normally, the equipotential lines in contour plots represent "heights" V_i that differ from one another by equal amounts. This means that regions of the plot where the gradient, $\nabla V(x', y')$, is "steep" (or the force is large) would exhibit closely packed contour lines. Regions where the gradient is "flat" (or the force approaches zero) would exhibit sparsely packed contour lines. We have not adhered to this convention in Figure 7.4.4. We have decreased the "step size" between contour heights that pass near the five Lagrangian points to illuminate those positions more clearly.

You might guess that it would be possible for a tertiary to remain at any one of these five points, synchronously locked to the two primaries as they rotate about their center of mass. It turns out that this never happens in nature for any tertiary located at L_1–L_3. These are points of unstable equilibrium. If a body located at one of these points is perturbed ever so slightly, it moves toward one primary and away from the other, or away from both primaries.

Close examination of Figures 7.4.3 and 7.4.4 reveals that the effective potential is rather flat and broad around L_4 and L_5, suggesting that a reasonably extensive, almost force-free, region exists where a tertiary might comfortably sit, more or less balanced by the opposing action of the gravitational and centrifugal forces. Because L_4 and L_5 are locations of absolute maxima, however, you might also guess that no stable, synchronous orbit is possible at these points either. Remember, though, that all the forces acting on the tertiary are not derived from the gradient of $V(x', y')$. The velocity-dependent Coriolis force must be considered and it has a nonnegligible effect, particularly in any region where it dominates, which under certain conditions can be the case in the region surrounding L_4 and L_5. The Coriolis force always acts perpendicular to the velocity of a particle. Thus, it does not alter its kinetic energy because $\mathbf{F} \cdot \mathbf{v} = 0$. If a tertiary is nearly stationary in the $x'y'$ frame of reference, moving slowly in the proper direction near either L_4 or L_5, the Coriolis force might dominate the nearly balanced gravitational and centrifugal forces and simply redirect its velocity, causing the tertiary to circulate around L_4 or L_5. In fact, this can and does happen in nature. The Coriolis force creates an effective, quasi-elliptical barrier around the L_4 and L_5 points, thus, turning the maxima of the effective potential into small "wells" of stability. Given the right conditions, we might expect the tertiary to closely follow one of the equipotential contours around L_4 and L_5, both its kinetic and potential energies remaining fairly constant throughout its motion.

The situation just described is analogous to the circulation of air that occurs around high-pressure systems, or "bumps," in the Earth's atmosphere. Gravity tries to pull the air toward the Earth; centrifugal force tries to throw it out; as air spills down from the high, the Coriolis force causes it to circulate about the high-pressure bump, clockwise in the Northern Hemisphere. Such circulating systems in the atmosphere of Earth are only stable temporarily. They form and then dissipate. The Great Red Spot on Jupiter, however, is a high-pressure storm that is a permanent feature of its atmosphere—permanent in the sense that it has been there ever since Galileo saw it with his telescope about 400 years ago! Note that these circulatory patterns are "stationary" with respect to the rotating system. The same holds true for the orbit of a tertiary around L_4 and L_5.

The Trojan Asteroids

The Trojan asteroids are a particular group of asteroids in a 1:1 orbital resonance with Jupiter and whose centroids lie along the orbit of Jupiter, 60° ahead of it and 60° behind (see Figure 7.4.5). These are the L_4 and L_5 points in the Sun–Jupiter primary system. Notice that the Trojans are spread out somewhat diffusely about the L_4 and L_5 points. Each member of the group rotates with Jupiter about the Sun in a fixed frame of reference

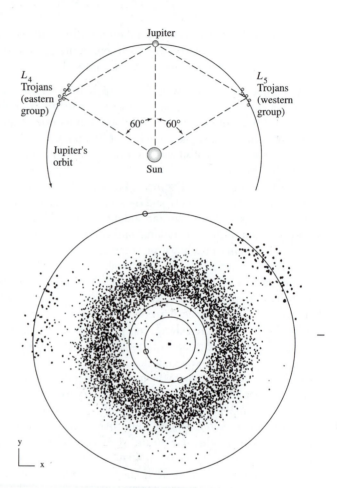

Figure 7.4.5 (a) The Trojan asteroids. (b) Trojan asteroids and the asteroid belt shown with orbits of Jupiter, Mars and Earth.

but is slowly circulating clockwise about L_4 and L_5, as viewed from above in the $x'y'$ frame of reference. In this section, we calculate some examples of the orbits of these asteroids.

First, we rewrite the equations of motion (Equations 7.4.7a and b) using the scaled coordinates we just introduced. Letting

$$r_1' = \sqrt{(x'-\alpha)^2 + y'^2} \qquad r_2' = \sqrt{(x'+1-\alpha)^2 + y'^2} \qquad (7.4.14)$$

Equations 7.4.7a and b become

$$\ddot{x}' = -(1-\alpha)\frac{(x'-\alpha)}{r_1'^3} - \alpha\frac{(x'+1-\alpha)}{r_2'^3} + x' + 2\dot{y}' \qquad (7.4.15a)$$

$$\ddot{y}' = -(1-\alpha)\frac{y'}{r_1'^3} - \alpha\frac{y'}{r_2'^3} + y' - 2\dot{x}' \qquad (7.4.15b)$$

In Example 4.3.2, we employed *Mathematica*'s numerical differential equation solver, *NDSolve*, to solve a set of coupled, second-order differential equations like the ones in Equations 7.4.15a and b. We employ the same technique here with one minor difference: we introduce two additional variables u' and v', such that

$$\dot{x}' = u' \tag{7.4.16a}$$

$$\dot{y}' = v' \tag{7.4.16b}$$

to convert the pair of second-order equations in Equations 7.4.15a and b into two first-order ones

$$\dot{u}' = -(1-\alpha)\frac{(x'-\alpha)}{r_1'^3} - \alpha\frac{(x'+1-\alpha)}{r_2'^3} + x' + 2v' \tag{7.4.16c}$$

$$\dot{v}' = -(1-\alpha)\frac{y'}{r_1'^3} \quad - \alpha\frac{y'}{r_2'^3} \quad + y' - 2u' \tag{7.4.16d}$$

This was the same trick we used in Section 3.8, where we solved for the motion of the self-limiting oscillator. The trick is a standard ploy used to convert n second-order differential equations into $2n$ first-order ones, making it possible to use Runge-Kutta techniques to solve the resulting equations. Most numerical differential equation solvers use this technique. *Mathcad* requires that the user input the $2n$ equations in first-order form. This is not a requirement in *Mathematica*, although it is still an option. We use the technique because it is so universally applicable. In the following section, we outline the specific call that we made to *NDSolve*. It is analogous to the one discussed in Example 4.3.2. We dropped the superfluous primes used to label the rotating coordinates because *Mathematica* uses primes in place of dots to denote the process of differentiation, that is, x' means $\dot{x}$. We urge you to remember that the variables x, y, u, and v used in *Mathematica* calls refer to the rotating coordinate system, and the number of primes beside a variable refer to the order of the derivative.

NDSolve [{equations, initial conditions}, {u, v, x, y}, {t, t_{min}, t_{max}}]

- {*equations, initial conditions*}
  Insert the four numerical differential equations and initial conditions using the following format

  {$x'[t] == u[t]$,

  $y'[t] == v[t]$,

  $u'[t] == -(1-\alpha)(x[t]-\alpha)/r_1(x[t], y[t])^3 - \alpha(x[t]+1-\alpha)/r_2(x[t], y[t])^3$
  $+ x[t] + 2v[t]$,

  $v'[t] == -(1-\alpha)y[t]/r_1(x[t], y[t])^3$
  $- \alpha y[t]/r_2(x[t], y[t])^3 + y[t] - 2\,u[t]$,

  $x[0] == x_0, y[0] == y_0, u[0] == u_0, v[0] == v_0$}

- {x, y, u, v}
  Insert the four dependent variables whose solutions are desired

  {x, y, u, v}

- {t, t_{min}, t_{max}}
  Insert the independent variable and its range over which the solution is to be evaluated {$t, 0, t_{max}$}

TABLE 7.4.1 Starting Conditions and Period of Orbits Around L_4

Parameter	Orbit 1	Orbit 2	Orbit 3	Orbit 4	Orbit 5
x_0	−0.509	−0.524	−0.524	−0.509	−0.532
y_0	0.883	0.909	0.920	0.883	0.920
u_0	0.0259	0.0647	0.0780	−0.0259	0.0780
v_0	0.0149	0.0367	0.0430	−0.049	0.0430
T (units)	80.3	118	210.5	80.3*	—
T (years)	152	223	397	152*	—

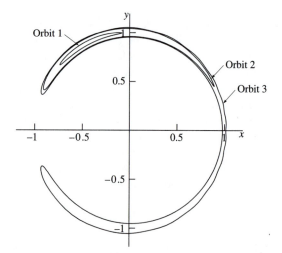

Figure 7.4.6 Orbits 1, 2, 3 of the Trojan asteroids corresponding to the conditions given in Table 7.4.1.

Note, the two functions $r_1[x, y]$ and $r_2[x, y]$ (see Equation 7.5.14) must be defined in *Mathematica* before the call to *NDSolve*. This is also true for the initial conditions x_0, y_0, u_0, and v_0 and the value of α. The value of α for the Sun–Jupiter system is 0.000953875.

We calculated orbits for five sets of initial conditions, in each case starting the terti-ary near L_4. The starting conditions and period of the resulting orbit (if the result is a stable orbit) are shown in Table 7.4.1.

As before (Example 4.3.2) we used *Mathematica's ParametricPlot* to generate plots of each of the orbits whose initial conditions are given in Table 7.4.1. Plots of the first three orbits are shown in Figure 7.4.6.

The unit of length is the mean distance between Jupiter and the Sun, $a + b = 5.203$ AU, or about 7.80×10^{11} m. The unit of time was defined such that one rotational period of the primary system, the orbital period of Jupiter ($T_J = 11.86$ years), equals 2π time units. Thus, one time unit equals $T_J/2\pi = 1.888$ years. Tertiaries that follow orbits 1 and 2 cir-culate slowly, clockwise, around L_4. Their calculated periods are 80.3 and 118 time units, respectively. Using the conversion factor gives us the periods of their orbits in years listed in the last row of Table 7.4.1. Orbit 3 is particularly interesting. The tertiary starts closer to Jupiter than do the other two and moves slowly over L_4 and back around the Sun, more or less along Jupiter's orbital path. It then slowly migrates toward Jupiter, passing under

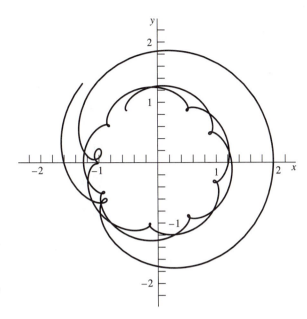

Figure 7.4.7   Trojan asteroids—
orbit 4 (see Table 7.4.1).

L_5 and approaching as close to Jupiter as when it started near L_4. Then it loops around L_5, passes back around the Sun, and moves back toward Jupiter, passing just under L_4 to the point where it started. The period of this orbit is 397 years.

Notice that the orbits closely follow the equipotential contours shown in Figure 7.4.4. This is not too surprising because as we remarked earlier, the Coriolis force does not change the kinetic energy of the tertiary. Thus, because the gravitational and centrifugal forces are more or less in balance, the orbits ought to follow the equipotential contours rather closely. The contours circulate around L_4 and L_5 individually, as do orbits 1 and 2, and a few contours circulate around L_4 and L_5 together, as does orbit 3. Given the shape of these orbits, it is easy to understand why the Trojan asteroids appear to be the rather loosely strung out cluster that you see in Figure 7.4.5.

In all cases, the orbits circulate in clockwise fashion like the air around high pressures in the Northern Hemisphere of Earth. The Coriolis force is directed "inward" for clockwise rotation and "outward" for counterclockwise rotation because of the sign of $\boldsymbol{\omega} \times \mathbf{v}$. Orbit 4, shown in Figure 7.4.7, reflects the consequences of a sign reversal in $\boldsymbol{\omega} \times \mathbf{v}$ if we try to set up a counterclockwise circulation about L_4. The orbit was generated with the same parameters as those of the stable orbit 1, except the sign of the initial velocity was reversed. The tertiary, after executing several loopty-loops, is soon thrown completely out of the region between Jupiter and the Sun. A velocity reversal like this would have no effect on the shape of a Keplerian orbit about a single, central gravitational force. The resulting stable orbit would simply be a reversed direction, retrograde orbit. Although most orbits in the solar system are prograde (counterclockwise as seen from above the plane of the ecliptic), retrograde (clockwise) orbits do occur, as, for example, Triton, Neptune's major moon. Reversed orbits are not possible around L_4 and L_5.

The conditions for the stability of these clockwise orbits around L_4 and L_5 have been studied in much more detail than can be presented here. The interested reader is referred

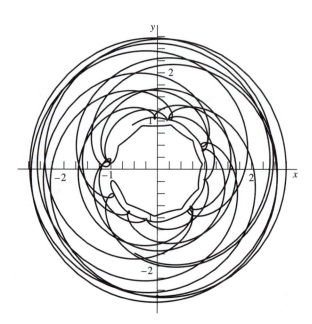

Figure 7.4.8 Trojan asteroids—
orbit 5 (see Table 7.4.1).

to the book by V. Szebehely referenced in footnote 4. Stable orbits are only possible for values of the mass parameter $\alpha_c = 0.03852$. The Jupiter–Sun system easily meets that condition, but the orbits of some of the Trojans are only marginally stable. This is particularly true for orbits such as orbit 3. Perturbations, if large enough, can have dramatic consequences for tertiaries in such orbits. Examine the starting conditions for orbit 5, which are virtually identical to those for orbit 3 except for the initial x' coordinate, which was changed by about 2%. The resultant "orbit" is shown in Figure 7.4.8. The trajectory of the tertiary was followed for 300 time units, or about 566 years. Eventually, as was the case for orbit 4, the asteroid was thrown completely out of the L_4–L_5 region, finally settling down in orbit about both primaries at a distance of about 3 units, or 15 AU, which places it somewhere between Saturn and Neptune. In fact, Jupiter is believed to have had just this effect on many of the asteroids that existed near it during the formative stages of the solar system.

Are there any other examples of objects orbiting primaries at the L_4 and L_5 points? A prime example is that of a number of Saturn's large supply of moons. Telesto and Calypso, two moons discovered by the *Voyager* mission, share an orbit with Tethys. Saturn and Tethys are the primaries, and Telesto is at L_4 and Calypso at L_5. Helene and Dione share another orbit that is 1.28 times farther from Saturn than the one occupied by Tethys, Telesto, and Calypso. Helene is located at the L_4 point of this orbit, and Dione is the primary. No moon is found for this orbit at L_5.

A number of space colony enthusiasts have argued that a large space colony could be deployed in a stable orbit at L_5 of the Earth–Moon primary system.[7] The mass parameter for the Earth–Moon system is $\alpha = 0.0121409$, which is certainly less than the critical

[7] G. K. O'Neill, "The Colonization of Space," *Phys. Today*, pp. 32–40 (September, 1974).

value α_c, so one might guess that orbits about L_5 would be stable. The Sun would exert perturbations on such an orbiting colony, however, and it is not obvious that its orbit would remain stable for long. This particular restricted four-body problem was only solved recently, in 1968. Quasi-elliptical orbits around L_5, with excursions limited to a few tenths of the Earth–Moon distance, were found to be stable.[8] If one adds the effects of Jupiter to the problem, however, long-term stability becomes problematical. The industrious student might want to tackle this problem numerically.

EXAMPLE 7.4.2

Calculate the coordinates of the L_1–L_3 collinear Lagrange points for the Earth–Moon system and the values of the effective potential function at those points.

Solution:

These three collinear Lagrange points all lie along the x'-axis, where $y' = 0$. These points represent extrema of the effective potential function, $V(x', y')$. Normally, we would find these points by searching for solutions of the equation

$$\frac{\partial}{\partial x'} V(x', y') \bigg|_{y'=0} = 0$$

Mathematica, however, has a tool, its *FindMinimum* function, that allows us to locate minima of functions directly, without first calculating their derivatives. *Mathematica* saves us a lot of work by effectively taking these derivatives for us. The Lagrange points, L_1–L_3, are located at the *maxima* of $V(x', y' = 0)$, however, so, to use *Mathematica's* *FindMinimum*, we need to pass to it a function $f(x') = -V(x', y' = 0)$ whose *minima* are the locations of L_1–L_3.

$$f(x') = -V(x', y') \bigg|_{y'=0} = \frac{1-\alpha}{|x'-\alpha|} + \frac{\alpha}{|x'-(\alpha-1)|} + \frac{x'^2}{2}$$

We have written the denominators in the preceding equation as absolute values to emphasize that they are positive definite quantities regardless of the value of x' relative to the critical values α and $\alpha - 1$. When we pass $f(x')$ to *Mathematica's* *FindMinimum* function, we need to ensure that: (1) *FindMinimum* can calculate the derivatives of $f(x')$ because that is one of the things it does in attempting to locate the minima and that (2) the values in the denominator remain positive definite regardless of any action that *FindMinimum* takes on $f(x')$. Thus, we need to remove the absolute values in the denominators of $f(x')$ to eliminate any possible pathologies in the derivative-taking process, but then we must replace their effect, for example, by multiplying the first two terms in the expression by a "step" function defined to take on the values ±1 depending on the value of x' relative to α and $\alpha - 1$. We call this "step" function sgn(x) and define it to equal −1 when its argument $x < 0$ and +1 when $x > 0$.

---

[8]R. Kolenkiewicz, L. Carpenter, "Stable Periodic Orbits About the Sun-Perturbed Earth-Moon Triangular Points," *AIAA J.* 6, 7, 1301 (1968).

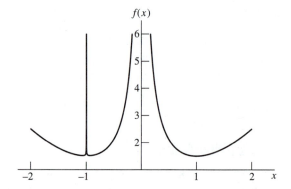

Figure 7.4.9   Regions of applicability for the sgn functions.

TABLE 7.4.2

Call	Lagrange Point	x_0	$\text{sgn}(x - \alpha)$	$\text{sgn}(x - (\alpha - 1))$	x_{min}	$f(x_{min})$
1	L_2	−1.2	−1	−1	−1.06883	1.51874
2	L_1	−0.8	−1	+1	−0.932366	1.51938
3	L_3	1.0	+1	+1	1.0004	1.50048

Inserting it into the expression above gives

$$f(x') = \text{sgn}(x' - \alpha)\frac{1-\alpha}{x'-\alpha} + \text{sgn}(x' - (\alpha-1))\frac{\alpha}{x'-(\alpha-1)} + \frac{x'^2}{2}$$

We can now pass the preceding function to *FindMinimum*. The sgn function takes on a value that insures that the terms in the equation always remain positive regardless of the region along the x'-axis that is being searched for one of the minima of $f(x')$. We also need to pass *FindMinimum* initial values of x' to begin the search. We plot $f(x')$ in Figure 7.4.9 to find approximate locations of the three minima we are using as these starting points. L_2 is the minimum located exterior to the singularity at $x' = -1$ that represents the location of Jupiter. Thus, $L_2 \approx -(1 + \epsilon)$. L_1 is located on the interior side of this singularity. Thus, $L_1 \approx -(1 - \epsilon)$ and L_3 is located just beyond the mirror image of Jupiter's singularity at $x' = +1$ opposite the Sun. Thus, $L_3 \approx +(1 + \epsilon)$. ϵ simply denotes some unknown small value. We now make three calls to *FindMinimum* to locate each of the three collinear Lagrange points.

Each call takes the form: *Find Minimum* [*function*, {*x, x_0*}] where the argument *function* means $f(x)$ as previously defined. Again, we drop the prime notation. x is the independent variable of the function, and x_0 is the value used to start the search. Table 7.4.2 lists the parameters input to each call. The output of the call are the locations x_{min} of the Lagrange points and the corresponding values of $f(x_{min})$. The values of x_0 were chosen to ensure that the search starts in the region in which the desired Lagrange point is located and fairly near to it.

7.5 | Collisions

Whenever two bodies undergo a collision, the force that either exerts on the other during the contact is an internal force, if the bodies are regarded together as a single system. The total linear momentum is unchanged. We can, therefore, write

$$\mathbf{p}_1 + \mathbf{p}_2 = \mathbf{p}_1' + \mathbf{p}_2' \tag{7.5.1a}$$

or, equivalently,

$$m_1\mathbf{v}_1 + m_2\mathbf{v}_2 = m_1\mathbf{v}_1' + m_2\mathbf{v}_2' \tag{7.5.1b}$$

The subscripts 1 and 2 refer to the two bodies, and the primes indicate the respective momenta and velocities after the collision. Equations 7.5.1a and b are quite general. They apply to any two bodies regardless of their shapes, rigidity, and so on.

With regard to the energy balance, we can write

$$\frac{p_1^2}{2m_1} + \frac{p_2^2}{2m_2} = \frac{p_1'^2}{2m_1} + \frac{p_2'^2}{2m_2} + Q \tag{7.5.2a}$$

or

$$\frac{1}{2}m_1v_1^2 + \frac{1}{2}m_2v_2^2 = \frac{1}{2}m_1v_1'^2 + \frac{1}{2}m_2v_2'^2 + Q \tag{7.5.2b}$$

Here the quantity Q is introduced to indicate the net loss or gain in kinetic energy that occurs as a result of the collision.

In the case of an *elastic* collision, no change takes place in the total kinetic energy, so that $Q = 0$. If an energy loss does occur, then Q is positive. This is called an *exoergic* collision. It may happen that an energy gain occurs. This would happen, for example, if an explosive was present on one of the bodies at the point of contact. In this case Q is negative, and the collision is called *endoergic*.

The study of collisions is of particular importance in atomic, nuclear, and high-energy physics. Here the bodies involved may be atoms, nuclei, or various elementary particles, such as electrons and quarks.

Direct Collisions

Let us consider the special case of a head-on collision of two bodies, or particles, in which the motion takes place entirely on a single straight line, the x-axis, as shown in Figure 7.5.1. In this case the momentum balance equation (Equation 7.5.1b) can be written

$$m_1\dot{x}_1 + m_2\dot{x}_2 = m_1\dot{x}_1' + m_2\dot{x}_2' \tag{7.5.3}$$

The direction along the line of motion is given by the signs of the $\dot{x}$'s.

To compute the values of the velocities after the collision, given the values before the collision, we can use the preceding momentum equation together with the energy balance equation (Equation 7.5.2b), if we know the value of Q. It is often convenient in this kind of problem to introduce another parameter ϵ called the *coefficient of restitution*.

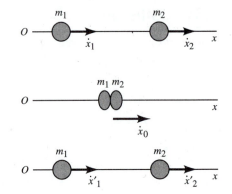

Figure 7.5.1   Head-on collision of two particles.

This quantity is defined as the ratio of the speed of separation v' to the speed of approach v. In our notation ϵ may be written as

$$\epsilon = \frac{|\dot{x}_2' - \dot{x}_1'|}{|\dot{x}_2 - \dot{x}_1|} = \frac{v'}{v} \tag{7.5.4}$$

The numerical value of ϵ depends primarily on the composition and physical makeup of the two bodies. It is easy to verify that in an elastic collision the value of $\epsilon = 1$. To do this, we set $Q = 0$ in Equation 7.5.2b and solve it together with Equation 7.5.3 for the final velocities. The steps are left as an exercise.

In the case of a *totally inelastic* collision, the two bodies stick together after colliding, so that $\epsilon = 0$. For most real bodies ϵ has a value somewhere between the two extremes of 0 and 1. For ivory billiard balls it is about 0.95. The value of the coefficient of restitution may also depend on the speed of approach. This is particularly evident in the case of a silicone compound known as Silly Putty. A ball of this material bounces when it strikes a hard surface at high speed, but at low speeds it acts like ordinary putty.

We can calculate the values of the final velocities from Equation 7.5.3 together with the definition of the coefficient of restitution (Equation 7.5.4). The result is

$$\dot{x}_1' = \frac{(m_1 - \epsilon m_2)\dot{x}_1 + (m_2 + \epsilon m_2)\dot{x}_2}{m_1 + m_2}$$

$$\dot{x}_2' = \frac{(m_1 + \epsilon m_1)\dot{x}_1 + (m_2 - \epsilon m_1)\dot{x}_2}{m_1 + m_2} \tag{7.5.5}$$

Taking the totally inelastic case by setting $\epsilon = 0$, we find, as we should, that $\dot{x}_1' = \dot{x}_2'$; that is, there is no rebound. On the other hand, in the special case that the bodies are of equal mass $m_1 = m_2$ and are perfectly elastic $\epsilon = 1$, we obtain

$$\dot{x}_1' = \dot{x}_2$$
$$\dot{x}_2' = \dot{x}_1 \tag{7.5.6}$$

The two bodies, therefore, just exchange their velocities as a result of the collision.

In the general case of a direct nonelastic collision, it is easily verified that the energy loss Q is related to the coefficient of restitution by the equation

$$Q = \tfrac{1}{2}\mu v^2 (1 - \epsilon^2) \tag{7.5.7}$$

in which $\mu = m_1 m_2/(m_1 + m_2)$ is the reduced mass, and $v = |\dot{x}_2 - \dot{x}_1|$ is the relative speed before impact. The derivation is left as an exercise (see Problem 7.9).

Impulse in Collisions

Forces of extremely short duration in time, such as those exerted by bodies undergoing collisions, are called *impulsive forces.* If we confine our attention to one body, or particle, the differential equation of motion is $d(m\mathbf{v})/dt = \mathbf{F}$, or in differential form $d(m\mathbf{v}) = \mathbf{F}\,dt$. Let us take the time integral over the interval $t = t_1$ to $t = t_2$. This is the time during which the force is considered to act. Then we have

$$\Delta(m\mathbf{v}) = \int_{t_1}^{t_2} \mathbf{F}\,dt \qquad (7.5.8a)$$

The time integral of the force is the impulse. It is customarily denoted by the symbol $\mathbf{P}$. Equation 7.5.8a is, accordingly, expressed as

$$\Delta(m\mathbf{v}) = \mathbf{P} \qquad (7.5.8b)$$

We can think of an *ideal impulse* as produced by a force that tends to infinity but lasts for a time interval that approaches zero in such a way that the integral $\int \mathbf{F}\,dt$ remains finite. Such an ideal impulse would produce an instantaneous change in the momentum and velocity of a body without producing any displacement.

EXAMPLE 7.5.1

Determining the Speed of a Bullet

A gun is fired horizontally, point-blank at a block of wood, which is initially at rest on a horizontal floor. The bullet becomes imbedded in the block, and the impact causes the system to slide a certain distance s before coming to rest. Given the mass of the bullet m, the mass of the block M, and the coefficient of sliding friction between the block and the floor μ_k, find the initial speed (muzzle velocity) of the bullet.

Solution:

First, from conservation of linear momentum, we can write

$$m\dot{x}_0 = (M + m)\dot{x}_0'$$

where $\dot{x}_0$ is the initial velocity of the bullet, and $\dot{x}_0'$ is the velocity of the system (block + bullet) immediately after impact. (The coefficient of restitution ϵ is zero in this case.) Second, we know that the magnitude of the retarding frictional force is equal to $(M + m)\,\mu_k g = (M + m)a$, where $a = -\ddot{x}$ is the deceleration of the system after impact, so $a = \mu_k g$. Now, from Chapter 2 we recall that $s = v_0^2/2a$ for the case of uniform acceleration in one dimension. Thus, in our problem

$$s = \frac{\dot{x}_0'^2}{2\mu_k g} = \left(\frac{m\dot{x}_0}{M + m}\right)^2 \left(\frac{1}{2\mu_k g}\right)$$

Solving for $\dot{x}_0$, we obtain

$$\dot{x}_0 = \left(\frac{M+m}{m}\right)(2\mu_k gs)^{1/2}$$

for the initial velocity of the bullet in terms of the given quantities.

As a numerical example, let the mass of the block be 4 kg, and that of the bullet 10 g = 0.01 kg (about that of a .38 calibre slug). For the coefficient of friction (wood-on-wood) let us take $\mu_k = 0.4$. If the block slides a distance of 15 cm = 0.15 m, then we find

$$\dot{x}_0 = \frac{4.01}{0.01}(2 \times 0.4 \times 9.8 \text{ ms}^{-2} \times 0.15 \text{ m})^{1/2} = 435 \text{ m/s}$$

7.6 | Oblique Collisions and Scattering: Comparison of Laboratory and Center of Mass Coordinates

We now turn our attention to the more general case of collisions in which the motion is not confined to a single straight line. Here the vectorial form of the momentum equations must be employed. Let us study the special case of a particle of mass m_1 with initial velocity $\mathbf{v}_1$ (the incident particle) that strikes a particle of mass m_2 that is initially at rest (the target particle). This is a typical problem found in nuclear physics. The momentum equations in this case are

$$\mathbf{p}_1 = \mathbf{p}_1' + \mathbf{p}_2' \tag{7.6.1a}$$

$$m_1\mathbf{v}_1 = m_1\mathbf{v}_1' + m_2\mathbf{v}_2' \tag{7.6.1b}$$

The energy balance condition is

$$\frac{p_1^2}{2m_1} = \frac{p_1'^2}{2m_1} + \frac{p_2'^2}{2m_2} + Q \tag{7.6.2a}$$

or

$$\tfrac{1}{2}m_1v_1^2 = \tfrac{1}{2}m_1v_1'^2 + \tfrac{1}{2}m_2v_2'^2 + Q \tag{7.6.2b}$$

Here, as before, the primes indicate the velocities and momenta after the collision, and Q represents the net energy that is lost or gained as a result of the impact. The quantity Q is of fundamental importance in atomic and nuclear physics, because it represents the energy released or absorbed in atomic and nuclear collisions. In many cases the target particle is broken up or changed by the collision. In such cases the particles that leave the collision are different from those that enter. This is easily taken into account by assigning different masses, say m_3 and m_4, to the particles leaving the collision. In any case, the law of conservation of linear momentum is always valid.

Consider the particular case in which the masses of the incident and target particles are the same. Then the energy balance equation (Equation 7.6.2a) can be written

$$p_1^2 = p_1'^2 + p_2'^2 + 2mQ \tag{7.6.3}$$

where $m = m_1 = m_2$. Now if we take the dot product of each side of the momentum equation (Equation 7.6.1a) with itself, we get

$$p_1^2 = (\mathbf{p}_1' + \mathbf{p}_2') \cdot (\mathbf{p}_1' + \mathbf{p}_2') = p_1'^2 + p_2'^2 + 2\mathbf{p}_1' \cdot \mathbf{p}_2' \tag{7.6.4}$$

Comparing Equations 7.6.3 and 7.6.4, we see that

$$\mathbf{p}_1' \cdot \mathbf{p}_2' = mQ \tag{7.6.5}$$

For an elastic collision ($Q = 0$) we have, therefore,

$$\mathbf{p}_1' \cdot \mathbf{p}_2' = 0 \tag{7.6.6}$$

so the two particles emerge from the collision at right angles to each other.

Center of Mass Coordinates

Theoretical calculations in nuclear physics are often done in terms of quantities referred to a coordinate system in which the center of mass of the colliding particles is at rest. On the other hand, the experimental observations on scattering of particles are carried out in terms of the laboratory coordinates. We, therefore, consider briefly the problem of conversion from one coordinate system to the other.

The velocity vectors in the laboratory system and in the center of mass system are illustrated diagrammatically in Figure 7.6.1. In the figure ϕ_1 is the angle of deflection of the incident particle after it strikes the target particle, and ϕ_2 is the angle that the line of motion of the target particle makes with the line of motion of the incident particle. Both ϕ_1 and ϕ_2 are measured in the laboratory system. In the center of mass system, because the center of mass must lie on the line joining the two particles at all times, both particles approach the center of mass, collide, and recede from the center of mass in opposite directions. The angle θ denotes the angle deflection of the incident particle in the center of mass system as indicated.

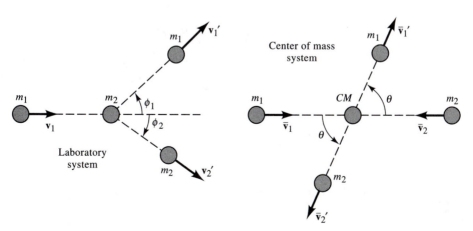

Figure 7.6.1   Comparison of laboratory and center of mass coordinates.

From the definition of the center of mass, the linear momentum in the center of mass system is zero both before and after the collision. Hence, we can write

$$\bar{\mathbf{p}}_1 + \bar{\mathbf{p}}_2 = 0 \tag{7.6.7a}$$

$$\bar{\mathbf{p}}_1' + \bar{\mathbf{p}}_2' = 0 \tag{7.6.7b}$$

The bars are used to indicate that the quantity in question is referred to the center of mass system. The energy balance equation reads

$$\frac{\bar{p}_1^2}{2m_1} + \frac{\bar{p}_2^2}{2m_2} = \frac{\bar{p}_1'^2}{2m_1} + \frac{\bar{p}_2'^2}{2m_2} + Q \tag{7.6.8}$$

We can eliminate $\bar{p}_2$ and $\bar{p}_2'$ from Equation 7.6.8 by using the momentum relations in Equations 7.6.7a and b. The result, which is conveniently expressed in terms of the reduced mass, is

$$\frac{\bar{p}_1^2}{2\mu} = \frac{\bar{p}_1'^2}{2\mu} + Q \tag{7.6.9}$$

The momentum relations, Equations 7.6.7a and b expressed in terms of velocities, read

$$m_1 \bar{\mathbf{v}}_1 + m_2 \bar{\mathbf{v}}_2 = 0 \tag{7.6.10a}$$

$$m_1 \bar{\mathbf{v}}_1' + m_2 \bar{\mathbf{v}}_2' = 0 \tag{7.6.10b}$$

The velocity of the center of mass is (see Equations 7.1.3 and 7.1.4)

$$\mathbf{v}_{cm} = \frac{m_1 \mathbf{v}_1}{m_1 + m_2} \tag{7.6.11}$$

Hence, we have

$$\bar{\mathbf{v}}_1 = \mathbf{v}_1 - \mathbf{v}_{cm} = \frac{m_2 \mathbf{v}_1}{m_1 + m_2} \tag{7.6.12}$$

The relationships among the velocity vectors $\mathbf{v}_{cm}$, $\mathbf{v}_1'$, and $\bar{\mathbf{v}}_1'$ are shown in Figure 7.6.2. From the figure, we see that

$$\begin{aligned} v_1' \sin \phi_1 &= \bar{v}_1' \sin \theta \\ v_1' \cos \phi_1 &= \bar{v}_1' \cos \theta + v_{cm} \end{aligned} \tag{7.6.13}$$

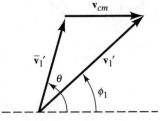

Figure 7.6.2   Velocity vectors in the laboratory system and the center of mass system.

Hence, by dividing, we find the equation connecting the scattering angles to be express-ible in the form

$$\tan \phi_1 = \frac{\sin \theta}{\gamma + \cos \theta} \tag{7.6.14}$$

in which γ is a numerical parameter whose value is given by

$$\gamma = \frac{v_{cm}}{\bar{v}_1'} = \frac{m_1 v_1}{\bar{v}_1'(m_1 + m_2)} \tag{7.6.15}$$

The last step follows from Equation 7.6.11.

Now we can readily calculate the value of $\bar{v}_1'$ in terms of the initial energy of the incident particle from the energy equation (Equation 7.6.9). This gives us the neces-sary information to find γ and, thus, determine the relationship between the scatter-ing angles. For example in the case of an elastic collision $Q = 0$, we find from the energy equation that $\bar{p}_1 = \bar{p}_1'$, or $\bar{v}_1 = \bar{v}_1'$. This result, together with Equation 7.6.12, yields the value

$$\gamma = \frac{m_1}{m_2} \tag{7.6.16}$$

for an elastic collision.

Two special cases of such elastic collisions are instructive to consider. First, if the mass m_2 of the target particle is very much greater than the mass m_1 of the incident particle, then γ is very small. Hence, $\tan \phi_1 \approx \tan \theta$, or $\phi_1 \approx \theta$. That is, the scattering angles as seen in the laboratory and in the center of mass systems are nearly equal.

The second special case is that of equal masses of the incident and target particles $m_1 = m_2$. In this case $\gamma = 1$, and the scattering relation reduces to

$$\tan \phi_1 = \frac{\sin \theta}{1 + \cos \theta} = \tan \frac{\theta}{2}$$

$$\phi_1 = \frac{\theta}{2} \tag{7.6.17}$$

That is, the angle of deflection in the laboratory system is just half that in the center of mass system. Furthermore, because the angle of deflection of the target particle is $\pi - \theta$ in the center of mass system, as shown in Figure 7.6.1, then the same angle in the labo-ratory system is $(\pi - \theta)/2$. Therefore, the two particles leave the point of impact at right angles to each other as seen in the laboratory system, in agreement with Equation 7.6.6.

In the general case of nonelastic collisions, it is left as a problem to show that γ is expressible as

$$\gamma = \frac{m_1}{m_2} \left[1 - \frac{Q}{T} \left(1 + \frac{m_1}{m_2} \right) \right]^{-1/2} \tag{7.6.18}$$

in which T is the kinetic energy of the incident particle as measured in the laboratory system.

EXAMPLE 7.6.1

In a nuclear scattering experiment a beam of 4-MeV alpha particles (helium nuclei) strikes a target consisting of helium gas, so that the incident and the target particles have equal mass. If a certain incident alpha particle is scattered through an angle of 30° in the laboratory system, find its kinetic energy and the kinetic energy of recoil of the target particle, as a fraction of the initial kinetic energy T of the incident alpha particle. (Assume that the target particle is at rest and that the collision is elastic.)

Solution:

For elastic collisions with particles of equal mass, we know from Equation 7.6.6 that $\phi_1 + \phi_2 = 90°$ (see Figure 7.6.1). Hence, if we take components parallel to and perpendicular to the momentum of the incident particle, the momentum balance equation (Equation 7.6.1a) becomes

$$p_1 = p_1' \cos \phi_1 + p_2' \sin \phi_1$$
$$0 = p_1' \sin \phi_1 - p_2' \cos \phi_1$$

in which $\phi_1 = 30°$. Solving the preceding pair of equations for the primed components, we find

$$p_1' = p_1 \cos \phi_1 = p_1 \cos 30° = \frac{\sqrt{3}}{2} p_1$$
$$p_2' = p_1 \sin \phi_1 = p_1 \sin 30° = \frac{1}{2} p_1$$

Therefore, the kinetic energies after impact are

$$T_1' = \frac{p_1'^2}{2m_1} = \frac{3}{4} \frac{p_1^2}{2m_1} = \frac{3}{4} T = 3 \text{ MeV}$$

$$T_2' = \frac{p_2'^2}{2m_2} = \frac{1}{4} \frac{p_1^2}{2m_1} = \frac{1}{4} T = 1 \text{ MeV}$$

EXAMPLE 7.6.2

What is the scattering angle in the center of mass system for Example 7.6.1?

Solution:

Here Equation 7.6.17 gives the answer directly, namely,

$$\theta = 2\phi_1 = 60°$$

EXAMPLE 7.6.3

(a) Show that, for the general case of elastic scattering of a beam of particles of mass m_1 off a stationary target of particles whose mass is m_2, the opening angle ψ in the lab is given by the expression

$$\psi = \phi_1 + \phi_2 = \frac{\pi}{2} + \frac{\phi_1}{2} - \frac{1}{2}\sin^{-1}\left(\frac{m_1}{m_2}\sin\phi_1\right)$$

(b) Suppose the beam of particles consists of protons and the target consists of helium nuclei. Calculate the opening angle for a proton scattered elastically at a lab angle $\phi_1 = 30°$.

Solution:

(a) Because particle 2 is at rest in the lab, its center of mass velocity $\bar{v}_2$ is equal in magnitude (and opposite in direction) to v_{cm}. For elastic collisions in the center of mass, momentum and energy conservation can be written as

$$\bar{p}_1 + \bar{p}_2 = \bar{p}_1' + \bar{p}_2' = 0$$

$$\frac{\bar{p}_1^2}{2m_1} + \frac{\bar{p}_2^2}{2m_2} = \frac{\bar{p}_1'^2}{2m_1} + \frac{\bar{p}_2'^2}{2m_2}$$

Solving for the magnitudes of the center of mass momenta of particle 1 in terms of particle 2, we obtain

$$\bar{p}_1 = \bar{p}_2 \qquad \bar{p}_1' = \bar{p}_2'$$

These expressions can be inserted into the energy conservation equation to obtain

$$\frac{\bar{p}_2^2}{2\mu} = \frac{\bar{p}_2'^2}{2\mu} \qquad \mu = \frac{m_1 m_2}{m_1 + m_2}$$

$$\therefore \bar{v}_2' = \bar{v}_2 = v_{cm}$$

Thus, in an elastic collision, the center of mass velocities of particle 2 are the same before and after the collision, and both are equal to the center of mass velocity. Moreover, the values of the center of mass velocities of particle 1 are also the same before and after the collision, and, from conservation of momentum in the center of mass, they are

$$\bar{v}_1' = \bar{v}_1 = \frac{m_2}{m_1}\bar{v}_2' = \frac{m_2}{m_1}v_{cm}$$

Shown below in Figure 7.6.3 is a vector diagram that relates the parameters of elastic scattering in the laboratory and center of mass frames of reference. From the geometry of Figure 7.6.3, we see that

$$\psi = \phi_1 + \phi_2$$

$$2\phi_2 = \pi - \theta$$

$$\phi_2 = \frac{\pi}{2} - \frac{\theta}{2}$$

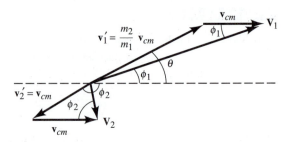

Figure 7.6.3   Velocity vectors in laboratory and center of mass frame for elastic scattering.

Now, applying the law of sines to the upper triangle of the figure, we obtain

$$\frac{(m_2/m_1)v_{cm}}{\sin\phi_1} = \frac{v_{cm}}{\sin(\theta - \phi_1)}$$

$$\sin(\theta - \phi_1) = \frac{m_1}{m_2}\sin\phi_1$$

$$\therefore \theta = \phi_1 + \sin^{-1}\left(\frac{m_1}{m_2}\sin\phi_1\right)$$

Finally, substituting this last expression for θ into the one preceding it for ϕ_2 and solving for the opening angle ψ, we obtain

$$\psi = \phi_1 + \phi_2 = \phi_1 + \left(\frac{\pi}{2} - \frac{\theta}{2}\right)$$

$$= \frac{\pi}{2} + \frac{\phi_1}{2} - \frac{1}{2}\sin^{-1}\left(\frac{m_1}{m_2}\sin\phi_1\right)$$

(b) For elastic scattering of protons off helium nuclei at $\phi_1 = 30°$, $m_1/m_2 = \frac{1}{4}$, and $\psi \approx 101°$.

(**Note:** In the case where $m_1 - m_2$, $\psi = 90°$ as derived in the text.)

7.7 | Motion of a Body with Variable Mass: Rocket Motion

Thus far, we have discussed only situations in which the masses of the objects under consideration remain constant during motion. In many situations this is not true. Raindrops falling though the atmosphere gather up smaller droplets as they fall, which increases their mass. Rockets propel themselves by burning fuel explosively and ejecting the resultant gasses at high exhaust velocities. Thus, they lose mass as they accelerate. In each case, mass is continually being added to or removed from the body in question, and this change in mass affects its motion. Here we derive the general differential equation that describes the motion of such objects.

So as not to get too confused with signs, we derive the equation by considering the case in which mass is added to the body as it moves. The equation of motion also applies to rockets, but in that case the rate of change of mass is a negative quantity. Examine Figure 7.7.1. A large mass is moving through some medium that is infested with small particles that stick to the mass as it strikes them. Thus, the larger body is continually gathering up mass as it moves through the medium. At some time t, its mass is $m(t)$ and its

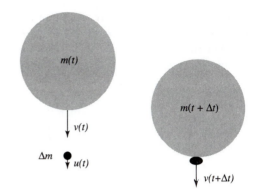

Figure 7.7.1 A mass m gathering up a small mass Δm as it moves through a medium.

velocity is $\mathbf{v}(t)$. The small particles are, in general, not at rest but are moving through the medium also with a velocity that we assume to be $\mathbf{u}(t)$. At time $t + \Delta t$, the large moving object has collided with some of these smaller particles and accumulated an additional small amount of mass Δm. Thus, its mass is now $m(t + \Delta t) = m(t) + \Delta m$ and its velocity has changed to $\mathbf{v}(t + \Delta t)$. In the small time interval Δt, the change (if any) in the total linear momentum of the system is

$$\Delta \mathbf{P} = (\mathbf{p}_{total})_{t+\Delta t} - (\mathbf{p}_{total})_{t} \tag{7.7.1}$$

This change can be expressed in terms of the masses and velocities before and after the collision

$$\Delta \mathbf{P} = (m + \Delta m)(\mathbf{v} + \Delta \mathbf{v}) - (m\mathbf{v} + \mathbf{u}\,\Delta m) \tag{7.7.2}$$

Because the velocity of Δm relative to m is $\mathbf{V} = \mathbf{u} - \mathbf{v}$, Equation 7.7.2 can be expressed as

$$\Delta \mathbf{P} = m\,\Delta \mathbf{v} + \Delta m\,\Delta \mathbf{v} - \mathbf{V}\,\Delta m \tag{7.7.3}$$

and on dividing by Δt we obtain

$$\frac{\Delta \mathbf{P}}{\Delta t} = (m + \Delta m)\frac{\Delta \mathbf{v}}{\Delta t} - \mathbf{V}\frac{\Delta m}{\Delta t} \tag{7.7.4}$$

In the limit as $\Delta t \to 0$, we have

$$\mathbf{F}_{ext} = \dot{\mathbf{P}} = m\dot{\mathbf{v}} - \mathbf{V}\dot{m} \tag{7.7.5}$$

The force $\mathbf{F}_{ext}$ represents any external force, such as gravity, air resistance, and so forth that acts on the system in addition to the impulsive force that results from the interaction between the masses m and Δm. If $\mathbf{F}_{ext} = 0$, then the total momentum $\mathbf{P}$ of the system is a constant of the motion and its net change is zero. This is the case for a rocket in deep space, beyond the gravitational influence of any planet or star, where $\mathbf{F}_{ext}$ is essentially zero.

We now apply this equation of motion to two special cases in which mass is added to or lost from the moving body. First, suppose that, as we have described, the body is falling through a fog or mist so that it collects mass as it goes, but assume that the small droplets of matter are suspended in the atmosphere such that their initial velocity prior to accretion is zero. In general, this will be a good approximation. Hence, $\mathbf{V} = -\mathbf{v}$, and we obtain

$$\mathbf{F}_{ext} = m\dot{\mathbf{v}} + \mathbf{v}\dot{m} = \frac{d}{dt}(m\mathbf{v}) \tag{7.7.6}$$

for the equation of motion. It applies only if the initial velocity of the matter that is being swept us is zero. Otherwise, the more general Equation 7.7.5, must be used.

For the second case, consider the motion of a rocket. The sign of $\dot{m}$ is negative because the rocket is losing mass in the form of ejected fuel. The term $\mathbf{V}\dot{m}$ in Equation 7.7.5 is called the thrust of the rocket, and its direction is opposite the direction of $\mathbf{V}$, the relative velocity of the exhaust products. Here, we solve the equation of motion for the simplest case of rocket motion in which the external force on it is zero; that is, the rocket is not subject to any force of gravity, air resistance, and so on. Thus, in Equation 7.7.5, $\mathbf{F}_{ext} = 0$, and we have

$$m\dot{\mathbf{v}} = \mathbf{V}\dot{m} \tag{7.7.7}$$

We can now separate the variables and integrate to find $\mathbf{v}$ as follows:

$$\int d\mathbf{v} = \int \frac{\mathbf{V}\,dm}{m} \tag{7.7.8}$$

If we assume that $\mathbf{V}$ is constant, then we can integrate between limits to find the *speed* as function of m:

$$\int_{v_0}^{v} dv = -V\int_{m_0}^{m} \frac{dm}{m}$$
$$v = v_0 + V \ln\frac{m_0}{m} \tag{7.7.9}$$

Here m_0 is the initial mass of the rocket plus unburned fuel, m is the mass at any time, and V is the speed of the ejected fuel relative to the rocket. Owing to the nature of the logarithmic function, the rocket must have a large fuel-to-payload ratio to attain the large speeds needed for launching satellites into space.

EXAMPLE 7.7.1

Launching an Earth Satellite from Cape Canaveral

We know from Example 6.5.3 that the speed of a satellite in a circular orbit near Earth is about 8 km/s. Satellites are launched toward the east to take advantage of Earth's rotation. For a point on the Earth near the equator the rotational speed is approximately $R_{Earth}\,\omega_{Earth}$, which is about 0.5 km/s. For most rocket fuels the effective ejection speed is of the order of 2 to 4 km/s. For example, if we take $V = 3$ km/s, then we find that the mass ratio calculated from Equation 7.7.9 is

$$\frac{m_0}{m} = \exp\left(\frac{v - v_0}{V}\right) = \exp\left(\frac{8.0 - 0.5}{3}\right) = e^{2.5} = 12.2$$

to achieve orbital speed from the ground. Thus, only about 8% of the total initial mass m_0 is payload.

Multi-Stage Rockets

Example 7.7.1 demonstrates that a large amount of fuel is necessary to put a small payload into low earth orbit (LEO) even if the effects of gravity and air resistance are absent. Neglecting air resistance is not a bad approximation because careful shaping of the rocket can greatly minimize its effect. However, as you most assuredly would suspect, we cannot ignore the effect of gravity because it greatly magnifies the problem of putting something into orbit.

The equation of motion of the rocket with gravity acting is given by Equation 7.7.5

$$m\frac{d\mathbf{v}}{dt} - \mathbf{V}\frac{dm}{dt} = m\mathbf{g} \tag{7.7.10}$$

Choosing the upward direction as positive and rearranging terms, we get

$$\frac{dv}{V} = -\frac{dm}{m} - \frac{g}{V}dt \tag{7.7.11}$$

For the rocket to achieve *liftoff*, the first term on the right of Equation 7.7.11 must exceed the second (remember, dm is negative); in other words, the rocket must eject a lot of matter, dm, at high exhaust velocity V. The reciprocal of the constant g/V in the second term is a "parameter of goodness" for a given type of rocket and has been given a special name, the *specific impulse* τ_s of the rocket engine.

$$\tau_s = \frac{V}{g} \tag{7.7.12}$$

It has the dimensions of time, and its value depends on the exhaust velocity of the rocket. This, in turn, depends primarily on the thermodynamics of what goes on inside the rocket's combustion chamber and the shape of the rocket nozzle. A well-designed chemical rocket that works by rapid oxidation of a fuel typically has an exhaust velocity of about 3000 m/s where the average molecular weight of the combustibles is about 30. Thus, $\tau_s = V/g \approx 300$ s.

We now integrate Equation 7.7.11 during the fuel burn up to the *time of burnout* τ_B to find the final velocity attained by the rocket.

$$\frac{1}{V}\int_0^{v_f} dv = -\int_{m_0}^{m_f} \frac{dm}{m} - \frac{1}{\tau_s}\int_0^{\tau_B} dt \tag{7.7.13a}$$

Completing the integration, we get

$$\frac{v_f}{V} = \ln\left[\frac{m_R + m_p + m_F}{m_R + m_p}\right] - \frac{\tau_B}{\tau_s} \tag{7.7.13b}$$

The masses in the above equation are m_R = mass of the rocket, m_p = mass of the payload, and m_F = mass of the fuel (plus oxidizer).

Solving Equation 7.7.13b for the mass ratio, we get

$$\left[\frac{m_R + m_p + m_F}{m_R + m_p}\right] = e^{\left(\frac{\tau_B}{\tau_s} + \frac{v_f}{V}\right)} \tag{7.7.14}$$

The question of interest here is how much fuel is needed to boost the rocket and payload into LEO? The final velocity of the rocket must be about 8 km/s. Solving for the mass of the fuel relative to the mass of the rocket and its payload, we get

$$\frac{m_F}{m_R + m_p} = e^{\left(\frac{v_f}{V} + \frac{\tau_B}{\tau_s}\right)} - 1 \tag{7.7.15}$$

The burnout time of a rocket lifting a payload into LEO is about 600 s. Putting the relevant numbers into Equation 7.7.15 yields the result

$$\frac{m_F}{m_R + m_p} = e^{(2.67+2.00)} - 1 \approx 105$$

In other words, it takes about 105 kg of fuel to place 1 kg of *stuff* into orbit! This ratio is larger than that which is typically required. For example, the liftoff weight of the *Saturn V* was about 3.2 million kg and it could put 100,000 kg into orbit. This is a ratio of about 32 kg of fuel for every kilogram of orbital stuff. Why is our result a factor of 3 larger?

Saturn V used a more efficient, two-stage rocket to launch a payload into LEO. The tanks that hold the fuel for the first stage are jettisoned after the first stage burn is completed; thus, this now useless mass is not boosted into orbit, which greatly reduces the overall fuel requirement. Let's take a look at Equation 7.7.14 to see how this works. We denote the mass ratio by the symbol, μ

$$\left[\frac{m_R + m_p + m_F}{m_R + m_p}\right] = \mu \tag{7.7.16}$$

We assume that the mass ratio of the first stage μ_1 is equal to that of the second μ_2 and that the burnout times τ_{B1} and τ_{B2} for each stage are identical. We can then calculate the final velocities achieved by each stage from Equation 7.7.13b

$$\frac{v_{f1}}{V} = \ln \mu - \frac{\tau_B}{\tau_s} \tag{7.7.17}$$

and

$$\frac{v_{f2} - v_{f1}}{V} = \ln \mu - \frac{\tau_B}{\tau_s} \tag{7.7.18}$$

Solving for v_{f2} gives

$$\frac{v_{f2}}{V} = 2\ln \mu - 2\frac{\tau_B}{\tau_s} \tag{7.7.19}$$

Solving for the fuel to rocket and payload mass ratio as before gives

$$\frac{m_F}{m_R + m_p} = e^{\left[\frac{v_{f2}}{2V} + \frac{\tau_B}{\tau_s}\right]} - 1 \tag{7.7.20}$$

Putting in the numbers, we get

$$\frac{m_F}{m_R + m_p} \approx 27 \tag{7.7.21}$$

Thus, it takes only about 27 kg of fuel to put 1 kg of stuff into orbit using a two-stage rocket. Clearly, there is an enormous advantage to staging as was demonstrated in Saturn V.

The Ion Rocket

Chemical rockets use the thermal energy released in the explosive oxidation of the fuel in the rocket motor chamber to eject the reactant products out the rear end of the rocket to propel it forward. In an ion rocket, such as NASA's *Deep Space 1*,[9] atoms of xenon gas are stripped of one of their electrons, and the resulting positive Xe^+ ions are accelerated by an electric field in the rocket motor. These ejected ions impart a forward momentum to the rocket exactly in the same way as described by Equation 7.7.7. There are two essential differences between ion and chemical rockets:

- The exhaust velocity of an ion rocket is about 10 times larger than that of a chemical rocket, which gives a larger specific impulse (see Equation 7.7.12)
- The mass ejected per unit time, $\dot{m}$, is much smaller in ion rockets, which gives a much smaller thrust (the term $V\dot{m}$ in Equation 7.7.7).

These differences crop up because, even though the electrostatic acceleration of ions is more efficient than thermal acceleration by chemical explosions, the density of ejected ions is much less than the density of the ejected gasses. The upshot is that an ion rocket is more efficient than a chemical rocket in the sense that it takes much less fuel mass to propel the rocket to some desired speed, but the acceleration of the rocket is quite gentle, so that it takes more time to attain that speed. This makes ion rockets more suitable for deep space missions to, say, comets and asteroids and, perhaps, ultimately, to nearby star systems! Indeed, one of the purposes of NASA's Deep Space 1 mission is to test this hypothesis. Here we discuss its propulsion system to see what has been achieved so far with this new technology.

The electrostatic potential Φ_e through which the Xe^+ ions were accelerated, was 1280 volts. The ions were ejected from the 0.3-meter thruster through a pair of focusing molybdenum grids. We can estimate the maximum possible escape velocity of these ions by noting that charged particles in an electrostatic potential Φ_e accelerate and gain kinetic energy by losing electrostatic potential energy $e\Phi_e$ where e is their electric charge. The electrostatic potential energy of a charged particle in an electric field is analogous to the gravitational potential energy $m\Phi$ (Equation 6.7.6) of a particle in a gravitational field. Thus, we have

$$\tfrac{1}{2}mV^2 = e\Phi_e \tag{7.7.22}$$

---

[9] See the website http://nmp.jpl.nasa.gov/ds1/tech/ionpropfaq.html for a discussion of NASA's New Millennium Project, Deep Space 1 mission.

where m is the mass of a Xe^+ ion. Solving for the escape velocity, we get

$$V = \sqrt{\frac{2e\Phi_e}{m}} \tag{7.7.23}$$

Putting in numbers:[10]

$$m \approx 131 \text{ AMU} = 131 \times 1.66 \times 10^{-27} \text{ kg} = 2.17 \times 10^{-25} \text{ kg}$$
$$e = 1.6 \times 10^{-19} \text{ C} \tag{7.7.24}$$
$$V = 4.3 \times 10^4 \text{ m/s}$$

Thus, the maximum possible specific impulse of the ion rocket is

$$\tau = \frac{V}{g} = \frac{4.3 \times 10^4 \text{ m/s}}{9.8 \text{ m/s}^2} = 4.4 \times 10^3 \text{ s} \tag{7.7.25}$$

In fact, the specific impulse of Deep Space I ranges between 1900 s and 3200 s depending upon throttle power. The maximum calculated here assumes that all the available power accelerates the ions with 100% efficiency and ejects them exactly in the backward direction out the rear end of the rocket, which is virtually impossible to do. The specific impulse of Deep Space I is about 10 times greater than that of Saturn V.

We now calculate the thrust of Deep Space I, again assuming that all available power is converted into the ejected ion beam with 100% efficiency. The maximum available power on Deep Space I is P = 2.5 kW. Thus, the rate, $\dot{N}$, at which Xe^+ ions are ejected can be calculated from the expression

$$P = \dot{E} = \dot{N}e\Phi_e \tag{7.7.26}$$

Because $e\Phi_e$ is the potential energy lost in accelerating a single ion, the power consumed is equal to the potential energy lost per unit time to all the accelerated ions. The rate at which mass is ejected, $\dot{m}$, is equal to the mass of each ion times $\dot{N}$. Thus,

$$\dot{m} = m\dot{N} = \frac{mP}{e\Phi_e} = \frac{(2.17 \times 10^{-25} \text{ kg})(2500 \text{ J/s})}{(1.6 \times 10^{-19} \text{ C})(1280 \text{ V})} = 2.6 \times 10^{-6} \text{ kg/s} \tag{7.7.27}$$

where we have used the fact that $1 \text{ C} \times 1 \text{ V} = 1 \text{ J}$. The maximum thrust of the ion rocket is thus,

$$\text{Thrust} = V\dot{m} = (4.3 \times 10^4 \text{ m/s})(2.6 \times 10^{-6} \text{ kg/s}) = 0.114 \text{ N}$$

In fact, the maximum thrust achieved by Deep Space I is 0.092 N. We can compare this with the thrust developed by Saturn V. Saturn V ejected about 11,700 kg/s. Thus,

$$\frac{\text{Thrust(Saturn V)}}{\text{Thrust(Deep Space I)}} = \frac{V\dot{m}\text{(Saturn V)}}{V\dot{m}\text{(Deep Space I)}} = \frac{(3000 \text{ m/s})(11,700 \text{ kg/s})}{0.092 \text{ N}} = 3.8 \times 10^8$$

---

[10] An AMU is an atomic mass unit. It is equal to 1.66×10^{-27} kg. The unit of electric charge is the Coulomb (C). The charge of the electron is -1.6×10^{-19} C; thus, the charge of a singly charged positive ion is $+1.6 \times 10^{-19}$ C.

We conclude that ion rockets are not useful for launching payloads from Earth but are suitable for deep space missions starting from Earth orbit in which efficient but gentle propulsion systems can be used.

Problems

7.1   A system consists of three particles, each of unit mass, with positions and velocities as follows:

$$\mathbf{r}_1 = \mathbf{i} + \mathbf{j} \qquad \mathbf{v}_1 = 2\mathbf{i}$$
$$\mathbf{r}_2 = \mathbf{j} + \mathbf{k} \qquad \mathbf{v}_2 = \mathbf{j}$$
$$\mathbf{r}_3 = \mathbf{k} \qquad \mathbf{v}_3 = \mathbf{i} + \mathbf{j} + \mathbf{k}$$

Find the position and velocity of the center of mass. Find also the linear momentum of the system.

7.2   (a) Find the kinetic energy of the system in Problem 7.1.
(b) Find the value of $m v_{cm}^2 / 2$.
(c) Find the angular momentum about the origin.

7.3   A bullet of mass m is fired from a gun of mass M. If the gun can recoil freely and the muzzle velocity of the bullet (velocity relative to the gun as it leaves the barrel) is v_0, show that the actual velocity of the bullet relative to the ground is $v_0/(1 + \gamma)$ and the recoil velocity for the gun is $-\gamma v_0/(1 + \gamma)$, where $\gamma = m/M$.

7.4   A block of wood rests on a smooth horizontal table. A gun is fired horizontally at the block and the bullet passes through the block, emerging with half its initial speed just before it entered the block. Show that the fraction of the initial kinetic energy of the bullet that is lost as frictional heat is $\frac{3}{4} - \frac{1}{4}\gamma$, where γ is the ratio of the mass of the bullet to the mass of the block ($\gamma < 1$).

7.5   An artillery shell is fired at an angle of elevation of 60° with initial speed v_0. At the uppermost part of its trajectory, the shell bursts into two equal fragments, one of which moves directly upward, relative to the ground, with initial speed $v_0/2$. What is the direction and speed of the other fragment immediately after the burst?

7.6   A ball is dropped from a height h onto a horizontal pavement. If the coefficient of restitution is ϵ, show that the total vertical distance the ball goes before the rebounds cease is $h(1 + \epsilon^2)/(1 - \epsilon^2)$. Find also the total length of time that the ball bounces.

7.7   A small car of a mass m and initial speed v_0 collides head-on on an icy road with a truck of mass $4m$ going toward the car with initial speed $\frac{1}{2}v_0$. If the coefficient of restitution in the collision is $\frac{1}{4}$, find the speed and direction of each vehicle just after colliding.

7.8   Show that the kinetic energy of a two-particle system is $\frac{1}{2}m v_{cm}^2 + \frac{1}{2}\mu v^2$, where $m = m_1 + m_2$, v is the relative speed, and μ is the reduced mass.

7.9   If two bodies undergo a direct collision, show that the loss in kinetic energy is equal to

$$\frac{1}{2}\mu v^2 (1 - \epsilon^2)$$

where μ is the reduced mass, v is the relative speed before impact, and ϵ is the coefficient of restitution.

7.10   A moving particle of mass m_1 collides elastically with a target particle of mass m_2, which is initially at rest. If the collision is head-on, show that the incident particle loses a fraction $4\mu/m$ of its original kinetic energy, where μ is the reduced mass and $m = m_1 + m_2$.

7.11   Show that the angular momentum of a two-particle system is

$$\mathbf{r}_{cm} \times m\mathbf{v}_{cm} + \mathbf{R} \times \mu\mathbf{v}$$

where $m = m_1 + m_2$, μ is the reduced mass, $\mathbf{R}$ is the relative position vector, and $\mathbf{v}$ is the relative velocity of the two particles.

7.12   The observed period of the binary system Cygnus X-1, presumed to be a bright star and a black hole, is 5.6 days. If the mass of the visible component is 20 $M_\odot$ and the black hole has a mass of 16 $M_\odot$, show that the semimajor axis of the orbit of the black hole relative to the visible star is roughly one-fifth the distance from Earth to the Sun.

7.13   (a) Using the coordinate convention given is Section 7.4 for the restricted three-body problem, find the coordinates (x', y') of the two Lagrangian points, L_4 and L_5.
    (b) Show that the gradient of the effective potential function $V(x', y')$ vanishes at L_4 and L_5.

7.14   A proton of mass m_p with initial velocity $\mathbf{v}_0$ collides with a helium atom, mass $4m_p$, that is initially at rest. If the proton leaves the point of impact at an angle of 45° with its original line of motion, find the final velocities of each particle. Assume that the collision is perfectly elastic.

7.15   Work Problem 7.14 for the case that the collision is inelastic and that Q is equal to one-fourth of the initial energy of the proton.

7.16   Referring to Problem 7.14, find the scattering angle of the proton in the center of mass system.

7.17   Find the scattering angle of the proton in the center-of-mass system for Problem 7.15.

7.18   A particle of mass m with initial momentum p_1 collides with a particle of equal mass at rest. If the magnitudes of the final momenta of the two particles are p_1' and p_2', respectively, show that the energy loss of the collision is given by

$$Q = \frac{p_1' p_2'}{m} \cos\psi$$

where ψ is the angle between the paths of the two particles after colliding.

7.19   A particle of mass m_1 with an initial kinetic energy T_1 makes an elastic collision with a particle of mass m_2 initially at rest. m_1 is deflected from its original direction with a kinetic energy T_1' through an angle ϕ_1 as in Figure 7.6.1. Letting $\alpha = m_2/m_1$ and $\gamma = \cos\phi_1$, show that the fractional kinetic energy lost by m_1, $\Delta T_1/T_1 = (T_1 - T_1')/T_1$, is given by

$$\frac{\Delta T_1}{T_1} = \frac{2}{1+\alpha} - \frac{2\gamma}{(1+\alpha)^2}\left(\gamma + \sqrt{\alpha^2 + \gamma^2 - 1}\right)$$

7.20   Derive Equation 7.6.18

7.21   A particle of mass m_1 scatters elastically from a particle of mass m_2 initially at rest as described in Problem 7.19. Find the curve $r(\phi_1)$ such that the time it takes the scattered particle to travel from the collision point to any point along the curve is a constant.

7.22   A uniform chain lies in a heap on a table. If one end is raised vertically with uniform velocity v, show that the upward force that must be exerted on the end of the chain is equal to the weight of a length $z + (v^2/g)$ of the chain, where z is the length that has been uncoiled at any instant.

7.23   Find the differential equation of motion of a raindrop falling through a mist collecting mass as it falls. Assume that the drop remains spherical and that the rate of accretion is proportional to the cross-sectional area of the drop multiplied by the speed of fall. Show that if the drop starts from rest when it is infinitely small, then the acceleration is constant and equal to $g/7$.

7.24 A uniform heavy chain of length a hangs initially with a part of length b hanging over the edge of a table. The remaining part, of length $a - b$, is coiled up at the edge of the table. If the chain is released, show that the speed of the chain when the last link leaves the end of the table is $[2g(a^3 - b^3)/3a^2]^{1/2}$.

7.25 A balloon of mass M containing a bag of sand of mass m_0 is filled with hot air until it becomes buoyant enough to rise ever so slightly above the ground, where it then hovers in equilibrium. Sand is then released at a constant rate such that all of it is dumped out in a time t_0. Find **(a)** the height of the balloon and **(b)** its velocity when all the sand has been released. Assume that the upward buoyancy force remains constant and neglect air resistance. **(c)** Assume that $\epsilon = m_0/M$ is very small, and find a power series expansion of your solutions for parts (a) and (b) in terms of this ratio. **(d)** Letting $M = 500$ kg, $m_0 = 10$ kg, and $t_0 = 100$ s, and keeping only the first-order term in the expansions obtained in part (c), find a numerical value for the height and velocity attained when all the sand has been released.

7.26 A rocket, whose total mass is m_0, contains a quantity of fuel, whose mass is ϵm_0 $(0 < \epsilon < 1)$. Suppose that, on ignition, the fuel is burned at a constant mass-rate k, ejecting gasses with a constant speed V relative to the rocket. Assume that the rocket is in a force-free environment. **(a)** Find the distance that the rocket has traveled at the moment it has burnt all the fuel. **(b)** What is the maximum possible distance that the rocket can travel during the burning phase? Assume that it starts from rest.

7.27 A rocket traveling through the atmosphere experiences a linear air resistance $-k\mathbf{v}$. Find the differential equation of motion when all other external forces are negligible. Integrate the equation and show that if the rocket starts from rest, the final speed is given by

$$v = V\alpha[1 - (m/m_0)^{1/\alpha}]$$

where V is the relative speed of the exhaust fuel, $\alpha = |\dot{m}/k| =$ constant, m_0 is the initial mass of the rocket plus fuel, and m is the final mass of the rocket.

7.28 Find the equation of motion for a rocket fired vertically upward, assuming g is constant. Find the ratio of fuel to payload to achieve a final speed equal to the escape speed v_e from the Earth if the speed of the exhaust gas is kv_e, where k is a given constant, and the fuel burning rate is $|\dot{m}|$. Compute the numerical value of the fuel–payload ratio for $k = \frac{1}{4}$, and $|\dot{m}|$ equal to 1% of the mass of the fuel per second.

7.29 Alpha Centauri is the nearest star system, about 4 light years from Earth. Assume that an ion rocket has been built to travel to Alpha Centauri. Suppose the exhaust velocity of the ions is one-tenth the speed of light. Let the initial mass of the fuel be twice that of the payload (ignore the mass of the rocket, itself). Also, assume that it takes about 100 hours to exhaust all the fuel of the rocket. How long does it take the rocket to reach Alpha Centauri? (The speeds are small enough that you can neglect the effects of special relativity.)

7.30 Consider the ion rocket described in Problem 7.29. Let's compare it to a chemical rocket whose exhaust velocity is 3 km/s. In the case of the ion rocket, 1 kg of fuel accelerates 1 kg of payload to a final velocity v_f. What fuel mass is required to accelerate the same payload to the same final velocity with the chemical rocket? (In each case, ignore the mass of the rocket.)

Computer Problems

C 7.1 Let two particles ($m_1 = m_2 = 1$ kg) repel each other with equal and opposite forces given by

$$\mathbf{F}_{12} = k\frac{b^2}{r^2}\mathbf{r}_{12} = -\mathbf{F}_{21}$$

where $b = 1$ m and $k = 1$ N. Assume that the initial positions of m_1 and m_2 are given by $(x_1, y_1)_0 = (-10, 0.5)$ m and $(x_2, y_2)_0 = (0, -0.5)$ m. Let the initial velocity of m_1 be 10 m/s in the $+x$ direction and m_2 be at rest. Numerically integrate the equations of motion for these two particles undergoing this two-dimensional "collision."

(a) Plot their trajectories up to a point where their distance of separation is 10 m.
(b) Measure the scattering angle of the incident particle and the recoil angle of the scattered particle. Is the sum of these two angles equal to 90°?
(c) What is the vector sum of their final momenta? Is it equal to the initial momentum of the incident particle?

C 7.2 Using a numerical optimization tool such as *Mathematica's FindMinimum* function, find the coordinates (x', y') of the Lagrange point L_4 in the restricted three-body problem. Do not assume, as you probably did in Problem 7.13, that L_4 is located at one of the corners of an equilateral triangle whose opposite base is formed by the two primaries. However, you should start the search for the coordinates of L_4 by using a point near the position of the suspected solution.

C 7.3 The total mass of a new experimental rocket, including payload, is 2×10^6 kg, and 90% of its mass is fuel. It burns fuel at a constant rate of 18,000 kg/s and exhausts the spent gasses at a speed of 3000 m/s. Assume that the rocket is launched vertically. Ignore the rotation of the Earth.

(a) Ignore air resistance, and assume that g, the acceleration due to gravity, is constant. Calculate the maximum altitude attained by the launched rocket.
(b) Repeat part (a), but include the effect of air resistance and the variable of g with altitude. Assume that the rocket presents a resistive surface to air that is equivalent to that of a sphere whose diameter is 0.5 m. Assume that the force of air resistance varies quadratically with speed and is given by

$$F(v) = -c_2 v |v|$$

and that c_2 scales with altitude y above Earth's surface as

$$c_2(y) = c_2(0) \exp^{-y/H}$$

where $c_2(0) = 0.22\, D^2$ as given in Chapter 2. H ($= 8$ km) is the scale height of the atmosphere, and the variation of g with altitude is

$$g(y) = \frac{9.8}{(1 + y/R_E)^2}\ \text{m/s}^2$$

as given in Computer Problem C 2.1.

8

Mechanics of Rigid Bodies: Planar Motion

"... centre of gravity implies the more restricted concept of a solid that is only heavy, while the centre of inertia is defined by means of the inertia alone, the forces to which the solid is subject being neglected. ... Euler also defines the moments of inertia—a concept which Huygens lacked and which considerably simplifies the language—and calculates these moments for Homogeneous bodies."

—Rene Dugas, *A History of Mechanics,* Editions du Griffon, Neuchatel, Switzerland, 1955; synopsis of Leonhard Euler's comments in Theoria motus corporum solidorum seu rigidorum, 1760

A rigid body may be regarded as a system of particles whose *relative* positions are fixed, or, in other words, the distance between any two particles is constant. This definition of a rigid body is idealized. In the first place, as pointed out in the definition of a particle, there are no true particles in nature. Second, real extended bodies are not strictly rigid; they become more or less deformed (stretched, compressed, or bent) when external forces are applied. For the present, we shall ignore such deformations. In this chapter we take up the study of rigid-body motion for the case in which the direction of the axis of rotation does not change. The general case, which involves more extensive calculation, is treated in the next chapter.

8.1 | Center of Mass of a Rigid Body

We have already defined the center of mass (Section 7.1) of a system of particles as the point (x_{cm}, y_{cm}, z_{cm}) where

$$x_{cm} = \frac{\sum_i x_i m_i}{\sum_i m_i} \qquad y_{cm} = \frac{\sum_i y_i m_i}{\sum_i m_i} \qquad z_{cm} = \frac{\sum_i z_i m_i}{\sum_i m_i} \tag{8.1.1}$$

For a rigid extended body, we can replace the summation by an integration over the volume of the body, namely,

$$x_{cm} = \frac{\int_v \rho x\, dv}{\int_v \rho\, dv} \qquad y_{cm} = \frac{\int_v \rho y\, dv}{\int_v \rho\, dv} \qquad z_{cm} = \frac{\int_v \rho z\, dv}{\int_v \rho\, dv} \qquad (8.1.2)$$

where ρ is the density and dv is the element of volume.

If a rigid body is in the form of a thin shell, the equations for the center of mass become

$$x_{cm} = \frac{\int_s \rho x\, ds}{\int_s \rho\, ds} \qquad y_{cm} = \frac{\int_s \rho y\, ds}{\int_s \rho\, ds} \qquad z_{cm} = \frac{\int_s \rho z\, ds}{\int_s \rho\, ds} \qquad (8.1.3)$$

where ds is the element of area and ρ is the mass per unit area, the integration extending over the area of the body.

Similarly, if the body is in the form of a thin wire, we have

$$x_{cm} = \frac{\int_l \rho x\, dl}{\int_l \rho\, dl} \qquad y_{cm} = \frac{\int_l \rho y\, dl}{\int_l \rho\, dl} \qquad z_{cm} = \frac{\int_l \rho z\, dl}{\int_l \rho\, dl} \qquad (8.1.4)$$

In this case, ρ is the mass per unit length and dl is the element of length.

For uniform homogeneous bodies, the density factors ρ are constant in each case and, therefore, may be canceled out in each of the preceding equations.

If a body is composite, that is, if it consists of two or more parts whose centers of mass are known, then it is clear, from the definition of the center of mass, that we can write

$$x_{cm} = \frac{x_1 m_1 + x_2 m_2 + \cdots}{m_1 + m_2 + \cdots} \qquad (8.1.5)$$

with similar equations for y_{cm} and z_{cm}. Here (x_1, y_1, z_1) is the center of mass of the part m_1, and so on.

Symmetry Considerations

If a body possesses symmetry, it is possible to take advantage of that symmetry in locating the center of mass. Thus, if the body has a plane of symmetry, that is, if each particle m_i has a mirror image of itself m_i' relative to some plane, then the center of mass lies in that plane. To prove this, let us suppose that the xy plane is a plane of symmetry. We have then

$$z_{cm} = \frac{\displaystyle\sum_i (z_i m_i + z_i' m_i')}{\displaystyle\sum_i (m_i + m_i')} \qquad (8.1.6)$$

But $m_i = m_i'$ and $z_i = -z_i'$. Hence, the terms in the numerator cancel in pairs, and so $z_{cm} = 0$; that is, the center of mass lies in the xy plane.

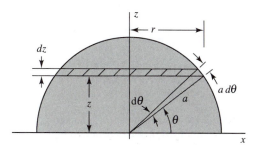

Figure 8.1.1 Coordinates for calculating the center of mass of a hemisphere.

Similarly, if the body has a line of symmetry, it is easy to show that the center of mass lies on that line. The proof is left as an exercise.

Solid Hemisphere

To find the center of mass of a solid homogeneous hemisphere of radius a, we know from symmetry that the center of mass lies on the radius that is normal to the plane face. Choosing coordinate axes as shown in Figure 8.1.1, we see that the center of mass lies on the z-axis. To calculate z_{cm} we use a circular element of volume of thickness dz and radius $= (a^2 - z^2)^{1/2}$, as shown. Thus,

$$dv = \pi(a^2 - z^2)dz \qquad (8.1.7)$$

Therefore,

$$z_{cm} = \frac{\int_0^a \rho\pi z(a^2 - z^2)dz}{\int_0^a \rho\pi(a^2 - z^2)dz} = \frac{3}{8}a \qquad (8.1.8)$$

Hemispherical Shell

For a hemispherical shell of radius a, we use the same axes as in Figure 8.1.1. Again, from symmetry, the center of mass is located on the z-axis. For our element of surface ds, we choose a circular strip of width $dl = a\,d\theta$. Hence,

$$ds = 2\pi r\,dl = 2\pi(a^2 - z^2)^{1/2}a\,d\theta$$

$$\theta = \sin^{-1}\left(\frac{z}{a}\right) \qquad d\theta = (a^2 - z^2)^{-1/2}\,dz \qquad (8.1.9)$$

$$\therefore ds = 2\pi a\,dz$$

The location of the center of mass is accordingly given by

$$z_{cm} = \frac{\int_0^a \rho 2\pi a z\,dz}{\int_0^a \rho 2\pi a\,dz} = \frac{1}{2}a \qquad (8.1.10)$$

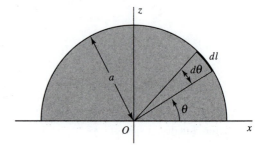

Figure 8.1.2 Coordinates for calculating the center of mass of a wire bent into the form of a semicircle.

Semicircle

To find the center of mass of a thin wire bent into the form of a semicircle of radius a, we use axes as shown in Figure 8.1.2. We have

$$dl = a\,d\theta \tag{8.1.11}$$

and

$$z = a\sin\theta \tag{8.1.12}$$

Hence,

$$z_{cm} = \frac{\int_0^\pi \rho(a\sin\theta)a\,d\theta}{\int_0^\pi \rho a\,d\theta} = \frac{2a}{\pi} \tag{8.1.13}$$

Semicircular Lamina

In the case of a uniform semicircular lamina, the center of mass is on the z-axis (Figure 8.1.2). As an exercise, the student should verify that

$$z_{cm} = \frac{4a}{3\pi} \tag{8.1.14}$$

Solid Cone of Variable Density: Numerical Integration

Sometimes we are confronted with the unfortunate prospect of having to find the center of mass of a body whose density is not uniform. In such cases, we must resort to numerical integration. Here we present a moderately complex case that we will solve numerically even though it can be solved analytically. We do this to illustrate how such a calculation can be easily carried out using the tools available in *Mathematica*.

Consider a solid, "unit" cone bounded by the conical surface $z^2 = x^2 + y^2$ and the plane $z = 1$ as shown in Figure 8.1.3, whose mass density function is given by

$$\rho(x,y,z) = \sqrt{x^2 + y^2} \tag{8.1.15}$$

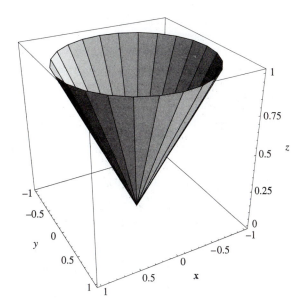

Figure 8.1.3   Solid cone whose surface is given by the curve $z^2 = x^2 + y^2$ and $z = 1$.

The center of mass of this cone can be calculated by solving the integrals given in Equation 8.1.2. The mass of the cone is given by

$$M = \int_{-1}^{1} \int_{-\sqrt{1-x^2}}^{\sqrt{1-x^2}} \int_{\sqrt{x^2+y^2}}^{1} \sqrt{x^2 + y^2}\ dz\, dy\, dx \tag{8.1.16}$$

Notice that the limits of integration over the variable y depend on x and that the limits of integration over z depend on both x and y. Because this integral is symmetric about both the x and y axes, it simplifies to

$$M = 4\int_{0}^{1} \int_{0}^{\sqrt{1-x^2}} \int_{\sqrt{x^2+y^2}}^{1} \sqrt{x^2 + y^2}\ dz\, dy\, dx \tag{8.1.17}$$

The first moments of the mass are given by the integrals

$$M_{yz} = \int_{-1}^{1} \int_{-\sqrt{1-x^2}}^{\sqrt{1-x^2}} \int_{\sqrt{x^2+y^2}}^{1} x\sqrt{x^2 + y^2}\ dz\, dy\, dx \tag{8.1.18a}$$

$$M_{xz} = \int_{-1}^{1} \int_{-\sqrt{1-x^2}}^{\sqrt{1-x^2}} \int_{\sqrt{x^2+y^2}}^{1} y\sqrt{x^2 + y^2}\ dz\, dy\, dx \tag{8.1.18b}$$

$$M_{xy} = \int_{-1}^{1} \int_{-\sqrt{1-x^2}}^{\sqrt{1-x^2}} \int_{\sqrt{x^2+y^2}}^{1} z\sqrt{x^2 + y^2}\ dz\, dy\, dx \tag{8.1.18c}$$

The location of the center of mass is then

$$(x_{cm}, y_{cm}, z_{cm}) = \left(\frac{M_{yz}}{M}, \frac{M_{xz}}{M}, \frac{M_{xy}}{M} \right) \tag{8.1.19}$$

The first moments of the mass about the x and y axes must vanish because, again, the mass distribution is symmetric about those axes. This is reflected in the fact that the integrals in Equations 8.1.18a and b are odd functions and, therefore, vanish. Thus, the only integrals we need evaluate are those in Equations 8.1.17 and 8.1.18c.

We performed these integrations numerically by invoking *Mathematica*'s *NIntegrate* function. The call to this function for a three-dimensional integral is

$$M(M_{xy}) = NIntegrate\ [Integrand,\ \{x,\ x_{min},\ x_{max}\},\ \{y,\ y_{min},\ y_{max}\},\ \{z,\ z_{min},\ z_{max}\}]$$

where the arguments are appropriate for M or M_{xy} and should be self-explanatory. The output of the two necessary calls yield the values: $M = 0.523599$ and $M_{xy} = 0.418888$. Thus, the coordinates of the center of mass are

$$(x_{cm},\ y_{cm},\ z_{cm}) = (0,0,0.800017) \tag{8.1.20}$$

We leave it as an exercise for the ambitious student to solve this problem analytically.

8.2 | Rotation of a Rigid Body about a Fixed Axis: Moment of Inertia

The simplest type of rigid-body motion, other than pure translation, is that in which the body is constrained to rotate about a fixed axis. Let us choose the z-axis of an appropriate coordinate system as the axis of rotation. The path of a representative particle m_i located at the point (x_i, y_i, z_i) is then a circle of radius $(x_i^2 + y_i^2)^{1/2} = r_i$ centered on the z-axis. A representative cross section parallel to the xy plane is shown in Figure 8.2.1.

The speed v_i of particle i is given by

$$v_i = r_i \omega = \left(x_i^2 + y_i^2\right)^{1/2} \omega \tag{8.2.1}$$

where ω is the angular speed of rotation. From a study of Figure 8.2.1, we see that the velocity has components as follows:

$$\dot{x}_i = -v_i \sin\phi_i = -\omega y_i$$
$$\dot{y}_i = v_i \cos\phi_i = \omega x_i \tag{8.2.2}$$
$$\dot{z}_i = 0$$

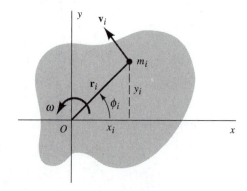

Figure 8.2.1   Cross section of a rigid body rotating about the z-axis. (The z-axis is out of the page.)

where ϕ_i is defined as shown in Figure 8.2.1. Equations 8.2.2 can also be obtained by extracting the components of the vector equation

$$\mathbf{v}_i = \boldsymbol{\omega} \times \mathbf{r}_i \tag{8.2.3}$$

where $\boldsymbol{\omega} = \mathbf{k}\omega$.

Let us calculate the kinetic energy of rotation of the body. We have

$$T_{rot} = \sum_i \tfrac{1}{2} m_i v_i^2 = \tfrac{1}{2}\left(\sum_i m_i r_i^2\right)\omega^2 = \tfrac{1}{2} I_z \omega^2 \tag{8.2.4}$$

where

$$I_z = \sum_i m_i r_i^2 = \sum_i m_i \left(x_i^2 + y_i^2\right) \tag{8.2.5}$$

The quantity I_z, defined by Equation 8.2.5, is called the *moment of inertia* about the z-axis.

To show how the moment of inertia further enters the picture, let us next calculate the angular momentum about the axis of rotation. Because the angular momentum of a single particle is, by definition, $\mathbf{r}_i \times m_i \mathbf{v}_i$, the z-component is

$$m_i(x_i \dot{y}_i - y_i \dot{x}_i) = m_i \left(x_i^2 + y_i^2\right)\omega = m_i r_i^2 \omega \tag{8.2.6}$$

where we have made use of Equations 8.2.2. The total z-component of the angular momentum, which we call L_z, is then given by summing over all the particles, namely,

$$L_z = \sum_i m_i r_i^2 \omega = I_z \omega \tag{8.2.7}$$

In Section 7.2 we found that the rate of change of angular momentum for any system is equal to the total moment of the external forces. For a body constrained to rotate about a fixed axis, taken here as the z-axis, then

$$N_z = \frac{dL_z}{dt} = \frac{d(I_z \omega)}{dt} \tag{8.2.8}$$

where N_z is the total moment of all the applied forces about the axis of rotation (the component of $\mathbf{N}$ along the z-axis). If the body is rigid, then I_z is constant, and we can write

$$N_z = I_z \frac{d\omega}{dt} \tag{8.2.9}$$

The analogy between the equations for translation and for rotation about a fixed axis is shown in the following table:

Translation along x-axis		Rotation about z-axis	
Linear momentum	$p_x = mv_x$	Angular momentum	$L_z = I_z \omega$
Force	$F_x = m\dot{v}_x$	Torque	$N_z = I_z \dot{\omega}$
Kinetic energy	$T = \tfrac{1}{2} mv^2$	Kinetic energy	$T_{rot} = \tfrac{1}{2} I_z \omega^2$

Thus, the moment of inertia is analogous to mass; it is a measure of the rotational inertia of a body relative to some fixed axis of rotation, just as mass is a measure of translational inertia of a body.

8.3 | Calculation of the Moment of Inertia

In calculations of the moment of inertia $\Sigma\, m_i r_i^2$ for extended bodies, we can replace the summation by an integration over the body, just as we did in calculation of the center of mass. Thus, we may write for any axis

$$I = \int r^2 dm \tag{8.3.1}$$

where the element of mass dm is given by a density factor multiplied by an appropriate differential (volume, area, or length), and r is the perpendicular distance from the element of mass to the axis of rotation.[1]

In the case of a composite body, from the definition of the moment of inertia, we may write

$$I = I_1 + I_2 + \cdots \tag{8.3.2}$$

where I_1, I_2, and so on, are the moments of inertia of the various parts about the particular axis chosen.

Let us calculate the moments of inertia for some important special cases.

Thin Rod

For a thin, uniform rod of length a and mass m, we have, for an axis perpendicular to the rod at one end (Figure 8.3.1a),

$$I_z = \int_0^a x^2\, \rho dx = \tfrac{1}{3}\rho a^3 = \tfrac{1}{3}ma^2 \tag{8.3.3}$$

The last step follows from the fact that $m = \rho a$.

If the axis is taken at the center of the rod (Figure 8.3.1b), we have

$$I_z = \int_{-a/2}^{a/2} x^2 \rho\, dx = \tfrac{1}{12}\rho a^3 = \tfrac{1}{12}ma^2 \tag{8.3.4}$$

Hoop or Cylindrical Shell

In the case of a thin circular hoop or cylindrical shell, for the central, or *symmetry*, axis, all particles lie at the same distance from the axis. Thus,

$$I_{axis} = ma^2 \tag{8.3.5}$$

where a is the radius and m is the mass.

---

[1] In Chapter 9, when we discuss the rotational motion of three-dimensional bodies, the distance between the mass element dm and the axis of rotation $r_\perp$ is designed to remind us that the relevant distance is the one perpendicular to the axis of rotation.

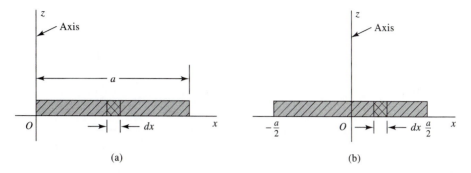

Figure 8.3.1   Coordinates for calculating the moment of inertia of a rod (a) about one end and (b) about the center of the rod.

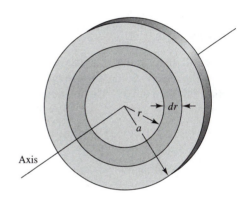

Figure 8.3.2   Coordinates for finding the moment of inertia of a disc.

Circular Disc or Cylinder

To calculate the moment of inertia of a uniform circular disc of radius a and mass m, we use polar coordinates. The element of mass, a thin ring of radius r and thickness dr, is given by

$$dm = \rho 2\pi r\, dr \tag{8.3.6}$$

where ρ is the mass per unit area. The moment of inertia about an axis through the center of the disc normal to the plane faces (Figure 8.3.2) is obtained as follows:

$$I_{axis} = \int_0^a r^2 \rho\, 2\pi r\, dr = 2\pi\rho\frac{a^4}{4} = \frac{1}{2}ma^2 \tag{8.3.7}$$

The last step results from the relation $m = \rho\pi a^2$.

   Equation 8.3.7 also applies to a uniform right-circular cylinder of radius a and mass m, the axis being the central axis of the cylinder.

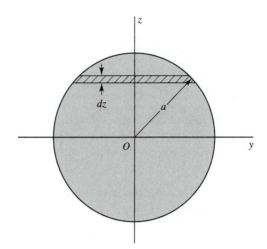

Figure 8.3.3   Coordinates for finding the moment of inertia of a sphere.

Sphere

Let us find the moment of inertia of a uniform solid sphere of radius a and mass m about an axis (the z-axis) passing through the center. We divide the sphere into thin circular discs, as shown in Figure 8.3.3. The moment of inertia of a representative disc of radius y, from Equation 8.3.7, is $\frac{1}{2}y^2 dm$. But $dm = \rho \pi y^2\, dz$; hence,

$$I_z = \int_{-a}^{a} \frac{1}{2}\pi\rho y^4\, dz = \int_{-a}^{a} \frac{1}{2}\pi\rho(a^2 - z^2)^2\, dz = \frac{8}{15}\pi\rho a^5 \tag{8.3.8}$$

The last step in Equation 8.3.8 should be filled in by the student. Because the mass m is given by

$$m = \frac{4}{3}\pi a^3 \rho \tag{8.3.9}$$

we have

$$I_z = \frac{2}{5}ma^2 \tag{8.3.10}$$

for a solid uniform sphere. Clearly also, $I_x = I_y = I_z$.

Spherical Shell

The moment of inertia of a thin, uniform, spherical shell can be found very simply by application of Equation 8.3.8. If we differentiate with respect to a, namely,

$$dI_z = \frac{8}{3}\pi\rho a^4\, da \tag{8.3.11}$$

the result is the moment of inertia of a shell of thickness da and radius a. The mass of the shell is $4\pi a^2\rho\, da$. Hence, we can write

$$I_z = \frac{2}{3}ma^2 \tag{8.3.12}$$

for the moment of inertia of a thin shell of radius a and mass m. The student should verify this result by direct integration.

EXAMPLE 8.3.1

Shown in Figure 8.3.4 is a uniform chain of length $l = 2\pi R$ and mass $m = M/2$ that is initially wrapped around a uniform, thin disc of radius R and mass M. One tiny piece of chain initially hangs free, perpendicular to the horizontal axis. When the disc is released, the chain falls and unwraps. The disc begins to rotate faster and faster about its fixed z-axis, without friction. (a) Find the angular speed of the disc at the moment the chain completely unwraps. (b) Solve for the case of a chain wrapped around a wheel whose mass is the same as that of the disc but concentrated in a thin rim.

Solution:

(a) Figure 8.3.4 shows the disc and chain at the moment the chain unwrapped. The final angular speed of the disc is ω. Energy was conserved as the chain unwrapped. Because the center of mass of the chain originally coincided with that of the disc, it fell a distance $l/2 = \pi R$, and we have

$$mg\frac{l}{2} = \frac{1}{2}I\omega^2 + \frac{1}{2}mv^2$$

$$\frac{l}{2} = \pi R \qquad v = \omega R \qquad I = \frac{1}{2}MR^2$$

Solving for ω^2 gives

$$\omega^2 = \frac{mg(l/2)}{\left[\left(\frac{1}{2}\right)(M/2) + \left(\frac{1}{2}\right)(m)\right]R^2} = \frac{mg\pi R}{\left[\left(\frac{1}{2}\right)m + \left(\frac{1}{2}\right)m\right]R^2}$$

$$= \pi\frac{g}{R}$$

(b) The moment of inertia of a wheel is $I = MR^2$. Substituting this into the preceding equation yields

$$\omega^2 = \pi\frac{2g}{3R}$$

Even though the mass of the wheel is the same as that of the disc, its moment of inertia is larger, because all its mass is concentrated along the rim. Thus, its angular acceleration and final angular velocity are less than that of the disc.

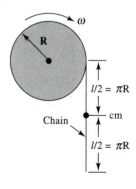

Figure 8.3.4   Falling chain attached to disc, free to rotate about a fixed z-axis.

Perpendicular-Axis Theorem for a Plane Lamina

Consider a rigid body that is in the form of a plane lamina of any shape. Let us place the lamina in the xy plane (Figure 8.3.5). The moment of inertia about the z-axis is given by

$$I_z = \sum_i m_i \left(x_i^2 + y_i^2 \right) = \sum_i m_i x_i^2 + \sum_i m_i y_i^2 \qquad (8.3.13)$$

The sum $\Sigma\, m_i x_i^2$ is just the moment of inertia I_y about the y-axis, because z_i is zero for all particles. Similarly, $\Sigma_i\, m_i y_i^2$ is the moment of inertia I_x about the x-axis. Equation 8.3.13 can, therefore, be written

$$I_z = I_x + I_y \qquad (8.3.14)$$

This is the perpendicular-axis theorem. In words:

The moment of inertia of any plane lamina about an axis normal to the plane of the lamina is equal to the sum of the moments of inertia about any two mutually perpendicular axes passing through the given axis and lying in the plane of the lamina.

As an example of the use of this theorem, let us consider a thin circular disc in the xy plane (Figure 8.3.6). From Equation 8.3.7 we have

$$I_z = \tfrac{1}{2} m a^2 = I_x + I_y \qquad (8.3.15)$$

In this case, however, we know from symmetry that $I_x = I_y$. We must, therefore, have

$$I_x = I_y = \tfrac{1}{4} m a^2 \qquad (8.3.16)$$

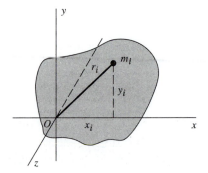

Figure 8.3.5 The perpendicular-axis theorem for a lamina.

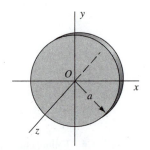

Figure 8.3.6 Circular disc.

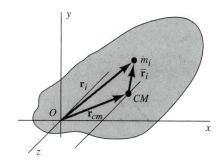

Figure 8.3.7   The parallel-axis theorem for any rigid body.

for the moment of inertia about any axis in the plane of the disc passing through the center. This result can also be obtained by direct integration.

Parallel-Axis Theorem for Any Rigid Body

Consider the equation for the moment of inertia about some axis, say the z-axis,

$$I_z = \sum_i m_i \left(x_i^2 + y_i^2 \right) \qquad (8.3.17)$$

Now we can express x_i and y_i in terms of the coordinates of the center of mass (x_{cm}, y_{cm}, z_{cm}) and the coordinates *relative* to the center of mass $(\bar{x}_i, \bar{y}_i, \bar{z}_i)$ (Figure 8.3.7) as follows:

$$x_i = x_{cm} + \bar{x}_i \qquad\qquad y_i = y_{cm} + \bar{y}_i \qquad (8.3.18)$$

We have, therefore, after substituting and collecting terms,

$$I_z = \sum_i m_i \left(\bar{x}_i^2 + \bar{y}_i^2 \right) + \sum_i m_i \left(x_{cm}^2 + y_{cm}^2 \right) + 2x_{cm} \sum_i m_i \bar{x}_i + 2y_{cm} \sum_i m_i \bar{y}_i \qquad (8.3.19)$$

The first sum on the right is just the moment of inertia about an axis parallel to the z-axis and passing through the center of mass. We call it I_{cm}. The second sum is equal to the mass of the body multiplied by the square of the distance between the center of mass and the z-axis. Let us call this distance l. That is, $l^2 = x_{cm}^2 + y_{cm}^2$.

Now, from the definition of the center of mass,

$$\sum_i m_i \bar{x}_i = \sum_i m_i \bar{y}_i = 0 \qquad (8.3.20)$$

Hence, the last two sums on the right of Equation 8.3.19 vanish. The final result may be written in the general form for any axis

$$I = I_{cm} + ml^2 \qquad (8.3.21)$$

This is the *parallel-axis* theorem. It is applicable to any rigid body, solid as well as laminar. The theorem states, in effect, that:

> The moment of inertia of a rigid body about any axis is equal to the moment of inertia about a parallel axis passing through the center of mass plus the product of the mass of the body and the square of the distance between the two axes.

We can use the parallel-axis theorem to calculate the moment of inertia of a uniform circular disc about an axis *perpendicular* to the plane of the disc and passing through an edge (see Figure 8.3.8a). Using Equations 8.3.7 and 8.3.21, we get

$$I = \tfrac{1}{2}ma^2 + ma^2 = \tfrac{3}{2}ma^2 \qquad (8.3.22)$$

We can also use the parallel-axis theorem to calculate the moment of inertia of the disc about an axis *in the plane* of the disc and *tangent* to an edge (see Figure 8.3.8b). Using Equations 8.3.16 and 8.3.21, we get

$$I = \tfrac{1}{4}ma^2 + ma^2 = \tfrac{5}{4}ma^2 \qquad (8.3.23)$$

As a second example, let us find the moment of inertia of a uniform circular cylinder of length b and radius a about an axis through the center and *perpendicular* to the central axis, namely I_x or I_y in Figure 8.3.9. For our element of integration, we choose a disc of thickness dz located a distance z from the xy plane. Then, from the previous result for a thin disc (Equation 8.3.16), together with the parallel-axis theorem, we have

$$dI_x = \tfrac{1}{4}a^2 dm + z^2 dm \qquad (8.3.24)$$

in which $dm = \rho\pi a^2\, dz$. Thus,

$$I_x = \rho\pi a^2 \int_{-b/2}^{b/2} \left(\tfrac{1}{4}a^2 + z^2\right)dz = \rho\pi a^2\left(\tfrac{1}{4}a^2 b + \tfrac{1}{12}b^3\right) \qquad (8.3.25)$$

But the mass of the cylinder is $m = \rho\pi a^2 b$, therefore,

$$I_x = I_y = m\left(\tfrac{1}{4}a^2 + \tfrac{1}{12}b^2\right) \qquad (8.3.26)$$

Radius of Gyration

Note the similarity of Equation 8.2.5, the expression for the moment of inertia I_z of a rigid body about the z-axis, to the expressions for center of mass developed in Section 8.1. If we were to divide Equation 8.2.5 by the total mass of the rigid body, we would obtain the mass-weighted average of the square of the positions of all the mass elements away from the z-axis.

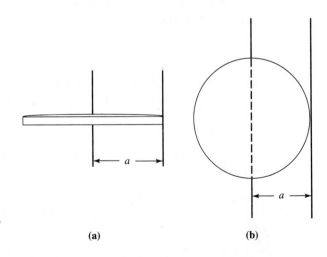

Figure 8.3.8   Moment of inertia of a uniform, thin disc about axes (a) perpendicular to the plane of the disc and through an edge and (b) in the plane of the disc and tangent to an edge.

(a)   (b)

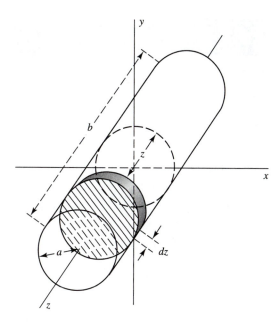

Figure 8.3.9   Coordinates for finding the moment of inertia of a circular cylinder.

Thus, moment of inertia is, in essence, the average of the squares of the radial distances away from the z-axis of all the mass elements making up the rigid body. You can understand physically why the moment of inertia must depend on the square (or, at least, some even power) of the distances away from the rotational axis; it could not be represented by a linear average over all the mass elements (or any average of the odd power of distance). If such were the case, then a body whose mass was symmetrically distributed about its rotational axis, such as a bicycle wheel, would have zero moment of inertia because of a term-by-term cancellation of the positive and negative weighted mass elements in the symmetrical distribution. An application of the slightest torque would spin up a bicycle wheel into an instantaneous frenzy, a condition that any bike racer knows is impossible.

We can formalize this discussion by defining a distance k, called the *radius of gyration*, to be this average, that is,

$$k^2 = \frac{I}{m} \qquad k = \sqrt{\frac{I}{m}} \tag{8.3.27}$$

Knowing the radius of gyration of any rigid body is equivalent to knowing its moment of inertia, but it better characterizes the nature of the averaging process on which the concept of moment of inertia is based.

For example, we find for the radius of gyration of a thin rod about an axis passing through one end (see Equation 8.3.3)

$$k = \sqrt{\frac{\left(\frac{1}{3}\right)ma^2}{m}} = \frac{a}{\sqrt{3}} \tag{8.3.28}$$

Moments of inertia for various objects can be tabulated simply by listing the squares of their radii of gyration (Table 8.3.1).

TABLE 8.3.1	Values of k^2 of Various Bodies (Moment of Inertia = Mass $\times k^2$)	
Body	**Axis**	k^2
Thin rod, lenght a	Normal to rod at its center	$\dfrac{a^2}{12}$
	Normal to rod at one end	$\dfrac{a^2}{3}$
Thin rectangular lamina, sides a and b	Through the center, parallel to side b	$\dfrac{a^2}{12}$
	Through the center, normal to the lamina	$\dfrac{a^2+b^2}{12}$
Thin circular disc, radius a	Through the center, in the plane of the disc	$\dfrac{a^2}{4}$
	Through the center, normal to the disc	$\dfrac{a^2}{2}$
Thin hoop (or ring), radius a	Through the center, in the plane of the hoop	$\dfrac{a^2}{2}$
	Through the center, normal to the plane of the hoop	a^2
Thin cylindrical shell, radius a, length b	Central longitudinal axis	a^2
Uniform solid right circular cylinder, radius a, length b	Central longitudinal axis	$\dfrac{a^2}{2}$
	Through the center, perpendicular to longitudinal axis	$\dfrac{a^2}{4}+\dfrac{b^2}{12}$
Thin spherical shell, radius a	Any diameter	$\dfrac{2}{3}a^2$
Uniform solid sphere, radius a	Any diameter	$\dfrac{2}{5}a^2$
Uniform solid rectangular parallelepiped, sides $a, b,$ and c	Through the center, normal to face ab, parallel to edge c	$\dfrac{a^2+b^2}{12}$

8.4| The Physical Pendulum

A rigid body that is free to swing under its own weight about a fixed horizontal axis of rotation is known as a *physical pendulum,* or *compound pendulum.* A physical pendulum is shown in Figure 8.4.1. Here *CM* is the center of mass, and *O* is the point on the axis of rotation that is in the vertical plane of the circular path of the center of mass.

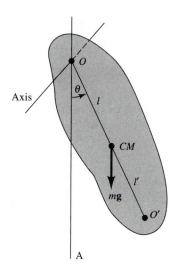

Figure 8.4.1 The physical pendulum.

Denoting the angle between the line OCM and the vertical line OA by θ, the moment of the gravitational force (acting at CM) about the axis of rotation is of magnitude

$$mgl \sin\theta$$

The fundamental equation of motion $N = I\dot{\omega}$ then takes the form $-mgl \sin\theta = I\ddot{\theta}$

$$\ddot{\theta} + \frac{mgl}{I}\sin\theta = 0 \qquad (8.4.1)$$

Equation 8.4.1 is identical in form to the equation of motion of a simple pendulum. For small oscillations, as in the case of the simple pendulum, we can replace $\sin\theta$ by θ:

$$\ddot{\theta} + \frac{mgl}{I}\theta = 0 \qquad (8.4.2)$$

The solution, as we know from Chapter 3, can be written

$$\theta = \theta_0 \cos(2\pi f_0 t - \delta) \qquad (8.4.3)$$

where θ_0 is the amplitude and δ is a phase angle. The frequency of oscillation is given by

$$f_0 = \frac{1}{2\pi}\sqrt{\frac{mgl}{I}} \qquad (8.4.4)$$

The period is, therefore, given by

$$T_0 = \frac{1}{f_0} = 2\pi\sqrt{\frac{I}{mgl}} \qquad (8.4.5)$$

(To avoid confusion, we have used the frequency f_0 instead of the angular frequency ω_0 to characterize the oscillation of the pendulum.) We can also express the period in terms of the radius of gyration k, namely,

$$T_0 = 2\pi \sqrt{\frac{k^2}{gl}} \qquad (8.4.6)$$

Thus, the period is the same as that of a simple pendulum of length k^2/l.

Consider as an example a thin uniform rod of length a swinging as a physical pendulum about one end: $k^2 = a^2/3$, $l = a/2$. The period is then

$$T_0 = 2\pi \sqrt{\frac{a^2/3}{ga/2}} = 2\pi \sqrt{\frac{2a}{3g}} \qquad (8.4.7)$$

which is the same as that of a simple pendulum of length $\frac{2}{3}a$.

Center of Oscillation

By use of the parallel-axis theorem, we can express the radius of gyration k in terms of the radius of gyration about the center of mass k_{cm}, as follows:

$$I = I_{cm} + ml^2 \qquad (8.4.8)$$

or

$$mk^2 = mk_{cm}^2 + ml^2 \qquad (8.4.9a)$$

Canceling the m's, we get

$$k^2 = k_{cm}^2 + l^2 \qquad (8.4.9b)$$

Equation 8.4.6 can, therefore, be written as

$$T_0 = 2\pi \sqrt{\frac{k_{cm}^2 + l^2}{gl}} \qquad (8.4.10)$$

Suppose that the axis of rotation of a physical pendulum is shifted to a different position O' at a distance l' from the center of mass, as shown in Figure 8.4.1. The period of oscillation T_0' about this new axis is given by

$$T_0' = 2\pi \sqrt{\frac{k_{cm}^2 + l'^2}{gl'}} \qquad (8.4.11)$$

The periods of oscillation about O and about O' are equal, provided

$$\frac{k_{cm}^2 + l^2}{l} = \frac{k_{cm}^2 + l'^2}{l'} \qquad (8.4.12)$$

Equation 8.4.12 readily reduces to

$$ll' = k_{cm}^2 \qquad (8.4.13)$$

The point O', related to O by Equation 8.4.13, is called the *center of oscillation* for the point O. O is also the center of oscillation for O'. Thus, for a rod of length a swinging about one end, we have $k_{cm}^2 = a^2/12$ and $l = a/2$. Hence, from Equation 8.4.13, $l' = a/6$, and so the rod has the same period when swinging about an axis located a distance $a/6$ from the center as it does for an axis passing through one end.

The "Upside-Down Pendulum": Elliptic Integrals

When the amplitude of oscillation of a pendulum is so large that the approximation $\sin \theta = \theta$ is not valid, the formula for the period (Equation 8.4.5) is not accurate. In Example 3.7.1 we obtained an improved formula for the period of a simple pendulum by using a method of successive approximations. That result also applies to the physical pendulum with l replaced by I/ml, but it is still an approximation and is completely erroneous when the amplitude approaches 180° (vertical position) (Figure 8.4.2).

To find the period for large amplitude, we start with the energy equation for the physical pendulum

$$\tfrac{1}{2}I\dot{\theta}^2 + mgh = E \tag{8.4.14}$$

where h is the vertical distance of the center of mass from the equilibrium position, that is, $h = l(1 - \cos \theta)$. Let θ_0 denote the amplitude of the pendulum's oscillation. Then $\dot{\theta} = 0$ when $\theta = \theta_0$, so that $E = mgl(1 - \cos \theta_0)$. The energy equation can then be written

$$\tfrac{1}{2}I\dot{\theta}^2 + mgl(1 - \cos \theta) = mgl(1 - \cos \theta_0) \tag{8.4.15}$$

Solving for $\dot{\theta}$ gives

$$\frac{d\theta}{dt} = \pm \left[\frac{2mgl}{I}(\cos \theta - \cos \theta_0) \right]^{1/2} \tag{8.4.16}$$

Thus, by taking the positive root, we can write

$$t = \sqrt{\frac{I}{2mgl}} \int_0^{\theta} \frac{d\theta}{(\cos \theta - \cos \theta_0)^{1/2}} \tag{8.4.17}$$

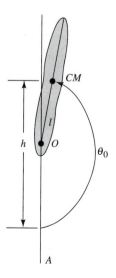

Figure 8.4.2   The upside-down pendulum.

from which we can, in principle, find t as a function of θ. Also, we note that θ increases from 0 to θ_0 in just one quarter of a complete cycle. The period T can, therefore, be expressed as

$$T = 4\sqrt{\frac{I}{2mgl}} \int_0^{\theta_0} \frac{d\theta}{(\cos\theta - \cos\theta_0)^{1/2}} \tag{8.4.18}$$

Unfortunately, the integrals in Equations 8.4.17 and 8.4.18 cannot be evaluated in terms of elementary functions. They can, however, be expressed in terms of special functions known as *elliptic integrals*. For this purpose it is convenient to introduce a new variable of integration ϕ, which is defined as follows:

$$\sin\phi = \frac{\sin(\theta/2)}{\sin(\theta_0/2)} = \frac{1}{k}\sin\left(\frac{\theta}{2}\right) \tag{8.4.19}$$

where[2]

$$k = \sin\left(\frac{\theta_0}{2}\right)$$

Thus, when $\theta = \theta_0$, we have $\sin\phi = 1$ and so $\phi = \pi/2$. The result of making these substitutions in Equations 8.4.17 and 8.4.18 yields

$$t = \sqrt{\frac{I}{mgl}} \int_0^{\phi} \frac{d\phi}{(1 - k^2\sin^2\phi)^{1/2}} \tag{8.4.20a}$$

$$T = 4\sqrt{\frac{I}{mgl}} \int_0^{\pi/2} \frac{d\phi}{(1 - k^2\sin^2\phi)^{1/2}} \tag{8.4.20b}$$

The steps are left as an exercise and involve use of the identity $\cos\theta = 1 - 2\sin^2\left(\frac{\theta}{2}\right)$.

Tabulated values of the integrals in the preceding expressions can be found in various handbooks and mathematical tables. The first integral

$$\int_0^{\phi} \frac{d\phi}{(1 - k^2\sin^2\phi)^{1/2}} = F(k, \phi) \tag{8.4.21}$$

is called the *incomplete elliptic integral of the first kind*. In our problem, given a value of the amplitude θ_0, we can find the relationship between θ and t through a series of steps involving the definitions of k and ϕ. We are more interested in finding the period of the pendulum, which involves the second integral

$$\int_0^{\pi/2} \frac{d\phi}{(1 - k^2\sin^2\phi)^{1/2}} = F\left(k, \frac{\pi}{2}\right) \tag{8.4.22}$$

---

[2] Note that k defined here is a parameter that characterizes elliptic integrals. It is not the k defined previously as the radius of gyration.

TABLE 8.4.1	Selected Values of the Complete Elliptic Integral and Corresponding Period of Oscillation of a Physical Pendulum[1]		
Amplitude, θ_0	$k = \sin\left(\dfrac{\theta_0}{2}\right)$	$F\left(k, \dfrac{\pi}{2}\right)$	Period, T
0°	0	1.5708 = π/2	T_0
10°	0.0872	1.5738	1.0019 T_0
45°	0.3827	1.6336	1.0400 T_0
90°	0.7071	1.8541	1.1804 T_0
135°	0.9234	2.4003	1.5281 T_0
178°	0.99985	5.4349	3.5236 T_0
179°	0.99996	5.2660	4.6002 T_0
180°	1	∞	∞

[1] For more extensive tables and other information on elliptic integrals, consult any treatise on elliptic functions, such as (1) H. B. Dwight, *Tables of Integrals and Other Mathematical Data,* The Macmillan Co., New York, 1961; and (2) M. Abramowitz and A. Stegun, *Handbook of Mathematical Functions,* Dover Publishing, New York, 1972.

known as the *complete elliptic integral of the first kind.* (It is also variously listed as $K(k)$ or $F(k)$ in many tables.) In terms of it, the period is

$$T = 4\sqrt{\frac{I}{mgl}}\, F\left(k, \frac{\pi}{2}\right) \tag{8.4.23}$$

Table 8.4.1 lists selected values of $F(k, \pi/2)$. Also listed is the period T as a factor multiplied by the period for zero amplitude: $T_0 = 2\pi(I/mgl)^{1/2}$.

Table 8.4.1 shows the trend as the amplitude approaches 180° at which value the elliptic integral diverges and the period becomes infinitely large. This means that, *theoretically,* a physical pendulum, such as a rigid rod, if placed exactly in the vertical position with absolutely zero initial angular velocity, would remain in that same unstable position indefinitely.

EXAMPLE 8.4.1

A physical pendulum, as shown in Figure 8.4.1, is hanging vertically at rest. It is struck a sudden blow such that its total energy after the blow is $E = 2mgl$, where m is the mass of the pendulum and l is the distance of its center of mass to the pivot point. (a) Calculate the angle of displacement θ away from the vertical as a function of time. (b) Does the pendulum reach the "upside-down" configuration, $\theta = \pi$? If so, use your result from part (a) to calculate how long it takes.

Solution:

(a) We begin by writing down the total energy of the pendulum as in Equation 8.4.15

$$\tfrac{1}{2}I\dot{\theta}^2 + mgl(1 - \cos\theta) = 2mgl$$

Solving for $\dot{\theta}^2$

$$\dot{\theta}^2 = \frac{2mgl}{I}(1 + \cos\theta) = \frac{4mgl}{I}\cos^2\frac{\theta}{2}$$

We introduce the following substitution, $y = \sin\theta/2$, to eliminate integrals involving trigonometric functions and obtain a moderately simple analytic solution.

As θ varies from 0 to π, y varies from 0 to 1. We now calculate $\dot{y}$

$$\dot{y} = \frac{1}{2}\left(\cos\frac{\theta}{2}\right)\dot{\theta} = \frac{1}{2}(1 - y^2)^{1/2}\dot{\theta}$$

where we have used the substitution $\cos\theta/2 = (1 - y^2)^{1/2}$

We now solve for $\dot{\theta}$ in terms of y and $\dot{y}$

$$\dot{\theta} = \frac{2\dot{y}}{(1 - y^2)^{1/2}} = 2\left(\frac{mgl}{I}\right)^{1/2}(1 - y^2)^{1/2}$$

We can now find a first-order differential equation describing the motion in terms of y

$$\dot{y} = \left(\frac{mgl}{I}\right)^{1/2}(1 - y^2)$$

The solution is

$$y = \tanh\left(\frac{mgl}{I}\right)^{1/2}t$$

(b) As $t \to \infty$, $y \to 1$, and $\theta \to \pi$ and the pendulum goes "upside-down"—eventually. Compare this result with the last line in Table 8.4.1.

8.5 | The Angular Momentum of a Rigid Body in Laminar Motion

Laminar motion takes place when all the particles that make up a rigid body move parallel to some fixed plane. In general, the rigid body undergoes both translational and rotational acceleration. The rotation takes place about an axis whose direction, but not necessarily its location, remains fixed in space. The rotation of a rigid body about a fixed axis is a special case of laminar motion, such as the physical pendulum discussed in the previous section. A cylinder rolling down an inclined plane is another example. We discuss motion of each of these types in the sections that follow, but as a prelude to these analyses, we first develop a theorem about the angular momentum of a rigid body in laminar motion.

We showed in Section 7.2 that the rate of change of the angular momentum of any system of particles is equal to the net applied torque

$$\frac{d\mathbf{L}}{dt} = \mathbf{N} \tag{8.5.1}$$

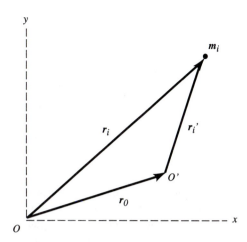

Figure 8.5.1 Vector position of a particle in a rigid body in laminar motion.

or

$$\frac{d}{dt}\sum_i \mathbf{r}_i \times m_i \mathbf{v}_i = \sum_i \mathbf{r}_i \times \mathbf{F}_i \qquad (8.5.2)$$

where all quantities are referred to an inertial coordinate system.

What happens, however, if we choose to describe the rotation of a rigid body (which is a system of particles whose relative positions are fixed) about an axis that might also be accelerating, such as that which takes place when a ball rolls down an inclined plane? To take into account such a possibility, we again consider a system of particles, as in Section 8.2, that is rotating about an axis whose direction is fixed in space. However, here we allow for the possibility that the axis might be accelerating. We begin by referring the position of a particle, m_i, to the origin, O, of an inertial frame of reference (see Figure 8.5.1). Let the point O' represent the origin of the axis in question, about which we wish to refer the rotation of the system of particles. The vectors, $\mathbf{r}_i$ and $\mathbf{r}'_i$, denote the position of the ith particle relative to the points O and O', respectively. We now calculate the total torque $\mathbf{N}'$ about the axis O'

$$\mathbf{N}' = \sum_i \mathbf{r}'_i \times \mathbf{F}_i \qquad (8.5.3)$$

From Figure 8.5.1, we see that

$$\mathbf{r}_i = \mathbf{r}_0 + \mathbf{r}'_i \qquad (8.5.4)$$

and

$$\mathbf{v}_i = \mathbf{v}_0 + \mathbf{v}'_i \qquad (8.5.5)$$

In the inertial frame of reference we have

$$\mathbf{F}_i = \frac{d}{dt}(m\mathbf{v}_i) \qquad (8.5.6)$$

Thus, Equation 8.5.4 becomes

$$\mathbf{N}' = \sum_i \mathbf{r}'_i \times \mathbf{F}_i = \sum_i \mathbf{r}'_i \times \frac{d}{dt} m_i(\mathbf{v}_0 + \mathbf{v}'_i) \tag{8.5.7a}$$

$$= -\dot{\mathbf{v}}_0 \times \sum_i m_i \mathbf{r}'_i + \sum_i \mathbf{r}'_i \times \frac{d}{dt} m_i \mathbf{v}'_i \tag{8.5.7b}$$

$$= -\dot{\mathbf{v}}_0 \times \sum_i m_i \mathbf{r}'_i + \frac{d}{dt} \sum_i \mathbf{r}'_i \times m_i \mathbf{v}'_i \tag{8.5.7c}$$

The step from Equation 8.5.7a to 8.5.7b follows because $\dot{\mathbf{v}}_0$ is not being summed and, therefore, may be extracted from the summation with impunity. The minus sign emerges because of the reversal of the order of the cross product. Extraction of the time derivative from inside the summation in Equation 8.5.7b to its position outside the summation in Equation 8.5.7c is permissible because it then generates a term, $\Sigma_i \mathbf{v}'_i \times m_i \mathbf{v}'_i$, that is the cross product of a vector with itself, which is zero.

The last term on the right in Equation 8.5.7c is the rate of change of the angular momentum, $\mathbf{L}'$, about the O' axis. Thus, we may rewrite this equation as

$$\mathbf{N}' = -\ddot{\mathbf{r}}_0 \times \sum_i m_i \mathbf{r}'_i + \frac{d}{dt} \mathbf{L}' \tag{8.5.8}$$

in which we have replaced $\dot{\mathbf{v}}_0$ with $\ddot{\mathbf{r}}_0$.

The equation of torque (8.5.1), thus, cannot be applied directly in its standard form to a system rotating about an axis that is undergoing acceleration. The correct equation (8.5.8) differs from Equation 8.5.1 by the presence of the extra term on the left.

However, this added term vanishes when any of three possible conditions are satisfied, as schematized in Figure 8.5.2a, b, and c:

1. The acceleration, $\ddot{\mathbf{r}}_0$, of the axis of rotation, O', vanishes (Figure 8.5.2a).

2. The point, O', is the center of mass of the system of particles that make up the rigid body. Under this condition, the term, $\Sigma_i m_i \mathbf{r}'_i = 0$ by definition (Figure 8.5.2b).

3. The O' axis passes through the point of contact between the cylinder and the plane. The vector, represented by the sum, $\Sigma_i m_i \mathbf{r}'_i$, passes through the center of mass. We can see this by noting that $\Sigma_i m_i \mathbf{r}'_i = M\mathbf{r}'_{cm}$ where $M = \Sigma_i m_i$ is the total mass and $\mathbf{r}'_{cm}$ is the vector position of the center of mass relative to O'. Therefore, if the vector $\ddot{\mathbf{r}}_0$ also passes through the center of mass, then their cross product will vanish (Figure 8.5.2c)

We will see in the next section that this last condition proves useful when solving problems involving rigid bodies that are rolling, but not sliding!

Condition 2 above should be emphasized. *The equation of torque for a rigid body undergoing laminar motion can always be expressed in the form given by Equation 8.5.1, if we take torques and calculate angular momentum about an axis that passes through the center of mass.* We write the equation here using appropriate notation to emphasize that it must be applied by summing torques about an axis that passes through the center of

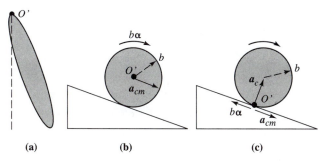

Figure 8.5.2   (a) A physical pendulum swinging about a fixed axis O'. The acceleration of the axis $\ddot{\mathbf{r}}_0$ is zero. (b) A cylinder rolling down an inclined plane. An axis O' through its center of mass is accelerating, but Equation 8.5.9 may be used to describe its rotational motion. (c) The same cylinder as in (b) but the axis O' through the point of contact between the cylinder and the plane is accelerating, even though it is instantaneously at rest (no slipping). (The net tangential acceleration of the axis is zero because $\mathbf{a}_{cm} = b\alpha$, where α is the angular acceleration of the cylinder and b is its radius. The net acceleration of the axis is, therefore, its centripetal acceleration, $\mathbf{a}_C$, directed toward the center of mass.)

mass of the rigid body

$$\mathbf{N}_{cm} = \frac{d}{dt}\mathbf{L}_{cm} = I_{cm}\dot{\boldsymbol{\omega}} \tag{8.5.9}$$

If in doubt, use this equation!

8.6 │ Examples of the Laminar Motion of a Rigid Body

To sum up, if a rigid body undergoes a laminar motion, the motion is most often specified as a translation of its center of mass and a rotation about an axis that passes through the center of mass and whose direction is fixed in space. Sometimes though, some other axis is a more appropriate choice. Such situations are usually obvious, as in the case of the physical pendulum, whose motion is a rotation about the fixed axis that passes through its pivot point.

The fundamental equation that governs the translation of a rigid body is

$$\mathbf{F} = m\ddot{\mathbf{r}}_{cm} = m\dot{\mathbf{v}}_{cm} = m\mathbf{a}_{cm} \tag{8.6.1}$$

where $\mathbf{F}$ is the vector sum of all the external forces acting on the body, m is its mass, and $\mathbf{a}_{cm}$ is the acceleration of its center of mass.

The fundamental equation that governs the rotation of the body about an axis O' that satisfies one of the conditions 1 to 3 given in Section 8.5 is

$$\mathbf{N}_{O'} = \frac{d}{dt}\mathbf{L}_{O'} = I_{O'}\,\dot{\boldsymbol{\omega}} \tag{8.6.2}$$

If an axis of rotation, other than that which passes through the center of mass, is chosen to describe the rotational motion, care should be taken in considering whether condition 1 or 3 is satisfied. If not, then the more general form of the equation of torque given by Equation 8.5.8 must be used instead.

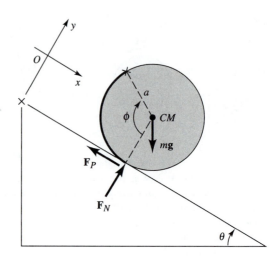

Figure 8.6.1   Body rolling down an inclined plane.

Body Rolling Down an Inclined Plane

As an illustration of laminar motion, we study the motion of a round object (cylinder, ball, and so on) rolling down an inclined plane. As shown in Figure 8.6.1, three forces are acting on the body. These are (1) the downward force of gravity, (2) the normal reaction of the plane $\mathbf{F}_N$, and (3) the frictional force parallel to the plane $\mathbf{F}_P$. Choosing axes as shown, the component equations of the translation of the center of mass are

$$m\ddot{x}_{cm} = mg \, \sin\theta - F_P \tag{8.6.3}$$
$$m\ddot{y}_{cm} = -mg \, \cos\theta + F_N \tag{8.6.4}$$

where θ is the inclination of the plane to the horizontal. Because the body remains in contact with the plane, we have

$$y_{cm} = \text{constant} \tag{8.6.5a}$$

Hence,

$$\ddot{y}_{cm} = 0 \tag{8.6.5b}$$

Therefore, from Equation 8.6.4,

$$F_N = mg \, \cos\theta \tag{8.6.6}$$

The only force that exerts a moment about the center of mass is the frictional force $\mathbf{F}_P$. The magnitude of this moment is $F_P a$, where a is the radius of the body. Hence, the rotational equation (Equation 8.6.2) becomes

$$I_{cm}\dot{\omega} = F_P a \tag{8.6.7}$$

To discuss the problem further, we need to make some assumptions regarding the contact between the plane and the body. We solve the equations of motion for two cases.

Motion with No Slipping

If the contact is very rough so that no slipping can occur, that is, if $F_P \le \mu_s F_N$, where μ_s is the coefficient of *static* friction, we have the following relations:

$$\dot{x}_{cm} = a\dot{\phi} = a\omega \tag{8.6.8a}$$

$$\ddot{x}_{cm} = a\ddot{\phi} = a\dot{\omega} \tag{8.6.8b}$$

where ϕ is the angle of rotation. Equation 8.6.7 can then be written

$$\frac{I_{cm}}{a^2}\ddot{x}_{cm} = F_P \tag{8.6.9}$$

Substituting this value for F_P into Equation 8.6.3 yields

$$m\ddot{x}_{cm} = mg\ \sin\theta - \frac{I_{cm}}{a^2}\ddot{x}_{cm} \tag{8.6.10}$$

Solving for $\dot{x}_{cm}$, we find

$$\ddot{x}_{cm} = \frac{mg\ \sin\theta}{m + (I_{cm}/a^2)} = \frac{g\ \sin\theta}{1 + \left(k_{cm}^2/a^2\right)} \tag{8.6.11}$$

where k_{cm} is the radius of gyration about the center of mass. The body, therefore, rolls down the plane with constant linear acceleration and with constant angular acceleration by virtue of Equations 8.6.8a and b.

For example, the acceleration of a uniform cylinder $(k_{cm}^2 = a^2/2)$ is

$$\ddot{x}_{cm} = \frac{g\ \sin\theta}{1 + \left(\frac{1}{2}\right)} = \frac{2}{3}g\ \sin\theta \tag{8.6.12}$$

whereas that of a uniform sphere $(k_{cm}^2 = 2a^2/5)$ is

$$\ddot{x}_{cm} = \frac{g\ \sin\theta}{1 + \left(\frac{2}{5}\right)} = \frac{5}{7}g\ \sin\theta \tag{8.6.13}$$

EXAMPLE 8.6.1

Calculate the center of mass acceleration of the cylinder rolling down the inclined plane in Figure 8.6.1 for the case of no slipping. Choose an axis O' that passes through the point of contact as in Figure 8.5.2c.

Solution:

As previously explained, the choice of this axis satisfies condition 3 given in Section 8.5 and we can use Equation 8.6.2 directly. The torque acting about O' is

$$N_{O'} = mg\ a\ \sin\theta$$

The moment of inertia of the cylinder about this point (see Equation 8.3.22) is

$$I_{O'} = \tfrac{3}{2}ma^2$$

Because there is no slipping, the relationship between the angular velocity of the cylinder about the axis O' and the center of mass velocity is

$$\dot{x}_{cm} = a\dot{\phi}$$

(**Note:** this is the same relationship that connects the angular velocity of the cylinder with the tangential velocity of any point on its surface *relative to the center of mass*.)

Therefore, the rotational equation of motion gives

$$mga \, \sin\theta = \tfrac{3}{2}ma^2\left(\frac{\ddot{x}_{cm}}{a}\right)$$

from which it immediately follows that

$$\ddot{x}_{cm} = \tfrac{2}{3}g \, \sin\theta$$

Energy Considerations

The preceding results can also be obtained from energy considerations. In a uniform gravitational field the potential energy V of a rigid body is given by the sum of the potential energies of the individual particles, namely,

$$V = \sum_i (m_i g h_i) = mgh_{cm} \tag{8.6.14}$$

where h_{cm} is the vertical distance of the center of mass from some (arbitrary) reference plane. Now if the forces, other than gravity, acting on the body do no work, then the motion is conservative, and we can write

$$T + V = T + mgh_{cm} = E = \text{constant} \tag{8.6.15}$$

where T is the kinetic energy.

In the case of the body rolling down the inclined plane (see Figure 8.6.1), the kinetic energy of translation is $\tfrac{1}{2}m\dot{x}_{cm}^2$ and that of rotation is $\tfrac{1}{2}I_{cm}\omega^2$, so the energy equation reads

$$\tfrac{1}{2}m\dot{x}_{cm}^2 + \tfrac{1}{2}I_{cm}\omega^2 + mgh_{cm} = E \tag{8.6.16}$$

But $\omega = \dot{x}_{cm}/a$ and $h_{cm} = -x_{cm} \sin\theta$. Hence,

$$\tfrac{1}{2}m\dot{x}_{cm}^2 + \tfrac{1}{2}mk_{cm}^2\frac{\dot{x}_{cm}^2}{a^2} - mgx_{cm} \, \sin\theta = E \tag{8.6.17}$$

In the case of pure rolling motion, the frictional force does not appear in the energy equation because no mechanical energy is converted into heat unless slipping occurs. Thus, the total energy E is constant. Differentiating with respect to t and collecting terms yields

$$m\dot{x}_{cm}\ddot{x}_{cm}\left(1+\frac{k_{cm}^2}{a^2}\right) - mg\dot{x}_{cm} \, \sin\theta = 0 \tag{8.6.18}$$

Canceling the common factor $\dot{x}_{cm}$ (assuming, of course, that $\dot{x}_{cm} \neq 0$) and solving for $\ddot{x}_{cm}$, we find the same result as that obtained previously using forces and moments (Equation 8.6.11).

Occurrence of Slipping

Let us now consider the case in which the contact with the plane is not perfectly rough but has a certain coefficient of *sliding* friction μ_k. If slipping occurs, then the magnitude of the frictional force $\mathbf{F}_P$ is given by

$$F_P = \mu_k F_N = \mu_k mg \cos\theta \qquad (8.6.19)$$

The equation of translation (Equation 8.6.3) then becomes

$$m\ddot{x}_{cm} = mg \sin\theta - \mu_k mg \cos\theta \qquad (8.6.20)$$

and the rotational equation (Equation 8.6.7) is

$$I_{cm}\dot{\omega} = \mu_k mga \cos\theta \qquad (8.6.21)$$

From Equation 8.6.20 we see that again the center of mass undergoes constant acceleration:

$$\ddot{x}_{cm} = g(\sin\theta - \mu_k \cos\theta) \qquad (8.6.22)$$

and, at the same time, the angular acceleration is constant:

$$\dot{\omega} = \frac{\mu_k mga \cos\theta}{I_{cm}} = \frac{\mu_k ga \cos\theta}{k_{cm}^2} \qquad (8.6.23)$$

Let us integrate these two equations with respect to t, assuming that the body starts from rest, that is, at $t = 0$, $\dot{x}_{cm} = 0, \dot{\phi} = 0$. We obtain

$$\dot{x}_{cm} = g(\sin\theta - \mu_k \cos\theta)t \qquad (8.6.24)$$

$$\omega = \dot{\phi} = g\left(\frac{\mu_k a \cos\theta}{k_{cm}^2}\right)t \qquad (8.6.25)$$

Consequently, the linear speed and the angular speed have a constant ratio, and we can write

$$\dot{x}_{cm} = \gamma a\omega \qquad (8.6.26)$$

where

$$\gamma = \frac{\sin\theta - \mu_k \cos\theta}{\mu_k a^2 \cos\theta/k_{cm}^2} = \frac{k_{cm}^2}{a^2}\left(\frac{\tan\theta}{\mu_k} - 1\right) \qquad (8.6.27)$$

Now $a\omega$ cannot be greater than $\dot{x}_{cm}$, so γ cannot be less than unity. The limiting case, that for which we have pure rolling, is given by $\dot{x}_{cm} = a\omega$, that is,

$$\gamma = 1$$

Solving for μ_k in Equation 8.6.27 with $\gamma = 1$, we find that the critical value of the coefficient of friction is given by

$$\mu_{crit} = \frac{\tan\theta}{1 + (a/k_{cm})^2} \tag{8.6.28}$$

(Actually this is the critical value for the coefficient of *static* friction μ_s.) If μ_s is greater than that given in Equation 8.6.28, then the body rolls without slipping.

For example, if a ball is placed on a 45° plane, it will roll without slipping, provided μ_s is greater than tan 45°/(1 + $\frac{5}{2}$) or $\frac{2}{7}$.

EXAMPLE 8.6.2

A small, uniform cylinder of radius R rolls without slipping along the inside of a large, fixed cylinder of radius $r > R$ as shown in Figure 8.6.2. Show that the period of small oscillations of the rolling cylinder is equivalent to that of a simple pendulum whose length is $3(r - R)/2$.

Solution:

A key to an easy solution hinges on the realization that the total energy of the rolling cylinder is a constant of the motion. There is no relative motion between the two surfaces because there is no slipping. In other words, O' and O coincide when the small cylinder is at the equilibrium position and the arc lengths $O'P$ and OP are identical. The force of friction $\mathbf{F}$, therefore, does not remove the energy from the rolling cylinder, nor does the normal force $\mathbf{N}$ do any work. It generates no torque because its line of action always passes through the center of mass, and it does not affect the translational kinetic energy

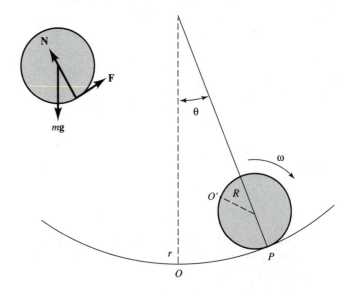

Figure 8.6.2 Small cylinder rolling without slipping on the inside of a large, fixed cylinder.

because it is always directed perpendicular to the motion of the center of the mass. The only force that does do work is the conservative force of gravity, $m\mathbf{g}$. Thus, the energy of the cylinder is conserved, and we can solve the problem by setting its time derivative equal to zero. The total energy of the cylinder is

$$E = T + V = \frac{1}{2} I_{cm} \omega^2 + \frac{1}{2} m v_{cm}^2 + mgh$$

where h is the height of the cylinder above that at its equilibrium position, v_{cm} is the speed of its center of mass, and I_{cm} is the moment of inertia about its center of mass (see Figure 8.6.2).

From the figure, we see that for small oscillations

$$h = (r - R)(1 - \cos\theta) \approx \frac{1}{2}(r - R)\theta^2$$

and because the cylinder rolls without slipping, we have

$$\omega = \frac{v_{cm}}{R} = \frac{(r - R)}{R}\dot{\theta}$$

Inserting these relations for h and ω into the energy equation gives

$$E = \frac{I_{cm}}{2R^2}(r - R)^2\dot{\theta}^2 + \frac{m}{2}(r - R)^2\dot{\theta}^2 + \frac{m}{2}g(r - R)\theta^2$$

On taking the derivative of the preceding equation and setting the result equal to zero, we obtain

$$\dot{E} = \frac{I_{cm}}{R^2}(r - R)^2\ddot{\theta}\dot{\theta} + m(r - R)^2\ddot{\theta}\dot{\theta} + mg(r - R)\theta\dot{\theta} = 0$$

and cancelling out common terms yields

$$\left(\frac{I_{cm}}{R^2} + m\right)(r - R)\ddot{\theta} + mg\theta = 0$$

The moment of inertia of the cylinder about its center of mass is $I_{cm} = mR^2/2$, and on substituting it into the preceding equation yields the equation of motion of the cylinder for small excursions about equilibrium

$$\ddot{\theta} + \frac{g}{\left(\frac{3}{2}\right)(r - R)}\theta = 0$$

This equation of motion is the same as that of a simple pendulum of length $3(r - R)/2$. Thus, their periods are identical.

(The student might wish to solve this problem using the method of forces and torques. The relevant forces acting on the rolling cylinder are shown in the insert in Figure 8.6.2.)

8.7 | Impulse and Collisions Involving Rigid Bodies

In the previous chapter we considered the case of an impulsive force acting on a particle. In this section we extend the notion of impulsive force to the case of laminar motion of a rigid body. First, we know that the translation of the body, assuming constant mass, is governed by the general equation $\mathbf{F} = m\,d\mathbf{v}_{cm}/dt$, so that if $\mathbf{F}$ is an impulsive type of force, the change of linear momentum of the body is given by

$$\int \mathbf{F}\,dt = \mathbf{P} = m\,\Delta\mathbf{v}_{cm} \tag{8.7.1}$$

Thus, the result of an impulse $\mathbf{P}$ is to produce a sudden change in the velocity of the center of mass by an amount

$$\Delta\mathbf{v}_{cm} = \frac{\mathbf{P}}{m} \tag{8.7.2}$$

Second, the rotational part of the motion of the body obeys the equation $N = \dot{L} = I\,d\omega/dt$, so the change in angular momentum is

$$\int N\,dt = I\,\Delta\omega \tag{8.7.3}$$

The integral $\int N\,dt$ is called the *rotational impulse*. Now if the primary impulse $\mathbf{P}$ is applied to the body in such a way that its line of action is a distance l from the reference axis about which the angular momentum is calculated, then $N = Fl$, and we have

$$\int N\,dt = Pl \tag{8.7.4}$$

Consequently, the change in angular velocity produced by an impulse $\mathbf{P}$ acting on a rigid body in laminar motion is given by

$$\Delta\omega = \frac{Pl}{I} \tag{8.7.5}$$

For the general case of free laminar motion, the reference axis must be taken through the center of mass, and the moment of inertia $I = I_{cm}$. On the other hand, if the body is constrained to rotate about a fixed axis, then the rotational equation alone suffices to determine the motion, and I is the moment of inertia about the fixed axis.

In collisions involving rigid bodies, the forces and, therefore, the impulses that the bodies exert on one another during the collision are always equal and opposite. Thus, the principles of conservation of linear and angular momentum apply.

Center of Percussion: The "Baseball Bat Theorem"

To illustrate the concept of center of percussion, let us discuss the collision of a ball of mass m, treated as a particle, with a rigid body (bat) of mass M. For simplicity we assume that the body is initially at rest on a smooth horizontal surface and is free to move in laminar-type motion. Let $\mathbf{P}$ denote the impulse delivered to the body by the ball. Then the equations for translation are

$$\mathbf{P} = M\mathbf{v}_{cm} \tag{8.7.6}$$

$$-\mathbf{P} = m\mathbf{v}_1 - m\mathbf{v}_0 \tag{8.7.7}$$

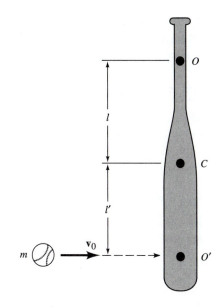

Figure 8.7.1   Baseball colliding with a bat.

where v_0 and v_1 are, respectively, the initial and final velocities of the ball and v_{cm} is the velocity of the mass center of the body after the impact. The preceding two equations imply conservation of linear momentum.

Because the body is initially at rest, the rotation about the center of mass, as a result of the impact, is given by

$$\omega = \frac{Pl'}{I_{cm}} \tag{8.7.8}$$

in which l' is the distance $O'C$ from the center of mass C to the line of action of $\boldsymbol{P}$, as shown in Figure 8.7.1. Let us now consider a point O located a distance l from the center of mass such that the line CO is the extension of $O'C$, as shown. The (scalar) velocity of O is obtained by combining the translational and rotational parts, namely,

$$v_O = v_{cm} - \omega l = \frac{P}{M} - \frac{Pl'}{I_{cm}}l = P\left(\frac{1}{M} - \frac{ll'}{I_{cm}}\right) \tag{8.7.9}$$

In particular, the velocity of O will be zero if the quantity in parentheses vanishes, that is, if

$$ll' = \frac{I_{cm}}{M} = k_{cm}^2 \tag{8.7.10}$$

where k_{cm} is the radius of gyration of the body about its center of mass. In this case the point O is the instantaneous center of rotation of the body just after impact. O' is called the *center of percussion* about O. The two points are related in the same way as the centers of oscillation, defined previously in our analysis of the physical pendulum (Equation 8.4.13).

Anyone who has played baseball knows that if the ball hits the bat in just the right spot there is no "sting" on impact. This "right spot" is just the center of percussion about the point at which the bat is held.

EXAMPLE 8.7.1

Shown in Figure 8.7.2 is a thin rod of length b and mass m suspended from an endpoint on a frictionless pivot. The other end of the rod is struck a blow that delivers a horizontal impulse P' to the rod. Calculate the horizontal impulse P delivered to the pivot by the suspended rod.

Solution:

First, we calculate the velocity of the center of mass after the blow by noting that the net horizontal impulse delivered to the rod is equal to its change in momentum.

$$P' - P = mv_{cm}$$

Now we consider the resulting rotation of the rod about the pivot point (the choice of this axis satisfies condition 1 in Section 8.5). The moment of inertia of the rod about an axis passing through that point is given by Equation 8.3.3

$$I = \tfrac{1}{3}mb^2$$

Now we calculate the angular velocity of the rod about the pivot using Equation 8.7.8

$$P'b = I\omega$$

But the velocity of the center of mass and the angular velocity are related according to

$$v_{cm} = \frac{b}{2}\omega$$

Thus, we can write

$$P' - P = m\frac{b}{2}\omega = m\frac{b}{2}\left[\frac{P'b}{I}\right] = \tfrac{3}{2}P'$$

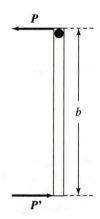

Figure 8.7.2   Thin rod suspended from frictionless pivot.

Therefore,

$$P = -\frac{1}{2}P'$$

The impulse delivered by the pivot to the rod is in the same direction as the impulse delivered by the horizontal blow, to the right in Figure 8.7.2. The impulse delivered by the rod to the pivot is in the opposite direction, to the left in the figure.

Problems

8.1 Find the center of mass of each of the following:
  (a) A thin wire bent into the form of a three-sided, block-shaped "⊔" with each segment of equal length b
  (b) A quadrant of a uniform circular lamina of radius b
  (c) The area bounded by parabola $y = x^2/b$ and the line $y = b$
  (d) The volume bounded by paraboloid of revolution $z = (x^2 + y^2)/b$ and the plane $z = b$
  (e) A solid uniform right circular cone of height b

8.2 The linear density of a thin rod is given by $\rho = cx$, where c is a constant and x is the distance measured from one end. If the rod is of length b, find the center of mass.

8.3 A solid uniform sphere of radius a has a spherical cavity of radius $a/2$ centered at a point $a/2$ from the center of the sphere. Find the center of mass.

8.4 Find the moments of inertia of each of the objects in Problem 8.1 about their symmetry axes.

8.5 Find the moment of inertia of the sphere in Problem 8.3 about an axis passing through the center of the sphere and the center of the cavity.

8.6 Show that the moment of inertia of a solid uniform octant of a sphere of radius a is $(\frac{2}{5})ma^2$ about an axis along one of the straight edges. (*Note:* This is the same formula as that for a solid sphere of the same radius.)

8.7 Show that the moments of inertia of a solid uniform rectangular parallelepiped, elliptic cylinder, and ellipsoid are, respectively, $(m/3)(a^2 + b^2)$, $(m/4)(a^2 + b^2)$, and $(m/5)(a^2 + b^2)$, where m is the mass, and $2a$ and $2b$ are the principal diameters of the solid at right angles to the axis of rotation, the axis being through the center in each case.

8.8 Show that the period of a physical pendulum is equal to $2\pi(d/g)^{1/2}$, where d is the distance between the point of suspension O and the center of oscillation O'.

8.9 (a) An idealized simple pendulum consists of a particle of mass M suspended by a thin massless rod of length a. Assume that an actual simple pendulum consists of a thin rod of mass m attached to a spherical bob of mass $M - m$. If the radius of the spherical bob is equal to b, and the length of the thin rod is equal to $a - b$, calculate the ratio of the period of the actual simple pendulum to the idealized simple one.
  (b) Calculate a value for this ratio if $m = 10$ g, $M = 1$ kg, $a = 1.27$ m, and $b = 5$ cm.

8.10 The period of a physical pendulum is 2 s. (Such a pendulum is called a "seconds" pendulum.) The mass of the pendulum is M, and its center of mass is 1 m below the axis of oscillation. A particle of mass m is attached to the bottom of the pendulum, 1.3 m below the axis, in line with the center of gravity. It is then found that the pendulum "loses" time at the rate of 20 s/day. Find the ratio of m to M.

8.11 A circular hoop of radius a swings as a physical pendulum about a point on the circumference. Find the period of oscillation for small amplitude if the axis of rotation is

(a) Normal to the plane of the hoop

(b) In the plane of the hoop

8.12 A uniform solid ball has a few turns of light string wound around it. If the end of the string is held steady and the ball is allowed to fall under gravity, what is the acceleration of the center of the ball? (Assume the string remains vertical.)

8.13 Two people are holding the ends of a uniform plank of length l and mass m. Show that if one person suddenly lets go, the load supported by the other person suddenly drops from $mg/2$ to $mg/4$. Show also that the initial downward acceleration of the free end is $\frac{3}{2}g$.

8.14 A uniform solid ball contains a hollow spherical cavity at its center, the radius of the cavity being $\frac{1}{2}$ the radius of the ball. Show that the acceleration of the ball rolling down a rough inclined plane is just $\frac{98}{101}$ of that of a uniform solid ball with no cavity. (*Note:* This suggests a method for nondestructive testing.)

8.15 Two weights of mass m_1 and m_2 are tied to the ends of a light inextensible cord. The cord passes over a rough pulley of radius a and moment of inertia I. Find the accelerations of the weights, assuming $m_1 > m_2$ and ignoring friction in the axle of the pulley.

8.16 A uniform right-circular cylinder of radius a is balanced on the top of a perfectly rough fixed cylinder of radius $b(b > a)$, the axes of the two cylinders being parallel. If the balance is slightly disturbed, show that the rolling cylinder leaves the fixed one when the line of centers makes an angle with the vertical of $\cos^{-1}\left(\frac{4}{7}\right)$.

8.17 A uniform ladder leans against a smooth vertical wall. If the floor is also smooth, and the initial angle between the floor and the ladder is θ_0, show that the ladder, in sliding down, will lose contact with the wall when the angle between the floor and the ladder is $\sin^{-1}\left(\frac{2}{3}\sin\theta_0\right)$.

8.18 At Cape Canaveral a Saturn V rocket stands in a vertical position ready for launch. Unfortunately, before firing, a slight disturbance causes the rocket to fall over. Find the horizontal and vertical components of the reaction on the launch pad as functions of the angle θ between the rocket and the vertical at any instant. Show from this that the rocket will tend to slide backward for $\theta < \cos^{-1}\left(\frac{2}{3}\right)$ and forward for $\theta > \cos^{-1}\left(\frac{2}{3}\right)$. (Assume the rocket to be a thin uniform rod.)

8.19 A ball is initially projected, without rotation, at a speed v_0 up a rough inclined plane of inclination θ and coefficient of sliding friction μ_k. Find the position of the ball as a function of time, and determine the position of the ball when pure rolling begins. Assume that μ_k is greater than $\frac{2}{7}\tan\theta$.

8.20 A billiard ball of radius a is initially spinning about a horizontal axis with angular speed ω_0 and with zero forward speed. If the coefficient of sliding friction between the ball and the billiard table is μ_k, find the distance the ball travels before slipping ceases to occur.

8.21 Figure P8.21 Illustrates two discs of radii a and b mounted inside a fixed, immovable circular track of radius c, such that $c = a + 2b$. The central disc A is mounted to a drive axle at point O. Disc B is sandwiched between disc A and track C and can roll without slipping when disc A is driven by an externally applied torque through its drive axle. Initially, the system is at rest such that the dashed lines denoting the spatial orientation of discs A and B line up horizontally in the figure. A constant torque K is applied for a time t_0 through the drive axle causing disc A to rotate, such that at time t_0 the dashed line denoting its spatial orientation makes an angle α with the horizontal. Disc B rolls between the track and disc A, and its orientation is denoted by the dashed line making an angle β with the direction toward O. Calculate the final angular speed of the two discs, ω_A and ω_B.

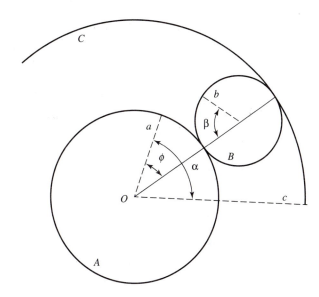

Figure P8.21

8.22 A thin uniform plank of length l lies at rest on a horizontal sheet of ice. If the plank is given a kick at one end in a direction normal to the plank, show that the plank will begin to rotate about a point located a distance $l/6$ from the center.

8.23 Show that the edge (cushion) of a billiard table should be at a height of $\frac{7}{10}$ of the diameter of the billiard ball in order that no reaction occurs between the table surface and the ball when the ball strikes the cushion.

8.24 A ballistic pendulum is made of a long plank of length l and mass m. It is free to swing about one end O and is initially at rest in a vertical position. A bullet of mass m' is fired horizontally into the pendulum at a distance l' below O, the bullet coming to rest in the plank. If the resulting amplitude of oscillation of the pendulum is θ_0, find the speed of the bullet.

8.25 Two uniform rods AB and BC of equal mass m and equal length l are smoothly joined at B. The system is initially at rest on a smooth horizontal surface, the points A, B, and C lying in a straight line. If an impulse $\boldsymbol{P}$ is applied at A at right angles to the rod, find the initial motion of the system. (*Hint*: Isolate the rods.)

Computer Problems

C 8.1 The table shown here displays the density of a 10-solar mass star versus radial distance from its core. The densities are given as $\log_{10} \rho/\rho_c$, where ρ is the density at the distance r and ρ_c is the core density of the star. The distances are given as fractions of the star's radius (r/R_*), where R_* is equal to 4 solar radii.

The mass of the sun is $M_\odot = 1.989 \times 10^{30}$ kg, and its radius is $R_\odot = 6.96 \times 10^5$ km.

(**a**) Using the data in the table, estimate the core density of the star by numerical integration.

(**b**) Find the distance R_3, such that the mass contained within R_3 is equal to $3M_\odot$.

Distance (r/R_*)	$\text{Log}_{10}\,(\rho/\rho_c)$
0.	0.
0.01130	−0.0007676
0.02373	−0.0032979
0.03740	−0.0081105
0.05244	−0.0159332
0.06898	−0.0275835
0.08718	−0.0440001
0.10720	−0.0666168
0.12921	−0.0966376
0.15343	−0.136117
0.18008	−0.187302
0.20938	−0.253082
0.24162	−0.338876
0.27708	−0.461671
0.31609	−0.607536
0.35900	−0.780852
0.40620	−0.949463
0.45812	−1.20746
0.51523	−1.46811
0.57805	−1.77071
0.64715	−2.12543
0.72316	−2.55734
0.80678	−3.11969
0.89876	−3.95562
1.00000	−6.28531

(c) Estimate the moment of inertia of the $3M_\odot$ portion of the star within R_3.

(d) Assuming that the star rotates as a rigid body once every 25 days, estimate the rotational angular momentum of the portion of the star within R_3.

(e) Suppose that this star "explodes" as a supernova and that when this happens, its outer $7M_\odot$ layer is blown off but its inner $3M_\odot$ collapses to form a solid, uniformly dense, spherical ball of radius 10 km. Calculate the density and period of rotation of this new, compact stellar object.

9

Motion of Rigid Bodies in Three Dimensions

"The body can no longer participate in the diurnal motion which actuates our sphere. Indeed, although because of its short length, its axis appears to preserve its original direction relatively to terrestrial objects, the use of a microscope is sufficient to establish an apparent and continuous motion which follows the motion of the celestial sphere exactly. . . . As the original direction of this axis is disposed arbitrarily in all azimuths about the vertical, the observed deviations can be, at will, given all the values contained between that of the total deviation and that of this total deviation as reduced by the sine of the latitude. . . . In one fell swoop, with a deviation in the desired direction, a new proof of the rotation of the Earth is obtained; this with an instrument reduced to small dimensions, easily transportable, and which mirrors the continuous motion of the Earth itself. . . . "

—J. B. L. Foucault, *Comptes rendus de l' Academie Sciences,* Vol 35, 27-Sep-1852

In the motion of a rigid body constrained either to rotate about a fixed axis or to move parallel to a fixed plane, the direction of the axis does not change. In the more general cases of rigid-body motion, which we take up in this chapter, the direction of the axis may vary. Compared with the previous chapter, the analysis here is considerably more involved. In fact, even in the case of a freely rotating body on which no external forces whatever are acting, the motion, as we shall see, is not simple.

9.1 | Rotation of a Rigid Body about an Arbitrary Axis: Moments and Products of Inertia—Angular Momentum and Kinetic Energy

We begin the study of the general motion of a rigid body with some mathematical preliminaries. First, we calculate the moment of inertia about an axis whose direction is arbitrary. The axis passes through a fixed point O, Figure 9.1.1a, taken as the origin of our

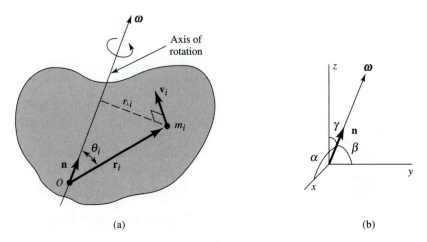

Figure 9.1.1   (a) The velocity vector of a representative particle of a rotating rigid body. (b) α, β, and γ are the angles that the angular velocity vector $\boldsymbol{\omega}$ (or the vector $\mathbf{n}$) makes with the x, y, and z axes, respectively.

coordinate system. We apply the fundamental definition

$$I = \sum_i m_i r_{\perp i}^2 \tag{9.1.1}$$

where $r_{\perp i}$ is the perpendicular distance from the particle of mass m_i to the axis of rotation. The direction of the axis of rotation is defined by the unit vector $\mathbf{n}$. Then

$$r_{\perp i} = |r_i \sin \theta_i| = |\mathbf{r}_i \times \mathbf{n}| \tag{9.1.2}$$

in which θ_i is the angle between $\mathbf{r}_i$ and $\mathbf{n}$, and

$$\mathbf{r}_i = \mathbf{i} x_i + \mathbf{j} y_i + \mathbf{k} z_i \tag{9.1.3}$$

is the position vector of the ith particle. Let the direction cosines of the axis be $\cos \alpha$, $\cos \beta$, and $\cos \gamma$ (see Figure 9.1.1b). Then

$$\mathbf{n} = \mathbf{i} \cos \alpha + \mathbf{j} \cos \beta + \mathbf{k} \cos \gamma \tag{9.1.4}$$

and so

$$
\begin{aligned}
r_{\perp i}^2 &= |\mathbf{r}_i \times \mathbf{n}|^2 \\
&= (y_i \cos \gamma - z_i \cos \beta)^2 + (z_i \cos \alpha - x_i \cos \gamma)^2 + (x_i \cos \beta - y_i \cos \alpha)^2
\end{aligned}
\tag{9.1.5a}
$$

On rearranging terms, we can write

$$
\begin{aligned}
r_{\perp i}^2 &= \left(y_i^2 + z_i^2\right)\cos^2\alpha + \left(z_i^2 + x_i^2\right)\cos^2\beta + \left(x_i^2 + y_i^2\right)\cos^2\gamma \\
&\quad - 2y_i z_i \cos \beta \cos \gamma - 2z_i x_i \cos \gamma \cos \alpha - 2x_i y_i \cos \alpha \cos \beta
\end{aligned}
\tag{9.1.5b}
$$

The moment of inertia about our general axis of rotation is then given by the rather lengthy expression

$$I = \sum_i m_i \left(y_i^2 + z_i^2 \right) \cos^2\alpha + \sum_i m_i \left(z_i^2 + x_i^2 \right) \cos^2\beta$$

$$+ \sum_i m_i \left(x_i^2 + y_i^2 \right) \cos^2\gamma - 2\sum_i m_i y_i z_i \cos\beta \cos\gamma \qquad (9.1.6)$$

$$- 2\sum_i m_i z_i x_i \cos\gamma \cos\alpha - 2\sum_i m_i x_i y_i \cos\alpha \cos\beta$$

As we shall see later, the formula can be simplified. First, we immediately recognize the sums involving the squares of the coordinates as the moments of inertia of the body about the three coordinate axes. We use a slightly modified notation for them as follows:

$$\sum_i m_i \left(y_i^2 + z_i^2 \right) = I_{xx} \qquad \textit{moment of inertia about the x-axis} \qquad (9.1.7a)$$

$$\sum_i m_i \left(z_i^2 + x_i^2 \right) = I_{yy} \qquad \textit{moment of inertia about the y-axis} \qquad (9.1.7b)$$

$$\sum_i m_i \left(x_i^2 + y_i^2 \right) = I_{zz} \qquad \textit{moment of inertia about the z-axis} \qquad (9.1.7c)$$

The sums involving the products of the coordinates are new to us. They are called *products of inertia*. We designate these quantities as follows:

$$-\sum_i m_i x_i y_i = I_{xy} = I_{yx} \qquad \textit{xy product of inertia} \qquad (9.1.8a)$$

$$-\sum_i m_i y_i z_i = I_{yz} = I_{zy} \qquad \textit{yz product of inertia} \qquad (9.1.8b)$$

$$-\sum_i m_i z_i x_i = I_{zx} = I_{xz} \qquad \textit{zx product of inertia} \qquad (9.1.8c)$$

Notice that our definition includes the minus sign. (In some textbooks the minus sign is not included.) Products of inertia have the same physical dimensions as moments of inertia, namely, mass $\times$ (length)^2, and their values are determined by the mass distribution and orientation of the body relative to the coordinate axes. In general, they appear because the axis of rotation points along an arbitrary direction, whereas in the previous chapter the axis of rotation pointed along one of the coordinate axes. For the moments and products of inertia to be constant quantities, it is generally necessary to employ a coordinate system that is fixed to the body and rotates with it.

In actually computing the moments and products of inertia of an extended rigid body, we replace the summations by integrations

$$I_{zz} = \int (x^2 + y^2)\,dm \qquad (9.1.9a)$$

$$I_{xy} = -\int xy\,dm \qquad (9.1.9b)$$

with similar expressions for the other I's. We have already found the moments of inertia for a number of cases in the previous chapter. It is important to remember that the values of the moments and products of inertia depend on the choice of the coordinate system.

Using the above notation, the general expression (Equation 9.1.6) for the moment of inertia about an arbitrary axis becomes

$$I = I_{xx} \cos^2 \alpha + I_{yy} \cos^2 \beta + I_{zz} \cos^2 \gamma + 2I_{yz} \cos \beta \cos \gamma$$
$$+ 2I_{zx} \cos \gamma \cos \alpha + 2I_{xy} \cos \alpha \cos \beta \tag{9.1.10}$$

Although Equation 9.1.10 seems rather cumbersome for obtaining the moment of inertia, it is nevertheless useful for certain applications. Furthermore, the calculation is included here to show how the products of inertia enter into the general problem of rigid-body dynamics.

At this point, we would like to express the moment of inertia of a rigid body in the more compact notational form of tensors or equivalent matrices. Such a representation conveys more than just economy and elegance. Expressions for certain kinematic variables are more easily remembered, and they provide us with more powerful techniques for solving complicated problems in rotational motion of rigid bodies.

If we examine the expressions we've developed for the moment of inertia of a rigid body about an arbitrary axis, they look like they could be written as the components I_{ij} of a symmetric, 3×3 matrix. Let us define the quantity $\mathbf{I}$ to be the *moment of inertia tensor*[1] whose components in matrix form are the values given in Equations 9.1.7a–9.1.8c.

$$\mathbf{I} = \begin{pmatrix} I_{xx} & I_{xy} & I_{xz} \\ I_{yx} & I_{yy} & I_{yz} \\ I_{zx} & I_{zy} & I_{zz} \end{pmatrix} \tag{9.1.11}$$

$I_{ij} = I_{ji}$ and the matrix is symmetric. Furthermore, because the vector $\mathbf{n}$ can be represented in matrix notation by the column vector

$$\mathbf{n} = \begin{pmatrix} \cos \alpha \\ \cos \beta \\ \cos \gamma \end{pmatrix} \tag{9.1.12}$$

we can express the moment of inertia about an axis aligned with the vector $\mathbf{n}$ in matrix notation as

$$I = \tilde{\mathbf{n}}\mathbf{I}\mathbf{n} = (\cos \alpha \quad \cos \beta \quad \cos \gamma) \begin{pmatrix} I_{xx} & I_{xy} & I_{xz} \\ I_{yx} & I_{yy} & I_{yz} \\ I_{zx} & I_{zy} & I_{zz} \end{pmatrix} \begin{pmatrix} \cos \alpha \\ \cos \beta \\ \cos \gamma \end{pmatrix} \tag{9.1.13}$$

where $\tilde{\mathbf{n}}$ means "$\mathbf{n}$ transpose." The transpose of a matrix is obtained by simply "flipping the matrix over about its diagonal," that is, exchanging the row elements for the column elements. In Equation 9.1.13 $\tilde{\mathbf{n}}$ is a row vector whose elements are equal to those of the column vector $\mathbf{n}$. Equation 9.1.13 is identical to Equation 9.1.10. Take note of its notational compactness and elegance, however.

---

[1] The notation used for tensors can be confusing. Tensors will be written throughout this text in **boldface** type—the same type used to denote a vector. The tensors defined in this chapter are tensors of the second rank. Vectors are tensors of the first rank. Hopefully, the distinction between the two will be clear from the context.

Angular Momentum Vector

In Chapter 7 we showed that the time rate of change of the angular momentum of a system of particles is equal to the total moment of all the external forces acting on the system. This rotational equation of motion is expressed by Equation 7.2.7

$$\frac{d\mathbf{L}}{dt} = \mathbf{N} \qquad (9.1.14)$$

The angular momentum of a system of particles about some coordinate origin is given by Equation 7.2.8

$$\mathbf{L} = \sum_i \mathbf{r}_i \times m_i \mathbf{v}_i \qquad (9.1.15)$$

These equations also apply to a rigid body, which is nothing other than a system of particles whose relative positions are fixed. Before we can apply the rotational equation of motion to a rigid body, however, we must be able to calculate its angular momentum about an arbitrary axis.

First, we note that the rotational velocity of any constituent particle of the rigid body is given by the cross product

$$\mathbf{v}_i = \boldsymbol{\omega} \times \mathbf{r}_i \qquad (9.1.16)$$

The total angular momentum of the rigid body is the sum of the angular momenta of each particle about the coordinate origin

$$\mathbf{L} = \sum_i [m_i \mathbf{r}_i \times \mathbf{v}_i] = \sum_i [m_i \mathbf{r}_i \times (\boldsymbol{\omega} \times \mathbf{r}_i)] \qquad (9.1.17)$$

This expression contains a vector triple product, which can be reduced to

$$[\mathbf{r}_i \times (\boldsymbol{\omega} \times \mathbf{r}_i)] = r_i^2 \boldsymbol{\omega} - \mathbf{r}_i (\mathbf{r}_i \cdot \boldsymbol{\omega}) \qquad (9.1.18)$$

Hence, the angular momentum of the rigid body can be written as

$$\mathbf{L} = \sum_i m_i r_i^2 \boldsymbol{\omega} - \sum_i m_i \mathbf{r}_i (\mathbf{r}_i \cdot \boldsymbol{\omega}) \qquad (9.1.19a)$$

We could easily evaluate the x, y, z components of the angular momentum vector using this equation. In keeping with the philosophy initiated earlier, however, we cast this equation into tensor form, defining a tensor in the process

$$\mathbf{L} = \left(\sum_i m_i r_i^2 \right) \boldsymbol{\omega} - \left(\sum_i m_i \mathbf{r}_i \mathbf{r}_i \right) \cdot \boldsymbol{\omega}$$

$$= \left[\left(\sum_i m_i r_i^2 \mathbf{1} \right) - \left(\sum_i m_i \mathbf{r}_i \mathbf{r}_i \right) \right] \cdot \boldsymbol{\omega} \qquad (9.1.19b)$$

where the vector $\boldsymbol{\omega}$ in the first term has been written as

$$\mathbf{1} \cdot \boldsymbol{\omega} = \boldsymbol{\omega} \qquad (9.1.20)$$

which may be viewed as the definition of the *unit tensor*

$$\mathbf{1} = \mathbf{ii} + \mathbf{jj} + \mathbf{kk} \qquad (9.1.21)$$

You can confirm the identity in Equation 9.1.21 by carrying out the dot product operation on the vector $\boldsymbol{\omega}$, that is,

$$
\begin{aligned}
\mathbf{1} \cdot \boldsymbol{\omega} &= (\mathbf{ii} + \mathbf{jj} + \mathbf{kk}) \cdot \boldsymbol{\omega} \\
&= \mathbf{i}(\mathbf{i} \cdot \boldsymbol{\omega}) + \mathbf{j}(\mathbf{j} \cdot \boldsymbol{\omega}) + \mathbf{k}(\mathbf{k} \cdot \boldsymbol{\omega}) \\
&= \mathbf{i}\omega_x + \mathbf{j}\omega_y + \mathbf{k}\omega_z \\
&= \boldsymbol{\omega}
\end{aligned}
\tag{9.1.22}
$$

Both the unit tensor and the second term in brackets in Equation 9.1.19b contain a product of vectors that we have never seen before. This type of vector product (for example, $\mathbf{ab}$) is called a *dyad product*. It is a tensor defined by its dot product operation on another vector $\mathbf{c}$ in the same way that we defined the unit tensor in Equations 9.1.21 and 9.1.22.

$$
(\mathbf{ab}) \cdot \mathbf{c} = \mathbf{a}(\mathbf{b} \cdot \mathbf{c})
\tag{9.1.23a}
$$

This dot product yields a vector. The operation may be expressed in matrix form as

$$
\begin{aligned}
(\mathbf{ab}) \cdot \mathbf{c} &=
\begin{pmatrix}
a_x b_x & a_x b_y & a_x b_z \\
a_y b_x & a_y b_y & a_y b_z \\
a_z b_x & a_z b_y & a_z b_z
\end{pmatrix}
\begin{pmatrix}
c_x \\ c_y \\ c_z
\end{pmatrix} \\[6pt]
&=
\begin{pmatrix}
a_x(b_x c_x + b_y c_y + b_z c_z) \\
a_y(b_x c_x + b_y c_y + b_z c_z) \\
a_z(b_x c_x + b_y c_y + b_z c_z)
\end{pmatrix}
= \mathbf{a}(\mathbf{b} \cdot \mathbf{c})
\end{aligned}
\tag{9.1.23b}
$$

If we were to "dot" Equation 9.1.23a from the left with another vector $\mathbf{d}$, the result would be a simple scalar, that is,

$$
\mathbf{d} \cdot (\mathbf{ab}) \cdot \mathbf{c} = (\mathbf{d} \cdot \mathbf{a})(\mathbf{b} \cdot \mathbf{c})
\tag{9.1.24}
$$

We leave it as an exercise for the reader to obtain this result using matrices.

In three dimensions a tensor has nine components. The components may be generated in the following way:

$$
T_{ij} = \mathbf{i} \cdot \mathbf{T} \cdot \mathbf{j}
\tag{9.1.25}
$$

You should be able to convince yourself that the components of the dyad product $\mathbf{ab}$ contained in the matrix of Equation 9.1.23b can be generated by the operation given in Equation 9.1.25.

Using these definitions, we can see that the term in brackets in Equation 9.1.19b

$$
\mathbf{I} = \sum_i m_i r_i^2 \mathbf{1} - \sum_i m_i \mathbf{r}_i \mathbf{r}_i
\tag{9.1.26}
$$

is the previously defined moment of inertia tensor whose components were given in Equation 9.1.6. We can demonstrate this equivalence by calculating the components as follows:

$$\mathbf{i} \cdot \mathbf{I} \cdot \mathbf{i} = \mathbf{i} \cdot \left\{ \sum_i \left[m_i r_i^2 (\mathbf{ii} + \mathbf{jj} + \mathbf{kk}) - m_i \mathbf{r}_i \mathbf{r}_i \right] \right\} \cdot \mathbf{i}$$

$$I_{xx} = \sum_i m_i r_i^2 - m_i x_i^2 = \sum_i m_i \left(y_i^2 + z_i^2 \right)$$

(9.1.27a)

$$\mathbf{i} \cdot \mathbf{I} \cdot \mathbf{j} = \mathbf{i} \cdot \left\{ \sum_i \left[m_i r_i^2 (\mathbf{ii} + \mathbf{jj} + \mathbf{kk}) - m_i \mathbf{r}_i \mathbf{r}_i \right] \right\} \cdot \mathbf{j}$$

$$I_{xy} = -\sum_i m_i x_i y_i, \text{ and so on}$$

(9.1.27b)

Thus, using tensor notation, the angular momentum vector can be written as

$$\mathbf{L} = \mathbf{I} \cdot \boldsymbol{\omega}$$

(9.1.28)

An important fact should now be apparent: *The direction of the angular momentum vector is not necessarily aligned along the axis of rotation;* **L** *and* $\boldsymbol{\omega}$ *are not necessarily parallel.* For example, let $\boldsymbol{\omega}$ be directed along the x-axis ($\omega_x = \omega$, $\omega_y = 0$, $\omega_z = 0$). In this case the preceding expression reduces to

$$\mathbf{L} = \mathbf{I} \cdot \boldsymbol{\omega}$$
$$= \mathbf{i}(\omega_x I_{xx} + \omega_y I_{xy} + \omega_z I_{xz}) + \mathbf{j}(\omega_x I_{yx} + \omega_y I_{yy} + \omega_z I_{yz})$$
$$+ \mathbf{k}(\omega_x I_{zx} + \omega_y I_{zy} + \omega_z I_{zz})$$
$$= \mathbf{i}\omega I_{xx} + \mathbf{j}\omega I_{xy} + \mathbf{k}\omega I_{xz}$$

(9.1.29)

Thus, **L** may have components perpendicular to the x-axis (axis of rotation). Note that the component of angular momentum along the axis of rotation is $L_x = \omega I_{xx}$, in agreement with the results of Chapter 8.

EXAMPLE 9.1.1

Find the moment of inertia of a uniform square lamina of side a and mass m about a diagonal.

Solution:

Let us choose coordinate axes as shown in Figure 9.1.2 with the lamina lying in the xy plane with a corner at the origin. Then, from the previous chapter, we have $I_{xx} = I_{yy} = ma^2/3$ and $I_{zz} = I_{xx} + I_{yy} = 2ma^2/3$. Now $z = 0$ for all points in the lamina; therefore, the xz and yz products of inertia vanish: $I_{xz} = I_{yz} = 0$. The xy product of inertia is found by integrating as follows:

$$I_{xy} = I_{yx} = -\int_0^a \int_0^a xy\rho \, dx \, dy = -\rho \int_0^a \frac{a^2}{2} y \, dy = -\rho \frac{a^4}{4}$$

where ρ is the mass per unit area; that is, $\rho = m/a^2$. We, therefore, get

$$I_{xy} = -\tfrac{1}{4}ma^2$$

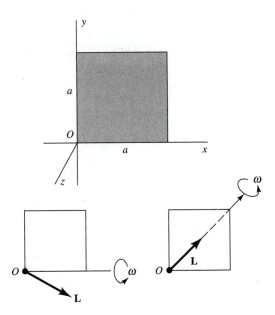

Figure 9.1.2   Square lamina.

We now have the essential ingredients necessary to construct the moment of inertia tensor of the square plate related to the xyz coordinate system indicated in Figure 9.1.2. It is

$$\mathbf{I} = \begin{pmatrix} \frac{ma^2}{3} & \frac{-ma^2}{4} & 0 \\ \frac{-ma^2}{4} & \frac{ma^2}{3} & 0 \\ 0 & 0 & \frac{2ma^2}{3} \end{pmatrix} = ma^2 \begin{pmatrix} \frac{1}{3} & -\frac{1}{4} & 0 \\ -\frac{1}{4} & \frac{1}{3} & 0 \\ 0 & 0 & \frac{2}{3} \end{pmatrix}$$

To calculate the moment of inertia of the plate about its diagonal according to Equation 9.1.13, we need the components of a unit vector $\mathbf{n}$ directed along the diagonal. The angles that this vector makes with the coordinate axes are $\alpha = \beta = 45°$ and $\gamma = 90°$. The direction cosines are then $\cos \alpha = \cos \beta = \frac{1}{\sqrt{2}}$ and $\cos \gamma = 0$. Equation 9.1.13 then gives

$$I = \tilde{\mathbf{n}} \mathbf{I} \mathbf{n} = \begin{pmatrix} \frac{1}{\sqrt{2}} & \frac{1}{\sqrt{2}} & 0 \end{pmatrix} \begin{pmatrix} \frac{1}{3} & -\frac{1}{4} & 0 \\ -\frac{1}{4} & \frac{1}{3} & 0 \\ 0 & 0 & \frac{2}{3} \end{pmatrix} \begin{pmatrix} \frac{1}{\sqrt{2}} \\ \frac{1}{\sqrt{2}} \\ 0 \end{pmatrix} ma^2$$

$$= \begin{pmatrix} \frac{1}{\sqrt{2}} & \frac{1}{\sqrt{2}} & 0 \end{pmatrix} \begin{pmatrix} \frac{1}{12\sqrt{2}} \\ \frac{1}{12\sqrt{2}} \\ 0 \end{pmatrix} ma^2 = \frac{1}{12} ma^2$$

for the moment of inertia about a diagonal. This result could also be obtained by direct integration. The student may wish to verify this as an exercise. See also Problem 9.2(c).

EXAMPLE 9.1.2

Find the angular momentum about the origin of the square plate in Example 9.1.1 when it is rotating with angular speed ω about (a) the x-axis and (b) the diagonal through the origin.

Solution:

(a) For rotation about the x-axis we have

$$\omega_x = \omega \qquad \omega_y = 0 \qquad \omega_z = 0$$

The total angular momentum is $\mathbf{L} = \mathbf{I} \cdot \boldsymbol{\omega}$, or, in matrix notation,

$$\mathbf{L} = \mathbf{I}\boldsymbol{\omega} = ma^2 \begin{pmatrix} \frac{1}{3} & -\frac{1}{4} & 0 \\ -\frac{1}{4} & \frac{1}{3} & 0 \\ 0 & 0 & \frac{2}{3} \end{pmatrix} \begin{pmatrix} \omega \\ 0 \\ 0 \end{pmatrix} = ma^2\omega \begin{pmatrix} \frac{1}{2} \\ -\frac{1}{4} \\ 0 \end{pmatrix}$$

with the final result expressed in matrix form as a column vector.

(b) The components of $\boldsymbol{\omega}$ about the diagonal are

$$\omega_x = \omega_y = \omega \cos 45° = \frac{\omega}{\sqrt{2}} \qquad \omega_z = 0$$

Therefore,

$$\mathbf{L} = ma^2 \begin{pmatrix} \frac{1}{3} & -\frac{1}{4} & 0 \\ -\frac{1}{4} & \frac{1}{3} & 0 \\ 0 & 0 & \frac{2}{3} \end{pmatrix} \begin{pmatrix} \frac{\omega}{\sqrt{2}} \\ \frac{\omega}{\sqrt{2}} \\ 0 \end{pmatrix} = \frac{ma^2\omega}{\sqrt{2}} \begin{pmatrix} \frac{1}{12} \\ \frac{1}{12} \\ 0 \end{pmatrix}$$

Notice that in case (a) the angular momentum vector $\mathbf{L}$ does not point in the same direction as the angular velocity vector $\boldsymbol{\omega}$ but points downward as shown in Figure 9.1.2. In case (b), however, the two vectors point in the same direction as shown in Figure 9.1.2.

The magnitude of the angular momentum vector for case (a) is given by $(\mathbf{L} \cdot \mathbf{L})^{1/2}$. Using matrix notation, we obtain

$$L^2 = \tilde{\mathbf{L}}\mathbf{L} = (ma^2\omega)^2 \begin{pmatrix} \frac{1}{3} & -\frac{1}{4} & 0 \end{pmatrix} \begin{pmatrix} \frac{1}{3} \\ -\frac{1}{4} \\ 0 \end{pmatrix} = (ma^2\omega)^2 \left[\frac{1}{9} + \frac{1}{16} \right]$$

$$= (ma^2\omega)^2 \frac{25}{144} \qquad \therefore L = ma^2\omega \frac{5}{12}$$

and for case (b)

$$L = ma^2\omega \frac{1}{12}$$

Rotational Kinetic Energy of a Rigid Body

We next calculate the kinetic energy of rotation of our general rigid body of Figure 9.1.1. As in our calculation of the angular momentum, we use the fact that the velocity of a representative particle is given by $\mathbf{v}_i = \boldsymbol{\omega} \times \mathbf{r}_i$. The rotational kinetic energy is, therefore, given by the summation

$$T_{rot} = \sum_i \frac{1}{2} m_i \mathbf{v}_i \cdot \mathbf{v}_i = \frac{1}{2} \sum_i (\boldsymbol{\omega} \times \mathbf{r}_i) \cdot m_i \mathbf{v}_i \qquad (9.1.30)$$

Now in any triple scalar product we can exchange the dot and the cross: $(\mathbf{A} \times \mathbf{B}) \cdot \mathbf{C} = \mathbf{A} \cdot (\mathbf{B} \times \mathbf{C})$. (See Section 1.7.) Hence,

$$T_{rot} = \frac{1}{2} \sum_i \boldsymbol{\omega} \cdot (\mathbf{r}_i \times m_i \mathbf{v}_i) = \frac{1}{2} \boldsymbol{\omega} \cdot \sum_i (\mathbf{r}_i \times m_i \mathbf{v}_i) \qquad (9.1.31)$$

But, by definition, the sum $\sum_i (\mathbf{r}_i \times m_i \mathbf{v}_i)$ is the angular momentum $\mathbf{L}$. Thus, we can write

$$T_{rot} = \frac{1}{2} \boldsymbol{\omega} \cdot \mathbf{L} \qquad (9.1.32)$$

for the kinetic energy of rotation of a rigid body. We recall from Chapter 7 that the translational kinetic energy of any system is equal to the expression $\frac{1}{2} \mathbf{v}_{cm} \cdot \mathbf{p}$, where $\mathbf{p} = m\mathbf{v}_{cm}$ is the linear momentum of the system and $\mathbf{v}_{cm}$ is the velocity of the mass center. For a rigid body the total kinetic energy is accordingly

$$T = T_{rot} + T_{trans} = \frac{1}{2} \boldsymbol{\omega} \cdot \mathbf{L} + \frac{1}{2} \mathbf{v}_{cm} \cdot \mathbf{p} \qquad (9.1.33)$$

where $\mathbf{L}$ is the angular momentum about the center of mass.

Using the results of the previous section, we can express the rotational kinetic energy of a rigid body in terms of the moment of the inertia tensor as

$$T_{rot} = \frac{1}{2} \boldsymbol{\omega} \cdot \mathbf{I} \cdot \boldsymbol{\omega} \qquad (9.1.34)$$

or we can carry out the calculation explicitly in matrix notation

$$T_{rot} = \frac{1}{2} \tilde{\boldsymbol{\omega}} \mathbf{I} \boldsymbol{\omega} = \frac{1}{2} \begin{pmatrix} \omega_x & \omega_y & \omega_z \end{pmatrix} \begin{pmatrix} I_{xx} & I_{xy} & I_{xz} \\ I_{yx} & I_{yy} & I_{yz} \\ I_{zx} & I_{zy} & I_{zz} \end{pmatrix} \begin{pmatrix} \omega_x \\ \omega_y \\ \omega_z \end{pmatrix} \qquad (9.1.35)$$

$$= \frac{1}{2} \left(I_{xx} \omega_x^2 + I_{yy} \omega_y^2 + I_{zz} \omega_z^2 + 2 I_{xy} \omega_x \omega_y + 2 I_{xz} \omega_x \omega_z + 2 I_{yz} \omega_y \omega_z \right)$$

EXAMPLE 9.1.3

Find the rotational kinetic energy of the square plate in Example 9.1.2.

Solution:

For the case (a) rotation about the x-axis, we have

$$T = \frac{1}{2} \tilde{\boldsymbol{\omega}} \mathbf{I} \boldsymbol{\omega} = \frac{1}{2} ma^2 \omega^2 \begin{pmatrix} 1 & 0 & 0 \end{pmatrix} \begin{pmatrix} \frac{1}{3} & -\frac{1}{4} & 0 \\ -\frac{1}{4} & \frac{1}{3} & 0 \\ 0 & 0 & \frac{2}{3} \end{pmatrix} \begin{pmatrix} 1 \\ 0 \\ 0 \end{pmatrix} = \frac{1}{6} ma^2 \omega^2$$

For case (b), rotation about a diagonal, we have

$$T = \frac{1}{2}ma^2\omega^2 \begin{pmatrix} \frac{1}{\sqrt{2}} & \frac{1}{\sqrt{2}} & 0 \end{pmatrix} \begin{pmatrix} \frac{1}{3} & -\frac{1}{4} & 0 \\ -\frac{1}{4} & \frac{1}{3} & 0 \\ 0 & 0 & \frac{2}{3} \end{pmatrix} \begin{pmatrix} \frac{1}{\sqrt{2}} \\ \frac{1}{\sqrt{2}} \\ 0 \end{pmatrix}$$

$$= \frac{1}{2}ma^2\omega^2 \begin{pmatrix} \frac{1}{\sqrt{2}} & \frac{1}{\sqrt{2}} & 0 \end{pmatrix} \begin{pmatrix} \frac{1}{12\sqrt{2}} \\ \frac{1}{12\sqrt{2}} \\ 0 \end{pmatrix} = \frac{1}{24}ma^2\omega^2$$

9.2 | Principal Axes of a Rigid Body

A considerable simplification in the previously derived mathematical formulas for rigid-body motion results if we employ a coordinate system such that the products of inertia all vanish. It turns out that such a coordinate system does in fact exist for any rigid body and for any point taken as the origin. The axes of this coordinate system are *principal axes* for the body at the point O, the origin of the coordinate system in question. (Often we choose O to be the center of mass.)

Explicitly, if the coordinate axes are principal axes of the body, then $I_{xy} = I_{xz} = I_{yz} = 0$. In this case we employ the following notation:

$$\begin{array}{lll} I_{xx} = I_1 & \omega_x = \omega_1 & \mathbf{i} = \mathbf{e}_1 \\ I_{yy} = I_2 & \omega_y = \omega_2 & \mathbf{j} = \mathbf{e}_2 \\ I_{zz} = I_3 & \omega_z = \omega_3 & \mathbf{k} = \mathbf{e}_3 \end{array} \tag{9.2.1}$$

The three moments of inertia I_1, I_2, and I_3 are the *principal moments* of the rigid body at the point O. In a coordinate system whose axes are aligned with the principal axes, the moment of inertia tensor takes on a particularly simple, diagonal form

$$\mathbf{I} = \begin{pmatrix} I_1 & 0 & 0 \\ 0 & I_2 & 0 \\ 0 & 0 & I_3 \end{pmatrix} \tag{9.2.2}$$

Thus, the problem of finding the principal axes of a rigid body is equivalent to the mathematical problem of diagonalizing a 3×3 matrix. The moment of inertia tensor is always expressible as a square, symmetric matrix, and any such matrix can always be diagonalized. *Thus, a set of principal axes exists for any rigid body at any point in space.*

The moment of inertia, angular momentum, and rotational kinetic energy of a rigid body about any arbitrary rotational axis all take on fairly simple forms in a coordinate system whose axes are aligned with the principal axes of the rigid body. Let $\mathbf{n}$ be a unit vector designating the direction of the axis of rotation of a rigid body, and let its components relative to the principal axes be given by the direction cosines $(\cos\alpha, \cos\beta, \cos\gamma)$.

The moment of inertia about that axis is

$$I = \tilde{\mathbf{n}}\mathbf{I}\mathbf{n} = (\cos\alpha \quad \cos\beta \quad \cos\gamma)\begin{pmatrix} I_1 & 0 & 0 \\ 0 & I_2 & 0 \\ 0 & 0 & I_3 \end{pmatrix}\begin{pmatrix} \cos\alpha \\ \cos\beta \\ \cos\gamma \end{pmatrix} \tag{9.2.3}$$

$$= I_1 \cos^2\alpha + I_2 \cos^2\beta + I_3 \cos^2\gamma$$

The angular velocity $\boldsymbol{\omega}$ points in the same direction as $\mathbf{n}$, and its components relative to the principal axes are $(\omega_1, \omega_2, \omega_3)$. The total angular momentum $\mathbf{L} = \mathbf{I} \cdot \boldsymbol{\omega}$, in this frame of reference, can be written in matrix notation as

$$\mathbf{L} = \mathbf{I}\boldsymbol{\omega} = \begin{pmatrix} I_1 & 0 & 0 \\ 0 & I_2 & 0 \\ 0 & 0 & I_3 \end{pmatrix}\begin{pmatrix} \omega_1 \\ \omega_2 \\ \omega_3 \end{pmatrix} = \begin{pmatrix} I_1\omega_1 \\ I_2\omega_2 \\ I_3\omega_3 \end{pmatrix} \tag{9.2.4}$$

$$= \mathbf{e}_1 I_1\omega_1 + \mathbf{e}_2 I_2\omega_2 + \mathbf{e}_3 I_3\omega_3$$

And, finally, the kinetic energy of rotation T_{rot} is

$$T_{rot} = \frac{1}{2}\tilde{\boldsymbol{\omega}}\mathbf{I}\boldsymbol{\omega} = \frac{1}{2}(\omega_1 \quad \omega_2 \quad \omega_3)\begin{pmatrix} I_1 & 0 & 0 \\ 0 & I_2 & 0 \\ 0 & 0 & I_3 \end{pmatrix}\begin{pmatrix} \omega_1 \\ \omega_2 \\ \omega_3 \end{pmatrix} \tag{9.2.5}$$

$$= \frac{1}{2}\left(I_1\omega_1^2 + I_2\omega_2^2 + I_3\omega_3^2\right)$$

Let us now investigate the question of finding the principal axes. First, if the body possesses some symmetry, then it is usually possible to choose a set of coordinate axes by inspection such that one or more of the three products of inertia consists of two parts of equal magnitude and opposite algebraic sign and, therefore, vanishes. For example, the rectangular block and the symmetric laminar body (Ping-Pong paddle) have the principal axes at O as indicated in Figure 9.2.1.

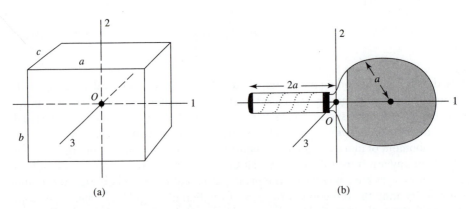

(a)   (b)

Figure 9.2.1   Principal axes for a (a) uniform rectangular block and (b) Ping-Pong paddle.

EXAMPLE 9.2.1

(a) For the rectangular block shown in Figure 9.2.1a, the principal moments at the center of mass O are those indicated in Table 8.3.1, namely,

$$I_1 = \frac{m}{12}(b^2 + c^2)$$

$$I_2 = \frac{m}{12}(a^2 + c^2)$$

$$I_3 = \frac{m}{12}(a^2 + b^2)$$

in which m is the mass of the block and a, b, and c are the edge lengths.

(b) To find the principal moments of the Ping-Pong paddle at the point O indicated in Figure 9.2.1b, we assume for simplicity that the paddle is a circular lamina of radius a and mass $m/2$ attached to a thin rod for a handle of mass $m/2$ and length $2a$. We borrow from the results of Section 8.3. The principal moments are each calculated by adding the moments about the respective axes for the two parts, namely,

$$I = I_{rod} + I_{disc}$$

$$I_1 = 0 + \frac{1}{4}\frac{m}{2}a^2 = \frac{1}{8}ma^2$$

$$I_2 = \frac{1}{3}\frac{m}{2}(2a)^2 + \frac{5}{4}\frac{m}{2}a^2 = \frac{31}{24}ma^2$$

$$I_3 = \frac{1}{3}\frac{m}{2}(2a)^2 + \frac{3}{2}\frac{m}{2}a^2 = \frac{17}{12}ma^2$$

We note that $I_3 = I_1 + I_2$ because the object is assumed to be laminar.

EXAMPLE 9.2.2

Find the moment of inertia about a diagonal of a uniform cube whose side length is a.

Solution:

The moment of inertia about any axis passing through the center of mass O of any uniform rectangular block is given by Equation 9.2.3. In particular, for a uniform cube ($a = b = c$), the principal moments at O are all equal. Thus, from Equation 9.2.3, we have

$$I = I_1 \cos^2\alpha + I_2 \cos^2\beta + I_3 \cos^2\gamma = I_1(\cos^2\alpha + \cos^2\beta + \cos^2\gamma) = I_1 = \frac{1}{6}ma^2$$

Because $\cos^2\alpha + \cos^2\beta + \cos^2\gamma = 1$, this is the moment of inertia about any axis that passes through the center of mass of the cube.

Dynamic Balancing

Suppose that a body is rotating about one of its principal axes, say the 1-axis. Then $\omega = \omega_1$, $\omega_2 = \omega_3 = 0$. The expression for the angular momentum (Equation 9.2.4) then reduces to just one term, namely,

$$\mathbf{L} = \mathbf{e}_1 I_1 \omega_1 \qquad\qquad (9.2.6a)$$

or, equivalently

$$\mathbf{L} = I_1 \boldsymbol{\omega} \qquad\qquad (9.2.6b)$$

Thus, in this circumstance the angular momentum vector is in the same direction as the angular velocity vector or axis of rotation. We have, therefore, the following important fact: *The angular momentum vector is either in the same direction as the axis of rotation, or is not, depending on whether the axis of rotation is, or is not, a principal axis.*

EXAMPLE 9.2.3

Suppose the Ping-Pong paddle of Example 9.2.1(b) is rotating with angular velocity $|\boldsymbol{\omega}| = \omega_3$ about its third principle axis (see Figure 9.2.1(b)). Its angular momentum $\mathbf{L}$ is then given by Equation 9.2.6a, with index 3 replacing index 1

$$\mathbf{L} = \mathbf{e}_3 I_3 \omega_3 = \mathbf{e}_3 \tfrac{17}{12} m a^2 \omega_3$$

The previous rule finds application in the case of a rotating device such as an automobile wheel or fan blade. If the device is *statically balanced,* the center of mass lies on the axis of rotation. To be *dynamically balanced* the axis of rotation must also be a principal axis so that, as the body rotates, the angular momentum vector $\mathbf{L}$ will lie along the axis. Otherwise, if the rotational axis is not a principal one, the angular momentum vector varies in direction: It describes a cone as the body rotates (Figure 9.2.2). Then, because $d\mathbf{L}/dt$ is equal to the applied torque, there must be a torque exerted on the body. The direction of this torque is at right angles to the axis. The result is a reaction on the bearings.

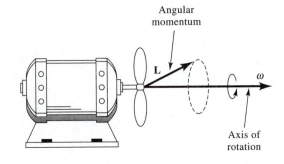

Figure 9.2.2   A rotating fan blade. The angular momentum vector $\mathbf{L}$ describes a cone about the axis of rotation if the blade is not dynamically balanced.

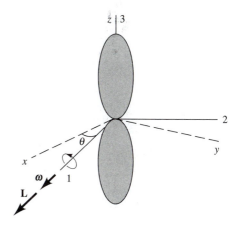

Figure 9.2.3 Determination of two principal axes (1 and 2) when the third one (z) is known.

Thus, in the case of a dynamically unbalanced wheel or rotator, there may be violent vibration and wobbling, even if the wheel is statically balanced.

Determination of the Other Two Principal Axes When One Is Known

In many instances a body possesses sufficient symmetry so that at least one principal axis can be found by inspection; that is, the axis can be chosen so as to make two of the three products of inertia vanish. If such is the case, then the other two principal axes can be determined as follows. Figure 9.2.3 is a front view of the fan blade in Figure 9.2.2. The z-axis is a symmetry axis and coincides with the third principal axis of the fan blade. Thus, for such cases, we have

$$I_{zx} = I_{zy} = 0 \qquad I_{zz} = I_3 \neq 0 \qquad (9.2.7)$$

The other two principal axes are each perpendicular to the z-axis. They must lie in the xy plane. Suppose the body is rotating about one of these two, as yet unknown, principal axes. If so, the rotating object is dynamically balanced as illustrated by the fan blade in Figure 9.2.3. The angular momentum vector $\mathbf{L}$ lies in the same direction as the angular velocity vector $\boldsymbol{\omega}$, thus

$$\mathbf{L} = I_1 \boldsymbol{\omega} = I_1 \begin{pmatrix} \omega_x \\ \omega_y \\ 0 \end{pmatrix} \qquad (9.2.8)$$

where I_1 is one of the two principal moments of inertia in question. Furthermore, in matrix notation, the angular momentum $\mathbf{L}$, in the xyz frame of reference, is given by

$$\mathbf{L} = \mathbf{I}\boldsymbol{\omega} = \begin{pmatrix} I_{xx} & I_{xy} & 0 \\ I_{xy} & I_{yy} & 0 \\ 0 & 0 & I_3 \end{pmatrix} \begin{pmatrix} \omega_x \\ \omega_y \\ 0 \end{pmatrix} \qquad (9.2.9)$$

(Remember, the products of inertia about the z-axis are zero.) Thus, equating components of the angular momentum given by these two expressions gives

$$I_{xx}\omega_x + I_{xy}\omega_y = I_1\omega_x \tag{9.2.10a}$$

$$I_{xy}\omega_x + I_{yy}\omega_y = I_1\omega_y \tag{9.2.10b}$$

Let θ denote the angle between the x-axis and the principal axis I_1 about which the body is rotating (see Figure 9.2.3). Then $\omega_y/\omega_x = \tan\theta$, so, on dividing by ω_x, we have

$$I_{xx} + I_{xy}\tan\theta = I_1 \tag{9.2.11a}$$

$$I_{xy} + I_{yy}\tan\theta = I_1\tan\theta \tag{9.2.11b}$$

Elimination of I_1 between the two equations yields

$$(I_{yy} - I_{xx})\tan\theta = I_{xy}(\tan^2\theta - 1) \tag{9.2.12}$$

from which θ can be found. In this calculation it is helpful to employ the trigonometric identity $\tan 2\theta = 2\tan\theta/(1 - \tan^2\theta)$. This gives

$$\tan 2\theta = \frac{2I_{xy}}{I_{xx} - I_{yy}} \tag{9.2.13}$$

In the interval $0°$ to $180°$ there are two values of θ, differing by $90°$, that satisfy Equation 9.2.13, and these give the directions of the two principal axes in the xy plane. In the case $I_{xx} = I_{yy}$, $\tan 2\theta = \infty$ so that the two values of θ are $45°$ and $135°$. (This is the case for the square lamina of Example 9.1.1 when the origin is at a corner.) Also, if $I_{xy} = 0$ the equation is satisfied by the two values $\theta = 0°$ and $\theta = 90°$; that is, the x- and y-axes are already principal axes.

EXAMPLE 9.2.4

Balancing a Crooked Wheel

Suppose an automobile wheel, through some defect or accident, has its axis of rotation (axle) slightly bent relative to the symmetry axis of the wheel. The situation can be remedied by use of counterbalance weights suitably located on the rim so as to make the axle a principal axis for the total system: wheel plus weights. For simplicity we shall treat the wheel as a thin uniform circular disc of radius a and mass m. Figure 9.2.4. We choose $Oxyz$ axes such that the disc lies in the yz plane, with the x-axis as the symmetry axis of the disc. The axis of rotation (axle) is taken as the 1-axis inclined by an angle θ relative to the x-axis and lying in the xy plane, as shown. Two balancing weights, each of mass m', are attached to the wheel by means of light supports of length b. The weights both lie in the xy plane, as indicated. The wheel is dynamically balanced if the 123 coordinate axes are principal axes for the total system.

Now, from symmetry relative to the xy plane, we see that the z-axis is a principal axis for the wheel plus weights: z is zero for the weights, and the xy plane divides the wheel into two equal parts having opposite signs for the products zx and zy. Consequently, we can use Equation 9.2.13 to find the relationship between θ and the other parameters.

From the previous chapter we know that for the wheel alone the moments of inertia about the x- and the y-axes are $\frac{1}{2}ma^2$, and $\frac{1}{4}ma^2$, respectively. Thus, for the wheel

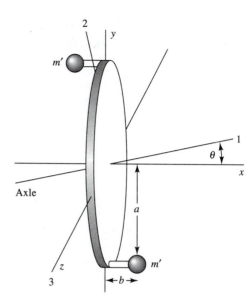

Figure 9.2.4 Principal axes for a bent
wheel with balancing weights.

plus weights

$$I_{xx} = \tfrac{1}{2}ma^2 + 2m'a^2$$

$$I_{yy} = \tfrac{1}{4}ma^2 + 2m'b^2$$

Now the xy product of inertia for the wheel alone is zero, and so we need consider only
the weights for finding I_{xy} for the system, namely,

$$I_{xy} = -\sum x_i y_i m_i' = -[(-b)am' + b(-a)m'] = 2abm'$$

Notice that this is a positive quantity for our choice of coordinate axes. Equation 9.2.13
then gives the inclination of the 1-axis:

$$\tan 2\theta = \frac{2I_{xy}}{I_{xx} - I_{yy}} = \frac{4abm'}{\tfrac{1}{4}ma^2 + 2m'(a^2 - b^2)}$$

If, as is typical, θ is very small, and m' is small compared with m, then we can express
the preceding relation in approximate form by ignoring the second term in the denominator and using the fact that $\tan u \approx u$ for small u. The result is

$$\theta \approx 8\frac{bm'}{am}$$

As a numerical example, let $\theta = 1° = 0.017$ rad, $a = 18$ cm, $b = 5$ cm, $m = 10$ kg. Solving
for m', we find

$$m' = \theta\frac{am}{8b} = 0.017\frac{18\,\text{cm} \times 10\,\text{kg}}{8 \times 5\,\text{cm}} = .076 \text{ kg} = 76 \text{ grams}$$

for the required balance weights.

Determining Principal Axes by Diagonalizing the Moment of Inertia Matrix

Suppose a rigid body has no axis of symmetry. Even so, the tensor that represents the moment of inertia of such a body, is characterized by a real, symmetric 3×3 matrix that can be *diagonalized* (see Equation 9.1.11). The resulting diagonal elements are the values of the principal moments of inertia of the rigid body. The axes of the coordinate system, in which this matrix is diagonal, are the principal axes of the body, because all products of inertia have vanished. Thus, finding the principal axes and corresponding moments of inertia of any rigid body, symmetric or not, is tantamount to diagonalizing its moment of inertia matrix.

There are a number of ways to diagonalize a real, symmetric matrix.[2] We present here a way that is quite standard. First, let's suppose that we have found the coordinate system in which all products of inertia vanish and the resulting moment of inertia tensor is now represented by a diagonal matrix whose diagonal elements are the principal moments of inertia. Let $\mathbf{e}_i$ be the unit vectors that represent this coordinate system, that is, they point along the direction along the three principal axes of the rigid body. If the moment of inertia tensor is "dotted" with one of these unit vectors, the result is equivalent to a simple multiplication of the unit vector by a scalar quantity

$$\mathbf{I} \cdot \mathbf{e}_i = \lambda_i \mathbf{e}_i \tag{9.2.14}$$

The scalar quantities λ_i are just the principal moments of inertia about their respective principal axes. The problem of finding the principal axes is one of finding those vectors $\mathbf{e}_i$ that satisfy the condition

$$(\mathbf{I} - \lambda \mathbf{1}) \cdot \mathbf{e}_i = 0 \tag{9.2.15}$$

In general this condition is not satisfied for any arbitrary set of orthonormal unit vectors $\mathbf{e}_i$. It is satisfied only by a set of unit vectors aligned with the principal axes of the rigid body. Any arbitrary xyz coordinate system can always be rotated such that the coordinate axes line up with the principal axes. The unit vectors specifying these coordinate axes then satisfy the condition in Equation 9.2.15. This condition is equivalent to the vanishing of the following determinant[3]

$$|\mathbf{I} - \lambda \mathbf{1}| = 0 \tag{9.2.16}$$

Explicitly, this equation reads

$$\begin{vmatrix} (I_{xx} - \lambda) & I_{xy} & I_{xz} \\ I_{yx} & (I_{yy} - \lambda) & I_{yz} \\ I_{zx} & I_{zy} & (I_{zz} - \lambda) \end{vmatrix} = 0 \tag{9.2.17}$$

---

[2] For example, see J. Mathews and R. L. Walker, *Mathematical Methods of Physics,* W. A. Benjamin, New York, 1970.

[3] *Ibid.*

It is a cubic in λ, namely,

$$-\lambda^3 + A\lambda^2 + B\lambda + C = 0 \qquad (9.2.18)$$

in which A, B, and C are functions of the I's. The three roots, λ_1, λ_2, and λ_3 are the three principal moments of inertia.

We now have the principal moments of inertia, but the task of specifying the components of the unit vectors representing the principal axes in terms of our initial coordinate system remains to be solved. Here we can make use of the fact that when the rigid body rotates about one of its principal axes, the angular momentum vector is in the same direction as the angular velocity vector. Let the angles of one of the principal axes relative to the initial xyz coordinate system be α_1, β_1, and γ_1, and let the body rotate about this axis. Therefore, a unit vector pointing in the direction of this principal axis has components ($\cos \alpha_1$, $\cos \beta_1$, $\cos \gamma_1$). The angular momentum is given by

$$\mathbf{L} = \mathbf{I} \cdot \boldsymbol{\omega} = \lambda_1 \boldsymbol{\omega} \qquad (9.2.19)$$

in which λ_1, the first principal moment of the three (λ_1, λ_2, and λ_3), is obtained by solving Equation 9.2.18. In matrix form Equation 9.2.19 reads

$$\begin{pmatrix} I_{xx} & I_{xy} & I_{xz} \\ I_{yx} & I_{yy} & I_{yz} \\ I_{zx} & I_{zy} & I_{zz} \end{pmatrix} \omega \begin{pmatrix} \cos \alpha_1 \\ \cos \beta_1 \\ \cos \gamma_1 \end{pmatrix} = \lambda_1 \omega \begin{pmatrix} \cos \alpha_1 \\ \cos \beta_1 \\ \cos \gamma_1 \end{pmatrix} \qquad (9.2.20)$$

We have extracted the common factor ω from each one of its components, thus, directly exposing the desired principal axis unit vector. The resultant equation is equivalent to the condition expressed by Equation 9.2.14, namely, that the dot product of the moment of inertia tensor with a principal axis unit vector is tantamount to multiplying that vector by the scalar quantity λ_1, that is,

$$\mathbf{I} \cdot \mathbf{e}_1 = \lambda_1 \mathbf{e}_1 \qquad (9.2.21)$$

This vector equation can be written in matrix form as

$$\begin{pmatrix} (I_{xx} - \lambda_1) & I_{xy} & I_{xz} \\ I_{yx} & (I_{yy} - \lambda_1) & I_{yz} \\ I_{zx} & I_{zy} & (I_{zz} - \lambda_1) \end{pmatrix} \begin{pmatrix} \cos \alpha_1 \\ \cos \beta_1 \\ \cos \gamma_1 \end{pmatrix} = 0 \qquad (9.2.22)$$

The direction cosines may be found by solving the above equations. The solutions are not independent. They are subject to the constraint

$$\cos^2 \alpha_1 + \cos^2 \beta_1 + \cos^2 \gamma_1 = 1 \qquad (9.2.23)$$

In other words the resultant vector $\mathbf{e}_1$ specified by these components is a unit vector. The other two vectors may be found by repeating the preceding process for the other two principal moments λ_2 and λ_3.

EXAMPLE 9.2.5

Find the principal moments of inertia of a square plate about a corner.

Solution:

We choose the same xyz system as initially chosen in Example 9.1.1. We have all the moments of inertia relative to those axes. They are the same as in Example 9.1.1. The vanishing of the determinant expressed by Equation 9.2.17 reads

$$\begin{vmatrix} \left(\frac{1}{3}-\lambda\right) & -\frac{1}{4} & 0 \\ -\frac{1}{4} & \left(\frac{1}{3}-\lambda\right) & 0 \\ 0 & 0 & \left(\frac{2}{3}-\lambda\right) \end{vmatrix} ma^2 = 0$$

(**Note:** We have extracted a common factor ma^2, which will leave us with only the desired numerical coefficients for each value of λ. We must then put the ma^2 factor back in to get the final values for the principal moments.)

Evaluating the preceding determinantal equation gives

$$\left[\left(\frac{1}{3}-\lambda\right)^2 - \left(\frac{1}{4}\right)^2\right]\left(\frac{2}{3}-\lambda\right) = 0$$

The second factor gives

$$\lambda_3 = \frac{2}{3}(ma^2)$$

The first factor gives

$$\frac{1}{3}-\lambda = \pm\frac{1}{4}$$
$$\lambda_1 = \frac{1}{12}(ma^2)$$
$$\lambda_2 = \frac{7}{12}(ma^2)$$

EXAMPLE 9.2.6

Find the directions of the principal axes of a square plate about a corner.

Solution:

Equations 9.2.22 give

$$\left(\frac{1}{3}-\lambda\right)\cos\alpha - \frac{1}{4}\cos\beta = 0$$
$$-\frac{1}{4}\cos\alpha + \left(\frac{1}{3}-\lambda\right)\cos\beta = 0$$
$$\left(\frac{2}{3}-\lambda\right)\cos\gamma = 0$$

We would guess that at least one of the principal axes (say, the third axis) is perpendicular to the plane of the square plate, that is, $\gamma_3 = 0°$, and $\alpha_3 = \beta_3 = 90°$. We would also guess from looking at the preceding equation that the principal moment about this axis would be $\lambda_3 = \frac{2}{3}(ma^2)$. Such choices would ensure that the third equation in the preceding group would automatically vanish as would the first two, because both $\cos\alpha_3$ and $\cos\beta_3$ would be identically zero. The remaining axes can be determined by inserting the other two principal moments into the preceding equations. Thus, if we set $\lambda_1 = \frac{1}{12}(ma^2)$, we obtain the conditions that

$$\cos\alpha_1 - \cos\beta_1 = 0 \qquad \cos\gamma_1 = 0$$

which can be satisfied only by $\alpha_1 = \beta_1 = 45°$ and $\gamma_1 = 90°$. Now, if we insert the final principal moment $\lambda_2 = \frac{7}{12}(ma^2)$ into the preceding equation, we obtain the conditions

$$\cos\alpha_2 + \cos\beta_2 = 0 \qquad \cos\gamma_2 = 0$$

that can be satisfied if $\alpha_2 = 135°$, $\beta_2 = 45°$, and $\gamma_2 = 90°$. Thus, two of the principal axes lie in the plane of the plate, one being the diagonal, the other perpendicular to the diagonal. The third is normal to the plate. The moment of inertia matrix in this coordinate representation is, thus,

$$\mathbf{I} = \begin{pmatrix} \frac{1}{12} & 0 & 0 \\ 0 & \frac{7}{12} & 0 \\ 0 & 0 & \frac{2}{3} \end{pmatrix} ma^2$$

and the corresponding principal axes in the original coordinate system are given by the vectors

$$\mathbf{e}_1 = \frac{1}{\sqrt{2}}\begin{pmatrix} 1 \\ 1 \\ 0 \end{pmatrix} \qquad \mathbf{e}_2 = \frac{1}{\sqrt{2}}\begin{pmatrix} -1 \\ 1 \\ 0 \end{pmatrix} \qquad \mathbf{e}_3 = \begin{pmatrix} 0 \\ 0 \\ 1 \end{pmatrix}$$

The principal axes can be obtained by a simple 45° rotation of the original coordinate system in the counterclockwise direction about the z-axis.

9.3 | Euler's Equations of Motion of a Rigid Body

We come now to what we may call the essential physics of the present chapter, namely, the actual three-dimensional rotation of a rigid body under the action of external forces. As we learned in Chapter 7, the fundamental equation governing the rotational part of the motion of any system, referred to an inertial coordinate system, is

$$\mathbf{N} = \frac{d\mathbf{L}}{dt} \qquad (9.3.1)$$

in which $\mathbf{N}$ is the net applied torque and $\mathbf{L}$ is the angular momentum. For a rigid body, we have seen that $\mathbf{L}$ is most simply expressed if the coordinate axes are principal axes for the body. Thus, in general, we must employ a coordinate system that is fixed in the body and rotates with it. That is, the angular velocity of the body and the angular velocity of the coordinate system are one and the same. (There is an exception: If two of the three principal moments I_1, I_2, and I_3 are equal to each other, then the coordinate axes need not be fixed in the body to be principal axes. This case if considered later.) In any case, our coordinate system is not an inertial one.

Referring to the theory of rotating coordinate systems developed in Chapter 5, we know that the time rate of change of the angular momentum vector in a fixed (inertial) system versus a rotating system is given by the formula (see Equations 5.2.9 and 5.2.10)

$$\left(\frac{d\mathbf{L}}{dt}\right)_{fixed} = \left(\frac{d\mathbf{L}}{dt}\right)_{rot} + \boldsymbol{\omega} \times \mathbf{L} \tag{9.3.2}$$

Thus, the equation of motion in the rotating system is

$$\mathbf{N} = \left(\frac{d\mathbf{L}}{dt}\right)_{rot} + \boldsymbol{\omega} \times \mathbf{L} \tag{9.3.3}$$

where

$$\dot{\mathbf{L}} = \mathbf{I} \cdot \dot{\boldsymbol{\omega}} \mathbf{L} \tag{9.3.4a}$$

$$\boldsymbol{\omega} \times \mathbf{L} = \boldsymbol{\omega} \times (\mathbf{I} \cdot \boldsymbol{\omega}) \tag{9.3.4b}$$

The latter cross product in Equation 9.3.4b can be written as the determinant

$$\boldsymbol{\omega} \times (\mathbf{I} \cdot \boldsymbol{\omega}) = \begin{vmatrix} \mathbf{e}_1 & \mathbf{e}_2 & \mathbf{e}_3 \\ \omega_1 & \omega_2 & \omega_3 \\ I_1\omega_1 & I_2\omega_2 & I_3\omega_3 \end{vmatrix} \tag{9.3.4c}$$

where the components of $\boldsymbol{\omega}$ are taken along the directions of the principal axes. Thus, Equation 9.3.3 can be written in matrix form as

$$\begin{pmatrix} N_1 \\ N_2 \\ N_3 \end{pmatrix} = \begin{pmatrix} I_1\dot{\omega}_1 \\ I_2\dot{\omega}_2 \\ I_3\dot{\omega}_3 \end{pmatrix} + \begin{pmatrix} \omega_2\omega_3(I_3 - I_2) \\ \omega_3\omega_1(I_1 - I_3) \\ \omega_1\omega_2(I_2 - I_1) \end{pmatrix} \tag{9.3.5}$$

These are known as Euler's equations for the motion of a rigid body in components along the principal axes of the body.

Body Constrained to Rotate About a Fixed Axis

As a first application of Euler's equations, we take up the special case of a rigid body that is constrained to rotate about a fixed axis with constant angular velocity. Then

$$\dot{\omega}_1 = \dot{\omega}_2 = \dot{\omega}_3 = 0 \tag{9.3.6}$$

and Euler's equations reduce to

$$\begin{aligned} N_1 &= \omega_2\omega_3(I_3 - I_2) \\ N_2 &= \omega_3\omega_1(I_1 - I_3) \\ N_3 &= \omega_1\omega_2(I_2 - I_1) \end{aligned} \tag{9.3.7}$$

These give the components of the torque that must be exerted on the body by the constraining support.

In particular, suppose that the axis of rotation is a principal axis, say the 1-axis. Then $\omega_2 = \omega_3 = 0$, $\omega = \omega_1$. In this case all three components of the torque vanish:

$$N_1 = N_2 = N_3 = 0 \tag{9.3.8}$$

That is, there is no torque at all. This agrees with our discussion concerning dynamic balancing in the previous section.

9.4 | Free Rotation of a Rigid Body: Geometric Description of the Motion

Let us consider the case of a rigid body that is free to rotate in any direction about a certain point O. No torques act on the body. This is the case of free rotation and is exemplified, for example, by a body supported on a smooth pivot at its center of mass. Another example is that of a rigid body moving freely under no forces or falling freely in a uniform gravitational field so that there are no torques. The point O in this case is the center of mass.

With zero torque the angular momentum of the body, as seen from the outside, must remain constant in direction and magnitude according to the general principle of conservation of angular momentum. With respect to rotating axes fixed in the body, however, the direction of the angular momentum vector may change, although its magnitude must remain constant. This fact can be expressed by the equation

$$\mathbf{L} \cdot \mathbf{L} = \text{constant} \tag{9.4.1a}$$

In terms of components referred to the principal axes of the body, Equation 9.4.1a reads

$$I_1^2 \omega_1^2 + I_2^2 \omega_2^2 + I_3^2 \omega_3^2 = L^2 = \text{constant} \tag{9.4.1b}$$

As the body rotates, the components of ω may vary, but they must always satisfy Equation 9.4.1b.

A second relation is obtained by considering the kinetic energy of rotation. Again, because there is zero torque, the total rotational kinetic energy must remain constant. This may be expressed as

$$\boldsymbol{\omega} \cdot \mathbf{L} = 2T_{rot} = \text{constant} \tag{9.4.2a}$$

or, equivalently in terms of components,

$$I_1 \omega_1^2 + I_2 \omega_2^2 + I_3 \omega_3^2 = 2T_{rot} = \text{constant} \tag{9.4.2b}$$

We now see that the components of $\boldsymbol{\omega}$ must simultaneously satisfy two different equations expressing the constancy of kinetic energy and of magnitude of angular momentum. (These two equations can also be obtained by use of Euler's equations. See Problem 9.7.) These are the equations of two ellipsoids whose principal axes coincide with the principal axes of the body. The first ellipsoid (Equation 9.4.1b) has principal diameters in the ratios $I_1^{-1} : I_2^{-1} : I_3^{-1}$. The second ellipsoid (Equation 9.4.2b) has principal diameters in the ratios $I_1^{-1/2} : I_2^{-1/2} : I_3^{-1/2}$. It is known as the *Poinsot ellipsoid*. As the body rotates, the

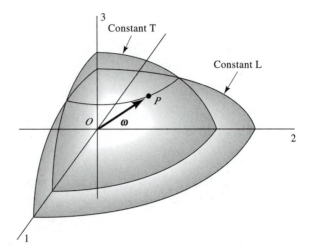

Figure 9.4.1 Intersecting ellipsoids of constant L and T for a rigid body undergoing torque-free rotation. (Only one octant is shown for clarity.)

extremity of the angular velocity vector, thus, describes a curve that is the intersection of the two ellipsoids. This is illustrated in Figure 9.4.1.

From the equations of the intersecting ellipsoids, when the initial axis of rotation coincides with one of the principal axes of the body, then the curve of intersection diminishes to a point. In other words, the two ellipsoids just touch at a principal diameter, and the body rotates steadily about this axis. This is true, however, only if the initial rotation is about the axis of either the largest or the smallest moment of inertia. If it is about the intermediate axis, say the 2-axis where $I_3 > I_2 > I_1$, then the intersection of the two ellipsoids is not a point, but a curve that goes entirely around both, as illustrated in Figure 9.4.2. In this case the rotation is unstable, because the axis of rotation precesses all around the body. See Problem 9.19. (If the initial axis of rotation is almost, but not exactly, along one of the two stable axes, then the angular velocity vector describes a tight cone about the corresponding axis.) These facts can easily be illustrated by tossing an oblong block, a book, or a Ping-Pong paddle into the air.

9.5 | Free Rotation of a Rigid Body with an Axis of Symmetry: Analytical Treatment

Although the geometric description of the motion of a rigid body given in the preceding section is helpful in visualizing free rotation under no torques, the method does not immediately give numerical values. We now proceed to augment that description with an analytical approach based on the direct integration of Euler's equations.

We shall solve Euler's equations for the special case in which the body possesses an axis of symmetry, so that two of the three principal moments of inertia are equal. An example of such an object is shown in Figure 9.5.1. Usually, one can see it being thrown around playfully by otherwise grown men every autumn weekend in large stadiums all around the country. The long, central axis of the object is its axis of symmetry. The object is a prolate spheroid, more commonly known as a football.

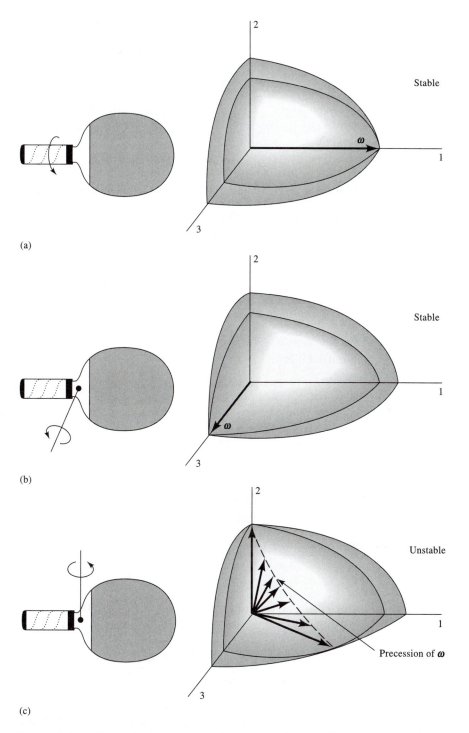

Figure 9.4.2   Ellipsoids of constant L and constant T for a rigid body rotating freely about the axis of the (a) least, (b) greatest, and (c) intermediate moment of inertia.

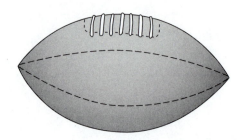

Figure 9.5.1 A prolate spheroid—whose long central axis is its axis of symmetry—more commonly known as a football.

Let us choose the 3-axis as the axis of symmetry. We introduce the following notation:

$I_s = I_3$ *moment of inertia about the symmetry axis*

$I = I_1 = I_2$ *moment about the axes normal to the symmetry axis*

For the case of zero torque, Euler's equations (9.3.5) then read

$$I\dot{\omega}_1 + \omega_2\omega_3(I_s - I) = 0 \tag{9.5.1a}$$

$$I\dot{\omega}_2 + \omega_3\omega_1(I - I_s) = 0 \tag{9.5.1b}$$

$$I_s\dot{\omega}_3 = 0 \tag{9.5.1c}$$

From the last equation it follows that

$$\omega_3 = \text{constant} \tag{9.5.2}$$

Let us now define a constant Ω as

$$\Omega = \omega_3 \frac{I_s - I}{I} \tag{9.5.3}$$

Equations 9.5.1a and b may be written

$$\dot{\omega}_1 + \Omega\omega_2 = 0 \tag{9.5.4a}$$

$$\dot{\omega}_2 - \Omega\omega_1 = 0 \tag{9.5.4b}$$

To separate the variables in Equations 9.5.4 a and b, we differentiate the first with respect to t and obtain

$$\ddot{\omega}_1 + \Omega\dot{\omega}_2 = 0 \tag{9.5.5}$$

Solving for $\dot{\omega}_2$ in Equation 9.5.4b and inserting the result into Equation 9.5.5, we find

$$\ddot{\omega}_1 + \Omega^2\omega_1 = 0 \tag{9.5.6}$$

This is the equation for simple harmonic motion. A solution is

$$\omega_1 = \omega_0 \cos \Omega t \tag{9.5.7a}$$

in which ω_0 is a constant of integration. To find ω_2, we differentiate Equation 9.5.7a with respect to t and insert the result into Equation 9.5.4a. We can then solve for ω_2 to obtain

$$\omega_2 = \omega_0 \sin \Omega t \tag{9.5.7b}$$

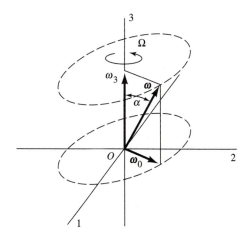

Figure 9.5.2   Angular velocity vector and its components for the free precession of a body with an axis of symmetry.

Thus, ω_1 and ω_2 vary harmonically in time with angular frequency Ω, and their phases differ by $\pi/2$. It follows that the projection of $\boldsymbol{\omega}$ on the 1, 2 plane describes a circle of radius ω_0 at the angular frequency Ω (see Figure 9.5.2).

We can summarize the preceding results as follows: In the free rotation of a rigid body with an axis of symmetry, the angular velocity vector describes a conical motion (precesses) about the symmetry axis. An observer, therefore, in the frame of reference attached to the rotating body, would see $\boldsymbol{\omega}$ trace out a cone around the symmetry axis of the body (called the *body cone*). (See Figure 9.6.4.) The angular frequency of this precession is the constant Ω defined by Equation 9.5.3. Let α denote the angle between the symmetry axis (3-axis) and the axis of rotation (direction of $\boldsymbol{\omega}$) as shown in Figure 9.5.2. Then $\omega_3 = \omega \cos \alpha$, and so

$$\Omega = \left(\frac{I_s}{I} - 1\right)\omega \, \cos \alpha \tag{9.5.8}$$

giving the rate of precession of the angular velocity vector about the axis of symmetry. (Some specific examples are discussed at the end of this section.)

We can now see the connection between the preceding analysis of the torque-free rotation of a rigid body and the geometric description of the previous section. The circular path of radius ω_0 traced out by the extremity of the angular velocity vector is just the intersection of the two ellipsoids of Figure 9.4.1.

*Free Rotation of a Rigid Body with Three Different Principal Moments: Numerical Solution

In the previous section we discussed the free rotation of a rigid body that had a single axis of symmetry such that two axes perpendicular to it were principle axes whose moments of inertia were identical. In the case here, we shall relax this condition and discuss the free rotation of a rigid body with three unequal principal moments of inertia $I_1 < I_2 < I_3$. The rigid body we consider is an ellipsoid of uniform mass distribution whose surface is

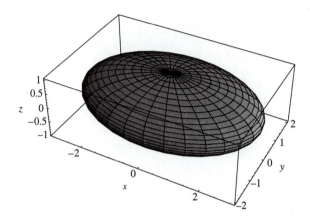

Figure 9.5.3 Ellipsoid with unequal semimajor axes $a > b > c$ and uniform mass distribution.

given by the equation

$$\frac{x^2}{a^2} + \frac{y^2}{b^2} + \frac{z^2}{c^2} = 1 \tag{9.5.9}$$

where $a > b > c$ are the semimajor axes of the ellipsoid (Figure 9.5.3). We solve Euler's equations (9.3.5) for this body numerically using *Mathematica*, and we see that even though the resulting motion is a little more complicated than that of the body with the axis of symmetry, many of the same general results are obtained.

First, we solve for the principal moments of inertia given by Equations 9.1.7a, b, and c converted suitably to integral form as in Equation 9.1.9a. Thus

$$I_3 = I_{zz} = 8\rho \int_0^a \int_0^{b(1-x^2/a^2)^{1/2}} \int_0^{c(1-x^2/a^2-y^2/b^2)^{1/2}} (x^2 + y^2)\, dz\, dy\, dx \tag{9.5.10}$$

with analogous expressions for the other two principal moments I_1 and I_2. The factor of 8 in Equation 9.5.10 arises because we have invoked symmetry to eliminate half of the integration about each of the three principal axes. Furthermore, there is no loss of generality if we set $8\rho = 1$, because it cancels out anyway when used in Euler's Equations, which are homogeneous in the absence of external torques.

We used *Mathematica*'s *NIntegrate* function (see Section 8.1) to evaluate the mass and three principal moments of inertia (remember, $8\rho = 1$) for an ellipsoid whose semimajor axes are $a = 3$, $b = 2$, and $c = 1$. The results are: $M = \pi$, $I_1 = \pi$, $I_2 = 2\pi$, and $I_3 = 8.168$. Euler's equations for free rotation of this ellipsoid then become

$$A_1 \omega_2 \omega_3 = \dot{\omega}_1$$
$$A_2 \omega_3 \omega_1 = \dot{\omega}_2 \tag{9.5.11}$$
$$A_3 \omega_1 \omega_2 = \dot{\omega}_3$$

where $A_1 = (I_2 - I_3)/I_1 = -0.6$, $A_2 = (I_3 - I_1)/I_2 = 0.8$, and $A_3 = (I_1 - I_2)/I_3 = -0.385$. We solved these three, first-order, coupled differential equations numerically with a call to *Mathematica*'s *NDSolve* (see Section 7.4, The Trojan Asteroids). The relevant call is

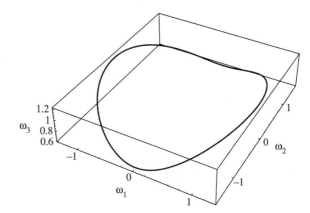

Figure 9.5.4 Free rotation of ellipsoid of Figure 9.5.3 plotted in phase space.

NDSolve [{Equations, Initial Conditions}, {$\omega_1, \omega_2, \omega_3$}, {$t, t_{min}, t_{max}$}]

We set the initial conditions to [$\omega_1(0), \omega_2(0), \omega_3(0)$] = (1, 1, 1)s^{-1}. The time interval [$0 - t_{max}$] was set to 3.3π, or about equal to one period of the resulting motion, which is shown in the phase space plot of Figure 9.5.4.

Note the similarity of the plot to Figure 9.5.2, which illustrates the precessional motion of an ellipsoid whose 3-axis is an axis of symmetry. In that case ω_1 and ω_2 precess around ω_3, which remains constant. Here the 3-axis is the principal axis about which the moment of inertia is the largest, and the angular velocity about that axis, although not constant, remains closer to constant than do ω_1 and ω_2. The motion is one in which ω_1 and ω_2 precess around ω_3, which itself wobbles slightly.

The analysis of Section 9.4 applies here as well. Both the kinetic energy T and the magnitude of the angular momentum, expressed as $\mathbf{L} \cdot \mathbf{L} = L^2$, are constants of the motion. As explained in Section 9.4, each of these values constrains the angular velocity vector to terminate on two ellipsoidal surfaces (see Equations 9.4.1b and 9.4.2b) given by the equations

$$\frac{\omega_1^2}{(2T/I_1)} + \frac{\omega_2^2}{(2T/I_2)} + \frac{\omega_3^2}{(2T/I_3)} = 1 \tag{9.5.12}$$

$$\frac{\omega_1^2}{(L/I_1)^2} + \frac{\omega_2^2}{(L/I_2)^2} + \frac{\omega_3^2}{(L/I_3)^2} = 1 \tag{9.5.13}$$

The values of the kinetic energy and the magnitude of the angular momentum are determined by the initial conditions for ω_1, ω_2, and ω_3 given earlier. The resultant ellipsoids of constant T and L^2 are shown in Figures 9.5.5a and b. Figure 9.5.5c shows the constant T and L^2 ellipsoids plotted together and their resultant intersection. The angular velocity vector is constrained to lie along this line to satisfy simultaneously the conditions that T and L^2 remain constant during the motion. Close examination of the intersection reveals that it is identical to the solution of Euler's equations plotted in Figure 9.5.4, as must be the case.

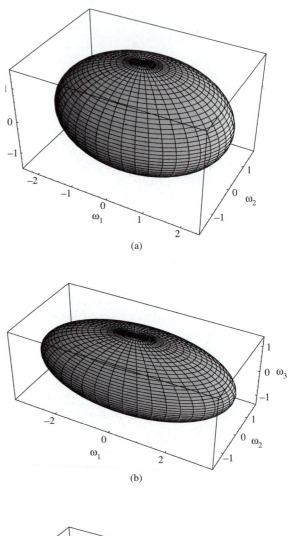

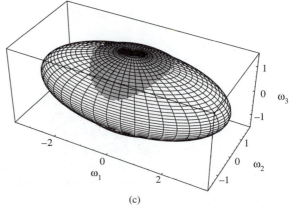

Figure 9.5.5 (a) Ellipsoid of constant T for the freely rotating mass ellipsoid of Figure 9.5.3. (b) Ellipsoid of constant L^2 for the freely rotating mass ellipsoid of Figure 9.5.3. (c) Constant T and L^2 ellipsoids plotted together showing the intersection that represents the possible phase path trajectory for the freely rotating ellipsoid of Figure 9.5.3.

9.6 | Description of the Rotation of a Rigid Body Relative to a Fixed Coordinate System: The Eulerian Angles

In the foregoing analysis of the free rotation of a rigid body, its precessional motion was described relative to a set of principal axes fixed in the body and rotating with it. To describe the motion relative to an observer outside the body, we must specify how the orientation of the body in space relative to a fixed coordinate system changes in time. There is no unique way to do this, but a commonly chosen scheme uses three angles ϕ, θ, and ψ to relate the direction of the three principal axes of the rigid body relative to those fixed in space. The scheme was first published by Leonhard Euler (1707–1783) in 1776 and is depicted in Figure 9.6.1. The coordinate system, labeled $O123$, is defined by the three principal axes fixed to the rigid body and rotates with it. The coordinate system fixed in space is labeled $Oxyz$. A third, rotating system, $Ox'y'z'$, providing a connection between the principal axes attached to the body and the axes fixed in space, is also shown and is defined as follows: The z'-axis coincides with the 3-axis of the body—its symmetry axis. The x'-axis is defined by the intersection of the body's 1-2 plane with the fixed xy plane; this axis is called the *line of nodes*. The angle between the x- and x'-axes is denoted by ϕ, and the angle between the z- and z'- (or 3-) axes is denoted by θ. The rotation of the body about its 3-axis is represented by the angle ψ between the 1-axis and the x'-axis. The three angles ϕ, θ, and ψ completely define the orientation of the body in space and are called the *Eulerian angles*.

The angular velocity $\boldsymbol{\omega}$ of the body is the vector sum of three angular velocities relating to the rates of change of the three Eulerian angles. We can see this by considering the infinitesimal change in the angular orientation of the body that occurs as it rotates in an infinitesimal amount of time. Assume that at time $t = 0$, the orientation of the body system (the $O123$ coordinate system) coincides with the fixed $Oxyz$ system. Some time dt later, the rigid body will have rotated through some infinitesimal angle $\boldsymbol{\omega}\, dt = d\boldsymbol{\beta}$.

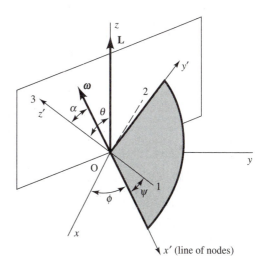

Figure 9.6.1 Diagram showing the relation of the Eulerian angles to the fixed and the rotating coordinate axes.

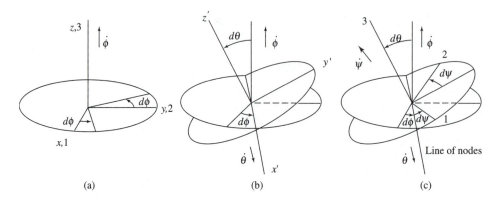

Figure 9.6.2   Generation of any arbitrary infinitesimal rotation of a rigid body as the vector sum of infinitesimal rotations through the free Eulerian angles, (a) $d\boldsymbol{\phi}$, (b) $d\boldsymbol{\theta}$, and (c) $d\boldsymbol{\psi}$.

We have written the infinitesimal angle of rotation as a vector. The angular velocity $\boldsymbol{\omega}$ of a rotating rigid body is a rotation through an infinitesimal angle in an infinitesimal amount of time and is a vector quantity.[4] A rotation through an infinitesimal angle is, therefore, also a vector that points in the same direction as its corresponding angular velocity vector. This direction is given by the conventional right-hand rule. The rotation through the angle $d\boldsymbol{\beta}$ can be "decomposed" into three successive infinitesimal rotations through the angles $d\boldsymbol{\phi}$, $d\boldsymbol{\theta}$, and $d\boldsymbol{\psi}$, respectively. First, starting at $t = 0$, rotate the $O123$-axes counterclockwise about the fixed z-axis through the angle $d\boldsymbol{\phi}$. This rotation is shown in Figure 9.6.2a. It brings the 1-axis into coincidence with the x'-axis shown in Figure 9.6.2b. The direction of $d\boldsymbol{\phi}$ is along the fixed z-axis—the direction a right-hand screw would advance as a result of the counterclockwise turning. Next rotate the $O123$-axes counterclockwise about the x'-axis through an angle $d\boldsymbol{\theta}$. This rotation is shown in Figure 9.6.2b. It brings the 3-axis into coincidence with the z'-axis. The direction of $d\boldsymbol{\theta}$ is along the x'-axis. Finally, rotate the $O123$-axes counterclockwise about the z'-axis through the angle $d\boldsymbol{\psi}$. This rotation is shown in Figure 9.6.2c and its direction is along the z'-axis. The resulting orientation of the $O123$ "body" coordinate system relative to the $Oxyz$ "fixed" coordinate system is what is shown in Figure 9.6.1, the only difference being that in this case the angles are infinitesimals. Clearly, this orientation is the result of the actual rotation of the body through an angle $d\boldsymbol{\beta}$ with angular velocity $\boldsymbol{\omega}$, where

$$d\boldsymbol{\beta} = \boldsymbol{\omega}\, dt = d\boldsymbol{\phi} + d\boldsymbol{\theta} + d\boldsymbol{\psi}$$

$$\therefore \boldsymbol{\omega} = \dot{\boldsymbol{\phi}} + \dot{\boldsymbol{\theta}} + \dot{\boldsymbol{\psi}}$$

(9.6.1)

Most problems involving a rotating rigid body prove to be more tractable when the equations of motion are written in terms of either the $O123$ or $Ox'y'z'$ rotating frames of reference. The rotational motion of the body in either of these two coordinate systems, however, can be related directly to a spatially fixed frame if we express the equations of

---

[4] Actually, it is a *pseudo-vector*. Vectors change sign under the parity operation. Pseudo-vectors do not. This difference in their behavior need not concern us here.

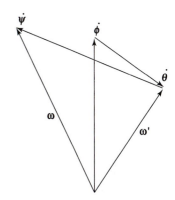

Figure 9.6.3 Vector diagram indicating the relationship among the angular velocities $\dot{\boldsymbol{\phi}}$, $\dot{\boldsymbol{\theta}}$, $\dot{\boldsymbol{\psi}}$, and $\boldsymbol{\omega}$ and $\boldsymbol{\omega}'$.

motion in terms of the *Eulerian angles* and their time derivatives. As a first step toward generating the equations of motion for a rotating body (we ignore its translational motion), we need to find the three components of the angular velocity $\boldsymbol{\omega}$ in each of these coordinate systems in terms of these variables. On examining Figures 9.6.1 and 9.6.2, we see that $\boldsymbol{\omega}$ consists of a rotation about the 3-axis with angular velocity $\dot{\boldsymbol{\psi}}$ superimposed on the rotation of the primed system whose angular velocity we will call $\boldsymbol{\omega}'$ (Figure 9.6.3). It should be clear that $\boldsymbol{\omega}'$ is the vector sum of $\dot{\boldsymbol{\phi}}$ and $\dot{\boldsymbol{\theta}}$ and is given by

$$\boldsymbol{\omega}' = \dot{\boldsymbol{\phi}} + \dot{\boldsymbol{\theta}} \tag{9.6.2}$$

We can now find the components of $\boldsymbol{\omega}'$ in the $Ox'y'z'$ system. First, we write down the relevant components of $\dot{\boldsymbol{\phi}}$, $\dot{\boldsymbol{\theta}}$, and $\dot{\boldsymbol{\psi}}$

$$
\begin{array}{lll}
\dot{\phi}_{x'} = 0 & \dot{\theta}_{x'} = \dot{\theta} & \dot{\psi}_{x'} = 0 \\
\dot{\phi}_{y'} = \dot{\phi}\sin\theta & \dot{\theta}_{y'} = 0 & \dot{\psi}_{y'} = 0 \\
\dot{\phi}_{z'} = \dot{\phi}\cos\theta & \dot{\theta}_{z'} = 0 & \dot{\psi}_{z'} = \dot{\psi}
\end{array}
\tag{9.6.3}
$$

Thus

$$
\begin{aligned}
\omega'_{x'} &= \dot{\phi}_{x'} = \dot{\theta} \\
\omega'_{y'} &= \dot{\phi}_{y'} = \dot{\phi}\sin\theta \\
\omega'_{z'} &= \dot{\phi}_{z'} = \dot{\phi}\cos\theta
\end{aligned}
\tag{9.6.4}
$$

Now $\boldsymbol{\omega}$ differs from $\boldsymbol{\omega}'$ only in the rate of rotation $\dot{\boldsymbol{\psi}}$ about the z'-axis. The components of $\boldsymbol{\omega}$ in the primed system are, therefore,

$$
\begin{aligned}
\omega_{x'} &= \dot{\theta} \\
\omega_{y'} &= \dot{\phi}\sin\theta \\
\omega_{z'} &= \dot{\phi}\cos\theta + \dot{\psi}
\end{aligned}
\tag{9.6.5}
$$

Next, we express the components of $\dot{\boldsymbol{\phi}}$, $\dot{\boldsymbol{\theta}}$, and $\dot{\boldsymbol{\psi}}$ in the $O123$ system

$$
\begin{array}{lll}
\dot{\phi}_1 = \dot{\phi}_{y'}\ \sin\psi = \dot{\phi}\sin\theta\sin\psi & \dot{\theta}_1 = \dot{\theta}_{x'}\cos\psi = \dot{\theta}\cos\psi & \dot{\psi}_1 = 0 \\
\dot{\phi}_2 = \dot{\phi}_{y'}\ \cos\psi = \dot{\phi}\sin\theta\cos\psi & \dot{\theta}_2 = -\dot{\theta}_{x'}\sin\psi = -\dot{\theta}\sin\psi & \dot{\psi}_2 = 0 \\
\dot{\phi}_3 = \dot{\phi}\cos\theta & \dot{\theta}_3 = 0 & \dot{\psi}_3 = \dot{\psi}
\end{array}
\tag{9.6.6}
$$

and then use them to obtain the components of $\boldsymbol{\omega}$ in the O123 system

$$\omega_1 = \dot{\phi}\sin\theta\sin\psi + \dot{\theta}\cos\psi$$
$$\omega_2 = \dot{\phi}\sin\theta\cos\psi - \dot{\theta}\sin\psi \qquad (9.6.7)$$
$$\omega_3 = \dot{\phi}\cos\theta + \dot{\psi}$$

(We do not need to use Equations 9.6.7 at present but will refer to them later.)

Now in the present case in which there is zero torque acting on the body, the angular momentum vector $\mathbf{L}$ is constant in magnitude and direction in the fixed system $Oxyz$. Let us choose the z-axis to be the direction of $\mathbf{L}$. This is known as the *invariable line*. From Figure 9.6.1 we see that the components of $\mathbf{L}$ in the primed system are

$$L_{x'} = 0$$
$$L_{y'} = L\sin\theta \qquad (9.6.8)$$
$$L_{z'} = L\cos\theta$$

We again restrict ourselves to the case of a body with an axis of symmetry, the 3-axis. Because the x' and y'-axes lie in the 1, 2 plane, and the z'-axis coincides with the 3-axis, then the primed axes are also principal axes. In fact, the principal moments are the same: $I_1 = I_2 = I_{x'x'} = I_{y'y'} = I$ and $I_3 = I_{z'z'} = I_s$.

Now consider the first of Equations 9.6.5 and 9.6.8 giving the x' component of the angular velocity and angular momentum of the body, namely zero. From these we see that $\dot{\theta} = 0$. Hence, θ is constant, and $\boldsymbol{\omega}$, having no x'-component, must lie in the $y'z'$ plane as shown. Let α denote the angle between the angular velocity vector $\boldsymbol{\omega}$ and the z'-axis. Then, in addition to Equations 9.6.5 and 9.6.8, we also have the following:

$$\omega_{y'} = \omega\sin\alpha \qquad \omega_{z'} = \omega\cos\alpha \qquad (9.6.9a)$$
$$L_{y'} = I\omega\sin\alpha \qquad L_{z'} = I_s\omega\cos\alpha \qquad (9.6.9b)$$

Therefore

$$\frac{L_{y'}}{L_{z'}} = \tan\theta = \frac{I}{I_s}\tan\alpha \qquad (9.6.10)$$

giving the relation between the angles θ and α.

According to the above result, θ is less than or greater than α, depending on whether I is less than I_s or greater than I_s, respectively. In other words, the angular momentum vector lies between the symmetry axis and the axis of rotation in the case of a flattened body ($I < I_s$), whereas in the case of an elongated body ($I > I_s$), the axis of rotation lies between the axis of symmetry and the angular momentum vector. The two cases are illustrated in Figure 9.6.4. In either case, as the body rotates, an observer in the fixed coordinate system would see the axis of symmetry (z'-axis or 3-axis) trace out a cone as it precesses about the constant angular momentum vector $\mathbf{L}$. At the same time the observer would see the axis of rotation ($\boldsymbol{\omega}$ vector) precess about $\mathbf{L}$ with the same frequency. The surface traced out by $\boldsymbol{\omega}$ about $\mathbf{L}$ is called the *space cone*, as indicated. The precessional motion of $\boldsymbol{\omega}$ and the body symmetry axis can be visualized by the body cone rolling

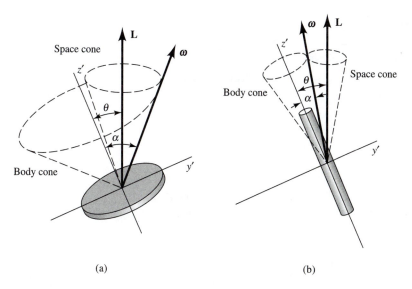

Figure 9.6.4 Free rotation of a (a) disc and (b) a rod. The space cones and body cones are shown dotted.

along the space cone (Figure 9.6.4). The central axis of the body cone is the symmetry axis of the body, and the line where the two cones touch defines the direction of $\boldsymbol{\omega}$.

Referring to Figure 9.6.1, we see that the angular speed of rotation of the $y'z'$ plane about the z-axis is equal to the time rate of change of the angle ϕ. Thus, $\dot{\phi}$ is the angular rate of precession of the symmetry axis and of the axis of rotation about the invariable line (**L** vector) as viewed from outside the body. This precession appears as a "wobble" such as that seen in an imperfectly thrown football or discus. From the second of Equations 9.6.5 and the first of Equations 9.6.9a, we have $\dot{\phi} \sin\theta = \omega \sin\alpha$ or

$$\dot{\phi} = \omega \frac{\sin\alpha}{\sin\theta} \tag{9.6.11}$$

for the rate of precession. Equation 9.6.11 can be put into a somewhat more useful form by using the relation between α and θ given by Equation 9.6.10. After a little algebra, we obtain

$$\dot{\phi} = \omega\left[1 + \left(\frac{I_s^2}{I^2} - 1\right)\cos^2\alpha\right]^{1/2} \tag{9.6.12}$$

for the wobble rate in terms of the angular speed ω of the body about its axis of rotation and the inclination α of the axis of rotation to the symmetry axis of the body.

To summarize our analysis of the free rotation of a rigid body with an axis of symmetry, there are three basic angular rates: the magnitude ω of the angular velocity, the precession of angular rate Ω of the axis of rotation (direction of $\boldsymbol{\omega}$) about the symmetry axis of the body, and the precession (wobble) of angular rate $\dot{\phi}$ of the symmetry axis about the invariable line (constant angular momentum vector).

EXAMPLE 9.6.1

Precession of a Frisbee

As an example of the above theory we consider the case of a thin disc, or any symmetric and fairly "flat" object, such as a china plate or a Frisbee. The perpendicular-axis theorem for principal axes is $I_1 + I_2 = I_3$, and, for a symmetric body $I_1 = I_2$, so that $2I_1 = I_3$. In our present notation this is $2I = I_s$, so the ratio

$$\frac{I_s}{I} = 2$$

to a good approximation. If our object is thrown into the air in such a way that the angular velocity $\boldsymbol{\omega}$ is inclined to the symmetry axis by an angle α, then Equation 9.5.8 gives

$$\Omega = \omega \cos \alpha$$

for the rate of precession of the rotational axis about the symmetry axis.

For the precession of the symmetry axis about the invariable line, the wobble as seen from the outside, Equation 9.6.12 yields

$$\dot{\phi} = \omega(1 + 3\cos^2\alpha)^{1/2}$$

In particular, if α is quite small so that $\cos\alpha$ is very nearly unity, then we have approximately

$$\Omega \approx \omega$$
$$\dot{\phi} \approx 2\omega$$

Thus, the wobble rate is very nearly twice the angular speed of rotation.

EXAMPLE 9.6.2

Free Precession of the Earth

In the motion of the Earth, it is known that the axis of rotation is very slightly inclined with respect to the geographic pole defining the axis of symmetry. The angle α is about 0.2 sec of arc (shown exaggerated in Figure 9.6.5). It is also known that the ratio of the moments of inertia I_s/I is about 1.00327 as determined from Earth's oblateness. From Equation 9.5.8 we have, therefore,

$$\Omega = 0.00327\omega$$

Then, because $\omega = 2\pi/\text{day}$, the period of the precession is calculated to be

$$\frac{2\pi}{\Omega} = \frac{1}{0.00327} \text{ days} = 305 \text{ days}$$

The observed period of precession of Earth's axis of rotation about the pole is about 440 days. The disagreement between the observed and calculated values is attributed to the fact that Earth is not perfectly rigid, nor is it in the shape of a perfectly symmetric oblate spheroid; it is shaped more like a slightly lumpy pear.

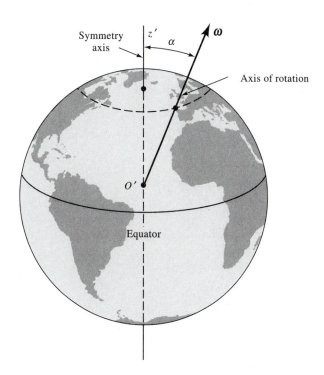

Figure 9.6.5 Showing the symmetry axis and the rotational axis of the Earth. (The angle α is greatly exaggerated.)

With regard to the precession of Earth's symmetry axis as viewed from space, Equation 9.6.12 gives

$$\dot{\phi} = 1.00327\omega$$

The associated period of Earth's wobble is, thus,

$$\frac{2\pi}{\dot{\phi}} = \frac{2\pi}{\omega}\frac{1}{1.00327} \approx 0.997 \ \ \text{day}$$

This precession is superimposed on the much longer precession of 26,000 years of Earth's rotational axis about an axis perpendicular to its orbital plane. This latter precession is caused by the torques exerted on the oblate Earth by the Sun and Moon. The fact that its period is so much longer than that of the free precession justifies ignoring the external torques in calculating the period of the free precession.

9.7 | Motion of a Top

In this section we study the motion of a symmetric rigid body that is free to turn about a fixed point and on which there is exerted a torque, instead of no torque, as in the case of free precession. The case is exemplified by a symmetric top, that is, a rigid body with $I_1 = I_2 \neq I_3$.

The notation for our coordinate axes is shown in Figure 9.7.1a. For clarity, only the z'-, y'-, and z-axes are shown in Figure 9.7.1b, the x'-axis being normal to the paper. The origin O is the fixed point about which the top turns.

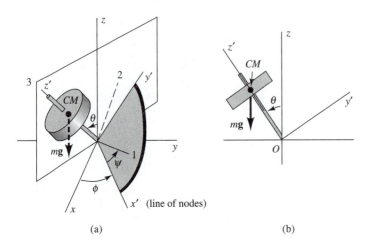

Figure 9.7.1 A symmetric top.

The torque about O resulting from the weight is of magnitude $mgl \sin \theta$, l being the distance from O to the center of mass CM. This torque is about the x'-axis, so that

$$
\begin{aligned}
N_{x'} &= mgl \sin \theta \\
N_{y'} &= 0 \\
N_{z'} &= 0
\end{aligned}
\tag{9.7.1}
$$

The components of the angular velocity of the top $\boldsymbol{\omega}$ are given by Equations 9.6.5. Hence, the angular momentum of the top has the following components in the primed system:

$$
\begin{aligned}
L_{x'} &= I_{x'x'} \omega_{x'} = I\dot{\theta} \\
L_{y'} - I_{y'y'} \omega_{y'} &= I\dot{\phi} \sin \theta \\
L_{z'} &= I_{z'z'} \omega_{z'} = I_s(\dot{\phi} \cos \theta + \dot{\psi}) = I_s S
\end{aligned}
\tag{9.7.2}
$$

Here we use the same notation for the moments of inertia as in the previous section, and in the last equation (9.7.2) we abbreviated the quantity $\omega_{z'} = \dot{\phi} \cos \theta + \dot{\psi}$ by the letter S, called the *spin*, which is simply the angular velocity about the top's axis of symmetry.

The fundamental equation of motion in the primed system (see Equation 5.2.10 or 9.3.3) is

$$
\mathbf{N} = \left(\frac{d\mathbf{L}}{dt} \right)_{rot} + \boldsymbol{\omega}' \times \mathbf{L}
\tag{9.7.3}
$$

in which the components of $\mathbf{N}$, $\mathbf{L}$, and $\boldsymbol{\omega}'$ are given by Equations 9.7.1, 9.7.2, and 9.6.4, respectively. Consequently, the component equations of motion are found to be the following:

$$
mgl \sin \theta = I\ddot{\theta} + I_s S\dot{\phi} \sin \theta - I\dot{\phi}^2 \cos \theta \sin \theta
\tag{9.7.4a}
$$

$$
0 = I \frac{d}{dt}(\dot{\phi} \sin \theta) - I_s S\dot{\theta} + I\dot{\theta}\dot{\phi} \cos \theta
\tag{9.7.4b}
$$

$$
0 = I_s \dot{S}
\tag{9.7.4c}
$$

The last equation (9.7.4c) shows that S, the spin of the top about its symmetry axis, remains constant. Also, of course, the component of the angular momentum along that axis is constant

$$L_{z'} = I_s S = \text{constant} \tag{9.7.5}$$

The second equation is then equivalent to

$$0 = \frac{d}{dt}(I\dot{\phi}\sin^2\theta + I_s S \cos\theta) \tag{9.7.6}$$

so that

$$I\dot{\phi}\sin^2\theta + I_s S \cos\theta = L_z = \text{constant} \tag{9.7.7}$$

This last constant is the component of the angular momentum along the fixed z-axis (see Problem 9.23). It is easy to see that the two angular momenta $L_{z'}$ and L_z must be constant. The gravitational torque is always directed along the x'-axis, or the *line of nodes*. Because both the body symmetry axis (the z'-axis) and the spatially fixed z-axis are perpendicular to the line of nodes, there can be no component of torque along either of them.

Steady Precession

At this point we discuss a simple special case of the motion of a top, namely that of steady precession in a horizontal plane. This is the common "demonstration" case in which the spin axis remains horizontal and precesses at a constant rate around a vertical line, the z-axis in our notation. Then we have $\theta = 90° = \text{constant}$, $\dot{\theta} = \ddot{\theta} = 0$. Equation 9.7.4a then reduces to the simple relation

$$mgl = I_s S \dot{\phi} \tag{9.7.8}$$

Now it is easy to see that the quantity mgl is just the (scalar) torque about the x'-axis. Furthermore, the horizontal (vector) component of the angular momentum has a magnitude of $I_s S$, and it describes a circle in the horizontal plane. Consequently, the extremity of the $\mathbf{L}$ vector has a velocity (time rate of change) of magnitude $I_s S \dot{\phi}$ and a direction that is parallel to the x'-axis. Thus, Equation 9.7.8 is simply a statement of the general relation $\mathbf{N} = d\mathbf{L}/dt$ for the special case in point (see Figure 9.7.2).

The more general case of steady precession in which the angle θ is constant but has a value other than 90° is still handled by use of Equation 9.7.4a, which gives, on setting $\ddot{\theta} = 0$ and canceling the common factor $\sin\theta$,

$$mgl = I_s S \dot{\phi} - S\dot{\phi}^2 \cos\theta \tag{9.7.9}$$

This is a quadratic equation in the unknown $\dot{\phi}$. Solving it yields two roots

$$\dot{\phi} = \frac{I_s S \pm \left(I_s^2 S^2 - 4mglI\cos\theta\right)^{1/2}}{2I\cos\theta} \tag{9.7.10}$$

Thus, for a given value of θ, two rates of steady precession of the top are possible: a fast precession (plus sign) and a slow precession (minus sign). Which of the two occurs depends

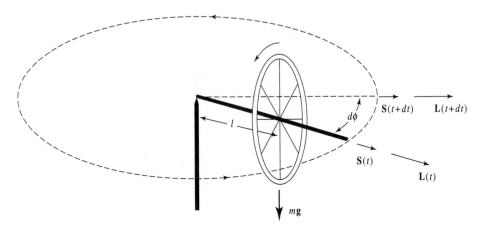

Figure 9.7.2   Bicycle wheel spinning with spin angular velocity **S** and spin angular momentum **L** about an axis pivoted at its end and executing steady precession in a horizontal plane.

on the initial conditions. Usually, it is the slower one that takes place in the motion of a simple top. In either case, the quantity in parentheses must be zero or positive for a physically possible solution, that is,

$$I_s^2 S^2 \geq 4mglI \cos\theta \qquad (9.7.11)$$

Sleeping Top

Anyone who has played with a top knows that if the top is set spinning sufficiently fast and is started in a vertical position, the axis of the top remains steady in the upright position, a condition called *sleeping*. This corresponds to a constant value of zero for θ in the above equations. Because $\dot\phi$ must be real, we conclude that the criterion for stability of the sleeping top is given by

$$I_s^2 S^2 \geq 4mglI \qquad (9.7.12)$$

If the top slows down through friction so that the condition no longer holds, then it begins to fall and eventually topples over.

EXAMPLE 9.7.1

A toy gyroscope has a mass of 100 g and is made in the form of a uniform disc of radius $a = 2$ cm fastened to a light spindle, the center of the disc being 2 cm from the pivot. If the gyroscope is set spinning at a rate of 20 revolutions per second, find the period for steady horizontal precession.

Solution:

Using cgs units we have $I_s = \frac{1}{2}ma^2 = \frac{1}{2} \times 100$ g $\times$ (2 cm)$^2 = 200$ g cm^2. For the spin we must convert revolutions per second to radians per second, that is, $S = 20 \times 2\pi$ rad/s.

Equation 9.7.8 then gives the precession rate

$$\dot{\phi} = \frac{mgl}{I_s S} = \frac{100 \text{ g} \times 980 \text{ cm s}^{-2} \times 2 \text{ cm}}{200 \text{ g cm}^2 \times 40 \times 3.142 \text{ s}^{-1}} = 7.8 \text{ s}^{-1}$$

in radians per second. The associated period is then

$$\frac{2\pi}{\dot{\phi}} = \frac{2 \times 3.142}{7.8 \text{ s}^{-1}} = 0.81 \text{ s}$$

EXAMPLE 9.7.2

Find the minimum spin of the gyroscope in Example 9.7.1 so that it can sleep in the vertical position.

Solution:

We need, in addition to the values in the previous example, the moment of inertia I about the x'- or y'-axes. By the parallel-axis theorem, we have $I = I_{x'x'} = I_{y'y'} = \frac{1}{4}ma^2 + ml^2 = \frac{1}{4}100 \text{ g} \times (2 \text{ cm})^2 + 100 \text{ g} \times (2 \text{ cm})^2 = 500 \text{ g cm}^2$. From Equation 9.7.12, we can then write

$$S \geq \frac{2}{I_s}(mglI)^{1/2} = \tfrac{2}{200}(200 \times 980 \times 2 \times 500)^{1/2} \text{ s}^{-1} = 140 \text{ s}^{-1}$$

or, in revolutions per second, the minimum spin is

$$S = \frac{140}{2\pi} = 22.3 \text{ rps}$$

9.8 | The Energy Equation and Nutation

Usually buried away in some dark labyrinth deep within the bowels of most university physics departments lies a veritable treasure of relatively old, cast-off experimental equipment no longer in active use. Squirreled away over the years, it is occasionally resurrected for duty as lecture demonstration apparatus after suitable modification. Some of it, though, is relatively new and was actually purchased intentionally for the expressed use of lecture demonstration. One of the more delightful devices of this type is the air gyroscope. It is periodically brought out and used by professors teaching classes on mechanics to illustrate precessional motion (Figure 9.8.1). Air gyros have the virtue of being almost friction-free devices so that all the nuances of rotational motion can be observed before they die away due to frictional effects. The rotating part of the air gyro is a machined spherical ball with a thin cylindrical rod attached to it that sticks out radially away from the center of the sphere. This rod marks the body axis of symmetry. The sphere rests inside a polished, machined, inverted, hemispherical surface, whose radius of curvature is matched to fit the spherical portion of the gyro. A thin hole is drilled up into the hemispherical surface so that air can be blown into it, creating an almost frictionless cushion on which the spherical ball rests. Normally, the gyro is made to spin about its body symmetry axis, and the axis is then released in a variety of ways to initialize the motion.

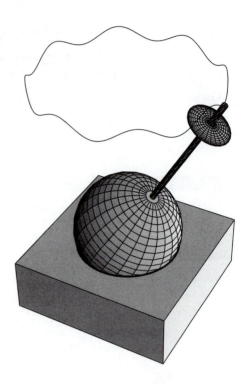

Figure 9.8.1   The air gyro.

In one of the more classic demonstrations with the air gyro, the professor makes it into a top by placing a small disc with a hole through its central axis about the thin cylindrical rod attached to the gyro, positioning it just below the tip of the rod, as shown in Figure 9.8.1. Then, to set the stage for the striking effects soon to follow, the instructor simply tilts the axis of the top somewhat away from the vertical and releases it from rest. Because the top is unbalanced by the weight of the disc mounted on the thin rod marking the axis, it topples over and strikes the surface of the table on which the air gyro rests— behavior that is completely expected by the on-looking students. Gravity creates a torque on the unbalanced top and it topples over. But now, the canny professor repeats the experiment, only this time initializing the motion of the top by again tilting it to some angle θ_1, but this time spinning it up about its symmetry axis before "dropping" it. Indeed, as before, the top starts to topple over, but it also begins to precess and its symmetry axis, to the great astonishment of the now-captivated student audience, does not fall all the way down to the surface. It seems to strike an impenetrable barrier at some angle θ_2 and, miraculously, it bounces back up, slowing down along the way, and finally halting momentarily at the same angle of tilt θ_1 at which it started. Then it falls, starts to precess again, and repeats this "up and down bobbing" motion over and over. The professor initialized this remarkable motion simply by holding its body symmetry axis steady at the angle of tilt θ_1, spinning the top and then releasing the axis. The bobbing motion, in which the symmetry axis oscillates up and down between the two angular limits, θ_1 and θ_2, is called *nutation*. As we shall see in the ensuing analysis, it is only one of several nutational modes that can be established, depending on how the motion is initialized.

In analyzing the motion of the top, we assume that there are no frictional forces; hence, the total energy $E = T_{rot} + V$ of the top remains constant

$$E = \tfrac{1}{2}\left(I\omega_x^2 + I\omega_y^2 + I_s\omega_z^2\right) + mgl\,\cos\theta = \text{constant} \tag{9.8.1}$$

or, in terms of the Eulerian angles,

$$E = \tfrac{1}{2}(I\dot\theta^2 + I\dot\phi^2\sin^2\theta + I_sS^2) + mgl\,\cos\theta \tag{9.8.2}$$

From Equation 9.7.7, we can solve for $\dot\phi$

$$\dot\phi = \frac{L_z - I_sS\cos\theta}{I\sin^2\theta} = \frac{L_z - L_{z'}\cos\theta}{I\sin^2\theta} \tag{9.8.3}$$

and substituting this expression into Equation 9.8.2, we obtain an expression for E

$$E = \tfrac{1}{2}I_sS^2 + \tfrac{1}{2}I\dot\theta^2 + \frac{(L_z - L_{z'}\cos\theta)^2}{2I\sin^2\theta} + mgl\,\cos\theta \tag{9.8.4}$$

in terms of θ and constants of the motion. Because the first term on the right-hand side of this equation, $I_sS^2/2$ (the "spin energy") is a constant of the motion, we can subtract it from the total energy of the top and define another constant

$$E' = E - \tfrac{1}{2}I_sS^2 \tag{9.8.5}$$

which essentially is the residual total energy of the top in the motion along the θ-coordinate. We now rewrite Equation 9.8.4 as

$$E' = \tfrac{1}{2}I\dot\theta^2 + V(\theta) \tag{9.8.6}$$

where we have defined the *effective potential*[5] of the top $V(\theta)$ to be

$$V(\theta) = \frac{(L_z - L_{z'}\cos\theta)^2}{2I\sin^2\theta} + mgl\,\cos\theta \tag{9.8.7}$$

This definition of the effective potential allows us to focus exclusively on the motion of the top along the θ-coordinate only by means of a relation that looks like a one-dimensional total energy equation; that is, $E' = T + V$. The definition of $V(\theta)$ here is analogous to the definition of $U(r)$ presented in Section 6.11, where we focused on the radial component of the orbital motion of an object subject to a central force field (such as planets in orbit about the Sun). In that particular instance, we saw that the effective potential constrained the object to lie within two radial limits ($r_1 \le r \le r_2$) called the turning points of the orbit. The same thing happens in this case for the motion along the θ coordinate. The moving object cannot pass into a region where its effective potential would exceed its total energy E'. Figure 9.8.2 depicts this situation. It shows the effective potential $V(\theta)$ plotted between 0 and π, the minimum and maximum possible values of θ, along with the constant value E'. The actual value of E' determines the lower and upper bounds for the motion of the top along the θ coordinate; clearly, θ cannot fall outside the two limits, or "turning points," θ_1 and θ_2.

---

[5] This quantity is really a potential energy, not a potential. Nonetheless, the terminology is standard and we hope that it doesn't cause too much consternation for those who strive for semantic rigor.

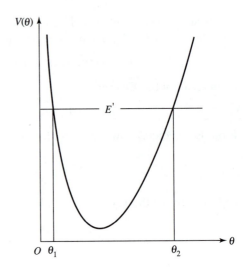

Figure 9.8.2   Energy plot.

Examining the figure shows us that the top reaches either of the turning points in its motion when $V(\theta) = E'$. We could find these values by solving for $\dot\theta$ in Equation 9.8.6

$$\dot\theta^2 = \frac{2}{I}[E' - V(\theta)] \tag{9.8.8}$$

and setting it equal to zero. This problem can be made a little more tractable by substituting $u = \cos\theta$ into Equation 9.8.8 and solving for $\dot u$

$$\dot u = -\dot\theta \sin\theta \tag{9.8.9a}$$

$$\dot\theta = -\frac{\dot u}{(1-u^2)^{1/2}} \tag{9.8.9b}$$

$$\dot u^2 = \frac{2}{I}(1-u^2)(E' - mglu) - \frac{1}{I^2}(L_z - L_{z'}u)^2 \tag{9.8.9c}$$

or

$$\dot u^2 = f(u) \tag{9.8.10}$$

The two turning points θ_1 and θ_2 correspond to the roots of the equation $f(u) = 0$. Note that $f(u)$ is a cubic polynomial; hence, it has three roots. One, though, is nonphysical.

In principle, we could solve for u (hence, θ) as a function of time by integration

$$t = \int \frac{du}{\sqrt{f(u)}} \tag{9.8.11}$$

We need not do this, however, if we merely want to investigate the general properties of the motion. In the region between the two turning points, in which motion is physically possible, $f(u)$ must be positive. Furthermore, because θ must lie between the absolute limits 0 and π, u is limited to values between -1 and $+1$. A plot of $f(u)$ is shown in Figure 9.8.3a for the case in which there are two roots u_1 and u_2 between 0 and $+1$,

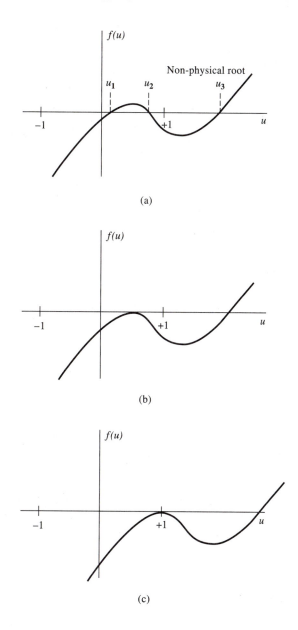

Figure 9.8.3   Effective potential
diagram for the spinning top.

a typical situation (the surface of the table on which the air gyro rests corresponds to
$\theta = \pi/2$ or $u = 0$). $f(u)$ is positive between those two roots and the region in between
is where the motion takes place. The values of θ corresponding to those two roots are
θ_1 and θ_2, the two turning points where θ reverses direction. The symmetry axis of
the top oscillates back and forth between these two values of θ as the top precesses
about the vertical. The overall motion is called *nutational precession*.

The nutational precession can take on any of the three modes indicated in Figures 9.8.4a,
b, and c. The paths shown there represent the projection of the body symmetry axis onto

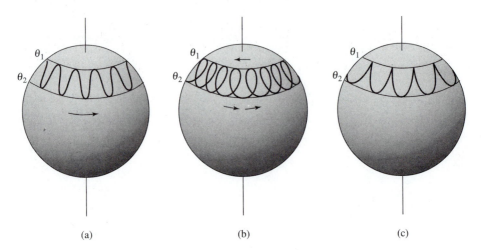

(a)   (b)   (c)

Figure 9.8.4   Nutational motion of the spinning top. (a) $\dot{\phi}$ never changes sign; (b) $\dot{\phi}$ changes sign; (c) $\dot{\phi} = 0$ at $\theta = \theta_1$ during a precessional period.

a sphere attached to the $Oxyz$ coordinate system fixed in space. The mode that actually occurs depends on whether or not the angular speed $\dot{\phi}$ (given by Equation 9.8.3) changes sign as θ varies between its two limits during the motion. This, in turn, depends on the values of the angular momentum about the z-axis L_z and the *spin angular momentum* $L_{z'} (= I_s S)$. If $\dot{\phi}$ does not change sign, the precession never reverses direction as the symmetry axis oscillates between the two angular limits θ_1 and θ_2. This is the mode depicted in Figure 9.8.4a. If $\dot{\phi}$ does change sign during the precessional nutation, it must have opposite signs at θ_1 and θ_2 (see Equation 9.8.3). In this case, the mode depicted by Figure 9.8.4b is obtained.

The third mode shown in Figure 9.8.4c is the common one previously described in which our canny professor initialized the motion by tilting the axis of the top, spinning it about this axis, and releasing it from rest. This had the effect of initializing the constants of motion L_z and $L_{z'}$, such that

$$\frac{L_z}{L_{z'}} = \cos\theta_1 \qquad\qquad (9.8.12)$$

and

$$\dot{\phi}\big|_{\theta=\theta_1} = 0 \qquad \dot{\theta}\big|_{\theta=\theta_1} = 0 \qquad\qquad (9.8.13)$$

The professor might then proceed to demonstrate the other two modes of precessional nutation simply by giving the tip of the body symmetry axis either a mild forward shove such that $\dot{\phi}\big|_{\theta=\theta_1} > 0$ (see Figure 9.8.4a) or a mild backward shove such that $\dot{\phi}\big|_{\theta=\theta_1} < 0$ (see Figure 9.8.4b).

There are two other possibilities for the motion of the top. If we have a double root as indicated by the plot of $f(u)$ in Figure 9.8.3b, that is, if $u_1 = u_2$, then there is no nutation, and the top precesses steadily. Our indubitable professor can demonstrate this mode by spinning the top quite rapidly and then slowly nudging the symmetry axis forward such that the angular velocity $\dot{\phi} \geq 0$ and then releasing it delicately with $\dot{\theta} = 0$ (see Problem 9.24). The final possibility is that of the sleeping top, discussed in the previous section. It corresponds to the plot in Figure 9.8.3c in which $f(u)$ has only the single physical root of $u_1 = 1$. Our intrepid professor can demonstrate this mode by aligning the axis of the top vertically, spinning it fast enough that the stability criterion set forth in Equation 9.7.12 is satisfied and then releasing it; if the professor fails to spin it fast enough, it will prove to be unstable and will topple over.

9.9 | The Gyrocompass

Let us consider the motion of a top that is mounted on a gimball support that constrains the spin axis to remain horizontal, but the axis is otherwise free to turn in any direction. The situation is diagrammed in Figure 9.9.1, which is taken from Figure 9.7.1 except that now $\theta = 90°$ and the unprimed axes are labeled to correspond to directions on the Earth's

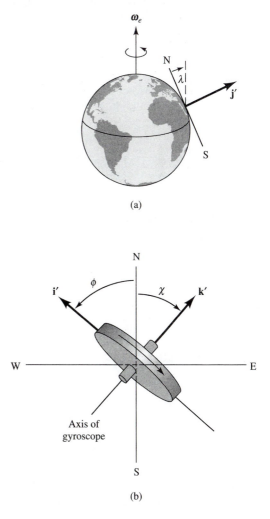

(a)

(b)

Figure 9.9.1   The gyrocompass.

surface as shown. The top is centrally mounted unlike the unbalanced tops just described, so that no gravitational torque acts upon it. In this case, the spinning top, although not completely torque-free, behaves somewhat like a gyroscope, which is a completely torque-free top.

We know from Chapter 5 that Earth's angular velocity, here denoted by $\boldsymbol{\omega}_e$, has components $\omega_e \cos \lambda$ (north) and $\omega_e \sin \lambda$ (vertical), where λ is the latitude. In the primed coordinate system we can then write

$$\boldsymbol{\omega}_e = \mathbf{i}'\omega_e \cos \lambda \cos \phi + \mathbf{j}'\omega_e \sin \lambda + \mathbf{k}'\omega_e \cos \lambda \sin \phi \qquad (9.9.1)$$

Now the primed system is turning about the vertical with angular rate $\dot{\phi}$ so the angular velocity of the primed system is

$$\boldsymbol{\omega}' = \boldsymbol{\omega}_e + \mathbf{j}'\dot{\phi} = \mathbf{i}'\omega_e \cos \lambda \cos \phi + \mathbf{j}'(\dot{\phi} + \omega_e \sin \lambda) + \mathbf{k}'\omega_e \cos \lambda \sin \phi \qquad (9.9.2)$$

Similarly, the gyrocompass itself is turning about the z'-axis at a rate $\dot{\psi}$ superimposed on the components in Equation 9.9.2 so that its angular velocity, referred to the primed system, is

$$\boldsymbol{\omega} = \boldsymbol{\omega}' + \mathbf{k}'\dot{\psi} = \mathbf{i}'\omega_e \cos \lambda \cos \phi + \mathbf{j}'(\dot{\phi} + \omega_e \sin \lambda) + \mathbf{k}'(\dot{\psi} + \omega_e \cos \lambda \sin \phi) \qquad (9.9.3)$$

The principal moments of inertia of the gyrocompass are, as before, $I_1 = I_2 = I$, $I_3 = I_s$; hence, the angular momentum can be expressed as

$$\mathbf{L} = \mathbf{i}'I\omega_e \cos \lambda \cos \phi + \mathbf{j}'I(\dot{\phi} + \omega_e \sin \lambda) + \mathbf{k}'I_s S \qquad (9.9.4)$$

where, in the last term, the total spin S is

$$S = \dot{\psi} + \omega_e \cos \lambda \sin \phi \qquad (9.9.5)$$

Now, because the gyrocompass is free to turn about both the vertical (y'-axis) and the spin or z'-axis, the applied torque required to keep the axis horizontal must be about the x'-axis: $\mathbf{N} = \mathbf{i}'N$. The equation of motion

$$\mathbf{N} = \left(\frac{d\mathbf{L}}{dt}\right)_{rot} + \boldsymbol{\omega}' \times \mathbf{L} \qquad (9.9.6)$$

thus, has components referred to the primed system as follows:

$$N = I\frac{d}{dt}(\omega_e \cos \lambda \cos \phi) + (\boldsymbol{\omega}' \times \mathbf{L})_{x'} \qquad (9.9.7a)$$

$$0 = I\frac{d}{dt}(\dot{\phi} + \omega_e \sin \lambda) + (\boldsymbol{\omega}' \times \mathbf{L})_{y'} \qquad (9.9.7b)$$

$$= I_s \frac{dS}{dt} + (\boldsymbol{\omega}' \times \mathbf{L})_{z'} \qquad (9.9.7c)$$

From the expressions for $\boldsymbol{\omega}'$ and $\mathbf{L}$, we find that $(\boldsymbol{\omega}' \times \mathbf{L})_{z'} = 0$, so Equation 9.9.7c becomes $dS/dt = 0$; thus, S is constant. Furthermore, we find that Equation 9.9.7b becomes

$$0 = I\ddot{\phi} + I\omega_e^2 \cos^2\lambda \cos\phi \sin\phi - I_s S\omega_e \cos\lambda \cos\phi \qquad (9.9.8)$$

It is convenient at this point to express the angle ϕ in terms of its complement $\chi = 90° - \phi$, so $\cos\phi = \sin\chi$ and $\ddot{\phi} = -\ddot{\chi}$. Furthermore, we can ignore the term involving ω_e^2 in Equation 9.9.8 because $S \gg \omega_e$ in the present case. Consequently, Equation 9.9.8 reduces to

$$\ddot{\chi} + \left(\frac{I_s S\omega_e \cos\lambda}{I}\right)\sin\chi = 0 \qquad (9.9.9)$$

This is similar to the differential equation for a pendulum. The variable χ oscillates about the value $\chi = 0$, and the presence of any damping causes the axis of the gyrocompass to "seek" and eventually settle down to a north–south direction. For small amplitude, the period of the oscillation is

$$T_0 = 2\pi \left(\frac{I}{I_s \omega_e S \cos\lambda}\right)^{1/2} \qquad (9.9.10)$$

Because the ratio I_s/I is very nearly 2 for any flat-type symmetric object, the period of oscillation is essentially independent of the mass and dimensions of the gyrocompass. Furthermore, because ω_e is very small, the spin S must be fairly large to have a reasonably small period. For example, let the gyrocompass spin at 60 Hz so that $S = 2\pi \times 60$ rad/s. Then, for a flat-type gyrocompass, we find for a latitude of 45° N

$$T_0 = 2\pi \left(\frac{24 \times 60 \times 60}{2 \times 2\pi \times 60 \times 2\pi \times 0.707}\right)^{1/2} \qquad s = \left(\frac{12 \times 60}{0.707}\right)^{1/2} \qquad s \qquad S = 31.9 \ s$$

or about $\frac{1}{2}$ min. In the preceding calculation we have used the fact that

$$\omega_e = 2\pi/(24 \times 60 \times 60) \text{ rad/s}$$

so that the factors 2π all cancel.

9.10 | Why Lance Doesn't Fall Over (Mostly)

In 2003, Lance Armstrong became only the second man in history to win the Tour de France five consecutive times. Armstrong's fifth win did not come easily. Usually, the winner of the Tour has been determined many stages before the last, when the riders enter Paris and make several circuits around the Champs-Elysees, the victor no longer challenged and cruising easily, well-protected by his team members in the pelaton. In 2003, Lance's victory was not secure until the finish of the 19th stage of the 20-stage race, a 49-km individual time trial. It was a rainy day. The road was wet and slippery, and the course was quite technical. Armstrong's chief rival, Jan Ullrich, was in second place overall, only a minute and several seconds behind Armstrong. Ullrich, a strong time trialist himself, had handed Armstrong a rare defeat in stage 12 by more than a minute, an earlier 47-km time trial. As he poised himself for the start of stage 19, he more than likely felt that he might assume the overall lead with another strong showing. Unfortunately, two-thirds of the way through the course, Ullrich, negotiating a curve too fast given the wet conditions, saw any chance of victory hopelessly dashed as his rear wheel skidded out from under him and he fell and slid along the wet road, crashing into hay bales along the side.

Armstrong, wearing the yellow jersey of the overall race leader, had the advantage of starting his trial three minutes after Ullrich. At the time he received word by radio of Ullrich's fall, he was neck and neck, time-wise, with the speedy German and more than likely would have remained well in front for the overall lead, even if Ullrich had not fallen. Given Ullrich's fall, however, Armstrong was able to slow down a bit from that point on, taking extreme caution to negotiate the slippery turns that remained. He cruised into the finish line in third place for the stage and one minute and sixteen seconds in front of Ullrich for the overall lead, his fifth consecutive victory for the Tour de France now assured. Ullrich, clearly shaken but recovering somewhat from his disastrous fall, finished fourth for the stage and second for the Tour.

In this section, we do not take it upon ourselves to anlyze the stability of cyclist and bicycle on a slippery road, clearly an exceedingly complicated issue. Instead, we take on the simpler problem of analyzing the stability of a disk bicycle wheel, of the sort typically used as the rear wheel of a time-trial bike, rolling along by itself on a perfectly rough horizontal road.

The motion of the wheel can be resolved into two parts: (1) translation of its center of mass and (2) rotation about its center of mass. The general equations describing the motion of the wheel were given by the force Equation 8.6.1 and the torque Equation 8.6.2. The external forces acting on the wheel are $m\mathbf{g}$ acting on its center of mass and $\mathbf{F}_P$ acting on it at its point of contact with the road (see Figure 9.10.1). The force equation, therefore, is

$$\mathbf{F}_P + m\mathbf{g} = m\frac{d\mathbf{v}_{cm}}{dt}$$

(9.10.1)

The torque equation is

$$\mathbf{r}_{OP} \times \mathbf{F}_P = \frac{d\mathbf{L}}{dt}$$

(9.10.2)

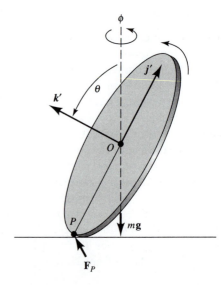

Figure 9.10.1   Coordinates for analyzing the motion of a rolling wheel.

where $\mathbf{r}_{OP}$ is the vector position of the point of contact relative to the center of mass. Solving for $\mathbf{F}_P$ in Equation 9.10.1 and substituting the result into Equation 9.10.2 yields

$$\mathbf{r}_{OP} \times \left(m\frac{d\mathbf{v}_{cm}}{dt} - m\mathbf{g} \right) = \frac{d\mathbf{L}}{dt} \tag{9.10.3}$$

If $\boldsymbol{\omega}$ is the angular velocity of the wheel, then the velocity of the mass center is given by

$$\mathbf{v}_{cm} = \boldsymbol{\omega} \times \mathbf{r}_{PO} = \boldsymbol{\omega} \times (-\mathbf{r}_{OP}) \tag{9.10.4}$$

If a is the radius of the wheel and θ is the inclination of its axis to the vertical as shown in Figure 9.10.1, then in the primed coordinate system, we have

$$\mathbf{r}_{OP} = -\mathbf{j}'a \tag{9.10.5}$$
$$\mathbf{g} = -g(\mathbf{j}' \sin\theta + \mathbf{k}' \cos\theta) \tag{9.10.6}$$

Substituting the preceding relations into Equation 9.10.3 gives

$$-a\mathbf{j}' \times \left[m\frac{d}{dt}(\boldsymbol{\omega} \times a\mathbf{j}') + mg(\mathbf{j}' \sin\theta + \mathbf{k}' \cos\theta) \right] = \frac{d\mathbf{L}}{dt} \tag{9.10.7}$$

Furthermore, the primed coordinate system rotates with angular velocity $\boldsymbol{\omega}'$ (remember— the $\mathbf{i}'$ direction always lies in a horizontal plane and the $\mathbf{j}'$ and $\mathbf{k}'$ vectors always lie in a vertical plane. See the discussion in Section 9.6). Thus, we have

$$\frac{d}{dt} = \left(\frac{d}{dt} \right)_{rot} + \boldsymbol{\omega}' \times \tag{9.10.8}$$

Substituting this operator relation into Equation 9.10.7 gives

$$-ma^2\mathbf{j}' \times \left[\left(\frac{d\boldsymbol{\omega}}{dt} \right)_{rot} \times \mathbf{j}' + \boldsymbol{\omega}' \times (\boldsymbol{\omega} \times \mathbf{j}') \right]$$
$$- mga[\mathbf{j}' \times (\mathbf{j}' \sin\theta + \mathbf{k}' \cos\theta)] = \left(\frac{d\mathbf{L}}{dt} \right)_{rot} + \boldsymbol{\omega}' \times \mathbf{L} \tag{9.10.9}$$

We do not attempt to solve the preceding general equation of motion. Rather, we wish to discuss a special case, namely, that for which the wheel stays very nearly vertical and the direction of the rolling motion is constant or nearly constant: *steady rolling*. This means that θ remains close to 90° so that the complement $\chi = 90° - \theta$ and the angle ϕ both remain small. Under these assumptions $\sin\chi = \chi$ and $\sin\phi = \phi$, approximately, and the general expressions for the components of $\boldsymbol{\omega}'$, $\boldsymbol{\omega}$, and $\mathbf{L}$ as given by Equations 9.6.4, 9.6.5, and 9.7.2 simplify to give

$$\boldsymbol{\omega}' = -\mathbf{i}'\dot{\chi} + \mathbf{j}'\dot{\phi} \tag{9.10.10a}$$
$$\boldsymbol{\omega} = -\mathbf{i}'\dot{\chi} + \mathbf{j}'\dot{\phi} + \mathbf{k}'S \tag{9.10.10b}$$
$$\mathbf{L} = -\mathbf{i}'I\dot{\chi} + \mathbf{j}'I\dot{\phi} + \mathbf{k}'I_s S \tag{9.10.10c}$$

Inserting these into Equation 9.10.9 and performing the indicated operations and dropping higher-order terms in the small quantities χ and ϕ, the following result is obtained:

$$ma^2(\mathbf{i}'\ddot{\chi} - \mathbf{k}'\dot{S} - \mathbf{i}'S\dot{\phi}) - mga\mathbf{i}'\chi = \mathbf{i}'(I\ddot{\chi} + T_s S\dot{\phi})$$
$$+ \mathbf{j}'(I\ddot{\phi} + I_s S\dot{\chi}) + \mathbf{k}'I_s\dot{S} \tag{9.10.11}$$

Equating the three components gives

$$ma^2(\ddot{\chi} - S\dot{\phi}) - mga\chi = -I\ddot{\chi} + I_s S\dot{\phi} \tag{9.10.12a}$$
$$0 = I\ddot{\phi} + I_s S\dot{\chi} \tag{9.10.12b}$$
$$0 = \dot{S}(I_s + ma^2) \tag{9.10.12c}$$

Equation 9.10.12c shows that $\dot{S} = 0$ so S is constant. Equation 9.10.12b can then be integrated to yield $I\dot{\phi} + I_s S\chi = 0$ provided we assume that $\chi = \dot{\phi} = 0$ for the initial condition. Then $\dot{\phi} = -I_s S\chi / I$, which, inserted into Equation 9.10.12a, gives the following separated differential equation for χ:

$$I(I + ma^2)\ddot{\chi} + [I_s(I_s + ma^2)S^2 - Imga]\chi = 0 \tag{9.10.13}$$

The assumed stable rolling then takes place if the quantity in brackets is positive. Thus, the stability criterion is

$$S^2 > \frac{Imga}{I_s(I_s + ma^2)} \tag{9.10.14}$$

EXAMPLE 9.10.1

How fast must a disc bicycle wheel roll to not topple over (assume that the wheel is a uniform thin, circular lamina and that its radius is $a = 0.35$ m)?

Solution:

The moments of inertia of a thin, uniform circular disk are $I_s = 2I = \frac{1}{2}ma^2$. The criterion for stable rolling is given by Equation 9.10.14

$$S^2 > \frac{mga}{2\left(\frac{1}{2}ma^2 + ma^2\right)} = \frac{g}{3a}$$

Because the rolling speed is $v = v_{cm} = aS$, the criterion for stable rolling becomes

$$v^2 > \frac{ga}{3}$$

Putting in numbers, we get

$$v > \left(\frac{9.80\ \text{ms}^{-2} \times 0.35\ \text{m}}{3}\right)^{1/2} = 1.07\ \text{m/s} = 3.85\ \text{km/hr}$$

So, if Lance rides fast in a relatively upright position on a dry road, he probably won't topple over unless an overly zealous spectator snares his handlebars with the straps of his ditty bag!

Problems

9.1   A thin uniform rectangular plate (lamina) is of mass m and dimensions $2a$ by a. Choose a coordinate system $Oxyz$ such that the plate lies in the xy plane with origin at a corner, the long dimension being along the x-axis. Find the following:
(a) The moments and products of inertia
(b) The moment of inertia about the diagonal through the origin
(c) The angular momentum about the origin if the plate is spinning with angular rate ω about the diagonal through the origin
(d) The kinetic energy in part (c)

9.2   A rigid body consists of three thin uniform rods, each of mass m and length $2a$, held mutually perpendicular at their midpoints. Choose a coordinate system with axes along the rods.
(a) Find the angular momentum and kinetic energy of the body if it rotates with angular velocity $\boldsymbol{\omega}$ about an axis passing through the origin and the point $(1, 1, 1)$.
(b) Show that the moment of inertia is the same for any axis passing through the origin.
(c) Show that the moment of inertia of a uniform square lamina is that given in Example 9.1.1 for any axis passing through the center of the lamina and lying in the plane of the lamina.

9.3   Find a set of principal axes for the lamina of Problem 9.1 in which the origin is
(a) At a corner
(b) At the center of the lamina

9.4   A uniform block of mass m and dimensions a by $2a$ by $3a$ spins about a long diagonal with angular velocity $\boldsymbol{\omega}$. Using a coordinate system with origin at the center of the block,
(a) Find the kinetic energy.
(b) Find the angle between the angular velocity vector and the angular momentum vector about the origin.

9.5   A thin uniform rod of length l and mass m is constrained to rotate with constant angular velocity $\boldsymbol{\omega}$ about an axis passing through the center O of the rod and making an angle α with the rod.
(a) Show that the angular momentum $\mathbf{L}$ about O is perpendicular to the rod and is of magnitude $(ml^2\omega/12)\sin\alpha$.
(b) Show that the torque vector $\mathbf{N}$ is perpendicular to the rod and to $\mathbf{L}$ and is of magnitude $(ml^2\omega^2/12)\sin\alpha\cos\alpha$.

9.6   Find the magnitude of the torque that must be exerted on the block in Problem 9.4 if the angular velocity $\boldsymbol{\omega}$ is constant in magnitude and direction.

9.7   A rigid body of arbitrary shape rotates freely under zero torque. By means of Euler's equations show that both the rotational kinetic energy and the magnitude of the angular momentum are constant, as stated in Section 9.4. (*Hint: For* $\mathbf{N}=0$, *multiply Euler's equations (Equation 9.3.5) by* ω_1, ω_2, *and* ω_3, *respectively, and add the three equations. The result indicates the constancy of kinetic energy. Next, multiply by* $I_1\omega_1$, $I_2\omega_2$, *and* $I_3\omega_3$, *respectively, and add. The result shows that* L^2 *is constant.*)

9.8   A lamina of arbitrary shape rotates freely under zero torque. Use Euler's equations to show that the sum $\omega_1^2 + \omega_2^2$ is constant if the 1, 2 plane is the plane of the lamina. This means that the projection of $\boldsymbol{\omega}$ on the plane of the lamina is constant in magnitude, although the component ω_3 normal to the plane is not necessarily constant. (*Hint: Use the perpendicular-axis theorem.*) What kind of lamina gives $\omega_3 = $ constant as well?

9.9   A square plate of side a and mass m is thrown into the air so that it rotates freely under zero torque. The rotational period $2\pi/\omega$ is 1 s. If the axis of rotation makes an angle of $45°$ with the symmetry axis of the plate, find the period of the precession of the axis of rotation about the symmetry axis and the period of wobble of the symmetry axis about the invariable line for two cases:

(a) A thin plate

(b) A thick plate of thickness $a/4$

9.10 A rigid body having an axis of symmetry rotates freely about a fixed point under no torques. If α is the angle between the axis of symmetry and the instantaneous axis of rotation, show that the angle between the axis of rotation and the invariable line (the **L** vector) is

$$\tan^{-1}\left[\frac{(I_s - I)\tan\alpha}{I_s + I\tan^2\alpha}\right]$$

where I_s (the moment of inertia about the symmetry axis) is greater than I (the moment of inertia about an axis normal to the symmetry axis).

9.11 Because the greatest value of the ratio $I_s/I = 2$ (symmetrical lamina), show from the result of Problem 9.10 that the angle between $\boldsymbol{\omega}$ and **L** cannot exceed $\tan^{-1}$ ($\frac{1}{\sqrt{8}}$) or about 19.5° and that the corresponding value of α is $\tan^{-1}\sqrt{2}$, or about 54.7°.

9.12 Find the angle between $\boldsymbol{\omega}$ and **L** for the two cases in Problem 9.9.

9.13 Find the same angle between $\boldsymbol{\omega}$ and **L** for Earth.

9.14 A space platform in the form of a thin circular disc of radius a and mass m (flying saucer) is initially rotating steadily with angular velocity $\boldsymbol{\omega}$ about its symmetry axis. A meteorite strikes the platform at the edge, imparting an impulse $\boldsymbol{P}$ to the platform. The direction of $\boldsymbol{P}$ is parallel to the axis of the platform, and the magnitude of $\boldsymbol{P}$ is equal to $ma\omega/4$. Find the resulting values of the precessional rate Ω, the wobble rate $\dot{\phi}$, and the angle α between the symmetry axis and the new axis of rotation.

9.15 A Frisbee is thrown into the air in such a way that it has a definite wobble. If air friction exerts a frictional torque $-c\boldsymbol{\omega}$ on the rotation of the Frisbee, show that the component of $\boldsymbol{\omega}$ in the direction of the symmetry axis decreases exponentially with time. Show also that the angle α between the symmetry axis and the angular velocity vector $\boldsymbol{\omega}$ decreases with time if I_s is larger than I, which is the case for a flat-type object. Thus, the degree of wobble steadily diminishes if there is air friction.

9.16 A simple top consists of a heavy circular disc of mass m and radius a mounted at the center of a thin rod of mass $m/2$ and length a. If the top is set spinning at a given rate S, and with the axis at an angle of 45° with the vertical, there are two possible values of the precession rate $\dot{\phi}$ such that the top precesses steadily at a constant value of $\theta = 45°$.

(a) Find the two numerical values of $\dot{\phi}$ when $S = 900$ rpm and $a = 10$ cm.

(b) How fast must the top spin to sleep in the vertical position? Express the results in revolutions per minute.

9.17 A pencil is set spinning in an upright position. How fast must the spin be for the pencil to remain in the upright position? Assume that the pencil is a uniform cylinder of length a and diameter b. Find the value of the spin in revolutions per second for $a = 20$ cm and $b = 1$ cm.

9.18 How fast must a penny (radius $a = 0.95$ cm) roll to remain upright?

9.19 A rigid body rotates freely under zero torque. By differentiating the first of Euler's equations with respect to t, and eliminating $\dot{\omega}_2$ and $\dot{\omega}_3$ by means of the second and third of

Euler's equations, show that the following result is obtained:

$$\ddot{\omega}_1 + K_1\omega_1 = 0$$

in which the function K_1 is given by

$$K_1 = -\omega_2^2\left[\frac{(I_3 - I_2)(I_2 - I_1)}{I_1 I_3}\right] + \omega_3^2\left[\frac{(I_3 - I_2)(I_3 - I_1)}{I_1 I_2}\right]$$

Two similar pairs of equations are obtained by cyclic permutations: $1 \rightarrow 2$, $2 \rightarrow 3$, $3 \rightarrow 1$. In the preceding expression for K_1 both quantities in brackets are *positive constants* if $I_1 < I_2 < I_3$, or if $I_1 > I_2 > I_3$. Discuss the question of the growth of ω_1 (stability) if initially ω_1 is very small and (a) $\omega_2 = 0$ and ω_3 is large: initial rotation is very nearly about the 3-axis, and (b) $\omega_3 = 0$ and ω_2 is large: initial rotation is nearly about the 2-axis. (*Note:* This is an analytical method of deducing the stability criteria illustrated in Figure 9.4.2.)

9.20 A rigid body consists of six particles, each of mass m, fixed to the ends of three light rods of length $2a$, $2b$, and $2c$, respectively, the rods being held mutually perpendicular to one another at their midpoints.
(a) Show that a set of coordinate axes defined by the rods are principal axes, and write down the inertia tensor for the system in these axes.
(b) Use matrix algebra to find the angular momentum and the kinetic energy of the system when it is rotating with angular velocity $\boldsymbol{\omega}$ about an axis passing through the origin and the point (a, b, c).

9.21 Work Problems 9.1 and 9.4 using matrix methods.

9.22 A uniform rectangular block of dimensions $2a$ by $2b$ by $2c$ and mass m spins about a long diagonal. Find the inertia tensor for a coordinate system with origin at the center of the block and with axes normal to the faces. Find also the angular momentum and the kinetic energy. Find also the inertia tensor for axes with origin at one corner.

9.23 Show that the z-component of angular momentum L_z of the simple top discussed in Section 9.7 is given by Equation 9.7.7.

9.24 If the top discussed in Sections 9.7 and 9.8 is set spinning very rapidly ($\dot{\psi} \gg 0$) its rate of precession will slow ($\dot{\phi} \approx 0$) and the angular difference $\theta_2 - \theta_1$ between the limits of its nutational motion will be small. Assuming this condition, show that the top can be made to precess without nutation if its motion is started with $\dot{\theta}|_{\theta=\theta_1} = 0$ and $\dot{\phi}|_{\theta=\theta_1} = mgl/L_{z'}$, where $L_{z'} = I_s S$.

9.25 Formaldehyde molecules (CH_2O) have been detected in outer space by the radio waves they emit when they change rotational states. Assume that the molecule is a rigid body, shaped like a regular tetrahedron whose faces make equilateral triangles. The masses of the oxygen, carbon, and hydrogen atoms are 16, 12, and 1 AMU, respectively.
(a) Show that a coordinate system whose 3-axis passes through the oxygen atom and its projection onto the face formed by the carbon and two hydrogen atoms, 1-axis passes through that point and the carbon atom, and 2-axis is parallel to the line connecting the two hydrogen atoms are principal axes of the molecule, as shown in Figure P9.25.
(b) Write down the inertia tensor for the molecule about these axes.
(c) Assume that the molecule rotates with angular velocity given by

$$\boldsymbol{\omega} = \omega_1\,\mathbf{e}_1 + \omega_2\,\mathbf{e}_2 + \omega_3\,\mathbf{e}_3$$

Show that motion is stable if the molecule rotates mostly about either the 3-axis or the 2-axis (i.e., if ω_1 and ω_2 are small compared with ω_3 or if ω_1 and ω_3 are small compared with ω_2) but not stable if it rotates mostly about the 1-axis.

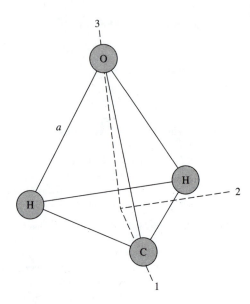

Figure P9.25

Computer Problems

C 9.1 Consider the spinning top discussed in Examples 9.7.1 and 9.7.2. Suppose that it is set spinning at 35 rev/s and initially its spin axis is held fixed at an angle $\theta_0 = 60°$. The axis is then released and the top starts to topple over. As it falls, its axis starts to precess as well as to nutate between two limiting polar angles θ_1 and θ_2.

(a) Calculate the two limits θ_1 and θ_2.

(b) Estimate the period of nutation analytically. *(Hint: Make approximations, where necessary, in the expression given in the text for $\dot{u}$ and then integrate.)*

(c) Estimate the average period of precession analytically. *(Hint: Make approximations in the expression for $\dot{\phi}$ given in the text.)*

(d) Find $\cos\theta(t)$ and $\phi(t)$ by numerically integrating the appropriate equations of motion over a time interval somewhat greater than one nutational period.

(e) Letting the function $x(t) = \cos\theta_1 - \cos\theta(t)$, where θ_1 is the smaller of the two angular limits of the nutational motion, plot $x(t)$ versus $\phi(t)$ over the same time interval as in part (d). From this plot, calculate both the nutational period and the average precessional period. Compare the results obtained from the plot with those from parts (a), (b), and (c).

C 9.2 (a) Reproduce the numerical calculation of the free rotation of the ellipsoid with three unequal principal moments of inertia given in Section 9.5. In particular, verify the values of the moments of inertia and the phase space plot given in that section.

(b) Calculate the phase space trajectory of that same ellipsoid by solving for the intersection of the constant T and L^2 ellipsoids.

(c) Find appropriate initial conditions for $[\omega_1(0), \omega_2(0), \omega_3(0)]$ that lead to precessional motion about an axis other than the 3-axis—if possible. *(Hint: The angular velocities about the 3-axis and either the 1- or 2-axis, but not both, should pass through zero during one period of the motion.)*

10

Lagrangian Mechanics

" . . . to reduce the theory of mechanics, and the art of solving the associated problems, to general formulae, whose simple development provides all the equations necessary for the solution of each problem. . . . to unite, and present from one point of view, the different principles which have, so far, been found to assist in the solution of problems in mechanics; by showing their mutual dependence and making a judgement of their validity and scope possible. . . . No diagrams will be found in this work. The methods that I explain in it require neither constructions nor geometrical or mechanical arguments, but only the algebraic operations inherent to a regular and uniform process. Those who love Analysis will, with joy, see mechanics become a new branch of it and will be grateful to me for having extended its field."

—Joseph Louis de Lagrange, *Avertissement for Mechanique Analytique*, 1788

Another way of looking at mechanics, other than from the Newtonian perspective, was developed in continental Europe somewhat contemporaneously with the efforts of Newton. This work was championed by Wilhelm von Leibniz (1646–1716) (with whom Newton was deeply embroiled in a bitter quarrel regarding who should get credit for the development of calculus). Leibniz's approach was based on mathematical operations with the scalar quantities of energy, as opposed to the vector quantities of force and acceleration. The development was to take more than a century to complete and would occupy the talents of many of the world's greatest minds. Following Leibniz, progress with the new mechanics was made chiefly by Johann Bernoulli (1667–1748). In 1717, he established the principle of "virtual work" to describe the equilibrium of static systems. This principle was extended by Jean LeRond D'Alembert (1717–1783) to include the motion of dynamical systems. The development culminated with the work of Joseph Louis de Lagrange (1736–1813), who used the virtual work principle and its D'Alembertian

417

extension as a foundation for the derivation of the dynamical equations of motion that, in his honor, now bear his name.[1]

Initially we do not take Lagrange's approach in developing his equations of motion. Instead, we take another approach, originally pursued with the goal of solving problems that run the entire gamut of physics, not merely those limited to the domain of classical mechanics. This approach stems from the deep philosophical belief that the physical universe operates according to laws of nature that are based on a principle of economy. They should be simple and elegant in form. This belief has gripped many of the most brilliant physicists and mathematicians throughout history, among them, Euler, Gauss, Einstein, Bernoulli, and Rayleigh to name a few. The basic idea is that "mother nature," given choices, always dictates that objects making up the physical universe follow paths through space and time based on extrema principles. For example, moving bodies "seek out" trajectories that are geodesics, namely, the shortest distance between two points on a given geometrical surface; a ray of light follows a path that minimizes (or, interestingly enough, maximizes) its transit time; and ensembles of particles assume equilibrium configurations that minimize their energy.

That such a hypothesis conveys some deep meaning about the workings of nature may or may not be true . This is an issue that provides fodder for philosophers and theologians alike. From the physicist's point of view, the proof, so to speak, is in the pudding. Elegant and beautiful though our laws of nature may be, we must insist ultimately on their experimental verification. The laws that we select to depict the reality of nature must stand up to scientific scrutiny. Failure to live up to the standards of this requirement relegated many an elegant hypothesis to the junk heap.

The hypothesis of global economy, however, which we introduce here, has withstood the assaults of all experimental battering rams (indeed, it leads to Newton's laws of motion). First announced in 1834 by the brilliant Irish mathematician Sir William Rowan Hamilton (1805–1865), it has proved to have such a far-reaching effect on the development of modern theoretical physics that most physicists have elevated the hypothesis to an even more fundamental status than that of Newton's laws. Thus, we have decided to use it as the fundamental postulate of mechanics in beginning the subject of this chapter.

In the sections that follow we

- Use this postulate, known as *Hamilton's variational principle,* to show that in the specific case of a body falling in a uniform gravitational field, it is equivalent to Newton's second law of motion.
- Use Hamilton's variational principle to derive Lagrange's equations of motion for a conservative system and demonstrate their use in several examples.
- Show how Lagrange's equations of motion need to be modified when *generalized forces of constraint* are a consideration.
- Present *D'Alembert's principle* and use it to derive Lagrange's equations for any physical system that involves any generalized force, including nonconservative ones,

---

[1] For a discussion of the principle of virtual work, see, for example, N. G. Chateau, *Theoretical Mechanics,* Springer-Verlag, Berlin, 1989. For the development of the Lagrange equations from D'Alembert's principle, see, for example, (1) H. Goldstein, *Classical Mechanics,* Addison-Wesley, Reading, MA, 1965; (2) F. A. Scheck, *Mechanics—From Newton's Laws to Deterministic Chaos,* Springer-Verlag, Berlin, 1990.

thereby completing a demonstration of the equivalence between the Newtonian and Lagrangian formulations of mechanics.

- Introduce the Hamiltonian formulation of mechanics and demonstrate its use in several examples.

10.1 | Hamilton's Variational Principle: An Example

Hamilton's variational principle states that the integral

$$J = \int_{t_1}^{t_2} L\, dt$$

taken along a path of the possible motion of a physical system is an extremum when evaluated along the path of motion that is the one actually taken. $L = T - V$ is the *Lagrangian* of the system, or the difference between its kinetic and potential energy. In other words, out of the myriad ways in which a system could change its configuration during a time interval $t_2 - t_1$, the actual motion that does occur is the one that either maximizes or minimizes the preceding integral. This statement can be expressed mathematically as

$$\delta J = \delta \int_{t_1}^{t_2} L\, dt = 0 \tag{10.1.1}$$

in which δ is an operation that represents a variation of any particular system parameter by an infinitesimal amount away from that value taken by the parameter when the integral in Equation 10.1.1 is an extremum. For example, the δ that occurs explicitly in Equation 10.1.1 represents a variation in the entire integral about its extremum value. Such a variation is obtained by varying the coordinates and velocities of a dynamical system away from the values actually taken as the system evolves in time from t_1 to t_2, under the constraint that the variation in all parameters is zero at the endpoints of the motion at t_1 and t_2. That is, the variation of the system parameters between t_1 and t_2 is completely arbitrary under the provisos that the motion must be completed during that time interval and that all system parameters must assume their unvaried values at the beginning and end of the motion.

Let us apply Hamilton's variational principle to the case of a particle dropped from rest in a uniform gravitational field. We will see that the integral in Equation 10.1.1 is an extremum when the path taken by the object is the one for which the particle obeys Newton's second law. Let the height of the particle above ground at any time t be denoted by y and its speed by $\dot{y}$. Then δy and $\delta \dot{y}$ represent small, virtual displacements of y and $\dot{y}$ away from the true position and speed of the particle at any time t during its actual motion. (Figure 10.1.1). The potential energy of the particle is mgy, and its kinetic energy is $m\dot{y}^2/2$. The Lagrangian is $L = m\dot{y}^2/2 - mgy$. The variation in the integral of the Lagrangian is given by

$$\delta J = \delta \int_{t_1}^{t_2} L\, dt = \delta \int_{t_1}^{t_2} \left[\frac{m\dot{y}^2}{2} - mgy \right] dt = \int_{t_1}^{t_2} (m\dot{y}\, \delta\dot{y} - mg\, \delta y)\, dt \tag{10.1.2}$$

The variation in the speed can be transformed into a coordinate variation by noting that

$$\delta\dot{y} = \frac{d}{dt}\, \delta y \tag{10.1.3a}$$

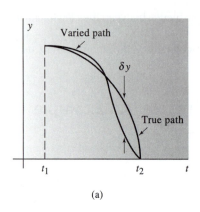

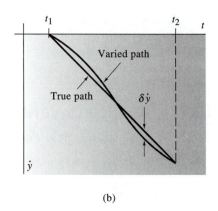

Figure 10.1.1 (a) Variation of the coordinate of a particle from its true path taken in free-fall. (b) Variation in the speed of a particle from the true value taken during free-fall.

Integrating the first term in Equation 10.1.2 by parts gives

$$\int_{t_1}^{t_2} m\dot{y}\,\delta\dot{y}\,dt = \int_{t_1}^{t_2} m\dot{y}\frac{d}{dt}\delta y\,dt = m\dot{y}\,\delta y\Big|_{t_1}^{t_2} - \int_{t_1}^{t_2} m\ddot{y}\,\delta y\,dt \qquad (10.1.3b)$$

The integrated term on the right-hand side is identically zero because the parameters of the admissible paths of motion do not vary at the endpoints of the motion. Hence, we obtain

$$\delta J = \delta \int_{t_1}^{t_2} L\,dt = \int_{t_1}^{t_2}(-m\ddot{y} - mg)\,\delta y\,dt = 0 \qquad (10.1.4)$$

Because δy represents a completely arbitrary variation of the parameter y away from its true value throughout the motion of the particle (except at the endpoints where the variation is constrained to be zero), the only way in which Equation 10.1.4 can be zero under such conditions is for the term in parentheses to be identically zero at all times. Thus,

$$-mg - m\ddot{y} = 0 \qquad (10.1.5)$$

which, as advertised, is Newton's second law of motion for a particle falling in a uniform gravitational field.

The solution to this equation of motion is $y(t) = -\frac{1}{2}gt^2$ (assuming that the object is dropped from rest at $y_0 = 0$). We now show that any solution that varies from this one yields an integral $J = \int L\,dt$ that is not an extremum. We can represent any possible variation in y away from the true solution by expressing it parametrically as $y(\alpha, t)$ such that

$$y(\alpha,t) = y(0,t) + \alpha\eta(t) \qquad (10.1.6)$$

When the parameter $\alpha = 0$, then $y = y(0, t) = y(t)$, the true solution. $\eta(t)$ is any function of time whose first derivative is continuous on the interval $[t_1, t_2]$ and whose values, $\eta(t_1)$

and $\eta(t_2)$, vanish, thus ensuring that $y(\alpha, t)$ attains its true value at those times regardless of the value of α. Because our choice of $\eta(t)$ is arbitrary, consistent only with the constraints expressed in Equation 10.1.6, the quantity $\alpha\eta(t)$ generates any variation $\delta y(t)$ taken away from the true dynamical path that we wish. An example of a possible variation away from the true dynamical path was depicted in Figure 10.1.1.

The integral J is now a function of the parameter α

$$J(\alpha) = \int_{t_1}^{t_2} L[y(\alpha,t), \dot{y}(\alpha,t); t]\, dt \tag{10.1.7}$$

We now proceed to calculate this integral for the case of the falling body. The expression for $\dot{y}$ in terms of the parameter α is

$$\dot{y}(\alpha,t) = \dot{y}(0,t) + \alpha\dot{\eta}(t) \tag{10.1.8}$$

where $\dot{y}(0,t) = -gt$. The kinetic and potential energies of the falling body are

$$T = \frac{1}{2}m\dot{y}^2 = \frac{1}{2}m[-gt + \alpha\dot{\eta}(t)]^2 \tag{10.1.9a}$$

$$V = mgy = mg\left[-\frac{1}{2}gt^2 + \alpha\eta(t)\right] \tag{10.1.9b}$$

The integral $J(\alpha)$ is, thus,

$$J(\alpha) = \int_{t_1}^{t_2} m\left(\frac{\dot{y}^2}{2} - gy\right) dt \tag{10.1.10}$$

$$= \int_{t_1}^{t_2} m\left\{g^2t^2 - \alpha g[t\dot{\eta}(t) + \eta(t)] + \frac{1}{2}\alpha^2\dot{\eta}^2(t)\right\} dt$$

The integral of the term linear in α, in the center square brackets in Equation 10.1.10, is

$$\int_{t_1}^{t_2} [t\dot{\eta}(t) + \eta(t)]\, dt = t\eta(t)\Big|_{t_2}^{t_1} - \int_{t_1}^{t_2} \eta(t)\, dt + \int_{t_1}^{t_2} \eta(t)\, dt \equiv 0 \tag{10.1.11}$$

The first term in Equation 10.1.11 vanishes because $\eta(t_1)$ and $\eta(t_2) = 0$. Thus, the term linear in α completely vanishes, and we obtain

$$J(\alpha) = \frac{1}{3}g^2\left(t_2^3 - t_1^3\right) + \frac{1}{2}\alpha^2\int_{t_1}^{t_2} \dot{\eta}^2(t)\, dt \tag{10.1.12}$$

The last term in Equation 10.1.12 is quadratic in the parameter α, and because the integral of $\dot{\eta}^2(t)$ must be positive for any $\dot{\eta}(t)$, $J(\alpha)$ exhibits the behavior pictured in Figure 10.1.2. The value of the integral is a minimum when

$$\frac{\partial J(\alpha)}{\partial \alpha}\bigg|_{\alpha=0} = 0 \tag{10.1.13}$$

which occurs at $\alpha = 0$.

Even though this result was based on a specific example, it is true for any integral J of a function *of the function y* (and its first derivative) that has the parametric form given by Equation 10.1.6. The resulting $J(\alpha)$ does not depend on α to first order, and its partial derivative vanishes at $\alpha = 0$, making the integral an extremum only when y is equal to the solution obtained from Newton's second law of motion.

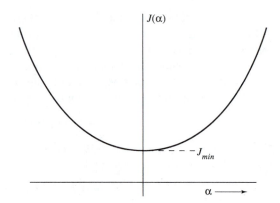

Figure 10.1.2 $J(\alpha) = a + b\alpha^2$ is a minimum at $\alpha = 0$.

EXAMPLE 10.1.1

Assume that a particle travels along a sinusoidal path from a point $x = 0$ to $x = x_1$ during a time interval Δt in a *force-field free region of space*. Use Hamilton's principle to show that the amplitude of the assumed sinusoidal path is zero, implying that the path the particle takes is really a straight line between the two points.

Solution:

Shown in Figure 10.1.3 are several possible paths that follow a sine curve between 0 and x_1 along with the presumed correct straight line path.

The motion that a particle actually carries out in a force-field–free region of space is given by the expression $x = v_x t$. Any other possible motion is constrained to be completed during the time interval

$$\Delta t = x_1/v_x$$

Thus, we can vary the possible varied sinusoidal paths according to

$$x = v_x t \qquad \text{and} \qquad y = \pm\eta \sin \pi v_x t/x_1$$

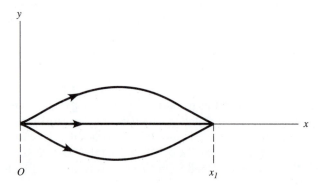

Figure 10.1.3 Possible paths for a particle in a force-field region.

where η is a parameter that can be varied to change the amplitude of the sinusoidal path of the particle. This variation insures that no matter which path the particle takes, it ends up at x_1 in the time Δt. The Lagrangian for the varied path is

$$L = T - V = \frac{1}{2} m \left[v_x^2 + \left(\frac{\eta \pi v_x}{x_1} \right)^2 \cos^2 \frac{\pi v_x t}{x_1} \right] - V$$

where V is the potential energy of the particle, which is constant because the motion is in a force-field–free region of space. Thus, we have

$$J = \int_0^{x_1/v_x} L dt = \frac{m v_x x_1}{2} + \frac{m v_x \eta^2 \pi^2}{4 x_1} - V \frac{x_1}{v_x}$$

We vary the path by varying η so that

$$\delta J = \left(\frac{\pi^2 m v_x}{2 x_1} \right) \eta \delta \eta = 0$$

Because $\delta \eta$ is not zero, η must be zero to satisfy Hamilton's principle, suggesting that the path represented by the equation $x = v_x t$ represents the actual path of the particle. Obviously, to demonstrate this conclusively, we would have to show that the same result is obtained for all possible variations of the path—a task we do not undertake here.

10.2 | Generalized Coordinates

Coordinates are used to define the position in space of an ensemble of particles. In general we can select any set of coordinates to describe the motion of a physical system. Certain choices, however, prove to be more economical than others because of the existence of geometrical constraints that restrict the allowable configuration of any system.

For example, consider the motion of the pendulum in Figure 10.2.1. It is constrained to move in the xy plane along an arc of radius r. We could choose to describe the configuration of the pendulum by means of its position vector

$$\mathbf{r} = x\mathbf{i} + y\mathbf{j} + z\mathbf{k} \tag{10.2.1}$$

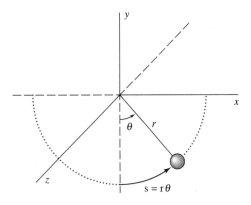

Figure 10.2.1 Pendulum swinging in the xy plane.

Clearly, such a choice is tantamount to lunacy because it ignores the two conditions of constraint to which the pendulum must adhere, namely

$$z = 0 \qquad r^2 - (x^2 + y^2) = 0 \tag{10.2.2}$$

Only one scalar coordinate is really needed to specify the position of the pendulum. At first sight either x or y might work, but to resolve a possible left–right ambiguity, we would probably choose x to define the position uniquely. We would then assume that the implied value of y is always negative. Given the x-value of the pendulum, y and z would then be determined completely by the conditions of constraint. This choice would still prove awkward for describing the configuration of the pendulum.

A better choice would be the arc length displacement s ($= r\theta$) or, equivalently, the angular displacement θ of the pendulum away from the vertical. Either of these choices is better, because only a single number is needed to tell us the whereabouts of the pendulum. The important point here is that the pendulum really has only *one degree of freedom;* that is, it can move only one way and that way is along an arc of radius r. There exists only a single, independent coordinate necessary to depict its configuration uniquely. *Generalized coordinates* are any collection of independent coordinates q_i (not connected by any equations of constraint) that just suffice to specify uniquely the configuration of a system of particles. The required number of generalized coordinates is equal to the system's number of degrees of freedom. If fewer than this number is chosen to describe the system's configuration, the result is indeterminate; if a greater number is chosen, then some of the coordinates must be determinable from the others by conditions of constraint.

For example, a single particle able to move freely in three-dimensional space exhibits three degrees of freedom and requires three coordinates to specify its configuration. There exist no equations of constraint connecting the coordinates of a single free particle. Two free particles would require six coordinates to specify the configuration completely, but two particles connected by a rigid straight line like a dumbbell (see Figure 10.2.2) would require only five coordinates. Let us see why. The position of particle 1 could be specified by the coordinates (x_1, y_1, z_1), whereas the position of particle 2 could be specified by the coordinates (x_2, y_2, z_2) as in the free particle case. An equation of constraint exists, however, that connects the coordinates

$$d^2 - [(x_1 - x_2)^2 + (y_1 - y_2)^2 + (z_1 - z_2)^2] = 0 \tag{10.2.3}$$

namely, the distance between the two particles is fixed and equal to d.

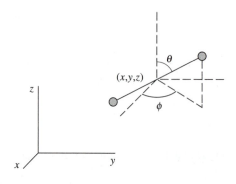

Figure 10.2.2   Generalized coordinates for two particles connected by an infinitesimally thin, rigid rod.

Suppose, in specifying the position of the system, we picked the above six coordinates, one by one. We would not have complete freedom of choice in the selection process, because the choice of the sixth coordinate would be forced on us after the first five had been made. It would make more sense to choose initially only five independent coordinates, say (X, Y, Z, θ, ϕ), unconnected by any equations of constraint, where (X, Y, Z) are the coordinates of the center of mass and (θ, ϕ) are the zenith and azimuthal angles, which describe the orientation of the dumbbell relative to the vertical ($\theta = 0°$ when particle 1 is directly above particle 2, and $\phi = 0°$ when the projection of the line from particle 2 to particle 1 onto the xy plane points parallel to the x-axis).

As a final example, consider the situation of a particle constrained to move along the surface of a sphere. Again, the coordinates (x, y, z) do not constitute an independent set. They are connected by the constraint

$$R^2 - (x^2 + y^2 + z^2) = 0 \qquad (10.2.4)$$

where R is the radius of the sphere. The particle has only two degrees of freedom available for its motion, and two independent coordinates are needed to specify completely its position on the sphere.

These coordinates could be taken as latitude and longitude, which denote positions on the spherical surface relative to an equator and a prime meridian as for Earth (see Figure 10.2.3), or we could choose the polar and azimuthal angles θ and ϕ as in the dumbbell example.

In general, if N particles are free to move in three-dimensional space but their $3N$ coordinates are connected by m conditions of constraint, then there exist $n = 3N - m$ independent generalized coordinates sufficient to describe uniquely the position of the N particles and n independent degrees of freedom available for the motion, provided the constraints are of the type described in the preceding examples. Such constraints are called *holonomic*. They must be expressible as equations of the form

$$f_j(x_i, y_i, z_i, t) = 0 \qquad i = 1, 2, \ldots, N \qquad j = 1, 2, \ldots, m \qquad (10.2.5)$$

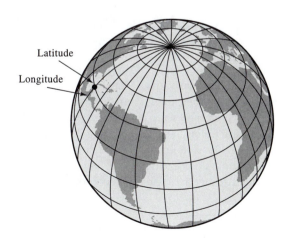

Figure 10.2.3   Coordinate of a point on Earth marked by latitude and longitude.

These equations are equalities, they are integrable in form, and they may or may not be explicitly time-dependent.

Constraints that cannot be expressed as equations of equality or that are nonintegrable in form are called *nonholonomic,* and the equations representing such constraints cannot be used to eliminate from consideration any dependent coordinates describing the configuration of the system. As an example of such a constraint, consider a particle constrained to remain outside the surface of a sphere. (Humans on Earth capable of going to the moon but incapable of going more than a few miles underground represent a reasonable approximation to this situation.) This condition of constraint is given by the inequality

$$(x^2 + y^2 + z^2) - R^2 \geq 0 \qquad (10.2.6)$$

Clearly, this equation cannot be used to reduce below three the required number of independent coordinates of the particle when it lies outside the sphere. Inside the sphere is another matter. In this case the single constraint reduces the degrees of freedom to zero. Because it is difficult to handle such situations using Lagrangian mechanics, we ignore them in this text.

Perhaps the classic example of a nonholonomic constraint in which the representative equation is nonintegrable appears in the case of a ball rolling along a rough, level surface without slipping. The "rolling" condition connects the coordinates. A change in the orientation of the ball cannot occur without an accompanying change in its position on the plane. The equation of constraint, however, represents a condition on velocities, not coordinates. The ball's point of contact with the surface is instantaneously at rest. The desired constraint on coordinates can be generated only by integrating the equation representing the velocity constraint. This cannot be done unless the ball's trajectory is known. Unfortunately, this is the very problem we wish to solve. Hence, the constraint equation is nonintegrable and, as mentioned previously, cannot be used to eliminate dependent coordinates from the problem. In contrast to the nonholonomic constraint represented by inequality conditions on the coordinates, however, this type of nonholonomic constraint can be handled tractably by the Lagrangian technique via the use of the method of Lagrange multipliers.[2] Again, we ignore such situations here.

10.3 | Calculating Kinetic and Potential Energies in Terms of Generalized Coordinates: An Example

The Lagrangian $L = T - V$ must be expressed as a function of the generalized coordinates and time derivatives (generalized velocities) appropriate for a given physical situation. (Sometimes, the Lagrangian may also be an explicit function of time, although we are not concerned with such cases here.) We need to find out how to generate such an expression before deriving Lagrange's equations of motion from Hamilton's variational principle. It is not obvious a priori just how to do this. Almost always, the kinetic energy of an ensemble of particles can be written as a quadratic form in the velocities of the particles related to a Cartesian coordinate system.

---

[2] For example, see pages 38–44 of H. Goldstein, *Classical Mechanics,* Addison-Wesley, Reading, MA, 1965.

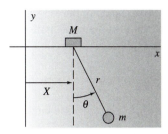

Figure 10.3.1   Simple pendulum attached to a movable support.

Because Cartesian coordinates are orthogonal, there are no cross-coupled terms in such an expression. This is usually not true, however, when the kinetic energy is expressed in terms of the generalized coordinates; that is, the coordinates chosen may lead to cross-coupled velocity terms of the form $\alpha \dot{q}_i \dot{q}_j$. There is no equivalent generalization regarding the expression for the potential energy of a system. In some cases, the expression contains cross terms even when expressed in Cartesian coordinates. Usually, however, the potential energy is expressible as some function of just one of the generalized coordinates, and it is easy to see just how it depends on that coordinate. One usually exploits this feature in choosing the generalized coordinates for any particular situation. Unfortunately, such choices almost invariably lead to cross terms in the expression for the kinetic energy of the system.

As a specific example, let us take the fairly nasty case, depicted in Figure 10.3.1, of a pendulum of mass m attached to a support of mass M that is free to move in a single dimension along a frictionless, horizontal surface. First, let us see just how many generalized coordinates are needed to specify the system configuration uniquely. Each mass needs three Cartesian coordinates, but there are four holonomic constraints

$$Z = 0 \qquad Y = 0$$
$$z = 0 \qquad [(x - X)^2 + y^2] - r^2 = 0 \tag{10.3.1}$$

The first two constraints ensure that the motion of the mass M lies along the x-axis. The second two constraints ensure that the pendulum swings in the xy plane along an arc of radius r, relative to its movable support. There are two degrees of freedom for the motion and two generalized coordinates necessary to describe the configuration of this system. We have chosen those coordinates to be X, which denotes the horizontal position mass of M, and θ, the angular displacement of the pendulum away from the vertical.

The potential and kinetic energies of this system can be expressed in terms of Cartesian coordinates and velocities as

$$T = \tfrac{1}{2} M \dot{X}^2 + \tfrac{1}{2} m(\dot{x}^2 + \dot{y}^2) \tag{10.3.2a}$$
$$V = mgy \tag{10.3.2b}$$

Obtaining the potential energy in terms of generalized coordinates requires a transformation of coordinates

$$x = X + r\sin\theta \qquad y = -r\cos\theta \qquad X = X \tag{10.3.3a}$$

while obtaining the kinetic energy as a function of generalized velocities can be effected by differentiating Equation 10.3.3a.

$$\dot{x} = \dot{X} + r\dot{\theta}\cos\theta \qquad \dot{y} = r\dot{\theta}\sin\theta \qquad \dot{X} = \dot{X} \qquad (10.3.3b)$$

Substituting these transformations into Equations 10.3.2a and b yields

$$T = \frac{M}{2}\dot{X}^2 + \frac{m}{2}[(\dot{X} + r\dot{\theta}\cos\theta)^2 + (r\dot{\theta}\sin\theta)^2]$$

$$= \frac{M}{2}\dot{X}^2 + \frac{m}{2}[\dot{X}^2 + (r\dot{\theta})^2 + 2\dot{X}r\dot{\theta}\cos\theta] \qquad (10.3.4a)$$

$$V = -mgr\cos\theta \qquad (10.3.4b)$$

Several features of Equations 10.3.4a and b illustrate our comments:

1. The kinetic energy is expressible as a quadratic form in the generalized velocities, including a cross term.
2. The potential energy is dependent on a single, generalized coordinate, in this case, the cosine of an angle.

Although it might have been easy to see how to write the potential energy directly in terms of a single, generalized coordinate, it might take some thought to write the kinetic energy directly in terms of the generalized velocities. It is worth doing, however, so let us give it a try.

The velocity of mass M relative to our fixed inertial frame of reference is

$$\mathbf{V}_M = \mathbf{i}\dot{X} \qquad (10.3.5)$$

The velocity of mass m can be expressed as the velocity of mass M plus the velocity of mass m relative to that of mass M, that is,

$$\mathbf{v}_m = \mathbf{V}_M + \mathbf{v}_{m(rel)} \qquad (10.3.6a)$$

where

$$\mathbf{v}_{m(rel)} = \mathbf{e}_\theta r\dot{\theta} \qquad (10.3.6b)$$

and $\mathbf{e}_\theta$ is a unit vector tangent to the arc along which the pendulum swings. Hence, the velocity of m can now be written in component form directly as a function of the generalized coordinates X and θ

$$\dot{x} = \dot{X} + r\dot{\theta}\cos\theta \qquad \dot{y} = r\dot{\theta}\sin\theta \qquad (10.3.7)$$

Plugging these expressions for the Cartesian velocities of the two masses into Equation 10.3.2a yields the kinetic energy in terms of the generalized coordinates as expressed in Equation 10.3.4a.

An even more direct way to generate the correct expression for the kinetic energy can be obtained by noting that

$$T = \frac{1}{2}M\mathbf{V}_M \cdot \mathbf{V}_M + \frac{1}{2}m\mathbf{v}_m \cdot \mathbf{v}_m \qquad (10.3.8)$$

where

$$\mathbf{V}_m = \mathbf{i}\dot{X} \qquad\qquad \mathbf{v}_m = \mathbf{i}\dot{X} + \mathbf{e}_\theta r\dot{\theta} \qquad\qquad (10.3.9a)$$

$$\mathbf{V}_M \cdot \mathbf{V}_M = \dot{X}^2 \qquad\qquad \mathbf{v}_m \cdot \mathbf{v}_m = \dot{X}^2 + r^2\dot{\theta}^2 + 2\dot{X}r\dot{\theta}\,\cos\theta \qquad\qquad (10.3.9b)$$

To obtain a correct expression for the kinetic energy of any given system, one rarely goes wrong by following the first outlined procedure (Equations 10.3.2a–10.3.4b), namely, write the kinetic energy in terms of velocities relative to Cartesian coordinates, find the transformation relating Cartesian to generalized coordinates, and then differentiate. In many cases, however, it is easier to use the last demonstrated procedure (Equations 10.3.8–10.3.9b), particularly if one is able to visualize just what form the generalized velocities take in terms of the unit vectors corresponding to the selected generalized coordinates.

When problems involve only holonomic constraints, there always exist transformation equations that relate the Cartesian coordinates of an ensemble of particles to their generalized coordinates, and, thus, the required generalized velocities can be obtained by differentiation. For example, in the case of a single particle, we have:

Three degrees of freedom—unconstrained motion in space

$$x = x(q_1, q_2, q_3)$$
$$y = y(q_1, q_2, q_3) \qquad\qquad (10.3.10)$$
$$z = z(q_1, q_2, q_3)$$

Two degrees of freedom—motion constrained to a surface

$$x = x(q_1, q_2)$$
$$y = y(q_1, q_2) \qquad\qquad (10.3.11)$$
$$z = z(q_1, q_2)$$

One degree of freedom—motion constrained to a line

$$x = x(q)$$
$$y = y(q) \qquad\qquad (10.3.12)$$
$$z = z(q)$$

And, as we did in Equations 10.3.3a and b, we can obtain the velocity transformations by simply differentiating the coordinate transformations:

$$\dot{x} = \sum_{i=1}^{n} \frac{\partial x}{\partial q_i}\dot{q}_i$$

$$\dot{y} = \sum_{i=1}^{n} \frac{\partial y}{\partial q_i}\dot{q}_i \qquad\qquad (10.3.13)$$

$$\dot{z} = \sum_{i=1}^{n} \frac{\partial z}{\partial q_i}\dot{q}_i$$

where n is the number of degrees of freedom.

10.4 | Lagrange's Equations of Motion for Conservative Systems

We are now ready to derive Lagrange's equations from Hamilton's variational principle. First, we should point out that all our examples thus far have consisted only of conservative systems whose motion is either unconstrained or, at worst, subject only to time-independent holonomic constraints. We continue to confine our analysis to such systems. The interested reader who wishes to endure the agony of dealing with either nonconservative systems or systems suffering nonholonomic constraints is urged to seek out the many excellent presentations contained in other, more advanced, texts on this subject.[3]

Hamilton's principle is expressed by Equation 10.1.1 We proceed from that point by carrying out the same variational procedure as we did in Section 10.1 for the specific case of an object freely falling in a gravitational field. Only this time we carry out the process for any general conservative system. We begin by assuming that the Lagrangian is a known function of the generalized coordinates q_i and velocities $\dot{q}_i$. The variation in its time integral is

$$\delta J = \delta \int_{t_1}^{t_2} L\, dt = \int_{t_1}^{t_2} \delta L\, dt = \int_{t_1}^{t_2} \sum_i \left(\frac{\partial L}{\partial q_i} \delta q_i + \frac{\partial L}{\partial \dot{q}_i} \delta \dot{q}_i \right) dt = 0 \qquad (10.4.1)$$

The q_i are functions of time; they change as the system "evolves" from t_1 to t_2. Therefore, the variation δq_i is equal to the difference between two slightly differing functions of time t. The $\delta \dot{q}_i$ can then be expressed as

$$\delta \dot{q}_i = \frac{d}{dt} \delta q_i \qquad (10.4.2)$$

This result can be substituted into the last term of Equation 10.4.1, which can then be integrated by parts to obtain

$$\int_{t_1}^{t_2} \sum_i \frac{\partial L}{\partial \dot{q}_i} \frac{d}{dt}(\delta q_i)\, dt = \sum_i \frac{\partial L}{\partial \dot{q}_i} \delta q_i \Big|_{t_1}^{t_2} - \int_{t_1}^{t_2} \sum_i \frac{d}{dt}\left(\frac{\partial L}{\partial \dot{q}_i} \right) \delta q_i\, dt \qquad (10.4.3)$$

The integrated term in brackets vanishes because the variation $\delta q_i = 0$ at the endpoints t_1 and t_2. Thus, we obtain

$$\delta \int_{t_1}^{t_2} L\, dt = \int_{t_1}^{t_2} \sum_i \left[\frac{\partial L}{\partial q_i} - \frac{d}{dt}\left(\frac{\partial L}{\partial \dot{q}_i} \right) \right] \delta q_i\, dt = 0 \qquad (10.4.4)$$

Each generalized coordinate q_i is independent of the others, as is each variation δq_i. Moreover, the actual value of each variation δq_i is completely arbitrary. In other words, we can vary each coordinate in any way we so choose as long as we make sure that its variation vanishes at the endpoints of the path. Consequently, the only way that we can ensure that the preceding integral vanishes, given all the infinite varieties of possible values for the δq_i's, is to demand that each bracketed term in the integrand of Equation 10.4.4

---

[3] See Footnote 1.

vanish separately; that is,

$$\frac{\partial L}{\partial q_i} - \frac{d}{dt}\left(\frac{\partial L}{\partial \dot{q}_i}\right) = 0 \qquad (i = 1, 2, \ldots, n) \qquad (10.4.5)$$

These are the desired Lagrangian equations of motion for a conservative system subject to, at worst, only holonomic constraints.

10.5 | Some Applications of Lagrange's Equations

Here we illustrate the great utility of Lagrange's equations of motion by using them to obtain the differential equations of motion for several different systems. The general problem-solving strategy proceeds as follows:

1. Select a suitable set of generalized coordinates that uniquely specifies the system configuration.
2. Find the equations of transformation relating the dependent Cartesian coordinates to the independent generalized coordinates.
3. Find the kinetic energy as a function of the generalized coordinates and velocities. If possible, use the prescription $T = m\mathbf{v} \cdot \mathbf{v}/2$, with $\mathbf{v}$ expressed in terms of unit vectors appropriate to the selected generalized coordinates. If necessary, express the kinetic energy in terms of Cartesian coordinates, then differentiate the coordinate transformations and plug the resulting velocity transformations into the kinetic energy expression.
4. Find the potential energy as a function of the generalized coordinates using, if necessary, the coordinate transformations.

EXAMPLE 10.5.1

The Harmonic Oscillator

Consider the case of a one-dimensional harmonic oscillator. Let x be the displacement coordinate. Step 2 is not explicitly necessary. The single Cartesian coordinate x is obviously the single generalized coordinate. The Lagrangian is

$$L(x, \dot{x}) = T - V = \tfrac{1}{2}m\dot{x}^2 - \tfrac{1}{2}kx^2$$

We have ignored the other two coordinates y and z, because they are both constrained to be zero. Now carry out the Lagrange operations of Equation 10.4.5

$$\frac{\partial L}{\partial \dot{x}} = m\dot{x} \qquad \frac{\partial L}{\partial x} = -kx$$

$$\frac{d}{dt}\left(\frac{\partial L}{\partial \dot{x}}\right) - \frac{\partial L}{\partial x} = m\ddot{x} + kx = 0$$

This is the equation of motion of the undamped harmonic oscillator discussed in Chapter 3.

EXAMPLE 10.5.2

Single Particle in a Central Force Field

The problem here is to use the Lagrange equations to generate the differential equations of motion for a particle constrained to move in a plane subject to a central force. The single constraint is given by $z = 0$, so we need two generalized coordinates. We choose plane polar coordinates: $q_1 = r$, $q_2 = \theta$. The transformation equations and resultant velocities are

$$x = r \cos\theta \qquad\qquad y = r \sin\theta$$
$$\dot{x} = \dot{r} \cos\theta - r\dot{\theta} \sin\theta \qquad\qquad \dot{y} = \dot{r} \sin\theta + r\dot{\theta} \cos\theta$$

Thus, the Lagrangian is

$$T = \tfrac{1}{2} m(\dot{x}^2 + \dot{y}^2) = \tfrac{1}{2} m(\dot{r}^2 + r^2\dot{\theta}^2) \qquad V = V(r)$$
$$L = \tfrac{1}{2} m(\dot{r}^2 + r^2\dot{\theta}^2) - V(r)$$

We could have obtained the kinetic energy term more directly by expressing the velocity vector in terms of radial and tangential unit vectors

$$\mathbf{v} = \mathbf{e}_r \dot{r} + \mathbf{e}_\theta r\dot{\theta}$$

and, thus, the square of the particle's velocity is

$$\mathbf{v} \cdot \mathbf{v} = \dot{r}^2 + r^2\dot{\theta}^2$$

which is just what we obtained using the coordinate transformations.

The relevant partial derivatives needed to implement the Lagrangian equations are

$$\frac{\partial L}{\partial \dot{r}} = m\dot{r} \qquad \frac{\partial L}{\partial r} = mr\dot{\theta}^2 - \frac{\partial V}{\partial r} = mr\dot{\theta}^2 + f(r)$$

$$\frac{\partial L}{\partial \theta} = 0 \qquad \frac{\partial L}{\partial \dot{\theta}} = mr^2\dot{\theta}$$

The equations of motion are, thus,

$$\frac{d}{dt}\frac{\partial L}{\partial \dot{r}} = \frac{\partial L}{\partial r} \qquad\qquad \frac{d}{dt}\frac{\partial L}{\partial \dot{\theta}} = \frac{\partial L}{\partial \theta} = 0$$

$$m\ddot{r} = mr\dot{\theta}^2 + f(r) \qquad\qquad \frac{d}{dt}(mr^2\dot{\theta}) = 0$$

For future reference we might note that because the time derivative of the quantity $mr^2\dot{\theta}$ is zero, it is a constant of the motion. This quantity is the angular momentum of the particle. We can easily see that its constancy arises naturally in the Lagrangian formalism because the θ coordinate is missing from the Lagrangian function.

EXAMPLE 10.5.3

Atwood's Machine

Atwood's machine consists of two weights of mass m_1 and m_2 connected by an ideal massless, inextensible string of length l that passes over a frictionless pulley of radius a and moment of inertia I (Figure 10.5.1). The system has only one degree of freedom—one mass moves either up or down while the other is constrained to move in the opposite sense, always separated from the first by the length of string. The pulley rotates appropriately. There exist five holonomic constraints. Four prevent motion in either the y or z direction, whereas the fifth expresses the previously mentioned constraint

$$(x_1 + \pi a + x_2) - l = 0$$

where x_1 and x_2 are the vertical positions of each mass relative to the center of the pulley. The Lagrangian is

$$T = \frac{1}{2}m_1\dot{x}^2 + \frac{1}{2}m_2\dot{x}^2 + \frac{1}{2}I\frac{\dot{x}^2}{a^2}$$

$$V = -m_1gx - m_2g(l - \pi a - x)$$

$$L = \frac{1}{2}\left(m_1 + m_2 + \frac{I}{a^2}\right)\dot{x}^2 + (m_1 - m_2)gx + m_2g(l - \pi a)$$

where x is the single generalized coordinate of the system. The Lagrange equations of motion are

$$\frac{d}{dt}\frac{\partial L}{\partial \dot{x}} = \frac{\partial L}{\partial x}$$

$$\left(m_1 + m_2 + \frac{I}{a^2}\right)\ddot{x} = (m_1 - m_2)g$$

$$\ddot{x} = \frac{(m_1 - m_2)g}{[m_1 + m_2 + I/a^2]}$$

giving the final acceleration of the system. If $m_1 > m_2$, then m_1 falls with constant acceleration while m_2 rises with the very same acceleration. The converse is true if $m_2 > m_1$. If $m_1 = m_2$, each mass remains at rest (or moves at constant velocity). The effect of the moment of inertia of the pulley is to reduce the acceleration of the system. The reader probably recalls this result from analysis presented in more elementary physics classes.

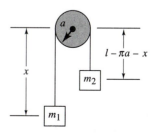

Figure 10.5.1 An Atwood machine.

EXAMPLE 10.5.4

The Double Atwood Machine

Consider the system shown in Figure 10.5.2. We have replaced one of the weights in the simple Atwood's machine by a second, simple Atwood's machine. There are two degrees of freedom for the motion of this system. Loosely speaking, they are the freedom of mass 1 (and the attached movable pulley) to move up and down about the fixed pulley and the freedom of mass 2 (and the attached mass 3) to move up and down about the movable pulley. No other motion is permissible. Thus, there must be 10 holonomic constraints. Eight of those constraints limit the motion of all three masses plus the movable pulley to only a single dimension. Two holonomic constraints connect the x coordinates

$$(x_p + x_1) - l = 0 \qquad (2x_1 + x_2 + x_3) - (2l + l') = 0$$

where x_i and x_p are the vertical positions of the masses and movable pulley relative to the center of the fixed pulley. (The student should verify these equations of constraint.) The implication of these constraints regarding the reduction to the two selected generalized coordinates x and x' is indicated in Figure 10.5.2. In this case we have assumed that each pulley is massless. We can, therefore, ignore the effects of moments of inertia. We have also assumed that the radii of the pulleys are essentially zero (or small compared with l and l', the lengths of the constraining string). This assumption allowed us to simplify the preceding equation of the constraint by ignoring the length of string that goes around each pulley. We can now write down the kinetic and potential energies for this system as well as its resultant Lagrangian

$$T = \tfrac{1}{2}m_1\dot{x}^2 + \tfrac{1}{2}m_2(-\dot{x} + \dot{x}')^2 + \tfrac{1}{2}m_3(-\dot{x} - \dot{x}')^2$$
$$V = -m_1 gx - m_2 g(l - x + x') - m_3 g(l - x + l' - x')$$
$$L = \tfrac{1}{2}m_1\dot{x}^2 + \tfrac{1}{2}m_2(-\dot{x} + \dot{x}')^2 + \tfrac{1}{2}m_3(\dot{x} + \dot{x}')^2$$
$$\qquad + (m_1 - m_2 - m_3)gx + (m_2 - m_3)gx' + \text{constant}$$

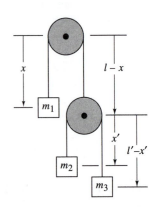

Figure 10.5.2 A compound Atwood machine.

The equations of motion are, thus,

$$\frac{d}{dt}\frac{\partial L}{\partial \dot{x}} = \frac{\partial L}{\partial x} \qquad\qquad \frac{d}{dt}\frac{\partial L}{\partial \dot{x}'} = \frac{\partial L}{\partial x'}$$

$$m_1\ddot{x} + m_2(\ddot{x} - \ddot{x}') + m_3(\ddot{x} + \ddot{x}') = (m_1 - m_2 - m_3)g$$

$$m_2(-\ddot{x} + \ddot{x}') + m_3(\ddot{x} + \ddot{x}') = (m_2 - m_3)g$$

The accelerations can be obtained from an algebraic solution of the preceding equations.

EXAMPLE 10.5.5

Euler's Equations for the Free Rotation of a Rigid Body

In this example, we use Lagrange's method to derive Euler's equations for the motion of a rigid body. We consider the case of torque-free rotation. No potential energy is involved, so the Lagrangian is equal to the kinetic energy

$$L = T = \frac{1}{2}\left(I_1\omega_1^2 + I_2\omega_2^2 + I_3\omega_3^2\right)$$

where the ω's are referred to principal axes of the body. In Equation 9.6.7 we expressed the ω's in terms of the Eulerian angles θ, ϕ, and ψ as follows:

$$\omega_1 = \dot{\theta}\cos\psi + \dot{\phi}\sin\theta\sin\psi$$

$$\omega_2 = -\dot{\theta}\sin\psi + \dot{\phi}\sin\theta\cos\psi$$

$$\omega_3 = \dot{\psi} + \dot{\phi}\cos\theta$$

Regarding the Eulerian angles as the generalized coordinates, the equations of motion are

$$\frac{d}{dt}\frac{\partial L}{\partial \dot{\theta}} = \frac{\partial L}{\partial \theta}$$

$$\frac{d}{dt}\frac{\partial L}{\partial \dot{\phi}} = \frac{\partial L}{\partial \phi}$$

$$\frac{d}{dt}\frac{\partial L}{\partial \dot{\psi}} = \frac{\partial L}{\partial \psi}$$

Now, by the chain rule,

$$\frac{\partial L}{\partial \dot{\psi}} = \frac{\partial L}{\partial \omega_3}\frac{\partial \omega_3}{\partial \dot{\psi}} = I_3\omega_3$$

so

$$\frac{d}{dt}\frac{\partial L}{\partial \dot{\psi}} = I_3\dot{\omega}_3$$

Again, using the chain rule, we have

$$\frac{\partial L}{\partial \psi} = I_1 \omega_1 \frac{\partial \omega_1}{\partial \psi} + I_2 \omega_2 \frac{\partial \omega_2}{\partial \psi}$$
$$= I_1 \omega_1 (-\dot\theta \sin\psi + \dot\phi \sin\theta \cos\psi) + I_2 \omega_2 (-\dot\theta \cos\psi - \dot\phi \sin\theta \sin\psi)$$
$$= I_1 \omega_1 \omega_2 - I_2 \omega_2 \omega_1$$

Consequently, the Lagrangian equation of motion in the coordinate ψ reduces to

$$I_3 \dot\omega_3 = \omega_1 \omega_2 (I_1 - I_2)$$

which, as we showed in Section 9.3, is the third of Euler's equations for the rotation of a rigid body under zero torque. The other two of Euler's equations can be obtained by cyclic permutation of the subscripts: $1 \to 2, 2 \to 3, 3 \to 1$. This is valid because we have not designated any particular principal axis as being preferred.

EXAMPLE 10.5.6

Particle Sliding on a Movable Inclined Plane

Consider the case of a particle of mass m free to slide along a smooth inclined plane of mass M. The inclined plane is not fixed but is free to slide along a smooth horizontal surface, as shown in Figure 10.5.3. There are only two degrees of freedom, because each object is constrained to move along a single dimension. We can most easily specify the position of the inclined plane by choosing its generalized coordinate to be x, the displacement of the plane relative to some fixed reference point. We can then complete the specification of the system's configuration by choosing x', the displacement down the plane relative to the top of the plane, to be the generalized coordinate of the particle.

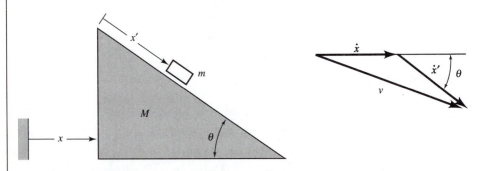

Figure 10.5.3 A block sliding down a movable wedge or inclined plane.

We can calculate the kinetic energy of each mass in terms of the dot product of its respective velocity vector, with each velocity vector specified in terms of unit vectors directed along the relevant generalized coordinate.

$$\mathbf{V} = \mathbf{i}\dot{x} \qquad \mathbf{v} = \mathbf{i}\dot{x} + \mathbf{e}_\theta \dot{x}'$$

The unit vector $\mathbf{e}_\theta$ is directed down the plane at an elevation angle θ relative to the horizontal (see Figure 10.5.3).

Thus, the kinetic energy is

$$T_M = \tfrac{1}{2}M\mathbf{V}\cdot\mathbf{V} = \tfrac{1}{2}M\dot{x}^2$$
$$T_m = \tfrac{1}{2}m\mathbf{v}\cdot\mathbf{v} = \tfrac{1}{2}m(\mathbf{i}\dot{x}+\mathbf{e}_\theta\dot{x}')\cdot(\mathbf{i}\dot{x}+\mathbf{e}_\theta\dot{x}')$$
$$= \tfrac{1}{2}m(\dot{x}^2 + \dot{x}'^2 + 2\dot{x}\dot{x}'\cos\theta)$$
$$T = T_M + T_m$$

The potential energy of the system depends only on the vertical position of the particle of mass m. We can choose it to be zero when the particle is at the top of the plane.

$$V = -mgx'\sin\theta$$

Therefore, the Lagrangian of the system is

$$L = \tfrac{1}{2}M\dot{x}^2 + \tfrac{1}{2}m(\dot{x}^2 + \dot{x}'^2 + 2\dot{x}\dot{x}'\cos\theta) + mgx'\sin\theta$$

and the equations of motion are

$$\frac{d}{dt}\frac{\partial L}{\partial \dot{x}} - \frac{\partial L}{\partial x} = 0 \qquad\qquad \frac{d}{dt}\frac{\partial L}{\partial \dot{x}'} = \frac{\partial L}{\partial x'}$$

$$\frac{d}{dt}[m(\dot{x}+\dot{x}'\cos\theta)+M\dot{x}] = 0 \qquad \frac{d}{dt}(\dot{x}'+\dot{x}\cos\theta) = g\sin\theta$$

(**Note:** The time derivative of the first term is zero. The first term in brackets is, therefore, a constant of the motion. This situation occurs because the Lagrangian is independent of the coordinate x, similar to the situation that occurred in Example 10.5.2. Close examination of this term reveals that it is the total linear momentum of the system in the x direction.)

The term $[m(\dot{x}+\dot{x}'\cos\theta)+M\dot{x}]$ being a constant of the motion also reflects the fact that from the Newtonian viewpoint there is no net force on the system in the x direction. Notice how naturally this result seemingly falls out of the Lagrangian formalism. We have more to say about this in the following section.

We carry out the preceding time derivatives, obtaining

$$m(\ddot{x} + \ddot{x}'\cos\theta) + M\ddot{x} = 0 \qquad\qquad \ddot{x}' + \ddot{x}\cos\theta = g\sin\theta$$

Solving for $\ddot{x}$ and $\ddot{x}'$ we find

$$\ddot{x} = \frac{-g\sin\theta\,\cos\theta}{(m+M)/m - \cos^2\theta} \qquad\qquad \ddot{x}' = \frac{g\sin\theta}{1 - m\cos^2\theta/(m+M)}$$

This particular example illustrates nicely the ease with which complicated problems fall apart when attacked with the Lagrangian formalism. One could certainly solve the problem using Newtonian methods, but such an attempt would require a great deal more thought and physical insight than demanded in the Lagrangian "turn the crank" method shown here.

10.6 | Generalized Momenta: Ignorable Coordinates

A key feature of Examples 10.5.2 and 10.5.5 was the emergence of a momentum, conserved along the direction of a generalized coordinate not explicitly contained in the Lagrangian of the system. We would like to explore this situation in a little more detail. Perhaps the simplest example that illustrates such a condition is a free particle moving in a straight line, say, along the x-axis. Its kinetic energy is

$$T = \tfrac{1}{2} m \dot{x}^2 \tag{10.6.1}$$

where m is the mass of the particle and $\dot{x}$ is its velocity. The Lagrangian for this system assumes the particularly simple form $L = T$. The Lagrangian equation of motion is, thus,

$$\frac{d}{dt}\frac{\partial L}{\partial \dot{x}} = \frac{d}{dt}\frac{\partial T}{\partial \dot{x}} = 0 \qquad \frac{d}{dt}(m\dot{x}) = 0 \tag{10.6.2}$$

$$\therefore m\dot{x} = \text{constant}$$

As occurred in Examples 10.5.2 and 10.5.5, when the Lagrangian is independent of a coordinate, a solution to the equation of motion leads to the constancy of a quantity that can be identified with the momentum of the system referred to that missing coordinate. In this case we see that the constant quantity is exactly equal to the product $m\dot{x}$, the "Newtonian" linear momentum p_x of the free particle. Hence, we make the formal definition that

$$p_x = \frac{\partial L}{\partial \dot{x}} = m\dot{x} \tag{10.6.3}$$

is the momentum of the particle. In the case of a system described by the generalized coordinates $q_1, q_2, \ldots, q_k, \ldots, q_n$, the quantities p_k defined by

$$p_k = \frac{\partial L}{\partial \dot{q}_k} \tag{10.6.4}$$

are called the *generalized momenta conjugate to the generalized coordinate* q_k. Lagrange's equations for a conservative system can then be written as

$$\dot{p}_k = \frac{\partial L}{\partial q_k} \tag{10.6.5}$$

It is now readily apparent that if the Lagrangian does not explicitly contain the coordinate q_k, then

$$\dot{p}_k = \frac{\partial L}{\partial q_k} = 0 \tag{10.6.6}$$

$$\therefore p_k = \text{constant} \tag{10.6.7}$$

The missing coordinate is ignorable, and its conjugate momentum is a constant of the motion.

EXAMPLE 10.6.1

Pendulum Attached to a Movable Support

Let us now continue the analysis of a pendulum attached to a movable support as outlined in Section 10.3 (see Figure 10.3.1). We have already calculated the kinetic and potential energies for this system in terms of the generalized coordinates X, the position of the movable support, and the angle θ that the pendulum makes with the vertical. They are given by Equation 10.3.4a. and b. The Lagrangian for this system is

$$L = \tfrac{1}{2}(M + m)\dot{X}^2 + \tfrac{1}{2}m(r^2\dot{\theta}^2 + 2\dot{X}r\dot{\theta} \, \cos\theta) + mgr \, \cos\theta$$

The equations of motion are

$$\dot{p}_X = \frac{d}{dt}\frac{\partial L}{\partial \dot{X}} = \frac{\partial L}{\partial X} = 0 \qquad \dot{p}_\theta = \frac{d}{dt}\frac{\partial L}{\partial \dot{\theta}} = \frac{\partial L}{\partial \theta}$$

$$\frac{d}{dt}[(M + m)\dot{X} + mr\dot{\theta} \, \cos\theta] = 0$$

$$\frac{d}{dt}[m(r^2\dot{\theta} + \dot{X}r \, \cos\theta)] = -m(\dot{X}r\dot{\theta} + gr) \sin\theta$$

$$\ddot{\theta} + \frac{\ddot{X}}{r} \cos\theta + \frac{g}{r} \sin\theta = 0$$

(**Note:** The Lagrangian is independent of the generalized coordinate X. It is an ignorable coordinate. Its conjugate momentum is the first term in brackets in the preceding equation: the total linear momentum of the system in the X direction. This momentum is a constant of the motion. It must be a conserved quantity, since the potential energy of the system is independent of this coordinate (that is why the Lagrangian is missing that coordinate) and, therefore, no net external forces are acting in this direction.)

Sometimes the differential equations of motion look complicated. One might wonder whether or not the differential equations are infested with errors. Such could be the situation on consideration of the second equation of motion derived earlier. As a check of the validity of such equations demand that the system adhere to certain limiting conditions, and then look closely at just what the derived equation of motion implies, given the imposed conditions. For instance, suppose that in the preceding problem, we "nail down" the movable support; that is, we fix it firmly to the track along which it was previously allowed to move without friction. We, thus, reduce the example to that of a simple pendulum, and the equation of motion had better reflect this fact. We see that it does. The central term containing the acceleration $\ddot{X}$ goes to zero, because we have eliminated any X-direction motion of the mass M. The resultant equation of motion becomes

$$\ddot{\theta} + \frac{g}{r} \sin \theta = 0$$

that of a simple pendulum.

So far, so good. Still, we might continue to be bothered about the $\ddot{X}$ term. The previous condition led to its elimination from the equation of motion. That is not a good way to see whether its presence makes any sense. Let us play another trick similar to the one used earlier, but this time with the angular acceleration and velocity terms $\ddot{\theta}$ and $\dot{\theta}$. Can we imagine a scenario in which they might be zero? What kind of motion would the system exhibit, given such a restriction? If those two terms are zero, the equation of motion reduces to a solution for θ in terms of the horizontal acceleration $\ddot{X}$ and g

$$\tan \theta = \frac{-\ddot{X}}{g}$$

In other words, if we uniformly accelerate the support toward the right, the solution in the preceding equation describes a pendulum that hangs "motionless" at an angle θ to the "left" of the vertical relative to its support; that is, the system becomes a simple "linear accelerometer." Indeed, this is a possible scenario for the motion of this system. Such an analysis was presented in Chapter 5 and is also presented in several beginning primers on basic physics.[4] You see, it all makes sense.

EXAMPLE 10.6.2

The Spherical Pendulum, or Bar of Soap in a Bowl

A classic problem in mechanics is that of a particle constrained to stay on a smooth spherical surface under gravity, such as a small mass sliding around inside a smooth spherical bowl. The case is also illustrated by a simple pendulum that is free to swing in any direction, Figure 10.6.1. This is the so-called spherical pendulum, mentioned in Section 5.6

There are two degrees of freedom, and we use generalized coordinates θ and ϕ, as shown. These are actually equivalent to spherical coordinates with $r = l = $ constant, in which l is the length of the pendulum cord. The two components of the velocity are $v_\theta = l\dot{\theta}$ and

---

[4] For example, see Example 6.8 in Serway and Jewett, *Physics for Scientists and Engineers,* 6th ed, Brooks/ Cole Thomson—Learning, Belmont, CA, 2004.

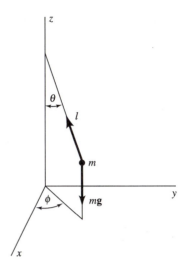

Figure 10.6.1 The spherical pendulum.

$v_\phi = l\dot{\phi} \sin\theta$. The height of the bob, measured from the xy plane, is $l - l\cos\theta$, so the Lagrangian function is

$$L = \tfrac{1}{2}ml^2(\dot{\theta}^2 + \dot{\phi}^2 \sin^2\theta) - mgl(1 - \cos\theta)$$

The coordinate ϕ is ignorable, so we have immediately

$$p_\phi = \frac{\partial L}{\partial \dot{\phi}} = ml^2\dot{\phi}\sin^2\theta = \text{constant}$$

This is the angular momentum about the vertical, or z-axis. We are left with just the equation in θ:

$$\frac{d}{dt}\frac{\partial L}{\partial \dot{\theta}} = \frac{\partial L}{\partial \theta}$$

which reads

$$ml^2\ddot{\theta} = ml^2\dot{\phi}^2 \sin\theta\cos\theta - mgl\sin\theta$$

Let us introduce the constant S, defined by

$$S = \dot{\phi}\sin^2\theta = \frac{p_\phi}{ml^2} \tag{10.6.8}$$

(This is the angular momentum divided by ml^2.) The differential equation of motion for θ then becomes

$$\ddot{\theta} + \frac{g}{l}\sin\theta - S^2\frac{\cos\theta}{\sin^3\theta} = 0 \tag{10.6.9}$$

It is instructive to consider some special cases at this point. First, if the angle ϕ is constant, then $\dot{\phi} = 0$, and so $S = 0$. Consequently, Equation 10.6.9 reduces to

$$\ddot{\theta} + \frac{g}{l} \sin \theta = 0$$

which, of course, is just the differential equation of the simple pendulum. The motion takes place in the plane $\phi = \phi_0 = $ constant.

The second special case is that of the *conical pendulum*. Here the bob describes a horizontal circle, so $\theta = \theta_0 = $ constant. In this case $\dot{\theta} = 0$ and $\ddot{\theta} = 0$, so Equation 10.6.9 reduces to

$$\frac{g}{l} \sin \theta_0 - S^2 \frac{\cos \theta_0}{\sin^3 \theta_0} = 0$$

or

$$S^2 = \frac{g}{l} \sin^4 \theta_0 \, \sec \theta_0 \tag{10.6.10}$$

Inserting the value of S given by Equation 10.6.8 into Equation 10.6.10 yields

$$\dot{\phi}_0^2 = \frac{g}{l} \sec \theta_0 \tag{10.6.11}$$

as the condition for conical motion of the pendulum.

Let us now consider the case in which the motion is almost conical; that is, the value of θ remains close to the value θ_0. If we insert the expression for S^2 given in Equation 10.6.10 into Equation 10.6.9, the result is

$$\ddot{\theta} + \frac{g}{l} \left(\sin \theta - \frac{\sin^4 \theta_0}{\cos \theta_0} \frac{\cos \theta}{\sin^3 \theta} \right) = 0$$

It is convenient at this point to introduce the new variable ξ defined as

$$\xi = \theta - \theta_0$$

The expression in parentheses, which we call $f(\xi)$, may be expanded as a power series in ξ according to the standard formula

$$f(\xi) = f(0) + f'(0)\xi + f''(0)\frac{\xi^2}{2!} + \cdots$$

We find, after performing the indicated operations, that $f(0) = 0$ and $f'(0) = 3\cos\theta_0 + \sec\theta_0$. Because we are concerned with the case of small values of ξ, we shall ignore higher powers of ξ than the first, and so we can write

$$\ddot{\xi} + \frac{g}{l}(3\cos\theta_0 + \sec\theta_0)\xi = 0$$

Thus, ξ oscillates harmonically about $\xi = 0$, or equivalently, θ oscillates harmonically about the value θ_0 with a period

$$T_1 = 2\pi \sqrt{\frac{l}{g(3\cos\theta_0 + \sec\theta_0)}}$$

Now the value of $\dot{\phi}$ does not vary greatly from the value given by the purely conical motion $\dot{\phi}_0$, so ϕ increases steadily during the oscillation of θ about θ_0. During one complete oscillation of θ, the value of the azimuth angle ϕ increases by the amount

$$\phi_1 \approx \dot{\phi}_0 T_1$$

From the values of $\dot{\phi}_0$ and T_1 given in the preceding equation, we find

$$\phi_1 = 2\pi(3\cos^2\theta_0 + 1)^{-1/2}$$

Now the quantity in parentheses is less than 4, for nonzero θ_0, so ϕ_1 is greater than π (180°). The excess $\Delta\phi$ is shown in Figure 10.6.2, which is a plot of the projection of the path of the pendulum bob on the xy plane. As the pendulum swings, it precesses in the direction of increasing ϕ, as indicated.

Finally, for the general case we can go back to the differential equation of motion (Equation 10.6.9) and integrate once with respect to θ by using the fact that $\ddot{\theta} = \dot{\theta}\,d\dot{\theta}/d\theta = \frac{1}{2}\,d\dot{\theta}^2/d\theta$. The result is

$$\tfrac{1}{2}\dot{\theta}^2 = \frac{g}{l}\cos\theta - \frac{S^2}{2\sin^2\theta} + C = -U(\theta) + C$$

in which C is the constant of integration and

$$U(\theta) = -\frac{g}{l}\cos\theta + \frac{S^2}{2\sin^2\theta}$$

is the effective potential. Actually, the integrated equation of motion is just the energy equation in which the total energy $E = C\,ml^2$. For a given initial condition, the motion

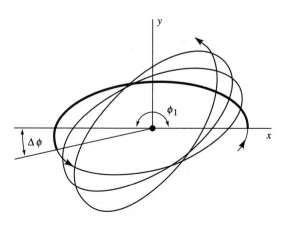

Figure 10.6.2   Projection on the xy plane of the path of motion of the spherical pendulum.

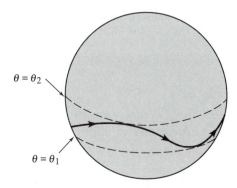

$\theta = \theta_2$

$\theta = \theta_1$

Figure 10.6.3   Illustrating the limits of the motion of the spherical pendulum.

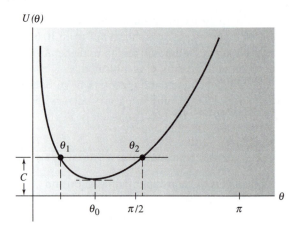

$U(\theta)$

θ_1 θ_2

C

θ_0 $\pi/2$ π θ

Figure 10.6.4   Graph of the effective potential for the spherical pendulum.

of the pendulum is such that the bob oscillates between two horizontal circles. These circles define the turning points of the θ-motion for which $\dot{\theta} = 0$ or $U(\theta) = C$. This is illustrated in Figures 10.6.3 and 10.6.4.

10.7 | Forces of Constraint: Lagrange Multipliers

Notice that even though the physical system discussed in several of our previous examples has been subject to holonomic constraints, nowhere in our calculations have we had to consider specifically the forces that result from those constraints. This is one of the great virtues of the Lagrangian method: direct inclusion of such forces of constraint in a solution for the motion of the constrained body is superfluous and, therefore, ignored. Suppose for some reason or another, however, we wish to know the values of those forces. For example, an engineer might wish to know the value of the normal force exerted by a curved bridge on a heavily weighted truck traveling over it—or Tarzan might wish to know the tension in a vine when swinging across a crocodile-infested river. Clearly, in such instances, ignorance is not necessarily bliss.

We can explicitly include the forces of constraint in the Lagrangian formulation if we so choose. In essence, this can be accomplished by not immediately invoking any equation(s) of constraint to reduce the number of degrees of freedom in a problem. Thus right from the outset we keep all the generalized coordinates in which the kinetic and potential energies of the system in question are expressed. This leads to more Lagrange equations for the problem, one for each additional generalized coordinate not eliminated by an equation of constraint. But because the coordinates are not independent, the resulting Lagrange equations as previously derived cannot be independent either. They can, however, be made independent through the technique of *Lagrange multipliers*.

For the sake of simplicity, we consider a system described by only two generalized coordinates q_1 and q_2 that are connected by a single equation of constraint

$$f(q_1, q_2, t) = 0 \tag{10.7.1}$$

We start with the Hamilton variational principle as given in Equation 10.4.4

$$\delta \int_{t_1}^{t_2} L \, dt = \int_{t_1}^{t_2} \sum_i \left[\frac{\partial L}{\partial q_i} - \frac{d}{dt}\left(\frac{\partial L}{\partial \dot{q}_i} \right) \right] \delta q_i \, dt = 0 \tag{10.7.2}$$

Only now, the δq_i are not independent; a variation in q_1 leads to a variation in q_2 consistent with the constraint given by Equation 10.7.1, *which*, because it is fixed at any instant of time t, obeys the condition

$$\delta f = \left(\frac{\partial f}{\partial q_1} \delta q_1 + \frac{\partial f}{\partial q_2} \delta q_2 \right) = 0 \tag{10.7.3}$$

Solving for δq_2 in terms of δq_1

$$\delta q_2 = -\left(\frac{\partial f / \partial q_1}{\partial f / \partial q_2} \right) \delta q_1 \tag{10.7.4}$$

and on substituting into Equation 10.7.2, we obtain

$$\int_{t_1}^{t_2} \left[\left(\frac{\partial L}{\partial q_1} - \frac{d}{dt}\frac{\partial L}{\partial \dot{q}_1} \right) - \left(\frac{\partial L}{\partial q_2} - \frac{d}{dt}\frac{\partial L}{\partial \dot{q}_2} \right)\left(\frac{\partial f / \partial q_1}{\partial f / \partial q_2} \right) \right] \delta q_1 \, dt = 0 \tag{10.7.5}$$

Only a single coordinate, q_1, is varied in this expression, and because it can be varied at will, the term in brackets must vanish. Thus,

$$\frac{(\partial L / \partial q_1) - (d/dt)(\partial L / \partial \dot{q}_1)}{(\partial f / \partial q_1)} = \frac{(\partial L / \partial q_2) - (d/dt)(\partial L / \partial \dot{q}_2)}{(\partial f / \partial q_2)} \tag{10.7.6}$$

The term on the left is a function only of the generalized coordinate q_1 and its derivative, while that on the right is a function only of q_2 and its derivative. Furthermore, each term is implicitly time-dependent through these variables and possibly explicitly as well. The only way they can be equal at all times throughout the motion is if they are equal to a single function of time, which we call $-\lambda(t)$. Thus, we have

$$\frac{\partial L}{\partial q_i} - \frac{d}{dt}\frac{\partial L}{\partial \dot{q}_i} + \lambda(t)\frac{\partial f}{\partial q_i} = 0 \qquad \{i = 1, 2 \tag{10.7.7}$$

We now have a problem with three unknown functions of time: $q_1(t)$, $q_2(t)$, and $\lambda(t)$, and we have three independent equations needed to solve for them: two Lagrangian equations of motion (10.7.7) and an equation of constraint (10.7.1). Thus, even though the two coordinates are connected by an equation of constraint and we could have immediately used it to reduce the degrees of freedom in the problem so that the motion could be characterized by a single, "multiplier-free" equation, we did not do so. The reason is that the Lagrange multiplier terms

$$Q_i = \lambda(t)\frac{\partial f}{\partial q_i} \qquad \{i = 1, 2 \qquad\qquad (10.7.8)$$

that appear in the two Lagrangian equations of motion (10.7.7) are the forces of constraint that we desire. They appear in the problem only because we did not initially invoke the equation of constraint to reduce the degrees of freedom. These terms Q_i are called *generalized forces of constraint*. They are identical to forces if their corresponding generalized coordinate q_i is a spatial coordinate. As we shall see, however, they are torques if their corresponding coordinate q_i is an angular coordinate.

The more general problem is describable in terms of n generalized coordinates connected by m equations of constraint. Thus, if knowledge of the constraining forces is desired, then the system is described by n Lagrange equations of motion of the form

$$\frac{\partial L}{\partial q_i} - \frac{d}{dt}\frac{\partial L}{\partial \dot{q}_i} + \sum_j \lambda_j(t)\frac{\partial f_j}{\partial q_i} \qquad \begin{cases} i = 1, 2, \ldots n \\ j = 1, 2, \ldots m \end{cases} \qquad (10.7.9)$$

Note, though, that there are $m + n$ unknowns—the n $q_i(t)$ and the m $\lambda_j(t)$. The additional information necessary to solve the problem comes from the m equations of constraint usually known in one—or some combination of—the following forms:

$$f_j(q_i, t) = 0 \qquad\qquad (10.7.10a)$$

$$\sum_i \frac{\partial f_j}{\partial q_i}dq_i + \frac{\partial f_j}{\partial t}dt = 0 \qquad \begin{cases} i = 1, 2, \ldots n \\ j = 1, 2, \ldots m \end{cases} \qquad (10.7.10b)$$

The second relation (Equation 10.7.10b) is the total differential of the first and so is equivalent to it. In many instances, the constraints are known in the following differential form in which there is no explicit time derivative:

$$\sum_i \frac{\partial f_j}{\partial q_i}dq_i = 0 \qquad \begin{cases} i = 1, 2, \ldots n \\ j = 1, 2, \ldots m \end{cases} \qquad (10.7.10c)$$

Constraints of this form, when used in Hamilton's variational principle to derive Lagrange's equations, are equivalent to the variation illustrated by Equation 10.7.3 and, thus, lead directly to the Lagrange equations 10.7.9. Even if the constraints depend explicitly on time as in Equations 10.7.10a and b, when used in Hamilton's variational principle, the time dependence has no effect on the Lagrangian equations of motion because *time is held fixed during the variational procedure*, a fact that we invoked in Equation 10.7.3. Thus, when one desires to know the forces of constraint and the constraints are of the type given by

Equations 10.7.10a, b or c, the Lagrangian equations of motion to use are the ones given by Equation 10.7.9.

EXAMPLE 10.7.1

Consider a disc that has a string wrapped around it with one end attached to a fixed support and allowed to fall with the string unwinding as it falls—as shown in Figure 10.7.1. (The situation is somewhat akin to that of a yo-yo whose string is attached to a finger and then allowed to drop, the finger held motionless as a fixed support.) Find the equations of motion of the falling disc and the forces of constraint.

Solution:

The kinetic energy of the falling disc is given by

$$T = \frac{1}{2}m\dot{y}^2 + \frac{1}{2}I_{cm}\dot{\phi}^2$$
$$= \frac{1}{2}m\dot{y}^2 + \frac{1}{4}ma^2\dot{\phi}^2$$

where m is the mass of the disc, a is its radius, and $I_{cm} = \frac{1}{2}ma^2$ is the moment of inertia of the disc about its central axis. The potential energy of the disc is

$$V = -mgy$$

where the reference level for the potential energy is located at $y = 0$, the point of suspension. The Lagrangian is, therefore,

$$L = T - V = \frac{1}{2}m\dot{y}^2 + \frac{1}{4}ma^2\dot{\phi}^2 + mgy$$

The equation of constraint is given by

$$f(y, \phi) = y - a\phi = 0$$

The system has one degree of freedom, and we could choose either y or ϕ as the generalized coordinate and then use the equation of constraint to eliminate the other from consideration. The preceding Lagrangian could then be transformed into a function

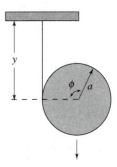

Figure 10.7.1   Falling disc unrolling from an attached string.

of the single generalized coordinate, and we could use Lagrange's equation (10.4.5) without Lagrange multipliers to solve for the equation of motion of the falling disc. This would not produce the desired forces of constraint, however. To find them, we must use the Lagrange equations (10.7.9) with the undetermined multipliers λ_j. In the case here, Equations 10.7.9 are

$$\frac{\partial L}{\partial y} - \frac{d}{dt}\frac{\partial L}{\partial \dot{y}} + \lambda\frac{\partial f}{\partial y} = 0$$

$$\frac{\partial L}{\partial \phi} - \frac{d}{dt}\frac{\partial L}{\partial \dot{\phi}} + \lambda\frac{\partial f}{\partial \phi} = 0$$

On carrying out the preceding prescription, we obtain the following equations of motion:

$$mg - m\ddot{y} + \lambda = 0$$

$$-\frac{1}{2}ma^2\ddot{\phi} - \lambda a = 0$$

Differentiating the equation of constraint gives

$$\ddot{\phi} = \frac{\ddot{y}}{a}$$

and inserting this into the second of the preceding equations of motion yields the value of λ

$$\lambda = -\frac{1}{3}mg$$

Inserting this into the first of the preceding equations of motion and using the differentiated equation of constraint yields

$$\ddot{y} = \frac{2}{3}g \qquad \ddot{\phi} = \frac{2}{3}\frac{g}{a}$$

Thus, we have the required equations of motion for the two, dependent, generalized coordinates.

If the disc were to fall freely, unconstrained by the attached string, the downward acceleration would be g. The upward tension in the string, the force of constraint, must reduce this by $\frac{1}{3}g$. Thus, λ must be equal to the tension in the string, one of the generalized forces of constraint. We can see that is so by calculating the values of the two generalized "forces"

$$Q_y = \lambda\frac{\partial f}{\partial y} = \lambda = -\frac{1}{3}mg$$

$$Q_\phi = \lambda\frac{\partial f}{\partial \phi} = -\lambda a = \frac{1}{3}mga$$

The generalized force Q_y is indeed the tension in the string that reduces the downward acceleration of the disc, while Q_ϕ is the torque on the disc that causes it to rotate about its center of mass.

10.8 | D'Alembert's Principle: Generalized Forces

In Section 10.1, we showed how Hamilton's variational principle led to Newton's laws of motion for the simple situation of a body falling freely in a uniform gravitational field. In Section 10.4, we used Hamilton's principle to derive the Lagrange equations of motion for conservative systems, thereby implying that the Newtonian and Lagrangian formulations of mechanics are equivalent, at least for that restricted class of problems. In Section 10.7 we showed how the Lagrangian formulation could be modified to include forces of constraint not derivable from a conservative potential function. Here we would like to show that, with a minor alteration, the Lagrangian formulation can be further extended to include those systems that might exhibit nonconservative forces as well. In doing so, we hope to eliminate any lingering doubt that the by-now somewhat overwhelmed student might harbor regarding the issue of the equivalence between the Newtonian and Lagrangian formulation of mechanics.

To do this we appeal to a principle first suggested by Bernoulli and formalized by D'Alembert. It is based on an extension of the fundamental condition of equilibrium, namely that all the forces acting on all the bodies that make up a physical system vanish when the system is in equilibrium

$$\sum_{i=1}^{N} \mathbf{F}_i = 0 \tag{10.8.1}$$

Here, $\mathbf{F}_i$ represents all the forces acting on the ith body, and N bodies comprise the system. This set of equations represents relationships among vector quantities. The formulation of mechanics developed in the late 17th and early 18th centuries, primarily in continental Europe, was based on relationships among scalar quantities, like energy and work. Bernoulli and D'Alembert realized that the net work done on bodies in equilibrium, if subjected to very small displacements from equilibrium, would also vanish. In essence, during such displacements, for every force that does positive work on the bodies, there is one that does an equal and opposite amount of negative work.

Using this idea, the method of finding the conditions necessary for equilibrium is carried out by *imagining that a system of bodies in a given configuration undergo small displacements away from their assumed positions, calculating the resultant work done on the system and then demanding that it sum to zero*. The displacements are imagined— bodies actually in equilibrium obviously do not undergo any real displacements—and they are assumed to be of infinitesimal extent. They are assumed to take place instantaneously (i.e., $\delta t = 0$), and they are assumed to take place in a way that is consistent with any imposed constraints. Such displacements are usually denoted by the symbol δ (as we previously did) and are called *virtual displacements*. The key point in this formulation is the realization that if the work vanishes for a system subjected to such virtual displacements, then the system *is* in equilibrium. Thus, we have a way of establishing the conditions necessary for equilibrium by dealing with the scalar quantity of work, rather than the vector quantity of force. This is the principle of *virtual work* mentioned at the beginning of the chapter, and it is expressed by the condition

$$\delta W = \sum_{i=1}^{N} \mathbf{F}_i \cdot \delta \mathbf{r}_i = 0 \tag{10.8.2}$$

However, problems in *dynamics* deal with bodies that are not in equilibrium. They must be solved using Newton's second law of motion

$$\mathbf{F}_i = \dot{\mathbf{p}}_i \qquad \{i = 1, 2, \ldots N \tag{10.8.3}$$

D'Alembert's insight was to realize that problems in dynamics could be cast in the same language as the principle of virtual work by including the inertial term $-\dot{\mathbf{p}}$ as a real force in Equation 10.8.2

$$\sum_{i=1}^{N} (\mathbf{F}_i - \dot{\mathbf{p}}_i) \cdot \delta\mathbf{r}_i = 0 \tag{10.8.4}$$

The condition expressed in Equation 10.8.4 is called *D'Alembert's principle,* and it is equivalent to Newton's second law of motion, written, however, as though the inertial term was a real force

$$\mathbf{F}_i - \dot{\mathbf{p}}_i = 0 \qquad \{i = 1, 2, \ldots N \tag{10.8.5}$$

(which we know, from our work in Chapter 5, is a valid way of analyzing motion in an appropriate noninertial frame of reference). In effect, D'Alembert reduced a problem in *dynamics* to one in *statics*. We now proceed to derive Lagrange's equations of motion from D'Alembert's principle, (as did Lagrange himself, by the way).

In what follows, however, we do not wish to cloud the derivation with myriad symbols such as double and triple sums and lots of indices, as sometimes happens in our zealous desire for completeness and rigor. Thus, we derive Lagrange's equations for the case of a single particle, assumed, however, to be subjected to some arbitrary number of forces. Presumably, the ambitious student could generalize the analysis to many forces acting on a many-particle system based on past material in the text through which we've already laboriously waded.

We begin with D'Alembert's principle, Equation 10.8.4, for a single particle expressed in three-dimensional Cartesian coordinates

$$\sum_{i=1}^{3} (F_i - \dot{p}_i)\, \delta x_i = 0 \tag{10.8.6}$$

The index i represents one of three Cartesian coordinates, and the F_i are the sum of all the force components acting on the body along the ith direction.

We assume that the motion of the particle is also describable in terms of its generalized coordinates q_j, which might or might not be connected by equations of constraint. For the moment, we assume that they are not and that the relationship between the three Cartesian coordinates and the three generalized coordinates then necessary to describe the motion of the particle is given by Equations 10.3.10 for an unconstrained system.

The first term in the sum of Equation 10.8.6 is the virtual work done on the system

$$\delta W = \sum_i F_i\, \delta x_i = \sum_j \left[\sum_i \left(F_i \frac{\partial x_i}{\partial q_j} \right) \right] \delta q_j = \sum_j Q_j\, \delta q_j \tag{10.8.7}$$

It is not equal to zero! The Q_j's, given by

$$Q_j = \sum_i F_i \frac{\partial x_i}{\partial q_j} \tag{10.8.8}$$

are called *generalized forces* corresponding to the generalized coordinates q_j. As was the case with the generalized forces of constraint discussed in Section 10.7, a generalized force Q_j is a force if the corresponding generalized coordinate q_j is a position; it is a torque if the corresponding generalized coordinate q_j is an angle. In each case, the product $Q_j \, \delta_j$ is work.

We now calculate the inertial term in Equation 10.8.6

$$\sum_i \dot{p}_i \, \delta x_i = \sum_i m \ddot{x}_i \, \delta x_i = \sum_j \left[\sum_i m \ddot{x}_i \frac{\partial x_i}{\partial q_j} \right] \delta q_j$$

$$= \sum_j \delta q_j \sum_i m \left[\frac{d}{dt}\left(\dot{x}_i \frac{\partial x_i}{\partial q_j} \right) - \dot{x}_i \frac{d}{dt}\left(\frac{\partial x_i}{\partial q_j} \right) \right]$$

(10.8.9)

Differentiating Equations 10.3.13, we see that

$$\frac{\partial \dot{x}_i}{\partial \dot{q}_j} = \frac{\partial x_i}{\partial q_j}$$

(10.8.10)

and we can insert this relation into the first term in the brackets of Equation 10.8.9 to obtain

$$\sum_i m \frac{d}{dt}\left(\dot{x}_i \frac{\partial x_i}{\partial q_j} \right) = \sum_i m \frac{d}{dt}\left(\dot{x}_i \frac{\partial \dot{x}_i}{\partial \dot{q}_j} \right)$$

$$= \frac{d}{dt}\left[\frac{\partial}{\partial \dot{q}_j}\left(\sum \tfrac{1}{2} m \dot{x}_i^2 \right) \right] = \frac{d}{dt}\left(\frac{\partial T}{\partial \dot{q}_j} \right)$$

(10.8.11)

To evaluate the second term in brackets in Equation 10.8.9 we need to take the time derivative of $\partial x_i / \partial q_j$. Because the time derivative of any general function of q_j and t is given by

$$\frac{d}{dt} f(q_j, t) = \sum_k \frac{\partial f}{\partial q_k} \dot{q}_k + \frac{\partial f}{\partial t}$$

(10.8.12)

on applying this operation we obtain

$$\sum_i m \dot{x}_i \frac{d}{dt}\left(\frac{\partial x_i}{\partial q_j} \right) = \sum_i m \dot{x}_i \left[\sum_k \frac{\partial^2 x_i}{\partial q_k \, \partial q_j} \dot{q}_k + \frac{\partial^2 x_i}{\partial q_j \, \partial t} \right] = \sum_i m \dot{x}_i \frac{\partial \dot{x}_i}{\partial q_j} \quad (10.8.13a)$$

This result is equivalent to reversing the order of differentiation

$$\frac{d}{dt} \frac{\partial x_i}{\partial q_j} = \frac{\partial}{\partial q_j} \frac{dx_i}{dt} = \frac{\partial \dot{x}_i}{\partial q_j}$$

(10.8.13b)

Thus, the second term becomes

$$\sum_i m \dot{x}_i \frac{d}{dt}\left(\frac{\partial x_i}{\partial q_j} \right) = \frac{\partial}{\partial q_j}\left[\sum_i \tfrac{1}{2} m \dot{x}_i^2 \right] = \frac{\partial T}{\partial q_j}$$

(10.8.13c)

The inertial term (Equation 10.8.9) is thus

$$\sum_i \dot{p}_i \, \delta x_i = \sum_j \left[\frac{d}{dt}\left(\frac{\partial T}{\partial \dot{q}_j} \right) - \frac{\partial T}{\partial q_j} \right] \delta q_j$$

(10.8.14)

Combining Equations 10.8.7 and 10.8.14, D'Alembert's principle becomes

$$\sum_i (F_i - \dot{p}_i)\,\delta x_i = \sum_j \left\{ Q_j - \left[\frac{d}{dt}\left(\frac{\partial T}{\partial \dot{q}_j} \right) - \frac{\partial T}{\partial q_j} \right] \right\} \delta q_j = 0 \qquad (10.8.15)$$

For the moment, we've assumed that there are no constraints; therefore, the q_j can be independently varied. Thus, the term in brackets must vanish, and we obtain

$$\frac{d}{dt}\left(\frac{\partial T}{\partial \dot{q}_j} \right) - \frac{\partial T}{\partial q_j} = Q_j \qquad (10.8.16)$$

If the external forces are conservative, they can be derived from a potential energy function, $F_i = -\nabla_i V$, and Equation 10.8.16 can be simplified even further

$$Q_j = \sum_i F_i \frac{\partial x_i}{\partial q_j} = -\sum_i \nabla_i V \frac{\partial x_i}{\partial q_j} \qquad (10.8.17a)$$

The latter expression in Equation 10.8.17a is the partial derivative of V with respect to q_j, thus

$$Q_j = -\frac{\partial V}{\partial q_j} \qquad (10.8.17b)$$

and it can be substituted into Equation 10.8.16, yielding

$$\frac{d}{dt}\left(\frac{\partial T}{\partial \dot{q}_j} \right) - \frac{\partial (T-V)}{\partial q_j} = 0 \qquad (10.8.18)$$

If the potential energy function V is independent of any generalized velocity $\dot{q}_j$, we can include that term along with the first term in Equation 10.8.18 as well

$$\frac{d}{dt}\left[\frac{\partial (T-V)}{\partial \dot{q}_j} \right] - \frac{\partial (T-V)}{\partial q_j} = \frac{d}{dt}\left(\frac{\partial L}{\partial \dot{q}_j} \right) - \frac{\partial L}{\partial q_j} = 0 \qquad (10.8.19)$$

where we have substituted the Lagrangian function $L = T - V$ introduced in Section 10.1. Equation 10.8.19 is the Lagrangian equation of motion for conservative systems that we derived previously from Hamilton's variational principle. Here, though, we have emphasized the restrictions to which the generalized forces and potential must adhere if this standard formalism is to apply. We emphasize this even further by noting the explicit functional dependencies permissible in the Lagrangian function used in Equation 10.8.19 [5]

$$L(q_j, \dot{q}_j, t) = T(q_j, \dot{q}_j, t) - V(q_j) \qquad (10.8.20)$$

---

[5] In some cases generalized forces can be derived from a velocity-dependent potential such as $V = V(q_j, \dot{q}_j, t)$ and

$$Q_j = \frac{d}{dt}\left(\frac{\partial V}{\partial \dot{q}_j} \right) - \frac{\partial V}{\partial q_j}$$

and the standard Lagrangian formalism still applies.

In general, for a system subject to nonconservative generalized forces and forces of constraint which we call Q'_j, Lagrange's equations of motion may be written as

$$\frac{d}{dt}\left[\frac{\partial L}{\partial \dot{q}_j}\right] - \frac{\partial L}{\partial q_j} = Q'_j \qquad \{j = 1, 2, \ldots n \qquad (10.8.21)$$

Although our derivation was for a single particle in which at most only three generalized coordinates are necessary to describe its motion, if N particles are involved, then $n = 3N$ is the number of generalized coordinates necessary to describe the system. The Lagrangian contains all conservative potentials, which, therefore, includes the effect of all conservative forces acting on the system. The forces of constraint comprising some of the Q'_j may be obtained using the method of Lagrange multipliers discussed in the previous section. The nonconservative generalized forces Q'_j, such as frictional or time-dependent forces, must be specifically included as known or unknown in Equation 10.8.21. If the nonconservative forces $\mathbf{F}_i$ are known, the Q'_j may be calculated immediately using Equation 10.8.8. If the $\mathbf{F}_i$ are unknown, then the Q'_j must be found from the process of solving Lagrange's equations. The Lagrangian equation of motion expressed by Equation 10.8.21 is thus completely general and equivalent to Newton's laws of motion from which it was derived using D'Alembert's principle.

Although earlier we derived Lagrange's equations from Hamilton's principle only for the case of conservative systems, it could also be used, as we just did using D'Alembert's principle, to derive Lagrange's equations for the more general case. One might, therefore, wonder about the possible equivalence of these two ways of formulating the laws of mechanics. They are equivalent. The generalized version of Hamilton's principle is in fact nothing other than an integral form of D'Alembert's principle.

Why the Lagrangian Method?

Given that both Hamilton's principle and D'Alembert's principle are equivalent to Lagrange's equations, why not use either of these two approaches as a beginning point for solving problems in mechanics? You could—and the method proves quite useful as an approximation technique for solving complicated problems in classical mechanics, but a discussion of such methods lies beyond the scope of this text.[6] Lagrange's equations, on the other hand, provide an incredibly consistent methodology and indeed an almost mind-numbingly mechanistic problem-solving strategy. In that characteristic, they seem to outrank even the universally applicable Newtonian approach. The question most students ask is then: "Why even use the Newtonian approach at all when the Lagrangian approach seems so much simpler, mechanistic, and powerful?" It seems to grant the practitioner omnipotent calculational prowess—enabling him or her to leap tall buildings with a single bound. . . .

The strength of the Lagrangian approach to solving problems is based on its ability to deal with scalar functions, whereas the Newtonian approach is based on the use of the vector quantities, forces, and momenta. When problems are not too complex, the Newtonian method is relatively straightforward to apply but as their complexity increases, solving problems with the Lagrangian approach begins to show its mettle. It is cast in the

---

[6] See, for example, C. G. Gray, G. Karl, and V. A. Novikov, Direct Use of Variational Principles as an Approximation Technique in Classical Mechanics, *Am. J. Phys.* **64**(9), 1177, 1996.

language of generalized coordinates that, except in the simplest of problems, are substantially easier to use as a way of describing the motion of many body systems or those shaped by complex constraints. The equations of motion are obtained exclusively from manipulations of scalar quantities in *configuration space* rather than vector operations in a rigid Cartesian coordinate system.

The Lagrangian approach is particularly powerful when dealing with a conservative system for which one only wishes to generate its equations of motion; forces of constraint are not an issue. Indeed, the standard Lagrangian formulation ignores them. If forces of constraint are of interest, then one must employ the method involving Lagrange multipliers to ferret them out. In such circumstances, it might be advisable to take the Newtonian approach instead. When nonconservative or velocity-dependent generalized forces rear their ugly heads, however, the Newtonian method is invariably the one of choice; in such cases, one really does need the sledgehammer to crack the walnut; delicate thrusts with the Lagrangian "rapier" will likely prove futile.

Finally, there is a real philosophical difference between the two approaches. The Newtonian method is differential; it has *cause and effect* embedded in it. The application of a force external to a body causes it to accelerate. Lagrange's equations are also differential, but they are cast in the language of kinetic and potential energies, scalar *"essences"* more intrinsic rather than extrinsic to a body. This distinction is particularly true for conservative systems, for nowhere in that Lagrangian formulation does the term *force* ever appear. Hamilton's variational principle represents the pinnacle of this point of view: it gives paramount importance to the energy concept at the expense of the concept of force. Indeed, in microscopic systems the concepts of force, cause, and effect lose their classic meaning entirely. In such systems, energies reign supreme, and it is no accident that the *Hamiltonian* and *Lagrangian* functions assume fundamental roles in a formulation of the theory of *quantum mechanics*.

The philosophical difference is accentuated even more if one considers Hamilton's variational principle to be the fundamental formulation of mechanics. It is an integral formulation as opposed to a differential one. The correct motion that a body takes through space is that which minimizes the time integral of the difference between the kinetic and potential energies. Such a perspective is a "global" one as opposed to the "local" one of the Newtonian formulation. Even theories of quantum mechanics exhibit a similar dichotomy of perspective, for instance, the path integral approach of Richard Feynman[7] (1918–1988) versus the differential approach of Erwin Schrödinger (1887–1961). Given that nature does seem to obey such global principles, it is not difficult to understand, as alluded to at the beginning of this chapter, why a number of philosophers have used them as a basis for the teleological argument that nature works to achieve some goal—that it has some purpose in mind—or that there was some purpose in mind for it. Indeed, Maupertuis argued that dynamical paths taken through space possessing nonminimal values of a mathematical quantity he called *action*[8] would actually be observed if nature exhibited less than perfect

[7] R. P. Feynman and A. R. Hibbs, *Quantum Mechanics and Path Integrals,* McGraw-Hill, New York, 1965.

[8] Pierre-Louis-Moreau de Maupertuis, *Essai de Cosmologie* (1751), in *Oeuvres,* Vol. 4, p. 3, Lyon, 1768, or see Henry Margenau, *The Nature of Physical Reality,* 2nd Ed., McGraw-Hill, New York, 1977. Announced by Maupertuis in 1747, and the first of the minimum principles, the *principle of least action* was the forerunner of Hamilton's principle.

laws of motion, thus arguing for the existence of a God who endowed our world with perfection. Such teleology smacks of the argument that eyes were designed so that we can see—as opposed to the view that they evolved slowly over time, each new stage of development emerging victorious by natural selection until ultimately there existed an eye that bestowed the survival advantage of sight on its owners.

We cast this philosophical issue aside because it has no bearing on the actual operational method we should choose when solving a problem in mechanics. Indeed, the actual motion a mechanical system exhibits is invariant under this choice and, therefore, cannot be used to decide which philosophy is the superior one to adopt. There is a superior method of choice when solving a problem, however, and that is the one that possesses those calculational advantages intrinsic to the particular problem under consideration.

10.9 | The Hamiltonian Function: Hamilton's Equations

Consider the following function of the generalized coordinates:

$$H = \sum_i \dot{q}_i p_i - L \tag{10.9.1}$$

For simple dynamic systems the kinetic energy T is a homogeneous quadratic function of the $\dot{q}$'s, and the potential energy V is a function of the q's alone, so that

$$L = T(q_i, \dot{q}_i) - V(q_i) \tag{10.9.2}$$

Now, from Euler's theorem for homogeneous functions,[9] we have

$$\sum_i \dot{q}_i p_i = \sum_i \dot{q}_i \frac{\partial L}{\partial \dot{q}_i} = \sum_i \dot{q}_i \frac{\partial T}{\partial \dot{q}_i} = 2T \tag{10.9.3}$$

Therefore,

$$H = \sum_i \dot{q}_i p_i - L = 2T - (T - V) = T + V \tag{10.9.4}$$

That is, the function H is equal to the total energy for the type of system we are considering.

Suppose we regard the n equations

$$p_i = \frac{\partial L}{\partial \dot{q}_i} \qquad (i = 1, 2, \ldots, n) \tag{10.9.5}$$

as solved for the $\dot{q}$'s in terms of the p's and the q's:

$$\dot{q}_i = \dot{q}_i(p_i, q_i) \tag{10.9.6}$$

---

[9] Euler's theorem states that for a homogeneous function f of degree n in the variables $x_1, x_2, \ldots, x_r$

$$x_1 \frac{\partial f}{\partial x_1} + x_2 \frac{\partial f}{\partial x_2} + \cdots + x_r \frac{\partial f}{\partial x_r} = nf$$

With these equations we can then express H as a function of the p's and the q's:

$$H(p_i, q_i) = \sum_i p_i \dot{q}_i(p_i, q_i) - L \qquad (10.9.7)$$

Let us calculate the variation of the function H corresponding to a variation δp_i, δq_i. We have

$$\delta H = \sum_i \left[p_i\, \delta\dot{q}_i + \dot{q}_i\, \delta p_i - \frac{\partial L}{\partial \dot{q}_i}\delta\dot{q}_i - \frac{\partial L}{\partial q_i}\delta q_i \right] \qquad (10.9.8a)$$

The first and third terms in the brackets cancel, because $p_i = \partial L/\partial \dot{q}_i$ by definition. Also, because Lagrange's equations can be written as $\dot{p}_i = \partial L/\partial q_i$, we can write

$$\delta H = \sum_i [\dot{q}_i\, \delta p_i - \dot{p}_i\, \delta q_i] \qquad (10.9.8b)$$

Now the variation of H must be given by the equation

$$\delta H = \sum_i \left[\frac{\partial H}{\partial p_i}\delta p_i + \frac{\partial H}{\partial q_i}\delta q_i \right] \qquad (10.9.8c)$$

It follows that

$$\frac{\partial H}{\partial p_i} = \dot{q}_i$$

$$\frac{\partial H}{\partial q_i} = -\dot{p}_i \qquad (10.9.9)$$

These are known as *Hamilton's canonical equations of motion.* They consist of $2n$ first-order differential equations, whereas Lagrange's equations consist of n second-order equations. We have derived Hamilton's equations for simple conservative systems. Equations 10.9.9 also hold for more general systems, for example, nonconservative systems, systems in which the potential energy function involves the $\dot{q}$'s, and systems in which L involves the time explicitly, but in these cases the total energy is no longer necessarily equal to H.

Those of you who survive a class in classical mechanics will encounter Hamilton's equations again when studying quantum mechanics (the fundamental theory of atomic phenomena). Hamilton's equations also find application in celestial mechanics. For further reading the student is referred to the Selected References (under *Advanced Mechanics*) at the end of the book.

EXAMPLE 10.9.1

Obtain Hamilton's equations of motion for a one-dimensional harmonic oscillator.

Solution:

We have

$$T = \tfrac{1}{2}m\dot{x}^2 \qquad V = \tfrac{1}{2}kx^2 \qquad L = T - V$$

$$p = \frac{\partial L}{\partial \dot{x}} = m\dot{x} \qquad \dot{x} = \frac{p}{m}$$

Hence,

$$H = T + V = \frac{1}{2m}p^2 + \frac{k}{2}x^2$$

The equations of motion

$$\frac{\partial H}{\partial p} = \dot{x} \qquad \frac{\partial H}{\partial x} = -\dot{p}$$

then read

$$\frac{p}{m} = \dot{x} \qquad kx = -\dot{p}$$

The first equation merely amounts to a restatement of the momentum–velocity relationship in this case. Using the first equation, the second can be written

$$kx = -\frac{d}{dt}(m\dot{x})$$

or, on rearranging terms,

$$m\ddot{x} + kx = 0$$

which is the familiar equation of the harmonic oscillator.

EXAMPLE 10.9.2

Find the Hamiltonian equations of motion for a particle in a central field.

Solution:

Here we have

$$T = \frac{m}{2}(\dot{r}^2 + r^2\dot{\theta}^2)$$
$$V = V(r)$$
$$L = T - V$$

in polar coordinates. Hence,

$$p_r = \frac{\partial L}{\partial \dot{r}} = m\dot{r} \qquad \dot{r} = \frac{p_r}{m}$$

$$p_\theta = \frac{\partial L}{\partial \dot{\theta}} = mr^2\dot{\theta} \qquad \dot{\theta} = \frac{p_\theta}{mr^2}$$

Consequently,

$$H = \frac{1}{2m}\left(p_r^2 + \frac{p_\theta^2}{r^2}\right) + V(r)$$

The Hamiltonian equations

$$\frac{\partial H}{\partial p_r} = \dot{r} \qquad \frac{\partial H}{\partial r} = -\dot{p}_r \qquad \frac{\partial H}{\partial p_\theta} = \dot{\theta} \qquad \frac{\partial H}{\partial \theta} = -\dot{p}_\theta$$

then read

$$\frac{p_r}{m} = \dot{r}$$

$$\frac{\partial V(r)}{\partial r} - \frac{p_\theta^2}{mr^3} = -\dot{p}_r$$

$$\frac{p_\theta}{mr^2} = \dot{\theta}$$

$$0 = -\dot{p}_\theta$$

The last two equations yield the constancy of angular momentum:

$$p_\theta = \text{constant} \qquad \text{and} \qquad mr^2\dot{\theta} = ml$$

from which the first two give

$$m\ddot{r} = \dot{p}_r = \frac{ml^2}{r^3} - \frac{\partial V(r)}{\partial r}$$

for the radial equation of motion. This, of course, is equivalent to that found earlier in Example 10.5.2.

EXAMPLE 10.9.3

Consider the Rutherford scattering problem discussed in Section 6.14 in which a particle of electric charge q and mass m is moving towards a scattering center, a heavy nucleus of charge Q assumed to be immovable and at rest. Initially the incoming particle is infinitely far from the scattering center and moving with speed v_0 along a straight line whose perpendicular distance to the scattering center (the impact parameter) is b (see Figure 6.14.1). Derive an integral expression for the scattering angle θ_S using Hamilton's equations.

Solution:

Two coordinates are necessary to describe the motion of the particle, which is confined to a plane in space. We choose polar coordinates θ and r. We choose the direction of the polar axis such that the initial position of the incoming particle is $r = \infty$ at $\theta = 0$. The Hamiltonian for this problem is the same as the one given in the previous example (10.9.2). It is

$$H = \frac{1}{2m}\left(p_r^2 + \frac{p_\theta^2}{r^2}\right) + V(r) = E\left(= \frac{1}{2}mv_0^2\right)$$

The Hamiltonian is equal to the total energy of the incident particle and is a constant of the motion. Furthermore, θ is an ignorable coordinate so that the angular momentum

$$p_\theta = mr^2\dot\theta = L \ (= mv_0b)$$

is also a constant of the motion.

The relevant Hamilton's equations are

$$\frac{\partial H}{\partial p_\theta} = \dot\theta \qquad \frac{\partial H}{\partial r} = -\dot p_r$$

Differentiating the first with respect to r, we get

$$\frac{\partial}{\partial r}\dot\theta = \frac{\partial}{\partial r}\frac{\partial H}{\partial p_\theta} = \frac{\partial}{\partial p_\theta}\frac{\partial H}{\partial r} = -\frac{\partial \dot p_r}{\partial p_\theta}$$

Integrating the above gives us an expression for $\dot\theta$

$$\dot\theta = -\int \frac{\partial \dot p_r}{\partial p_\theta}\,dr = -\frac{d}{dt}\int\frac{\partial p_r}{\partial p_\theta}\,dr \qquad \text{or} \qquad d\theta = -d\left[\int \frac{\partial p_r}{\partial p_\theta}\,dr\right]$$

where, for the moment, we have left out the limits of integration. Examining Figure 6.14.1, we see that θ_0 is the change of the angular direction of the particle as it moves from $r = \infty$ to $r = r_{min}$, the distance of closest approach to the nucleus. Because

$$\frac{\partial p_r}{\partial p_\theta} = 0 \qquad at \qquad r = \infty, \theta = 0$$

we get

$$\theta_0 = -\int_\infty^{r_{min}} \frac{\partial p_r}{\partial p_\theta}\,dr$$

We can solve for p_r using the expression for the Hamiltonian

$$p_r = \left[2m(E - V(r)) - \frac{p_\theta^2}{r^2}\right]^{1/2}$$

Because the scattering angle is $\theta_S = \pi - 2\theta_0$ (see Figure 6.14.1) we get

$$\theta_S = \pi + 2\int_\infty^{r_{min}} \frac{\partial}{\partial p_\theta}\left[2m(E - V(r)) - \frac{p_\theta^2}{r^2}\right]^{1/2}dr$$

We can take the expression one step further by carrying out the differentiation inside the integral. We get

$$\theta_S = \pi - 2\int_\infty^{r_{min}} \frac{\dfrac{L}{r^2}}{\left[2m\left(E - \dfrac{qQ}{r}\right) - \dfrac{L^2}{r^2}\right]^{1/2}}\,dr$$

in which we have replaced p_θ with the constant angular momentum L. We have also inserted the expression for the particle's potential energy in the field of the nucleus, $V(r) = qQ/r$. The industrious student should be able to show that carrying out the integral in the above equation yields the value θ_0, given by Equation 6.14.6.

Problems

Lagrange's method should be used in all of the following problems, unless stated otherwise.

10.1   Calculate the integral

$$J(\alpha) = \int_{t_1}^{t_2} L[x(\alpha,t), \dot{x}(\alpha,t), t]\,dt$$

for the simple harmonic oscillator. Follow the analysis presented in Section 10.1. Show that $J(\alpha)$ is an extremum at $\alpha = 0$.

10.2   Find the differential equations of motion of a projectile in a uniform gravitational field without air resistance.

10.3   Find the acceleration of a solid uniform sphere rolling down a perfectly rough, fixed inclined plane. Compare with the result derived earlier in Section 8.6.

10.4   Two blocks of equal mass m are connected by a flexible cord. One block is placed on a smooth horizontal table, the other block hangs over the edge. Find the acceleration of the blocks and cord assuming (a) the mass of the cord is negligible and (b) the cord is heavy, of mass m'.

10.5   Set up the equations of motion of a "double-double" Atwood machine consisting of one Atwood machine (with masses m_1 and m_2) connected by means of a light cord passing over a pulley to a second Atwood machine with masses m_3 and m_4. Ignore the masses of all pulleys. Find the accelerations for the case $m_1 = m$, $m_2 = 4m$, $m_3 = 2m$, and $m_4 = m$.

10.6   A ball of mass m rolls down a movable wedge of mass M. The angle of the wedge is θ, and it is free to slide on a smooth horizontal surface. The contact between the ball and the wedge is perfectly rough. Find the acceleration of the wedge.

10.7   A particle slides on a smooth inclined plane whose inclination θ is increasing at a constant rate ω. If $\theta = 0$, at time $t = 0$, at which time the particle starts from rest, find the subsequent motion of the particle.

10.8   Show that Lagrange's method automatically yields the correct equations of motion for a particle moving in a plane in a rotating coordinate system Oxy. (*Hint: $T = \frac{1}{2}\, m\mathbf{v}\cdot\mathbf{v}$, where $\mathbf{v} = \mathbf{i}(\dot{x} - \omega y) + \mathbf{j}(\dot{y} + \omega x)$, and $F_x = -\partial V/\partial x$, $F_y = -\partial V/\partial y$.*)

10.9   Repeat Problem 10.8 for motion in three dimensions.

10.10   Find the differential equations of motion for an "elastic pendulum": a particle of mass m attached to an elastic string of stiffness K and unstretched length l_0. Assume that the motion takes place in a vertical plane.

10.11   A particle is free to slide along a smooth cycloidal trough whose surface is given by the parametric equations

$$x = \frac{a}{4}(2\theta + \sin 2\theta)$$

$$y = \frac{a}{4}(1 - \cos 2\theta)$$

where $0 \le \theta \le \pi$ and a is a constant. Find the Lagrangian function and the equation of motion of the particle.

10.12 A simple pendulum of length l and mass m is suspended from a point on the circumference of a thin massless disc of radius a that rotates with a constant angular velocity ω about its central axis as shown in Figure P10.12. Find the equation of motion of the mass m.

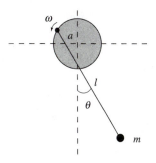

Figure P10.12

10.13 A bead of mass m is constrained to slide along a thin, circular hoop of radius l that rotates with constant angular velocity ω in a horizontal plane about a point on its rim as shown in Figure P10.13.

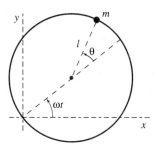

Figure P10.13

(a) Find Lagrange's equation of motion for the bead.
(b) Show that the bead oscillates like a pendulum about the point on the rim diametrically opposite the point about which the hoop rotates.
(c) What is the effective "length" of this "pendulum"?

10.14 The point of support of a simple pendulum is being elevated at a constant acceleration a, so that the height of the support is $\frac{1}{2}at^2$, and its vertical velocity is at. Find the differential equation of motion for small oscillations of the pendulum by Lagrange's method. Show that the period of the pendulum is $2\pi[l/(g + a)]^{1/2}$, where l is the length of the pendulum.

10.15 Work Problem 8.12 by using the method of Lagrange multipliers. (a) Show that the acceleration of the ball is $\frac{5}{7}g$. (b) Find the tension in the string.

10.16 A heavy elastic spring of uniform stiffness and density supports a block of mass m. If m' is the mass of the spring and k its stiffness, show that the period of vertical oscillations is

$$2\pi\sqrt{\frac{m + (m'/3)}{k}}$$

This problem shows the effect of the mass of the spring on the period of oscillation. (*Hint:* To set up the Lagrangian function for the system, assume that the velocity of any part of the spring is proportional to its distance from the point of suspension.)

10.17 Use the mathed of Lagrange multipliers to find the tensions in the two strings of the double Atwood machine of Example 10.5.4.

10.18 A smooth rod of length l rotates in a plane with a constant angular velocity ω about an axis fixed at one end of the rod and perpendicular to the plane of rotation. A bead of mass m is initially positioned at the stationary end of the rod and given a slight push such that its initial speed directed along the rod is ωl.
(a) Find the time it takes the bead to reach the other end of the rod.
(b) Use the method of Lagrange multipliers to find the reaction force **F** that the rod exerts on the bead.

10.19 A particle of mass m perched on top of a smooth hemisphere of radius a is disturbed ever so slightly, so that it begins to slide down the side. Find the normal force of constraint exerted by the hemisphere on the particle and the angle relative to the vertical at which it leaves the hemisphere. Use the method of Lagrange multipliers.

10.20 A particle of mass m_1 slides down the smooth circular surface of radius of curvature a of a wedge of mass m_2 that is free to move horizontally along the smooth horizontal surface on which it rests (Figure P10.20).

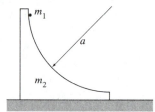

Figure P10.20

(a) Find the equations of motion for each mass.
(b) Find the normal force of constraint exerted by the wedge on the particle. Use the method of Lagrange multipliers.

10.21 (a) Find the general differential equations of motion for a particle in cylindrical coordinates· R, ϕ, z. Use the relation

$$v^2 = v_R^2 + v_\phi^2 + v_z^2$$
$$= \dot{R}^2 + R^2 \dot{\phi}^2 + \dot{z}^2$$

(b) Find the general differential equations of motion for a particle in spherical coordinates: r, θ, ϕ. Use the relation

$$v^2 = v_r^2 + v_\theta^2 + v_\phi^2$$
$$= \dot{r}^2 + r^2 \dot{\theta}^2 + r^2 \dot{\phi}^2 \sin^2\theta$$

(**Note:** Compare your results with the result derived in Chapter 1, Equations 1.12.3 and 1.12.14 by setting **F** $= m\mathbf{a}$ and taking components.)

10.22 Find the differential equations of motion in *three* dimensions for a particle in a central field using spherical coordinates.

10.23 A bar of soap slides in a smooth bowl in the shape of an inverted right circular cone of apex angle 2α. The axis of the cone is vertical. Treating the bar of soap as a particle of mass m, find the differential equations of motion using spherical coordinates with $\theta = \alpha = $ constant. As is the case with the spherical pendulum, Example 10.6.2, show that the particle, given an initial motion with $\dot{\phi}_0 \neq 0$, must remain between two horizontal circles on the cone.

(Hint: Show that $\dot{r}^2 = f(r)$, where $f(r) = 0$ has two roots that define the turning points of the motion in r.) What is the effective potential for this problem?

10.24 In Problem 10.23, find the value of $\dot{\phi}_0$ such that the particle remains on a *single* horizontal circle: $r = r_0$. Find also the period of small oscillations about this circle if $\dot{\phi}_0$ is not quite equal to the required value.

10.25 As stated in Section 4.5, the differential equation of motion of a particle of mass m and electric charge q moving with velocity $\mathbf{v}$ in a static magnetic field $\mathbf{B}$ is given by

$$m\ddot{\mathbf{r}} = q(\mathbf{v} \times \mathbf{B})$$

Show that the Lagrangian function

$$L = \tfrac{1}{2}mv^2 + q\mathbf{v} \cdot \mathbf{A}$$

yields the correct equation of motion where $\mathbf{B} = \nabla \times \mathbf{A}$. The quantity $\mathbf{A}$ is called the *vector potential*. *(Hint: In this problem it is necessary to employ the general formula $df(x,y,z)/dt = \dot{x}\,\partial f/\partial x + \dot{y}\,\partial f/\partial y + \dot{z}\,\partial f/\partial z$. Thus, for the part involving $\mathbf{v} \cdot \mathbf{A}$, we have*

$$\frac{d}{dt}\left[\frac{\partial(\mathbf{v} \cdot \mathbf{A})}{\partial \dot{x}}\right] = \frac{d}{dt}\left[\frac{\partial}{\partial \dot{x}}(\dot{x}A_x + \dot{y}A_y + \dot{z}A_z)\right] = \frac{d}{dt}(A_x)$$

$$= \dot{x}\frac{\partial A_x}{\partial x} + \dot{y}\frac{\partial A_x}{\partial y} + \dot{z}\frac{\partial A_x}{\partial z}$$

and similarly for the other derivatives.)

10.26 Write the Hamiltonian function and find Hamilton's canonical equations for the three-dimensional motion of a projectile in a uniform gravitational field with no air resistance. Show that these equations lead to the same equations of motion as found in Section 4.3.

10.27 Find Hamilton's canonical equations for
(a) A simple pendulum
(b) A simple Atwood machine
(c) A particle sliding down a smooth inclined plane

10.28 A particle of mass m is subject to a central, attractive force given by

$$\mathbf{F}(r, t) = -\mathbf{e}_r \frac{k}{r^2} \exp^{-\beta t}$$

where k and β are positive constants, t is the time, and r is distance to the center of force. (a) Find the Hamiltonian function for the particle. (b) Compare the Hamiltonian to the total energy of the particle. (c) Is the energy of the particle conserved? Discuss.

10.29 Two particles whose masses are m_1 and m_2 are connected by a massless spring of unstressed length l and spring constant k. The system is free to rotate and vibrate on top of a smooth horizontal plane that serves as its support. (a) Find the Hamiltonian of the system. (b) Find Hamilton's equations of motion. (c) What generalized momenta, if any, are conserved?

10.30 As we know, the kinetic energy of a particle in one-dimensional motion is $\tfrac{1}{2}m\dot{x}^2$. If the potential energy is proportional to x^2, say $\tfrac{1}{2}kx^2$, show by direct application of Hamilton's variational principle, $\delta \int L\, dt = 0$, that the equation of the simple harmonic oscillator is obtained.

10.31 The relativistic mass of a moving particle is given by the expression

$$m = \frac{m_0}{\sqrt{1 - v^2/c^2}}$$

where m_0 is its rest mass, v is its speed, and c is the speed of light.

(a) Show that the Lagrangian

$$L = -m_0 c^2 \sqrt{1 - v^2/c^2} - V$$

where the potential energy V is not velocity-dependent, provides the correct equation of motion of the particle.

(b) Find the generalized momentum of the particle and the Hamiltonian.

(c) If the relativistic kinetic energy of the particle is

$$T = \frac{m_0 c^2}{\sqrt{1 - v^2/c^2}}$$

show that $H = T + V$.

(d) Show that, except for an additive constant, the relativistic expression for the kinetic energy of slow moving particles reduces to the classical Newtonian expression.

Computer Problems

C 10.1 Assume that the spherical pendulum discussed in Section 10.6 is set into motion with the following initial conditions: $\phi_0 = 0$ rad, $\dot{\phi}_0 = 10.57$ rad/s, $\theta_0 = \pi/4$ rad, and $\dot{\theta}_0 = 0$ rad/s. Let the length of the pendulum be 0.284 m.

(a) Calculate θ_1 and θ_2, the polar angular limits of the motion.

$$\tfrac{1}{2}\dot{\theta}^2 = -U(\theta) + C = 0$$

(Hint: Solve the equation numerically for the condition of $\dot{\theta}_0 = 0$.)

(b) Solve the equations of motion of the pendulum numerically, and find the period of the θ-motion.

(c) Plot θ as a function of the azimuthal angle ϕ over two azimuthal cycles.

(d) Calculate the angle of precession $\Delta\phi$ that occurs during one complete cycle of θ.

C 10.2 A bead slides from rest down a smooth curve S in the xy plane from the point $(0, 2)$ to the point $(\pi, 0)$.

(a) Show that the curve S for which the total time of travel is a minimum is a cycloid, described by the parametric equations $x = \theta - \sin\theta$ and $y = 1 + \cos\theta$, where θ ranges from 0 to π. *(Hint: The time of travel is given by $T = \int ds/v$, where ds is a differential element of displacement and v is the speed of the bead along the curve S. Express ds in terms of $y' = dy/dx$ and dx. Express the resultant integrand as an explicit function of y, y', and possibly x. The integral is a minimum when the integrand satisfies Lagrange's equation. Find the differential equation of the curve generated by the Lagrange equation and solve it.)*

(b) Assume that the curve S can be approximated by a quadratic function, $y(x) = a_0 + a_1 x + a_2 x^2$. (This function must satisfy the boundary conditions given for the curve S.) Insert this function (and its derivative) into the integral in part (a) for the bead's time of travel. Find the constant coefficients a_i that minimize the integral.

(c) Estimate the minimum time of transit.

(d) Plot the function $y(x)$ obtained in part (b) along with the exact solution given by the equations representing a cycloid (from $x = 0$ to $x = \pi$). How well do the two solutions agree?

11

Dynamics of Oscillating Systems

The modern development of physics is continually enhancing Hamilton's name. His famous analogy between mechanics and optics virtually anticipated wave mechanics, which did not have to add much to his ideas, but only had to take them seriously—a little more seriously than he was able to take them, with the experimental knowledge of a century ago. The central conception of all modern theory in physics is "the Hamiltonian." If you wish to apply modern theory to any particular problem, you must start with putting the problem "in Hamiltonian form."

Thus Hamilton is one of the greatest men of science the world has produced."

—Erwin Schroedinger, *A Collection of Papers in Memory of Sir William Rowan Hamilton*, ed. D. E. Smith, *Scripta Mathematica Studies,* no. 2, N.Y., 1945

In the preceding chapters we studied simple systems that can oscillate about a configuration of equilibrium, including a simple pendulum, a particle suspended on an elastic spring, a physical pendulum, and so on. Each of these cases had only one degree of freedom characterized by a single frequency of oscillation. Here we consider more complicated systems—systems with several degrees of freedom that are characterized by several different frequencies of oscillation. The analysis is greatly simplified if we use generalized coordinates and Lagrange's method for finding the equations of motion in terms of those coordinates.

11.1 | Potential Energy and Equilibrium: Stability

Before we take up the study of motion of a system with many degrees of freedom about an equilibrium configuration, let us first examine just what is meant by the term *equilibrium*. As a way of introduction, let us recall the oscillatory motion of a mass on a spring about its equilibrium position. It is a conservative system, and its restoring force is derivable from a potential energy function

$$V(x) = \frac{1}{2}kx^2 \tag{11.1.1}$$

$$F(x) = -\frac{dV(x)}{dx} = -kx \tag{11.1.2}$$

The equilibrium position of the oscillator is at $x = 0$, the position where the restoring force vanishes or the derivative of the potential energy function is zero. If the oscillator was initially placed at rest at $x = 0$, it would remain there at rest.

Let us consider the motion of a simple pendulum of length r constrained to swing in a vertical plane (See Figure 10.2.1). As in our discussion of Section 10.2, let its position be described by the single generalized coordinate θ, the angle that it makes with the vertical. Taking the potential energy to be zero at $\theta = 0$, the potential energy function and the derived restoring "force" are given by

$$V = mgr(1 - \cos\theta) \tag{11.1.3}$$

$$N_\theta = -\frac{\partial V}{\partial\theta} = -mg(r\sin\theta) = -mgx \tag{11.1.4}$$

where x is the horizontal displacement of the pendulum bob from the vertical. The generalized coordinate of the pendulum is an angular variable, and the restoring force is actually a restoring torque N_θ. The pendulum is in its equilibrium position when the restoring torque is equal to zero. In each of these two cases, regardless of whether the potential energy is a function of either a positional or an angular coordinate, equilibrium corresponds to the configuration at which the derivative of the potential energy function vanishes.

Now let us generalize the above cases to a system with n degrees of freedom whose generalized coordinates $q_1, q_2, \ldots, q_n$ completely specify its configuration. The q's can be a mixture of both positional and angular variables. We assume that the system is conservative and that its potential energy function is a function of the q's alone:

$$V = V(q_1, q_2, \ldots, q_n) \tag{11.1.5}$$

All forces and torques acting on the system vanish when

$$\frac{\partial V}{\partial q_k} = 0 \qquad (k = 1, 2, \ldots, n) \tag{11.1.6}$$

This more complicated system is in equilibrium when Equation 11.1.6 holds true. These equations constitute a necessary condition for the system to remain at rest if it is initially at rest in such a configuration. If the system is given a small displacement from this configuration, however, it may or may not return to the equilibrium configuration. If it always tends to return to equilibrium, given a sufficiently small displacement, the equilibrium is *stable;* otherwise, it is *unstable.* (If the system has no tendency to move either toward or away from equilibrium, the equilibrium is *neutral.*)

A ball placed (1) at the bottom of a spherical bowl, (2) on top of a spherical cap, and (3) on a plane horizontal surface are examples of stable, unstable, and neutral equilibrium, respectively.

Intuition tells us that the potential energy must be a *minimum* in all cases for stable equilibrium. That this is so can be argued from energy considerations. If the system is conservative, the total energy $T + V$ is constant, so for a small change near equilibrium $\Delta T = -\Delta V$. Thus, T decreases if V increases; that is, the motion tends to slow down and return to the equilibrium position, given a small displacement. The reverse is true if the potential energy is *maximum*; that is, any displacement causes V to decrease and T to increase, so the system tends to move away from the equilibrium position at an ever-increasing rate.

Extended Criteria for Stable Equilibrium

We consider first a system with one degree of freedom. Suppose we expand the potential energy function $V(q)$ as a Taylor series about the point $q = 0$, namely,

$$V(q) = V_0 + qV_0' + \frac{q^2}{2!}V_0'' + \frac{q^3}{3!}V_0''' + \cdots + \frac{q_n}{n!}V_0^{(n)} + \cdots \tag{11.1.7a}$$

where we use the notation $V_0' = (dV/dq)_{q=0}$, and so on. Now if $q = 0$ is a position of equilibrium, then $V_0' = 0$. This eliminates the linear term in the expansion. Furthermore, the term V_0 is a constant whose value depends on the arbitrary choice of the zero of the potential energy, so without incurring any loss of generality we can set $V_0 = 0$. Consequently, the expression for $V(q)$ simplifies to

$$V(q) = \frac{q^2}{2}V_0'' + \cdots \tag{11.1.7b}$$

If V_0'' is not zero, then for a small displacement q from equilibrium the force is approximately linear in the displacement:

$$F(q) = -\frac{dV}{dq} = -qV_0'' \tag{11.1.7c}$$

This is of a restorative or stabilizing type if V_0'' is positive, whereas, if V_0'' is negative, the force is antirestoring and the equilibrium is unstable. If $V_0'' = 0$, then we must examine the first nonvanishing term in the expansion. If this term is of even order in n, then the equilibrium is again stable, or unstable, depending on whether the derivative $V_0^{(n)} = (d^n V/dq^n)_{q=0}$ is positive or negative, respectively. If the first nonvanishing derivative is of odd order in n, then the equilibrium is always unstable regardless of the sign of the derivative; this corresponds to the situation at point C in Figure 11.1.1. Clearly, if all derivatives vanish, then the potential energy function is a constant, and the equilibrium is neutral.

Similarly, for the case of a system with several degrees of freedom, we can effect a linear transformation so that $q_1 = q_2 = \cdots = q_n = 0$ is the configuration of the equilibrium, if an equilibrium configuration exists. The potential energy function can then be expanded in the form

$$V(q_1, q_2, \ldots, q_n) = \frac{1}{2}\left(K_{11}q_1^2 + 2K_{12}q_1q_2 + K_{22}q_2^2 + \cdots\right) \tag{11.1.8a}$$

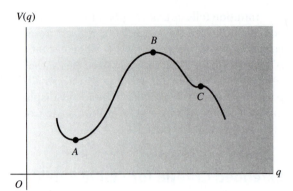

$V(q)$

Figure 11.1.1 Graph of a one-dimensional potential energy function. The point A is one of stable equilibrium. Points B and C are unstable.

where

$$K_{11} = \left(\frac{\partial^2 V}{\partial q_1^2}\right)_{q_1=q_2=\cdots=q_n=0}$$

$$K_{12} = \left(\frac{\partial^2 V}{\partial q_1 \, \partial q_2}\right)_{q_1=q_2=\cdots=q_n=0}$$

(11.1.8b)

and so on. We have arbitrarily set $V(0, 0, \ldots, 0) = 0$. The linear terms in the expansion are absent because the expansion is about an equilibrium configuration.

The expression in parentheses in Equation 11.1.8a is known as a *quadratic form*. If this quadratic form is positive definite,[1] that is, either zero or positive for all values of the q's, then the equilibrium configuration $q_1 = q_2 = \cdots = q_n = 0$ is stable.

EXAMPLE 11.1.1

Stability of Rocking Chairs, Pencils-on-End, and the Like

Let us examine the equilibrium of a body having a rounded (spherical or cylindrical) base that is balanced on a plane, horizontal surface. Let a be the radius of curvature of the base, and let the center of mass CM be a distance b from the initial point of contact, as shown in Figure 11.1.2a. In Figure 11.1.2b the body is shown in a displaced position, where θ is the angle between the vertical and the line OCM (O being the center of curvature), as shown. Let h denote the distance from the plane to the center of mass. Then the

---

[1] The necessary and sufficient conditions that the quadratic form in Equation 11.1.8a be positive definite are

$$K_{11} > 0 \qquad \begin{vmatrix} K_{11} & K_{12} \\ K_{21} & K_{22} \end{vmatrix} > 0 \qquad \begin{vmatrix} K_{11} & K_{12} & K_{13} \\ K_{21} & K_{22} & K_{23} \\ K_{31} & K_{32} & K_{33} \end{vmatrix} > 0 \qquad \text{and so on}$$

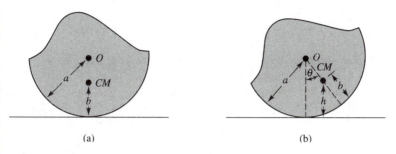

(a)   (b)

Figure 11.1.2   Coordinates for analyzing the stability of equilibrium of a round-bottomed object.

potential energy is given by

$$V = mgh = mg[a - (a - b)\cos\theta]$$

where m is the mass of the body. We have

$$V' = \frac{dV}{d\theta} = mg(a-b)\sin\theta$$

which gives, for $\theta = 0$,

$$V_0' = 0$$

Thus, $\theta = 0$ is a position of equilibrium. Furthermore, the second derivative is

$$V'' = mg(a-b)\cos\theta$$

so, for $\theta = 0$

$$V_0'' = mg(a-b)$$

Hence, the equilibrium is stable if $a > b$, that is, if the center of mass lies below the center of curvature O. If $a < b$, the second derivative is negative and the equilibrium is unstable, such as with a pencil standing on end. If $a = b$, the potential energy function is constant, and the equilibrium is neutral. In this latter case, the center of mass coincides with the center of curvature.

11.2 | Oscillation of a System with One Degree of Freedom about a Position of Stable Equilibrium

If a system has one degree of freedom, the kinetic energy may be expressed as

$$T = \tfrac{1}{2}M\dot{q}^2 \tag{11.2.1}$$

where the coefficient M may be a constant or a function of the generalized coordinate q. In any case if $q = 0$ is a position of equilibrium, we consider q small enough so that $M = M(0) = $ constant is a valid approximation. From the expression for the potential energy

(Equation 11.1.7b), we can write the Lagrangian function as

$$L = T - V = \tfrac{1}{2} M \dot{q}^2 - \tfrac{1}{2} V_0'' q^2 \tag{11.2.2}$$

Lagrange's equation of motion

$$\frac{d}{dt} \frac{\partial L}{\partial \dot{q}} - \frac{\partial L}{\partial q} = 0 \tag{11.2.3}$$

then becomes

$$M \ddot{q} + V_0'' q = 0 \tag{11.2.4}$$

Thus, if $q = 0$ is a position of stable equilibrium, that is, if $V_0'' > 0$, then the system oscillates harmonically about the equilibrium position with angular frequency

$$\omega = \sqrt{\frac{V_0''}{M}} \tag{11.2.5}$$

EXAMPLE 11.2.1

Consider the motion of the round-bottomed object discussed in Example 11.1.1 (see Figure 11.1.2). If the contact is perfectly rough, we have pure rolling, and the speed of the center of mass is approximately $b\dot{\theta}$ for small θ. The kinetic energy T is accordingly given by

$$T = \tfrac{1}{2} m (b\dot{\theta})^2 + \tfrac{1}{2} I_{cm} \dot{\theta}^2$$

where I_{cm} is the moment of inertia about the center of mass. Also, we can express the potential energy function V as follows:

$$V(\theta) = mg[a - (a - b)\cos\theta]$$

$$= mg\left[a - (a - b)\left(1 - \frac{\theta^2}{2!} + \frac{\theta^4}{2!} - \cdots\right)\right]$$

$$= \tfrac{1}{2} mg(a - b)\theta^2 + \text{constant} + \text{higher terms}$$

We can then write

$$L = \tfrac{1}{2}(mb^2 + I_{cm})\dot{\theta}^2 - \tfrac{1}{2}mg(a - b)\theta^2$$

ignoring constants and higher terms. Comparing with Equation 11.2.2, we see that

$$M = mb^2 + I_{cm}$$
$$V_0'' = mg(a - b)$$

The motion about the equilibrium position $\theta = 0$ is, therefore, approximately simple harmonic with angular frequency

$$\omega = \sqrt{\frac{mg(a - b)}{mb^2 + I_{cm}}}$$

EXAMPLE 11.2.2

Attitude Stability and Oscillation of an Orbiting Satellite

In this example we analyze the oscillatory motion of a nonspherical satellite traveling in a circular orbit. For simplicity, we consider the satellite to be a dumbbell consisting of two small spheres, of mass $m/2$ each, connected by a thin massless connecting cylinder of length $2a$, Figure 11.2.1. Polar coordinates r, θ specify the center of mass of the satellite, and the angle ϕ gives the "attitude" of the satellite axis relative to the radius vector $\mathbf{r}_0$. We treat the two end spheres as particles and assume that the motion is in a single plane, the plane of the orbit. For a circular orbit $r = r_0 =$ constant, and $\dot{\theta} = \omega_0 = v_{cm}/r_0 =$ constant.

The most important quantity to calculate in this example is the potential energy function of the satellite. It is given by

$$V = -\frac{GM_e m}{2}\left(\frac{1}{r_1} + \frac{1}{r_2}\right)$$

in which M_e is Earth's mass and r_1 and r_2 are the distances from the center of the Earth to the respective end spheres, as shown. From the law of cosines we have

$$r_{1,2} = \left(r_0^2 + a^2 \pm 2r_0 a \cos\phi\right)^{1/2} = \left(r_0^2 + a^2\right)^{1/2}(1 \pm \epsilon \cos\phi)^{1/2}$$

where $\epsilon = 2r_0 a/(r_0^2 + a^2)$. Now $a \ll r_0$, so ϵ is a very small quantity. We, therefore, express the potential energy function by use of the binomial series $(1+x)^{-1/2} = 1 - \frac{1}{2}x + \frac{3}{8}x^2 + \cdots$, where $x = \pm\epsilon \cos\phi$. The result, after collecting and canceling terms, is

$$V(\phi) = -\frac{GM_e m}{r_0}\left(1 + \frac{3a^2}{2r_0^2}\cos^2\phi + \cdots\right)$$

where we have ignored a^2 compared with r_0^2 in all terms involving the quantity $r_0^2 + a^2$.

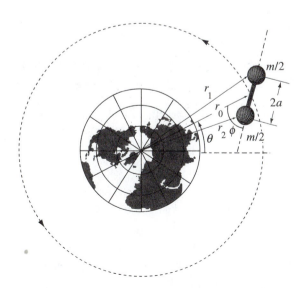

Figure 11.2.1 Dumbbell-shaped satellite in a circular orbit.

Taking the first and second derivatives with respect to ϕ, we find

$$V'(\phi) = \frac{GM_e m}{r_0^3} 3a^2 \sin\phi \cos\phi$$

$$V''(\phi) = \frac{GM_e m}{r_0^3} 3a^2 \cos(2\phi)$$

Thus, we have $\phi = 0$ and $\phi = \pi/2$ as two positions of equilibrium: $V'(0) = V'(\pi/2) = 0$. The first is stable, because $V''(0) > 0$. In this case the attitude of the satellite is such that the satellite's axis (line connecting the two masses) is along the radius vector $\mathbf{r}_0$. The second position is an unstable equilibrium because $V''(\pi/2) < 0$; here the axis is at right angles to the radius vector.

The rocking motion of the satellite about the position of stable equilibrium is given by Equation 11.2.4 with $q = \phi$, $M = I_{cm} = ma^2$, and $V_0'' = 3a^2 GM_e m/r_0^3$. Thus, the angular frequency of the oscillation is

$$\omega = \sqrt{\frac{V_0''}{I_{cm}}} = \sqrt{\frac{3GM_e}{r_0^3}}$$

(Note that this is independent of m and a.) Now the angular frequency of the circular orbit around Earth is given by $\omega_0^2 = v_{cm}^2/r_0^2 = GM_e/r_0^3$. (See Example 6.5.3.) Thus, we can write

$$\omega = \omega_0 \sqrt{3}$$

For a synchronous Earth satellite the orbital period $T_0 = 2\pi/\omega_0 = 23.934$ h.[2] Consequently, the rocking period of our dumbbell satellite in a synchronous orbit would be

$$\frac{2\pi}{\omega} = \frac{T_0}{\sqrt{3}} \, h = 13.818 \, h$$

11.3 | Coupled Harmonic Oscillators: Normal Coordinates

Before developing the general theory of oscillating systems with any number of degrees of freedom, we shall study a simple specific example, namely, a system consisting of two harmonic oscillators that are coupled together.

We use a model composed of particles attached to elastic springs, although any type of oscillator could be used. For simplicity we assume that the oscillators are identical and are restricted to move in a straight line (Figure 11.3.1). The coupling is represented by a spring of stiffness K' as shown. The system has two degrees of freedom. We choose

---

[2] The sidereal day, corresponding to one full 360° rotation of the Earth relative to the stars, is equal to 23 hr 56 m 3.44 s, about 4 m shorter than the mean solar day.

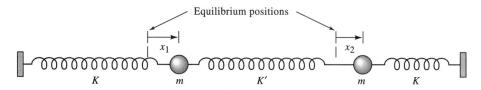

Figure 11.3.1 Model of two coupled harmonic oscillators.

coordinates x_1 and x_2, the displacements of the particles from their respective equilibrium positions, to represent the configuration of the system.

Before plunging into the mathematics describing the motion of this system, we should like to consider just what sort of behavior we might expect. We would guess that the actual motion would depend critically on the initial conditions of the system, whereas the vibrational frequency (or frequencies) would not. For example, suppose we held one mass at the position $x_1 = 0$ while we pulled the other mass a little to the right, say $x_2 = 1$, and then released them both from rest. Just after being released, m_2 is subject to a restoring force due to the compression of the right-hand spring and the stretching of the middle spring. m_1, even though at rest at $x_1 = 0$, is subject to a force due to the stretching of the middle spring. Hence, both masses start to move, m_1 away from $x_1 = 0$ and m_2 toward $x_2 = 0$. The resulting motion looks to be fairly complex, but one thing is certain: Overall energy is conserved. Thus, as m_1, initially at rest, moves away from $x_1 = 0$, it gains energy at the expense of that of m_2. As time goes by we might anticipate that m_1 will eventually be displaced to the left at $x_1 = -1$ while m_2 will be at $x_2 = 0$, both instantaneously at rest. This configuration is completely symmetrical to the initial one, with m_1 and m_2 having exchanged energies. The system should continue to repeat this motion, with m_1 and m_2 shuttling their energy back and forth through the coupling spring. The critical point here is that x_1 and x_2 are never simultaneously zero and that the coupling spring is never relaxed with m_1 and m_2 in that configuration. Hence, the two masses continue to exchange energy.

A second important feature of this motion is that each mass vibrates in a *multifrequency* fashion. This can be most readily seen by analyzing the cause of single-frequency motion. Such motion occurs when the acceleration (or force per unit mass) of a mass is proportional to the negative of its displacement. In the situation here, each mass is subject to two forces, one from each connecting spring. The middle spring generates a force on each mass that is proportional to the difference in their displacements. Thus, we might anticipate that the general motion of each mass would be a composite of two different frequencies, and we will soon see that this is the case.

Figure 11.3.2 shows the motion of the two previously described masses. The spring constants have values $K = 4$ and $K' = 1$ (arbitrary units), so this is a case of moderate coupling. The amplitude of oscillation of m_1 slowly builds up and then dies away in step with the dying away and buildup of the amplitude of oscillation of m_2. The motion has been plotted over one complete period. Each of these motions looks like a case of "beats" between two different single frequencies of the same amplitude. And that is exactly what they are.

The phenomenon of beats occurs when waves (or vibrations) of two different frequencies are added together. For example, let us assume that x_1 and x_2 can be represented

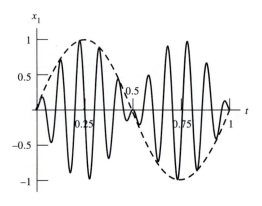

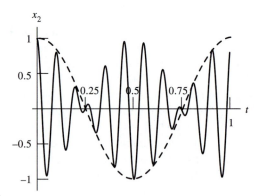

Figure 11.3.2   Displacement of two coupled harmonic oscillators.

by the sum (or difference) of two simple, equal-amplitude, harmonic motions whose frequencies are different. The resultant sum is equal to a product of sines (or cosines) of a sum and a difference of frequencies.

For example, suppose we define Q_1 and Q_2 as follows

$$Q_1 = \frac{1}{\sqrt{2}} \cos \omega_1 t \qquad Q_2 = \frac{1}{\sqrt{2}} \cos \omega_2 t \qquad (11.3.1)$$

(The factor $\frac{1}{\sqrt{2}}$ has been included in each of the definitions in Equation 11.3.1 solely for the purposes of normalization). Now, if we add Q_1 and Q_2 (again normalized with the factor $\frac{1}{\sqrt{2}}$), we get

$$\frac{1}{\sqrt{2}}(Q_1 + Q_2) = \frac{1}{2}(\cos \omega_1 t + \cos \omega_2 t)$$
$$= \cos\left[\frac{1}{2}(\omega_1 + \omega_2)t\right]\cos\left[\frac{1}{2}(\omega_1 - \omega_2)t\right] \qquad (11.3.2a)$$
$$= x_2$$

The resulting sum is equal to x_2 and is, in fact, the function that is plotted in Figure 11.3.2b. It has been properly normalized so that it satisfies the condition that $x_2(0) = 1$.

Now suppose we subtract Q_2 from Q_1

$$\frac{1}{\sqrt{2}}(Q_1 - Q_2) = \frac{1}{2}(\cos\omega_1 t - \cos\omega_2 t)$$
$$= \sin\left[\frac{1}{2}(\omega_1 + \omega_2)t\right]\sin\left[\frac{1}{2}(\omega_1 - \omega_2)t\right] \qquad (11.3.2b)$$
$$= x_1$$

The resulting difference is equal to x_1 and is the function that is plotted in Figure 11.3.2a. It satisfies the condition that $x_1(0) = 0$.

The fascinating thing here is that, although the coordinates x_1 and x_2 engage in this composite dance of energy exchange, Q_1 and Q_2 do not. They are functions only of the single frequencies ω_1 and ω_2. Because x_1 and x_2 are expressible as the difference and sum of Q_1 and Q_2 we can write them in matrix notation as

$$\mathbf{x} = \begin{pmatrix} x_1 \\ x_2 \end{pmatrix} = \frac{1}{\sqrt{2}}\begin{pmatrix} 1 & -1 \\ 1 & 1 \end{pmatrix}\begin{pmatrix} Q_1 \\ Q_2 \end{pmatrix} = \mathbf{AQ} \qquad (11.3.3a)$$

The relation can be inverted to obtain Q_1 and Q_2 as functions of x_1 and x_2

$$\mathbf{Q} = \begin{pmatrix} Q_1 \\ Q_2 \end{pmatrix} = \frac{1}{\sqrt{2}}\begin{pmatrix} 1 & 1 \\ -1 & 1 \end{pmatrix}\begin{pmatrix} x_1 \\ x_2 \end{pmatrix} = \mathbf{A}^{-1}\mathbf{x} \qquad (11.3.3b)$$

These matrix equations are equivalent to $\pm45°$ rotations of a two-dimensional coordinate system. This suggests that we can interpret x_1 and x_2 or Q_1 and Q_2 as components of a single vector $\mathbf{q}$ whose endpoint represents the instantaneous configuration of the coupled oscillators in either of two different coordinate systems. We show such a vector $\mathbf{q}$ in Figure 11.3.3.

As time goes on, the endpoint of $\mathbf{q}$ traces out a path in *configuration space* whose components are given by $x_1(t)$ and $x_2(t)$, the solutions to the equations of motion. Shown in Figure 11.3.4a is a plot of this path for the coupled oscillator. The trajectory is confined to a box whose boundaries make 45° lines with the coordinate axes. Because the Q_i-coordinate system is simply the x_i-coordinate system rotated through 45°, a similar plot made in that system ought to trace out a path confined to a box whose boundaries are parallel to the axes. The plot shown in Figure 11.3.4b demonstrates that this is the case.

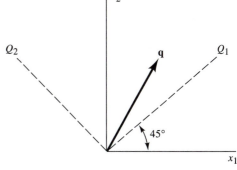

Figure 11.3.3 Vector whose components represent the displacements of two coupled oscillators. The Q_1, Q_2 coordinates are obtained by rotating the x_1, x_2 coordinates by 45°.

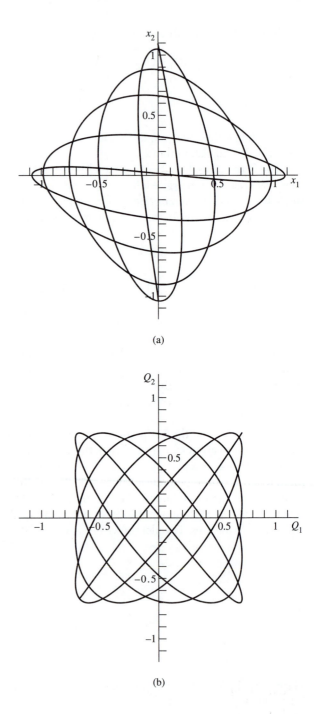

(a)

(b)

Figure 11.3.4 Motion of two coupled oscillators in configuration space. Coordinate axes are (a) the displacements x_1, x_2 of each mass from equilibrium (b) the linear sum $Q_1 = (x_1 + x_2)/\sqrt{2}$ and $Q_2 = (x_2 - x_1)/\sqrt{2}$.

That the boundaries lie parallel to the axes in the Q_i-coordinate system suggests that they might be a more suitable basis in which to express the equations of motion of this system.

The significance of the (Q_1, Q_2) coordinates becomes apparent if we ask ourselves whether or not the system can be started off such that the two masses vibrate at a single fixed frequency and never exchange energy. There are two ways to do this. First, suppose that we release the masses from rest after displacing them by equal amounts from their respective equilibrium positions. The initial conditions of the resulting motion are

$$x_1(0) = x_2(0) = 1 \qquad \dot{x}_1(0) = \dot{x}_2(0) = 0 \qquad (11.3.4)$$

If we examine Equations 11.3.3a and b, we see that $Q_2(0) = 0$ but $Q_1(0) = \sqrt{2}$. The central spring is neither stretched nor compressed during the initial displacement so no force tries to separate or pull the masses together any more than they already are. Furthermore, the two masses must return to their equilibrium positions at the same time, moving in the same direction with the same velocity because the restoring force on each mass is identical. But if the central spring always stays in a flaccid state, no energy can be passed back and forth between the two masses. Thus, at later times we have

$$x_1(t) = x_2(t) = \cos \omega_1 t$$
$$Q_1(t) = \sqrt{2} \, \cos \omega_1 t \qquad (11.3.5)$$
$$Q_2(t) = 0$$

The two masses vibrate back and forth as though they were completely independent simple harmonic oscillators with identical frequencies, $\omega_1 = \sqrt{K/m}$. Once the system is put in this mode of oscillation, it stays that way. The system is executing a *normal mode* of oscillation called the *symmetric mode,* which is pictured in Figures 11.3.5a–d.

The second way to get the two masses to oscillate at a single frequency is initially to displace them from their equilibrium positions by equal amounts but in opposite directions and then release them from rest. Thus

$$x_2(0) = -x_1(0) = 1 \qquad \dot{x}_1(0) = \dot{x}_2(0) = 0 \qquad (11.3.6)$$

If we now examine Equations 11.3.3a and b, we see that $Q_1(0) = 0$ but $Q_2(0) = \sqrt{2}$. This time, however, the central, connecting spring is initially stretched. But pay close attention to the central point on the connecting spring—it does not move. It is being pulled on by equal but oppositely directed forces. When the two masses are released, the central point does not move unless one mass moves closer to it than does the other, creating an imbalance in the two opposing forces. By symmetry, this cannot happen. Again, the restoring forces acting on each mass are equal and oppositely directed, and they remain that way throughout the motion. The central point is a *nodal point* in the vibration, and no energy can be transferred from one mass to the other across that point. As far as either mass is concerned, we could cut the central spring in half and attach each of the freed-up endpoints to a fixed, immobile boundary, exactly like the attachment of the end springs. The resulting effective spring constant for each mass is $K + 2K'$ (it is left to the student to show that this is true). Another way to look at it is to note that when the two masses pass through their respective equilibrium positions, the central spring is neither stretched nor compressed. No energy transfer is, therefore, possible, and each mass vibrates 180° out of phase with the other at the single frequency $\omega_2 = \sqrt{(K + 2K')/m}$. This normal mode

Symmetric mode

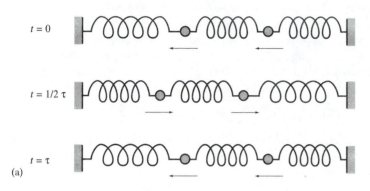

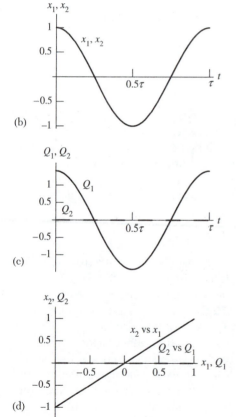

Figure 11.3.5   (a) Schematic of coupled oscillation in symmetric mode. (b) Displacements versus time. (c) Normal coordinates versus time. (d) Configuration space trajectory; generalized coordinates and normal coordinates.

of oscillation is called the *antisymmetric* mode or, for obvious reasons, the "breathing mode." Once placed in this mode, the system stays there. Its motion is described by the equations

$$x_2(t) = -x_1(t) = \cos\omega_2 t$$
$$Q_1(t) = 0 \qquad\qquad (11.3.7)$$
$$Q_2(t) = \sqrt{2}\,\cos\omega_2 t$$

and it is pictured in Figure 11.3.6a–d.

The Q_i-coordinates are called *normal coordinates*, and if the coupled oscillator is vibrating such that its configuration space vector $\mathbf{q}$ has (Q_1, Q_2) components given by

$$q = \begin{pmatrix} Q_1 \\ Q_2 \end{pmatrix} = \begin{pmatrix} 1 \\ 0 \end{pmatrix} B_1 \cos(\omega_1 t - \delta_1) \quad \text{or} \quad q = \begin{pmatrix} Q_1 \\ Q_2 \end{pmatrix} = \begin{pmatrix} 0 \\ 1 \end{pmatrix} B_2 \cos(\omega_2 t - \delta_2) \quad (11.3.8a)$$

or equivalently (x_1, x_2), components given by

$$q = \begin{pmatrix} x_1 \\ x_2 \end{pmatrix} = \begin{pmatrix} 1 \\ 1 \end{pmatrix} A_1 \cos(\omega_1 t - \delta_1) \quad \text{or} \quad q = \begin{pmatrix} x_1 \\ x_2 \end{pmatrix} = \begin{pmatrix} -1 \\ 1 \end{pmatrix} A_2 \cos(\omega_2 t - \delta_2) \quad (11.3.8b)$$

where A_i, B_i are amplitudes and δ_i are phase angles that depend on how the motion is initialized, then $\mathbf{q} =$ either $\mathbf{Q}_1$ or $\mathbf{Q}_2$, and the system is vibrating in one of its possible normal modes (see Figures 11.3.5c and 11.3.6c).

Method of Solution

Armed with this discussion of the coupled oscillator of Figure 11.3.1, let us now solve its equations of motion. The Lagrangian of the system is

$$L = T - V = \tfrac{1}{2}m\dot{x}_1^2 + \tfrac{1}{2}m\dot{x}_2^2 - \tfrac{1}{2}Kx_1^2 - \tfrac{1}{2}K'(x_2 - x_1)^2 - \tfrac{1}{2}Kx_2^2 \qquad (11.3.9)$$

Lagrange's equation then yields the equations of motion

$$\begin{aligned} m\ddot{x}_1 + (K+K')x_1 - \quad K'x_2 &= 0 \\ m\ddot{x}_2 - \quad K'x_1 + (K+K')x_2 &= 0 \end{aligned} \qquad (11.3.10a)$$

which we can write using matrix notation as

$$\mathbf{M\ddot{q} + Kq} = 0 \qquad (11.3.10b)$$

where $\mathbf{q}$ is the vector whose (x_1, x_2) components represent the configuration, or *state*, of the system. The matrix equation in component form is

$$\begin{pmatrix} m & 0 \\ 0 & m \end{pmatrix} \begin{pmatrix} \ddot{x}_1 \\ \ddot{x}_2 \end{pmatrix} + \begin{pmatrix} K+K' & -K' \\ -K' & K+K' \end{pmatrix} \begin{pmatrix} x_1 \\ x_2 \end{pmatrix} = 0 \qquad (11.3.10c)$$

As anticipated, the resulting equations of motion are coupled, as evidenced by the cross terms in Equation 11.3.10a or the nonzero, off-diagonal elements in the "**K**-matrix" of Equation 11.3.10c.

A completely general solution should yield a state vector $\mathbf{q}$ whose (x_1, x_2) components are functions of two frequencies ω_1 and ω_2. We know, however, that we can also search for particular single-frequency solutions that would correspond to the normal

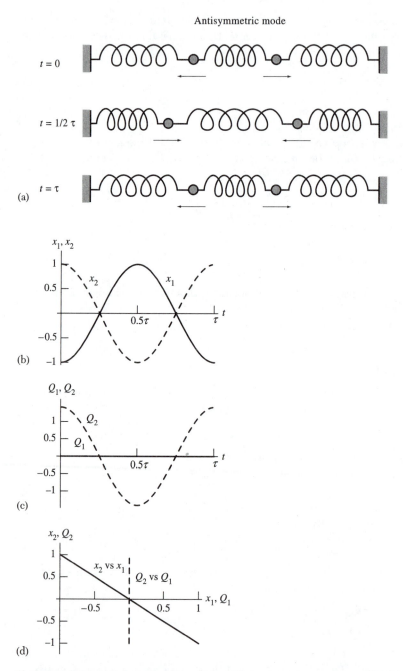

Figure 11.3.6 (a) Schematic of coupled oscillation in antisymmetric mode. (b) Displacements versus time. (c) Normal coordinates versus time. (d) Configuration space trajectory; generalized coordinates and normal coordinates.

modes of oscillation. In this case, the vector $\mathbf{q}$ would be one of the normal mode vectors $\mathbf{Q}_i$ whose direction points along one of the (Q_1, Q_2) coordinate axes. Once the solutions for the normal mode vectors are found, however, a solution for any general state vector $\mathbf{q}(x_1, x_2)$ can be assembled as a linear combination of the two normal mode vectors exactly as was the situation described by Equations 11.3.2a and b. We proceed then by searching for solutions of the form

$$\mathbf{q} = \mathbf{a} \cos(\omega t - \delta) \tag{11.3.11a}$$

whose components are, therefore,

$$x_1 = a_1 \cos(\omega t - \delta) \qquad x_2 = a_2 \cos(\omega t - \delta) \tag{11.3.11b}$$

Thus, each component has the same frequency and phase but a different amplitude. Plugging this assumed solution into Equation 11.3.10b yields

$$\mathbf{Ka} = \omega^2 \mathbf{Ma}$$

$$\begin{pmatrix} K + K' & -K' \\ -K' & K + K' \end{pmatrix} \begin{pmatrix} a_1 \\ a_2 \end{pmatrix} = \omega^2 \begin{pmatrix} m & 0 \\ 0 & m \end{pmatrix} \begin{pmatrix} a_1 \\ a_2 \end{pmatrix} \tag{11.3.12a}$$

The a_i (the amplitudes of the oscillation referred to the x_i coordinates) are components of a time-independent vector $\mathbf{a}$ that satisfies Equation 11.13.12a. This vector is called an *eigenvector*, and ω^2 is its *eigenvalue* or, in this case, its *eigenfrequency* (squared). A way in which the eigenvectors and eigenfrequencies can be found is by simultaneously diagonalizing the $\mathbf{K}$ and $\mathbf{M}$ matrices in Equation 11.3.12a.[3] We discuss such a method later. First, we present a method that is more direct and generally applicable but less physically intuitive.

Equation 11.3.12a is equivalent to a system of linear, homogeneous equations given by

$$\begin{pmatrix} K + K' - \omega^2 m & -K' \\ -K' & K + K' - \omega^2 m \end{pmatrix} \begin{pmatrix} a_1 \\ a_2 \end{pmatrix} = 0 \tag{11.3.12b}$$

These equations have a nontrivial solution (i.e., solutions other than $a_1 = a_2 = 0$) if and only if

$$\det |\mathbf{K} - \omega^2 \mathbf{M}| = 0 \tag{11.3.13a}$$

or

$$\begin{vmatrix} K + K' - \omega^2 m & -K' \\ -K' & K + K' - \omega^2 m \end{vmatrix} = 0 \tag{11.3.13b}$$

On expanding the determinant, we obtain

$$(K + K' - \omega^2 m)^2 - K'^2 = 0 \tag{11.3.13c}$$

which we can rearrange as

$$(\omega^2 m - K)[\omega^2 m - (K + 2K')] = 0 \tag{11.3.13d}$$

---

[3] This problem is equivalent to diagonalizing the moment of inertia matrix discussed in Section 9.2. For a treatment of this subject, the reader is referred to any text in mathematical physics such as: (1) J. Matthews and R. L. Walker, *Mathematical Methods of Physics*, W. A. Benjamin, New York (1970) or (2) S. I. Grossman and W. R. Derrick, *Advanced Engineering Mathematics*, HarperCollins Publ., New York (1988).

and the eigenfrequencies are now apparent.[4] They are the roots of Equation 11.3.13d and are given by

$$\omega_1^2 = \frac{K}{m} \qquad \omega_2^2 = \frac{K+2K'}{m} \tag{11.3.14}$$

These are the normal mode frequencies that we obtained previously, based on physical considerations alone.

We can now substitute each eigenfrequency back into Equation 11.3.12a or b to find the solutions for a_1 and a_2, the components of the eigenvectors. There are two eigenvectors, however, each with two components, so we rename these components a_{ij} meaning the ith component of the jth eigenvector.[5] Letting $\omega = \omega_1$ and inserting its value into Equation 11.3.12a, we obtain

$$\mathbf{K}\mathbf{a}_1 = \omega_1^2 \mathbf{M}\mathbf{a}_1$$

$$\begin{pmatrix} K+K' & -K' \\ -K' & K+K' \end{pmatrix}\begin{pmatrix} a_{11} \\ a_{21} \end{pmatrix} = \omega_1^2 \begin{pmatrix} m & 0 \\ 0 & m \end{pmatrix}\begin{pmatrix} a_{11} \\ a_{21} \end{pmatrix} \tag{11.3.15}$$

and using the first of the two equations in the matrix Equation 11.3.15

$$[(K+K') - \omega_1^2 m]a_{11} - K'a_{21} = 0 \tag{11.3.16}$$

yields the solution $a_{11} = a_{21}$.

Repeating the preceding process with the second eigenfrequency ω_2 yields components $a_{12} = -a_{22}$ for the second eigenvector $\mathbf{a}_2$. Thus, the (x_1, x_2) components of the two normal mode vectors are

$$\mathbf{Q}_1 = \begin{pmatrix} 1 \\ 1 \end{pmatrix}a_{11} \cos(\omega_1 t - \delta_1) \qquad \mathbf{Q}_2 = \begin{pmatrix} -1 \\ 1 \end{pmatrix}a_{12} \cos(\omega_2 t - \delta_2) \tag{11.3.17}$$

which can be compared with the (x_1, x_2) components of the normal mode vectors presented previously in Equation 11.3.8b.

The components $[x_1(t), x_2(t)]$, of any state vector $\mathbf{q}$ that represent the motion of the system in general, are linear combinations of the (x_1, x_2) components of the normal mode vectors $\mathbf{Q}_1$ and $\mathbf{Q}_2$. We show these components in the following table.

	$\mathbf{Q}_1$	$\mathbf{Q}_2$
x_1	$a_{11} \cos(\omega_1 t - \delta_1)$	$-a_{12} \cos(\omega_2 t - \delta_2)$
x_2	$a_{11} \cos(\omega_1 t - \delta_1)$	$a_{12} \cos(\omega_2 t - \delta_2)$

---

[4]We use the term eigenfrequency for either ω or ω^2. The one we mean should be clear from the context.

[5]When we describe a general n-dimensional eigenvector $\mathbf{a}$, we do so by denoting its components $a_1, a_2, \ldots, a_n$. There are n specific eigenvectors $\mathbf{a}_1, \mathbf{a}_2, \ldots, \mathbf{a}_n$, however, that are solutions to the equations of motion for a coupled oscillator system described by n generalized coordinates. Thus, when we describe one of these specific eigenvectors $\mathbf{a}_k$, we do so by denoting its components $a_{1k}, a_{2k}, \ldots, a_{nk}$. Do not confuse a specific eigenvector $\mathbf{a}_k$ with the scalar component a_k of some generalized eigenvector $\mathbf{a}$.

which makes it easy to see the general solutions for $x_1(t)$ and $x_2(t)$

$$x_1(t) = A_1 \cos(\omega_1 t - \delta_1) - A_2 \cos(\omega_2 t - \delta_2)$$
$$x_2(t) = A_1 \cos(\omega_1 t - \delta_1) + A_2 \cos(\omega_2 t - \delta_2)$$

(11.3.18)

To simplify notation, we have defined two new constants A_1 and A_2, such that $A_1/A_2 = a_{11}/a_{12}$. The four unknowns A_1, A_2, δ_1, and δ_2 can be determined from the initial values of the positions and velocities of each mass.

Initial Conditions

Let us now proceed to solve the specific problem that initiated this discussion, namely, mass m_2 is initially displaced one unit to the right, mass m_1 is held at $x_1 = 0$, and then they are simultaneously released from rest. These are the *initial conditions* for this problem. First, we derive relations for the constants in Equations 11.3.18 in terms of any general set of initial conditions, and then we invoke the specific initial conditions stated previously to solve for the constants in this particular problem. At time $t = 0$, Equation 11.3.18 becomes

$$x_1(0) = A_1 \cos\delta_1 - A_2 \cos\delta_2$$
$$x_2(0) = A_1 \cos\delta_1 + A_2 \cos\delta_2$$

(11.3.19a)

On differentiating Equation 11.3.18 and evaluating the result at $t = 0$, we obtain

$$\dot{x}_1(0) = \omega_1 A_1 \sin\delta_1 - \omega_2 A_2 \sin\delta_2$$
$$\dot{x}_2(0) = \omega_1 A_1 \sin\delta_1 + \omega_2 A_2 \sin\delta_2$$

(11.3.19b)

A little algebra yields the following solution for the amplitude constants

$$A_1^2 = \tfrac{1}{4}[x_1(0) + x_2(0)]^2 + \frac{1}{4\omega_1^2}[\dot{x}_1(0) + \dot{x}_2(0)]^2$$

$$A_2^2 = \tfrac{1}{4}[x_2(0) - x_1(0)]^2 + \frac{1}{4\omega_2^2}[\dot{x}_2(0) - \dot{x}_1(0)]^2$$

(11.3.20a)

and for the phase constants

$$\tan\delta_1 = \frac{\dot{x}_1(0) + \dot{x}_2(0)}{\omega_1[x_1(0) + x_2(0)]} \qquad \tan\delta_2 = \frac{\dot{x}_2(0) - \dot{x}_1(0)}{\omega_2[x_2(0) - x_1(0)]}$$

(11.3.20b)

Now, we insert the specific conditions for this problem

$$x_1(0) = 0 \qquad x_2(0) = 1 \qquad \dot{x}_1(0) = \dot{x}_2(0) = 0$$

(11.3.21)

into Equations 11.3.20a and b, and using Equation 11.3.19a to determine the signs of A_1 and A_2 gives

$$\delta_1 = \delta_2 = 0 \qquad A_1 = A_2 = \tfrac{1}{2}$$

(11.3.22)

Inserting these into Equation 11.3.18 yields the desired solutions

$$x_1(t) = \tfrac{1}{2}(\cos\omega_1 t - \cos\omega_2 t)$$
$$x_2(t) = \tfrac{1}{2}\cos(\omega_1 t + \cos\omega_2 t)$$

(11.3.23)

which should be compared with Equations 11.3.2a and b, which were put forth during our initial description of the motion that we anticipated for the coupled oscillator.

The Equations of Motion in Normal Coordinates

It is worth the effort to see if the Lagrangian can be expressed in a coordinate system such that no cross terms exist. The resulting equations of motion are not coupled when expressed in such coordinates. They break apart into two separate sets of linear, second-order differential equations whose solutions represent two decoupled simple harmonic oscillators. But this is exactly the characteristic of the normal modes. We suspect then, that transforming the Lagrangian to a function of the coordinates Q_1 and Q_2 ought to produce the desired decoupling.

In what follows, we carry out such a transformation using matrix methods. First we write the kinetic and potential energies of the coupled oscillator in (x_1, x_2) coordinates using matrix notation

$$T = \frac{1}{2}\tilde{\mathbf{x}}\mathbf{M}\dot{\mathbf{x}} = \frac{1}{2}(\dot{x}_1 \ \dot{x}_2)\begin{pmatrix} m & 0 \\ 0 & m \end{pmatrix}\begin{pmatrix} \dot{x}_1 \\ \dot{x}_2 \end{pmatrix}$$

$$= \frac{1}{2}m\dot{x}_1^2 + \frac{1}{2}m\dot{x}_2^2 \tag{11.3.24}$$

The potential energy is

$$V = \frac{1}{2}\tilde{\mathbf{x}}\mathbf{K}\mathbf{x} = \frac{1}{2}(x_1 \ x_2)\begin{pmatrix} K+K' & -K' \\ -K' & K+K' \end{pmatrix}\begin{pmatrix} x_1 \\ x_2 \end{pmatrix}$$

$$= \frac{1}{2}(K+K')x_1^2 + \frac{1}{2}(K+K')x_2^2 - K'x_1x_2 \tag{11.3.25}$$

Applying the transformation between the x_i coordinates and the Q_i coordinates in Equations 11.3.3a, we obtain

$$T = \frac{1}{2}\tilde{\dot{\mathbf{Q}}}\tilde{\mathbf{A}}\mathbf{M}\mathbf{A}\dot{\mathbf{Q}}$$

$$= \frac{1}{4}(\dot{Q}_1 \ \dot{Q}_2)\begin{pmatrix} 1 & 1 \\ -1 & 1 \end{pmatrix}\begin{pmatrix} m & 0 \\ 0 & m \end{pmatrix}\begin{pmatrix} 1 & -1 \\ 1 & 1 \end{pmatrix}\begin{pmatrix} \dot{Q}_1 \\ \dot{Q}_2 \end{pmatrix}$$

$$= \frac{1}{2}(\dot{Q}_1 \ \dot{Q}_2)\begin{pmatrix} m & 0 \\ 0 & m \end{pmatrix}\begin{pmatrix} \dot{Q}_1 \\ \dot{Q}_2 \end{pmatrix} \tag{11.3.26}$$

$$= \frac{1}{2}m\dot{Q}_1^2 + \frac{1}{2}m\dot{Q}_2^2$$

in which we have used the matrix identity $\mathbf{A}\tilde{\mathbf{Q}} = \tilde{\mathbf{Q}}\tilde{\mathbf{A}}$.

The potential energy is

$$V = \frac{1}{2}\tilde{\mathbf{Q}}\tilde{\mathbf{A}}\mathbf{K}\mathbf{A}\mathbf{Q}$$

$$= \frac{1}{4}(Q_1 \ Q_2)\begin{pmatrix} 1 & 1 \\ -1 & 1 \end{pmatrix}\begin{pmatrix} K+K' & -K' \\ -K' & K+K' \end{pmatrix}\begin{pmatrix} 1 & -1 \\ 1 & 1 \end{pmatrix}\begin{pmatrix} Q_1 \\ Q_2 \end{pmatrix}$$

$$= \frac{1}{2}(Q_1 \ Q_2)\begin{pmatrix} K & 0 \\ 0 & K+2K' \end{pmatrix}\begin{pmatrix} Q_1 \\ Q_2 \end{pmatrix} \tag{11.3.27}$$

$$= \frac{1}{2}KQ_1^2 + \frac{1}{2}(K+2K')Q_2^2$$

The Lagrangian is thus

$$L = \frac{1}{2} m \dot{Q}_1^2 + \frac{1}{2} m \dot{Q}_2^2 - \frac{1}{2} K Q_1^2 - \frac{1}{2}(K + 2K')Q_2^2 \tag{11.3.28}$$

As anticipated, it contains no cross terms, and the resulting equations of motion are

$$m\ddot{Q}_1 + KQ_1 = 0 \qquad m\ddot{Q}_2 + (K + 2K')Q_2 = 0 \tag{11.3.29a}$$

In matrix notation, they are

$$\begin{pmatrix} m & 0 \\ 0 & m \end{pmatrix}\begin{pmatrix} \ddot{Q}_1 \\ \ddot{Q}_2 \end{pmatrix} + \begin{pmatrix} K & 0 \\ 0 & K+2K' \end{pmatrix}\begin{pmatrix} Q_1 \\ Q_2 \end{pmatrix} = 0 \tag{11.3.29b}$$

These are the equations of motion of two uncoupled, simple harmonic oscillators whose solutions are

$$Q_1 = b_1 \cos(\omega_1 t - \epsilon_1) \qquad Q_2 = b_2 \cos(\omega_2 t - \epsilon_2) \tag{11.3.30a}$$

where

$$\omega_1^2 = \frac{K}{m} \qquad \omega_2^2 = \frac{K+2K'}{m} \tag{11.3.30b}$$

and b_1, b_2, ϵ_1, and ϵ_2 are constants of integration. In the Q_i-coordinate system, the normal mode vectors are

$$\mathbf{Q}_1 = \begin{pmatrix} 1 \\ 0 \end{pmatrix} b_1 \cos(\omega_1 t - \epsilon_1) \qquad \mathbf{Q}_2 = \begin{pmatrix} 0 \\ 1 \end{pmatrix} b_2 \cos(\omega_2 t - \epsilon_2) \tag{11.3.31}$$

which should be compared with those of Equation 11.3.8a.

A solution in terms of x_1 and x_2 can be obtained by transforming back to those coordinates using Equations 11.3.3a

$$
\begin{aligned}
x_1 &= \frac{1}{\sqrt{2}}(Q_1 - Q_2) \\
&= B_1 \cos(\omega_1 t - \epsilon_1) - B_2 \cos(\omega_2 t - \epsilon_2) \\
x_2 &= \frac{1}{\sqrt{2}}(Q_1 + Q_2) \\
&= B_1(\cos\omega_1 t - \epsilon_1) + B_2(\cos\omega_2 t - \epsilon_2)
\end{aligned}
\tag{11.3.32}
$$

where B_1, B_2, ϵ_1, and ϵ_2 are constants of integration. The $\frac{1}{\sqrt{2}}$ factor has been absorbed by the constants B_1 and B_2. The solutions are identical with those of Equation 11.3.18, obtained by solving the coupled equations of motion in x_i coordinates from the outset.

Diagonalizing the Lagrangian

Equations 11.3.29a are the equations of motion for two uncoupled, simple harmonic oscillators, and Equation 11.3.29b is its matrix representation. Both the **K** and **M** matrices in that equation are in diagonal form, that is, they have no nonzero, off-diagonal elements.

Each of those matrices was diagonalized by the *congruent transformation*

$$\mathbf{K}_{diag} = \tilde{\mathbf{A}}\mathbf{K}\mathbf{A} \qquad \mathbf{M}_{diag} = \tilde{\mathbf{A}}\mathbf{M}\mathbf{A} \qquad (11.3.33)$$

It is worth the effort to find such a matrix $\mathbf{A}$ in any given coupled oscillator problem for all one has to do to transform the problem to the simpler one of uncoupled oscillators is to diagonalize the $\mathbf{K}$ and $\mathbf{M}$ matrices via the transformation in Equation 11.3.33. The question is: what is the matrix $\mathbf{A}$? Close inspection of $\mathbf{A}$ in Equation 11.3.3a shows that *its columns are the* (x_1, x_2) *components of the eigenvectors* $\mathbf{a}_i$, that is,

$$\mathbf{A} = (\mathbf{a}_1 \; \mathbf{a}_2) = \begin{pmatrix} a_{11} & a_{12} \\ a_{21} & a_{22} \end{pmatrix} \qquad (11.3.34)$$

Each $\mathbf{a}_i$ in the preceding matrix is a column vector that was a solution to the equation of motion transformed to its eigenvector equivalent (Equation 11.3.12a) and written here as

$$\mathbf{K}\mathbf{a}_i = \omega_i^2 \mathbf{M}\mathbf{a}_i$$

$$\begin{pmatrix} K+K' & -K' \\ -K' & K+K' \end{pmatrix}\begin{pmatrix} a_{1i} \\ a_{2i} \end{pmatrix} = \omega_i^2 \begin{pmatrix} m & 0 \\ 0 & m \end{pmatrix}\begin{pmatrix} a_{1i} \\ a_{2i} \end{pmatrix} \qquad (11.3.35)$$

where a_{ji} refers to the jth component of the eigenvector $\mathbf{a}_i$ and ω_i^2 is its eigenfrequency. Thus, we see that

$$\tilde{\mathbf{a}}_i \mathbf{K}\mathbf{a}_i = \omega_i^2 \tilde{\mathbf{a}}_i \mathbf{M}\mathbf{a}_i \qquad (11.3.36a)$$

or

$$\frac{\tilde{\mathbf{a}}_i \mathbf{K}\mathbf{a}_i}{\tilde{\mathbf{a}}_i \mathbf{M}\mathbf{a}_i} = \omega_i^2 \qquad (11.3.36b)$$

Thus, the matrix formed with each eigenvector as one of the columns is the desired matrix $\mathbf{A}$ that transforms the generalized coordinates into normal coordinates and diagonalizes the $\mathbf{M}$ and $\mathbf{K}$ matrices that make up the Lagrangian of the system. Furthermore, note that the eigenfrequencies ω_1^2 and ω_2^2, corresponding to the normal mode eigenvectors, are simply the ratios of the elements of the diagonal matrices, $\mathbf{K}_{diag}$ and $\mathbf{M}_{diag}$. As an example, examination of the $\mathbf{K}_{diag}$ and $\mathbf{M}_{diag}$ matrices in Equations 11.3.26 and 11.3.27 reveals that the ratios of their diagonal elements are indeed equal to ω_1^2 and ω_2^2.

However, finding the transformation matrix $\mathbf{A}$ that diagonalizes $\mathbf{K}$ and $\mathbf{M}$ and, thus, solves the problem, means *first* solving the coupled equations of motion represented by Equations 11.3.10b or its equivalent 11.3.12a (or 11.3.35) for the eigenfrequencies ω_1 and ω_2, and *second*—using those solutions to solve for the corresponding normal mode eigenvectors $\mathbf{a}_1$ and $\mathbf{a}_2$. In other words, we must *first* solve the coupled equations of motion *before* finding the normal modes that we need to decouple them, a classic "catch-22" situation. How do we obtain the normal modes first, other than by "educated" guesswork? There is no general way to do this, but one method works quite well in many situations. It exploits the fact that *the Lagrangian is invariant under certain symmetry operations and the characteristics of the symmetry operators can be used to obtain the coordinates of the normal modes.* Let's see how this works.

The Lagrangian for any two-component coupled oscillator has the following general form in matrix notation, analogous to Equation 11.3.9

$$L = T - V = \frac{1}{2}\tilde{\mathbf{x}}\mathbf{M}\dot{\mathbf{x}} - \frac{1}{2}\tilde{\mathbf{x}}\mathbf{K}\mathbf{x}$$

$$= \frac{1}{2}(\dot{x}_1\,\dot{x}_2)\begin{pmatrix} M_{11} & M_{12} \\ M_{12} & M_{22} \end{pmatrix}\begin{pmatrix} \dot{x}_1 \\ \dot{x}_2 \end{pmatrix} - \frac{1}{2}(x_1\,x_2)\begin{pmatrix} K_{11} & K_{12} \\ K_{12} & K_{22} \end{pmatrix}\begin{pmatrix} x_1 \\ x_2 \end{pmatrix} \qquad (11.3.37a)$$

The $\mathbf{M}$ and $\mathbf{K}$ matrices are always real and symmetric. Hence, $K_{21} = K_{12}$ and $M_{21} = M_{12}$, facts that we have used in writing down the off-diagonal matrix elements in Equation 11.3.37a. Carrying out the matrix multiplication, L is found to be

$$L = \frac{1}{2}M_{11}\dot{x}_1^2 + \frac{1}{2}M_{22}\dot{x}_2^2 + M_{12}\dot{x}_1\dot{x}_2 - \frac{1}{2}K_{11}x_1^2 - \frac{1}{2}K_{22}x_2^2 - K_{12}x_1x_2 \qquad (11.3.37b)$$

Now suppose, in the preceding Lagrangian, we replace x_2 with $\pm\alpha x_1$ and x_1 with $\pm x_2/\alpha$. If the parameter α has the right value, the Lagrangian stays the same. It is invariant under this *exchange* operation. Carrying out the exchange ($\alpha x_1 \to x_2$ and $x_2/\alpha \to x_1$) transforms the Lagrangian to

$$L' = \frac{1}{2}(\dot{x}_2/\alpha\;\;\alpha\dot{x}_1)\begin{pmatrix} M_{11} & M_{12} \\ M_{12} & M_{22} \end{pmatrix}\begin{pmatrix} \dot{x}_2/\alpha \\ \alpha\dot{x}_1 \end{pmatrix} - \frac{1}{2}(x_2/\alpha\;\;\alpha x_1)\begin{pmatrix} K_{11} & K_{12} \\ K_{12} & K_{22} \end{pmatrix}\begin{pmatrix} x_2/\alpha \\ \alpha x_1 \end{pmatrix}$$

$$= \frac{1}{2}(\dot{x}_1\,\dot{x}_2)\begin{pmatrix} \alpha^2 M_{22} & M_{12} \\ M_{12} & M_{11}/\alpha^2 \end{pmatrix}\begin{pmatrix} \dot{x}_1 \\ \dot{x}_2 \end{pmatrix} - \frac{1}{2}(x_1\,x_2)\begin{pmatrix} \alpha^2 K_{22} & K_{12} \\ K_{12} & K_{11}/\alpha^2 \end{pmatrix}\begin{pmatrix} x_1 \\ x_2 \end{pmatrix} \qquad (11.3.38a)$$

or, after carrying out the matrix multiplication

$$L' = \frac{1}{2}M_{11}\frac{\dot{x}_2^2}{\alpha^2} + \frac{1}{2}M_{22}\alpha^2\dot{x}_1^2 + M_{12}\dot{x}_2\dot{x}_1 - \frac{1}{2}K_{11}\frac{x_2^2}{\alpha^2} + \frac{1}{2}K_{22}\alpha^2 x_1^2 - K_{12}x_2x_1 \qquad (11.3.38b)$$

The two cross terms in L' are identical to those in L, and the transformed Lagrangian L' equals the original Lagrangian L if

$$\frac{M_{11}}{M_{22}} = \frac{K_{11}}{K_{22}} \qquad (11.3.39a)$$

$$\alpha^2 = \frac{M_{11}}{M_{22}}\left(=\frac{K_{11}}{K_{22}}\right) \qquad (11.3.39b)$$

The first condition must be a property of the Lagrangian for the system under consideration. The second condition determines the ratio of the x components that must be used in the exchange process if the Lagrangian is to be invariant under that exchange. Because this latter condition is imposed on α^2, an exchange using $-\alpha$ as the parameter also satisfies the condition.

This suggests that the two eigenvectors $\mathbf{a}_1$ and $\mathbf{a}_2$ that make up the desired transformation matrix $\mathbf{A}$ have (x_1, x_2) components in which the second is $\pm\alpha$ times that of the first, that is

$$\mathbf{a}_1 = \begin{pmatrix} 1 \\ \alpha \end{pmatrix} \qquad \mathbf{a}_2 = \begin{pmatrix} -1 \\ \alpha \end{pmatrix} \qquad (11.3.40a)$$

The **A** matrix that generates the transformation from generalized to normal coordinates is thus

$$\mathbf{A} = (\mathbf{a}_1\, \mathbf{a}_2) = \begin{pmatrix} 1 & -1 \\ \alpha & \alpha \end{pmatrix} \tag{11.3.40b}$$

The Lagrangian $L' = \tilde{\mathbf{A}}L\mathbf{A}$ can have no cross terms if it is to remain invariant under each of the aforementioned exchange operations. Note what happens to the two eigenvectors $\mathbf{a}_i$ during an exchange: (1) for the exchange $\alpha x_1 \rightarrow x_2$ and $x_2/\alpha \rightarrow x_1$, we find that $\mathbf{a}_1 \rightarrow +\mathbf{a}_1$ but $\mathbf{a}_2 \rightarrow -\mathbf{a}_2$, but (2) for the exchange $\alpha x_1 \rightarrow -x_2$ and $x_2/\alpha \rightarrow -x_1$ we find that $\mathbf{a}_2 \rightarrow +\mathbf{a}_2$ but $\mathbf{a}_1 \rightarrow -\mathbf{a}_1$. In each case, one of the normal mode vectors changes sign, but the other stays the same. If the transformed Lagrangian had any cross terms, they would change signs but the squared terms would not. Hence, the transformed Lagrangian L' would no longer be invariant under the exchange. We conclude that the transformed Lagrangian L' must be a function of normal coordinates with no cross terms if the $\mathbf{a}_i$'s really are the desired eigenvectors and the corresponding matrix **A** is the desired transformation matrix.

It is worth reexamining the previous example in light of this discussion. Look at the diagonal elements of the **K** and **M** matrices of Equation 11.3.10c. They satisfy condition 11.3.39a. Condition 11.3.39b can be satisfied if $\alpha = 1$. The matrix **A** that diagonalizes **K** and **M** is, therefore, given by Equation 11.3.40b and is

$$\mathbf{A} = (\mathbf{a}_1\, \mathbf{a}_2) = \begin{pmatrix} 1 & -1 \\ 1 & 1 \end{pmatrix} \tag{11.3.41}$$

K and **M** are diagonalized according to the transformations given in Equation 11.3.33

$$\begin{aligned} \mathbf{K}_{diag} = \tilde{\mathbf{A}}\mathbf{K}\mathbf{A} &= \begin{pmatrix} 1 & 1 \\ -1 & 1 \end{pmatrix} \begin{pmatrix} K+K' & -K' \\ -K' & K+K' \end{pmatrix} \begin{pmatrix} 1 & -1 \\ 1 & 1 \end{pmatrix} \\ &= 2\begin{pmatrix} K & 0 \\ 0 & K+2K' \end{pmatrix} \end{aligned} \tag{11.3.42a}$$

and

$$\begin{aligned} \mathbf{M}_{diag} = \tilde{\mathbf{A}}\mathbf{M}\mathbf{A} &= \begin{pmatrix} 1 & 1 \\ -1 & 1 \end{pmatrix} \begin{pmatrix} m & 0 \\ 0 & m \end{pmatrix} \begin{pmatrix} 1 & -1 \\ 1 & 1 \end{pmatrix} \\ &= 2\begin{pmatrix} m & 0 \\ 0 & m \end{pmatrix} \end{aligned} \tag{11.3.42b}$$

The ratio of the diagonal elements of $\mathbf{K}_{diag}$ and $\mathbf{M}_{diag}$ yield the eigenfrequencies ω_1^2 and ω_2^2 obtained previously in Equation 11.3.14. (The multiplicative factor of 2 that occurs in Equations 11.3.42a and b cancels out in these ratios and is, therefore, irrelevant. It could be eliminated by normalizing the eigenvectors $\mathbf{a}_1$ and $\mathbf{a}_2$ by the factor $\frac{1}{\sqrt{2}}$.)

EXAMPLE 11.3.1

The Double Pendulum (Two Rock Climbers Dangling on a Single Rope)

Let us consider the motion of a double pendulum that consists of two simple pendula, each of mass m and length l. The first one is attached to a fixed support, and the second one is attached to the mass of the first, as shown in Figure 11.3.7a. Assuming that the

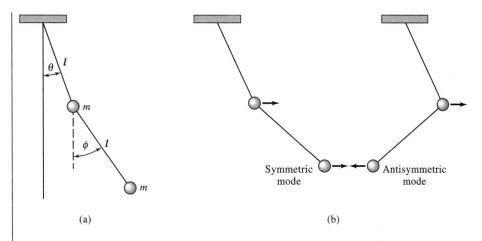

Figure 11.3.7 The double pendulum and its normal modes.

pendulum executes small oscillations confined to a single plane, find the normal modes and the corresponding normal mode frequencies.

Solution 1: "Guessing" the Normal Modes

Let us specify the configuration of the system by the two angles θ and ϕ as shown in the figure. The kinetic energy of the double pendulum is given by

$$T = \tfrac{1}{2} m \mathbf{v}_1 \cdot \mathbf{v}_1 + \tfrac{1}{2} m \mathbf{v}_2 \cdot \mathbf{v}_2 \qquad (11.3.43)$$

The velocities of each mass $\mathbf{v}_1$ and $\mathbf{v}_2$ can be expressed in terms of the angular velocities $\dot{\theta}$ and $\dot{\phi}$

$$\mathbf{v}_1 = \mathbf{e}_\theta l \dot{\theta} \qquad \mathbf{v}_2 = \mathbf{v}_1 + \mathbf{e}_\phi l \dot{\phi} \qquad (11.3.44)$$

where the second term in the latter expression is the velocity of the second mass relative to that of the first.

The kinetic energy is, thus,

$$
\begin{aligned}
T &= \tfrac{1}{2} m l^2 \dot{\theta}^2 + \tfrac{1}{2} m (\mathbf{e}_\theta l \dot{\theta} + \mathbf{e}_\phi l \dot{\phi}) \cdot (\mathbf{e}_\theta l \dot{\theta} + \mathbf{e}_\phi l \dot{\phi}) \\
&\approx \tfrac{1}{2} m l^2 \dot{\theta}^2 + \tfrac{1}{2} m l^2 (\dot{\theta} + \dot{\phi})^2 \\
&= \tfrac{1}{2} m l^2 (2\dot{\theta}^2 + \dot{\phi}^2 + 2\dot{\theta}\dot{\phi})
\end{aligned}
\qquad (11.3.45)
$$

in which we have used $\mathbf{e}_\theta \cdot \mathbf{e}_\phi \approx 1$ because these two unit vectors remain approximately parallel as long as the angular displacements are small. We can calculate the $\mathbf{M}$ matrix by comparing Equation 11.3.45 with Equations 11.3.37a and b

$$\mathbf{M} = m l^2 \begin{pmatrix} 2 & 1 \\ 1 & 1 \end{pmatrix} \qquad (11.3.46)$$

The sum of the potential energies of the two masses relative to their equilibrium position is

$$V = mgl(1 - \cos\theta) + mgl[2 - (\cos\theta + \cos\phi)]$$
$$\approx \frac{1}{2}mgl(2\theta^2 + \phi^2)$$

(11.3.47)

in which we used the small angle approximation for both cosine functions. We can calculate the **K** matrix by compairing Equation 11.3.47 with Equations 11.3.37a and b

$$\mathbf{K} = mgl\begin{pmatrix} 2 & 0 \\ 0 & 1 \end{pmatrix}$$

(11.3.48)

The condition given in Equation 11.3.39a is automatically satisfied and that given in Equations 11.3.39b is satisfied if $\alpha^2 = 2$. We, therefore, "guess" that the two eigenvectors for this coupled system, according to Equation 11.3.40a, are

$$\mathbf{a}_1 = \begin{pmatrix} 1 \\ \sqrt{2} \end{pmatrix} \qquad \mathbf{a}_2 = \begin{pmatrix} -1 \\ \sqrt{2} \end{pmatrix}$$

(11.3.49a)

The **A** matrix is then

$$\mathbf{A} = \begin{pmatrix} 1 & -1 \\ \sqrt{2} & \sqrt{2} \end{pmatrix}$$

(11.3.49b)

which we then use to diagonalize **M**

$$\mathbf{M}_{diag} = \tilde{\mathbf{A}}\mathbf{M}\mathbf{A} = ml^2\begin{pmatrix} 1 & \sqrt{2} \\ -1 & \sqrt{2} \end{pmatrix}\begin{pmatrix} 2 & 1 \\ 1 & 1 \end{pmatrix}\begin{pmatrix} 1 & -1 \\ \sqrt{2} & \sqrt{2} \end{pmatrix}$$
$$= 2ml^2\begin{pmatrix} 2+\sqrt{2} & 0 \\ 0 & 2-\sqrt{2} \end{pmatrix}$$

(11.3.50a)

then **K**

$$\mathbf{K}_{diag} = \tilde{\mathbf{A}}\mathbf{K}\mathbf{A} = mgl\begin{pmatrix} 1 & \sqrt{2} \\ -1 & \sqrt{2} \end{pmatrix}\begin{pmatrix} 2 & 0 \\ 0 & 1 \end{pmatrix}\begin{pmatrix} 1 & -1 \\ \sqrt{2} & \sqrt{2} \end{pmatrix}$$
$$= 4mgl\begin{pmatrix} 1 & 0 \\ 0 & 1 \end{pmatrix}$$

(11.3.50b)

The Lagrangian, expressed in terms of the normal coordinates, is

$$L = T - V = \frac{1}{2}\dot{\tilde{\mathbf{Q}}}\mathbf{M}_{diag}\dot{\mathbf{Q}} - \frac{1}{2}\tilde{\mathbf{Q}}\mathbf{K}_{diag}\mathbf{Q}$$
$$= ml^2(\dot{Q}_1\,\dot{Q}_2)\begin{pmatrix} 2+\sqrt{2} & 0 \\ 0 & 2-\sqrt{2} \end{pmatrix}\begin{pmatrix} \dot{Q}_1 \\ \dot{Q}_2 \end{pmatrix} - 2mgl(Q_1\,Q_2)\begin{pmatrix} 1 & 0 \\ 0 & 1 \end{pmatrix}\begin{pmatrix} Q_1 \\ Q_2 \end{pmatrix}$$
$$= ml^2\left[(2+\sqrt{2})\dot{Q}_1^2 + (2-\sqrt{2})\dot{Q}_2^2\right] - 2mgl\left(Q_1^2 + Q_2^2\right)$$

(11.3.51)

and it has no cross terms. The ratio of the diagonal elements of $\mathbf{K}_{diag}$ and $\mathbf{M}_{diag}$ yields the eigenfrequencies ω_1^2 and ω_2^2

$$\omega_1^2 = \frac{2mgl}{ml^2}\left(\frac{1}{2+\sqrt{2}}\right) = (2-\sqrt{2})\frac{g}{l} \qquad symmetric\ mode$$

$$\omega_2^2 = \frac{2mgl}{ml^2}\left(\frac{1}{2-\sqrt{2}}\right) = (2+\sqrt{2})\frac{g}{l} \qquad antisymmetric\ mode$$

(11.3.52a)

The ratio of the two normal mode frequencies is independent of all the parameters m, l, and g and is equal to

$$\frac{\omega_2}{\omega_1} = \left[\frac{(2+\sqrt{2})}{(2-\sqrt{2})}\right]^{1/2} = 2.414 \qquad (11.3.52b)$$

so the oscillation in the faster, antisymmetric mode has a frequency about two and one-half times that of the slower, symmetric mode.

Solution 2: The General Method

The equations of motion can be written in matrix form (using the $\mathbf{M}$ and $\mathbf{K}$ matrices found in Equations 11.3.46 and 11.3.48) as

$$\mathbf{M\ddot{q} + Kq} = 0$$

$$ml^2\begin{pmatrix} 2 & 1 \\ 1 & 1 \end{pmatrix}\begin{pmatrix} \ddot{\theta} \\ \ddot{\phi} \end{pmatrix} + mgl\begin{pmatrix} 2 & 0 \\ 0 & 1 \end{pmatrix}\begin{pmatrix} \theta \\ \phi \end{pmatrix} = 0 \qquad (11.3.53)$$

where the generalized coordinates (θ, ϕ) represent components of the generalized coordinate state vector $\mathbf{q}$. As before, let us assume that a solution for a normal mode exists that takes a form analogous to those found for the coupled oscillator discussed in the previous section and represented by Equations 11.3.11a and b.

$$\mathbf{q} = \mathbf{a}\cos\omega t$$

$$= \begin{pmatrix} a_1 \\ a_2 \end{pmatrix}\cos\omega t \qquad (11.3.54)$$

(For simplicity, we used only a cosine term with no phase angle.)

Plugging the assumed single frequency solution for a normal mode into the equations of motion yields the following

$$\begin{pmatrix} -2\omega^2 + 2 & -\omega^2 \\ -\omega^2 & -\omega^2 + 1 \end{pmatrix}\begin{pmatrix} a_1 \\ a_2 \end{pmatrix} = 0 \qquad (11.3.55)$$

in which we further simplified subsequent algebraic manipulation by omitting the factors g and l. As before, a nontrivial solution ($a_1, a_2 \neq 0$) exists only if the determinant of the matrix in Equation 11.3.55 is zero, that is

$$\begin{vmatrix} -2\omega^2 + 2 & -\omega^2 \\ -\omega^2 & -\omega^2 + 1 \end{vmatrix} = 0 \qquad (11.3.56a)$$

or

$$\omega^4 - 4\omega^2 + 2 = 0 \tag{11.3.56b}$$

The solution to this quadratic (in ω^2) equation yields the two eigenfrequencies ω_1^2 and ω_2^2, already obtained in Equations 11.3.52a. We can now calculate the ratio a_1/a_2 for the two eigenvectors by inserting these two frequencies, one by one, into either of the homogeneous equations contained in the matrix Equation 11.3.55. For example, the first equation is

$$(-2\omega^2 + 2)a_1 = \omega^2 a_2$$

$$\frac{a_1}{a_2} = \frac{\omega^2}{-2\omega^2 + 2} \tag{11.3.57a}$$

Inserting the two eigenfrequencies ω_1^2 and ω_2^2 from Equations 11.3.52a into Equation 11.3.57 yields the following conditions on the coordinates of the eigenvectors

$$\frac{a_1}{a_2} = +\frac{1}{\sqrt{2}} \; (\omega = \omega_1) \qquad \frac{a_1}{a_2} = -\frac{1}{\sqrt{2}} \; (\omega = \omega_2) \tag{11.3.57b}$$

We arbitrarily set $a_1 = 1$, which we are free to do because the solution yields only the ratio a_1/a_2.

Thus, we obtain

$$\theta = \cos\omega_1 t \qquad \phi = +\sqrt{2}\,\cos\omega_1 t \qquad \mathbf{a}_1 = \begin{pmatrix} 1 \\ \sqrt{2} \end{pmatrix} \qquad \textit{symmetric mode}$$

$$\tag{11.3.58}$$

$$\theta = -\cos\omega_2 t \qquad \phi = +\sqrt{2}\,\cos\omega_2 t \qquad \mathbf{a}_2 = \begin{pmatrix} -1 \\ \sqrt{2} \end{pmatrix} \qquad \textit{antisymmetric mode}$$

which are precisely the eigenvectors that we originally guessed.

EXAMPLE 11.3.2

A pendulum of mass m and length r is attached to a support, also of mass m, that is able to move along a frictionless, horizontal track. A spring, of force constant k, is attached between the support and an adjacent wall (Figure 11.3.8). The values of the mass m, spring constant k, and pendulum length r are such that $2mg = kr$, that is, if the spring were used to support the weight of the two masses, it would be stretched a distance equal to the length of the pendulum. Find the normal mode frequencies.

Solution:

We calculated the kinetic and potential energies of a system like this one (minus the spring) in Section 10.3. The kinetic energy of this system is the same. It is

$$T = \tfrac{1}{2}m\dot{X}^2 + \tfrac{1}{2}m[\dot{X}^2 + (r\dot{\theta})^2 + 2\dot{X}(r\dot{\theta})\cos\theta]$$

The potential energy (defined to be zero at the equilibrium configuration) is

$$V = mgr(1 - \cos\theta) + \tfrac{1}{2}kX^2$$

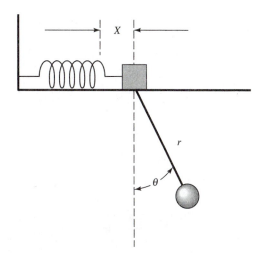

Figure 11.3.8 Pendulum attached to movable support which is attached to a wall with a spring.

Assuming that θ, $\dot{\theta}$ are small, we can approximate the above expressions to give

$$T \approx \frac{1}{2}m\dot{X}^2 + \frac{1}{2}m[\dot{X}^2 + (r\dot{\theta})^2 + 2\dot{X}(r\dot{\theta})]$$

$$V \approx mgr\frac{\theta^2}{2} + k\frac{X^2}{2}$$

We can further simplify these expressions by setting $q_1 = X$, $q_2 = r\theta$ and, using the conditions given in the problem, setting $\omega_0^2 \equiv g/r = k/2m$. Then, we have

$$T = \frac{1}{2}m\left[2\dot{q}_1^2 + \dot{q}_2^2 + 2\dot{q}_1\dot{q}_2\right]$$

$$V \approx \frac{1}{2}m\omega_0^2\left[2q_1^2 + q_2^2\right]$$

These last two expressions are identical in form to Equations 11.3.45 and 11.3.47 in Example 11.3.1; therefore, the solution is identical to the one we obtained there. The resulting normal mode frequencies are

$$\omega_1^2 = (2 - \sqrt{2})\omega_0^2 \qquad \text{and} \qquad \omega_2^2 = (2 + \sqrt{2})\omega_0^2$$

Does this seem reasonable to you?

11.4 | General Theory of Vibrating Systems

We turn now to a system with n degrees of freedom. In Section 10.3 we showed that the kinetic energy T is a homogeneous quadratic function of the generalized velocities. It can be written in matrix form as

$$T = \frac{1}{2}\tilde{\mathbf{q}}\mathbf{M}\dot{\mathbf{q}} = \frac{1}{2}\sum_{j,k}^{n} M_{jk}\,\dot{q}_j\dot{q}_k \qquad (11.4.1)$$

provided there are no moving constraints. M_{jk} are the elements of the real, symmetric $n \times n$ matrix $\mathbf{M}$, and $\dot{q}_j$ are the n generalized velocities that are the components of the vector $\dot{\mathbf{q}}$. Because we are concerned with motion about an equilibrium configuration, we assume, as in Section 11.2, that the M_{jk}'s are constant and equal to their values at the equilibrium configuration. We further assume that the origin of the n-dimensional coordinate system $(q_1, q_2, \ldots, q_n)$ has been chosen such that the equilibrium configuration is given by

$$q_1 = q_2 = \cdots = q_n = 0 \tag{11.4.2}$$

Accordingly, the potential energy V, from Equation 11.1.8a, is given by

$$V = \frac{1}{2} \tilde{\mathbf{q}} \mathbf{K} \mathbf{q} = \frac{1}{2} \sum_{j,k} K_{jk} q_j q_k \tag{11.4.3}$$

K_{jk} are the elements of the real, symmetric $n \times n$ matrix $\mathbf{K}$, and q_j are the n generalized coordinates in which the state vector $\mathbf{q}$ is expressed.

The Lagrangian function then assumes the form

$$
\begin{aligned}
L &= \frac{1}{2} \tilde{\dot{\mathbf{q}}} \mathbf{M} \dot{\mathbf{q}} - \frac{1}{2} \tilde{\mathbf{q}} \mathbf{K} \mathbf{q} \\
&= \frac{1}{2} \sum_{j,k}^{n} (M_{jk} \dot{q}_j \dot{q}_k - K_{jk} q_j q_k)
\end{aligned} \tag{11.4.4}
$$

and the resulting equations of motion are

$$\frac{d}{dt}\left(\frac{\partial L}{\partial \dot{q}_k}\right) - \frac{\partial L}{\partial q_k} = 0 \qquad (k = 1, 2, \ldots, n) \tag{11.4.5}$$

which are equal to

$$\sum_{j}^{n} (M_{jk} \ddot{q}_j + K_{jk} q_j) \qquad (k = 1, 2, \ldots, n) \tag{11.4.6}$$

or in matrix form

$$\mathbf{M}\ddot{\mathbf{q}} + \mathbf{K}\mathbf{q} = 0 \tag{11.4.7}$$

If a solution of the form

$$\mathbf{q} = \mathbf{a} \cos \omega t \tag{11.4.8}$$

exists, where $\mathbf{a}$ is a vector with n components a_j, the following equation(s) must be satisfied:

$$(\mathbf{K} - \omega^2 \mathbf{M})\mathbf{a} = 0 \tag{11.4.9}$$

This is a matrix representation of a homogeneous set of linear equations for the n components of the eigenvector $\mathbf{a}$

$$
\begin{pmatrix}
K_{11} - \omega^2 M_{11} & K_{12} - \omega^2 M_{12} & \cdots \\
K_{21} - \omega^2 M_{21} & K_{22} - \omega^2 M_{22} & \cdots \\
& \cdots & \cdots
\end{pmatrix}
\begin{pmatrix}
a_1 \\
a_2 \\
\cdots
\end{pmatrix}
=
\begin{pmatrix}
0 \\
0 \\
\cdots
\end{pmatrix} \tag{11.4.10}
$$

A nontrivial solution requires that the determinant of the coefficients of the **a** vanishes, that is

$$\det (\mathbf{K} - \omega^2 \mathbf{M}) = 0 \tag{11.4.11}$$

or

$$\begin{vmatrix} K_{11} - \omega^2 M_{11} & K_{12} - \omega^2 M_{12} & \cdots \\ K_{21} - \omega^2 M_{21} & K_{22} - \omega^2 M_{22} & \cdots \\ \cdots & \cdots & \cdots \end{vmatrix} = 0 \tag{11.4.12}$$

The preceding secular equation (11.4.12) is an equation of the nth degree in ω^2. The n roots are the eigenvalues or eigenfrequencies of the system.

Thus, if a given system has n degrees of freedom, there are, in general, n different possible eigenfrequencies of oscillation about the equilibrium configuration, each characterized by its own eigenvector that corresponds to a normal mode. Finding the roots, or eigenfrequencies, of the secular equation often entails the tedious task of solving a high-order polynomial, cubic for $n = 3$, quartic for $n = 4$, and so on. In some special situations the roots of the secular equation are either repeated or zero or both. In such cases, the problem of finding the roots might not be too difficult. Such an example is given at the end of this section. In the next section, we present a method of determining the normal mode frequencies for a linear array of coupled oscillators for which n may have any value.

As in the previous case of two coupled oscillators, the complete solution to Equation 11.4.12 gives the n eigenfrequencies ω_k^2, which can then be used in Equation 11.4.9 or 11.4.10 to calculate the components a_{ik} of the n eigenvectors $\mathbf{a}_k$. These eigenvectors and eigenfrequencies can then be used to construct the n normal mode vectors $\mathbf{Q}_k$, which is given by

$$\mathbf{Q}_k(t) = \mathbf{a}_k \cos(\omega_k t - \delta_k) \qquad \mathbf{a}_k = \begin{pmatrix} a_{1k} \\ a_{2k} \\ \cdots \\ \cdots \\ a_{nk} \end{pmatrix} \qquad (k = 1, 2, \ldots, n) \tag{11.4.13}$$

The amplitudes, or components a_{ik}, are not independent but are related because for each eigenfrequency ω_k they satisfy the homogeneous eigenvector Equation 11.4.9 or 11.4.10. This allows us to determine only the ratios of the eigenvector components $a_{1k} : a_{2k} : \ldots : a_{nk}$. We have freedom to normalize them as we choose. For the sake of simplicity, this is frequently done by setting the first component equal to one.

The displacement of each oscillator from its equilibrium position is represented by its generalized coordinate q_k, one of the components of the state vector $\mathbf{q}$ for the system. This is the general solution we desire. Each of these components is a linear combination of the components of the normal mode vector $\mathbf{Q}_k$, which oscillate at their respective frequencies ω_k. The components of these normal mode vectors are given in Equation 11.4.13, and they can be used to construct a table like the one that led to the solutions given in Equation

11.3.18 for the coupled oscillator. The table shows the linear combinations that make up the desired solutions q_k.

	$\mathbf{Q}_1$	$\mathbf{Q}_2$	$\cdots$ $\cdots$	$\mathbf{Q}_n$
q_1	$a_{11}\cos(\omega_1 t - \delta_1)$	$a_{12}\cos(\omega_2 t - \delta_2)$	$\cdots$ $\cdots$	$a_{1n}\cos(\omega_n t - \delta_n)$
q_2	$a_{21}\cos(\omega_1 t - \delta_1)$	$a_{22}\cos(\omega_2 t - \delta_2)$	$\cdots$ $\cdots$	$a_{2n}\cos(\omega_n t - \delta_n)$
$\cdots$	$\cdots$	$\cdots$	$\cdots$ $\cdots$	$\cdots$
$\cdots$	$\cdots$	$\cdots$	$\cdots$ $\cdots$	$\cdots$
q_n	$a_{n1}\cos(\omega_1 t - \delta_1)$	$a_{n2}\cos(\omega_2 t - \delta_2)$	$\cdots$ $\cdots$	$a_{nn}\cos(\omega_n t - \delta_n)$

The general solution for each generalized coordinate q_k is, thus,

$$q_k = \sum_{i=1}^{n} a_{ki}\cos(\omega_i t - \delta_i) \qquad (k = 1, 2, \dots, n) \qquad (11.4.14)$$

We emphasize that particularly when $n > 2$, it is well worth the effort to try to guess the normal mode vectors because constructing the transformation matrix $\mathbf{A}$ is then straightforward and the complete solution to the problem can be found with relatively simple operations using this matrix.

EXAMPLE 11.4.1

Linear Motion of a Triatomic Molecule

Let us consider the motion of a three-particle system in which all the particles lie in a straight line. An example of such a collinear system is the carbon dioxide molecule CO_2, which has the structure O–C–O. We consider motion only in one dimension, along the x-axis (Figure 11.4.1). The two end particles, each of mass m, are bound to the central particle, mass M, via a potential function that is equivalent to that of two springs of stiffness K, as shown in Figure 11.4.1. The coordinates expressing the displacements of each mass are x_1, x_2, and x_3.

Solution:

In this problem we can easily guess the normal modes. They are pictured in Figures 11.4.1(a)–(c). If you think about it a while you should realize that what's going on here is that the center of mass of the molecule is not accelerating. In mode (c) the central mass is vibrating 180° out of phase with the two end masses. The ratio of the vibrational amplitudes is such that the center of mass remains at rest. Mode (b) obeys the same condition. The central mass remains at rest while the two equal end masses vibrate 180° out of phase with each other, with equal amplitudes, again fixing the center of mass. Mode (a) depicts overall translation of the center of mass at constant velocity.

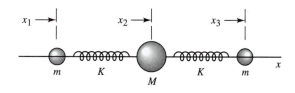

Figure 11.4.1   Model of a triatomic molecule and its three normal modes for motion in a single line.

We could go on and solve the problem using this guess. We do not do so, however. We solve it using the general method introduced in the previous example, in which we assume that the normal modes are unknown. We ultimately generate a secular equation that, in this example, is of third order in ω^2. (There are three coordinates, hence, three normal modes and frequencies in the solution.) It turns out that this particular third-order equation is very easy to solve. On obtaining the frequencies of each normal mode, we then insert them into any one of the equations relating the amplitudes of the displacement coordinates to one another (the matrix equivalent of the secular equation in ω^2), thus, obtaining the normal modes.

The Lagrangian of the system is

$$L = T - V$$
$$= \left(\frac{m}{2}\dot{x}_1^2 + \frac{M}{2}\dot{x}_2^2 + \frac{m}{2}\dot{x}_3^2 \right) - \left[\frac{K}{2}(x_2 - x_1)^2 + \frac{K}{2}(x_3 - x_2)^2 \right] \qquad (11.4.15)$$

and Lagrange's three equations of motion read

$$
\begin{aligned}
m\ddot{x}_1 + Kx_1 \quad\;\; -Kx_2 \qquad\qquad\qquad\;\; &= 0 \\
-Kx_1 \quad +M\ddot{x}_2 + 2Kx_2 \quad\;\; -Kx_3 \quad &= 0 \\
-Kx_2 \quad\;\; +m\ddot{x}_3 + Kx_3 \;\; &= 0
\end{aligned}
\qquad (11.4.16)
$$

If a solution of the form $x_1 = a_1 \cos \omega t$, $x_2 = a_2 \cos \omega t$, $x_3 = a_3 \cos \omega t$ exists, then

$$
\begin{pmatrix}
K - m\omega^2 & -K & 0 \\
-K & 2K - M\omega^2 & -K \\
0 & -K & K - m\omega^2
\end{pmatrix}
\begin{pmatrix}
a_1 \\ a_2 \\ a_3
\end{pmatrix}
= 0
\qquad (11.4.17)
$$

The secular equation is thus,

$$
\begin{vmatrix}
K - m\omega^2 & -K & 0 \\
-K & 2K - M\omega^2 & -K \\
0 & -K & K - m\omega^2
\end{vmatrix} = 0
\qquad (11.4.18a)
$$

which, on expanding the determinant and collecting terms, fortuitously becomes the product of three factors

$$
\omega^2(-m\omega^2 + K)(-mM\omega^2 + KM + 2Km) = 0
\qquad (11.4.18b)
$$

Equating each of the three factors to zero gives the three normal frequencies of the system:

$$
\omega_1 = 0 \qquad \omega_2 = \left(\frac{K}{m}\right)^{1/2} \qquad \omega_3 = \left(\frac{K}{m} + 2\frac{K}{M}\right)^{1/2}
\qquad (11.4.19)
$$

Let us discuss the modes corresponding to these three roots.

1. The first mode is no oscillation at all but is *pure translation* of the system as a whole. If we set $\omega = 0$ in Equations 11.4.17, we find that $a_1 = a_2 = a_3$ for this mode.

2. Setting $\omega = \omega_2$ in Equations 11.4.17 gives $a_2 = 0$ and $a_1 = -a_3$. In this mode the center particle is at rest while the two end particles vibrate in opposite directions (antisymmetrically) with the same amplitude.

3. Finally, setting $\omega = \omega_3$ in Equations 11.4.17 we obtain the following relations: $a_1 = a_3$ and $a_2 = -2a_1(m/M) = -2a_3(m/M)$. Thus, in this mode the two end particles vibrate in unison while the center particle vibrates oppositely with a different amplitude. The three modes are illustrated in Figure 11.4.1.

The ratio ω_3/ω_2 is independent of the constant K, namely,

$$
\frac{\omega_3}{\omega_2} = \left(1 + 2\frac{m}{M}\right)^{1/2}
$$

In the carbon dioxide molecule the mass ratio m/M is very nearly 16:12 for ordinary CO_2 (C_{12} and O_{16} atoms). Thus, the frequency ratio

$$
\frac{\omega_3}{\omega_2} = \left(1 + 2 \times \tfrac{16}{12}\right)^{1/2} = \left(\tfrac{11}{3}\right)^{1/2} = 1.915
$$

11.5 | Vibration of a Loaded String or Linear Array of Coupled Harmonic Oscillators

Any real, solid system contains many particles, each bound to a small region of space by atomic potentials, not just two or three particles coupled together by springs. The binding potential "felt" by each particle, however, is well represented by a quadratic function of the difference between each particle's displacement from its equilibrium position and the corresponding displacement of its immediate neighbor. Thus, such a system is essentially one of many coupled oscillators. Its analysis can lead to a description of the oscillations of a continuous medium, the propagation of waves through a continuous medium, or the vibrations of a crystalline lattice. In this section, we take the first step toward arming you with the theoretical weaponry necessary to attack such problems. We consider the motion of a simple mechanical system consisting of a light elastic string, clamped at both ends and loaded with n particles, each of mass m, equally spaced along the length of the string. Before proceeding with the analysis, however, we make a brief historical digression on this subject.[6]

An analysis of the dynamics of a line of interconnected masses was first attempted by Newton, himself. Two of his successors, the remarkable Bernoullis (John and his son Daniel), were the ones who had ultimate success with the problem. They demonstrated that a system of n masses has exactly n independent modes (for one-dimensional motion only). In 1753 Daniel Bernoulli (1700–1782) demonstrated that the general motion of this vibrating system is describable as a superposition of its normal modes. According to Leon Brillouin, a major contributor to the theory of vibrations of a crystalline lattice[7]:

> This investigation by the Bernoullis may be said to form the beginning of theoretical physics as distinct from mechanics, in the sense that it is the first attempt to formulate laws for the motion of a system of particles rather than for that of a single particle. The principle of superposition is important, as it is a special case of a Fourier series, and in time it was extended to become a statement of Fourier's theorem.

Strong words, these! Let us now begin the analysis.

Let us label the displacements of the various particles from their equilibrium positions by the coordinates $q_1, q_2, \ldots, q_n$. Actually, two types of displacement can occur, namely, a longitudinal displacement in which the particle moves along the direction of the string, and a transverse displacement in which the particle moves at right angles to the length of the string. These are illustrated in Figure 11.5.1. For simplicity we assume that the motion is either purely longitudinal or purely transverse, although in the actual physical situation a combination of the two could occur. The kinetic energy of the system is then given by

$$T = \frac{m}{2}\left(\dot{q}_1^2 + \dot{q}_2^2 + \cdots + \dot{q}_n^2\right) \tag{11.5.1}$$

If we use the letter k to denote any given particle, then, in the case of longitudinal motion, the stretch of the section of string between particle k and particle $k + 1$ is

$$q_{k+1} - q_k \tag{11.5.2}$$

Hence, the potential energy of this section of the string is

$$\tfrac{1}{2}K(q_{k+1} - q_k)^2 \tag{11.5.3}$$

in which K is the elastic stiffness coefficient of the section of string connecting the two adjacent particles.

---

[6] See, for example, A. P. French, *Vibrations and Waves*, The MIT Introductory Physics Series, Norton, New York (1971).

[7] L. Brillouin, *Wave Propagation in Periodic Structures*, Dover, New York (1953).

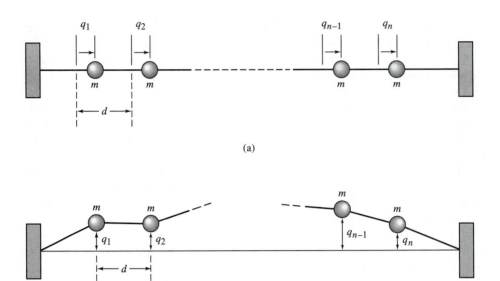

(a)

(b)

Figure 11.5.1   Linear array of vibrating particles or the loaded string. (a) Longitudinal motion. (b) Transverse motion.

For the case of transverse motion, the distance between particle k and $k + 1$ is

$$[d^2 + (q_{k+1} - q_k)^2]^{1/2} = d + \frac{1}{2d}(q_{k+1} - q_k)^2 + \cdots \qquad (11.5.4)$$

in which d is the equilibrium distance between two adjacent particles. The stretch of the section of string connecting the two particles is then approximately

$$\Delta l = \frac{1}{2d}(q_{k+1} - q_k)^2 \qquad (11.5.5)$$

Thus, if F is the force of tension in the string, the potential energy of the section under consideration is given by

$$F\Delta l = \frac{F}{2d}(q_{k+1} - q_k)^2 \qquad (11.5.6)$$

It follows that the total potential energy of the system in either the longitudinal or the transverse type of motion is expressible as a quadratic function of the form

$$V = \frac{K}{2}\left[q_1^2 + (q_2 - q_1)^2 + \cdots + (q_n - q_{n-1})^2 + q_n^2\right] \qquad (11.5.7)$$

in which

$$K = \frac{F}{d} = \frac{\text{tension}}{\text{separation}} \qquad transverse\ vibration$$

or

$$K = \text{elastic constant} \qquad \textit{longitudinal vibration}$$

The Lagrangian function for either case is, thus, given by

$$L = \tfrac{1}{2} \sum_k \left[m\dot{q}_k^2 - K(q_{k+1} - q_k)^2 \right] \tag{11.5.8}$$

The Lagrangian equations of motion

$$\frac{d}{dt}\frac{\partial L}{\partial \dot{q}_k} = \frac{\partial L}{\partial q_k} \tag{11.5.9}$$

then become

$$m\ddot{q}_k = -K(q_k - q_{k-1}) + K(q_{k+1} - q_k) \tag{11.5.10}$$

where $k = 1, 2, \ldots, n$.

To solve the preceding system of n equations, we use a trial solution in which the q's are assumed to vary harmonically with time:

$$q_k = a_k \cos \omega t \tag{11.5.11a}$$

where a_k is the amplitude of vibration of the kth particle. Substitution of the trial solution (Equation 11.5.11a) into the differential equations (Equations 11.5.10) yields the following recursion formula for the amplitudes:

$$-m\omega^2 a_k = K(a_{k-1} - 2a_k + a_{k+1}) \tag{11.5.11b}$$

This formula includes the endpoints of the string if we set

$$a_0 = a_{n+1} = 0 \tag{11.5.11c}$$

The secular determinant is, thus,

$$\begin{vmatrix} 2K - m\omega^2 & -K & 0 & \cdots & 0 \\ -K & 2K - m\omega^2 & -K & \cdots & 0 \\ 0 & -K & 2K - m\omega^2 & \cdots & 0 \\ \cdots & \cdots & \cdots & \cdots & \cdots \\ 0 & 0 & 0 & \cdots & 2K - m\omega^2 \end{vmatrix} = 0 \tag{11.5.12}$$

The determinant is of the nth order, and there are, thus, n values of ω that satisfy the equation. Rather than find these n roots by algebra, however it turns out that we can find them by working directly with the recursion relation (Equation 11.5.11b).

To this end we define a quantity ϕ related to the amplitudes a_k by the following equation:

$$a_k = A \sin k\phi \tag{11.5.13}$$

Direct substitution into the recursion formula (11.5.11b) then yields

$$-m\omega^2 A \sin(k\phi) = KA[\sin(k\phi - \phi) - 2\sin(k\phi) + \sin(k\phi + \phi)] \tag{11.5.14a}$$

which reduces to

$$m\omega^2 = K(2 - 2\cos\phi) = 4K\sin^2\frac{\phi}{2} \qquad (11.5.14b)$$

or

$$\omega = 2\omega_0 \sin\frac{\phi}{2} \qquad (11.5.14c)$$

in which

$$\omega_0 = \left(\frac{K}{m}\right)^{1/2} \qquad (11.5.14d)$$

Equation 11.5.14c gives the normal frequencies in terms of the quantity ϕ, which we have not, as yet, determined. Now, as a matter of fact, the same relation would have been obtained by any of the following substitutions for the amplitude a_k: $A\cos k\phi$, $Ae^{ik\phi}$, $Ae^{-ik\phi}$, or any linear combination of these. Only the substitution $a_k = A\sin(k\phi)$ satisfies the end condition $a_0 = 0$, however. To determine the actual value of the parameter ϕ, and, thus, find the normal frequencies of the vibrating string, we use the other end condition, namely, $a_{n+1} = 0$. This condition is met if we set

$$(n+1)\phi = N\pi \qquad (11.5.15)$$

in which N is an integer, because we then have

$$a_{n+1} = A\sin N\pi = 0 \qquad (11.5.16)$$

Having found ϕ, we can now calculate the normal frequencies. They are given by

$$\omega_N = 2\omega_0 \sin\left(\frac{N\pi}{2n+2}\right) \qquad (11.5.17)$$

Furthermore, from Equations 11.5.13 and 11.5.15 we see that the amplitudes for the normal modes are given by

$$a_k = A\sin\left(\frac{N\pi k}{n+1}\right) \qquad (11.5.18)$$

Here the value of $k = 1, 2, \ldots, n$ denotes a particular particle in the linear array, and the value of $N = 1, 2, \ldots, n$ refers to the normal mode in which the system is oscillating.

The different normal modes are illustrated graphically by plotting the amplitudes as given by Equation 11.5.18. These fall on a sine curve as shown in Figure 11.5.2, which shows the case of three particles $n = 3$. The actual motion of the system, when it is vibrating in a single pure mode, is given by the equation

$$q_k = a_k \cos\omega_N t = A\sin\left(\frac{\pi N k}{n+1}\right)\cos\omega_N t \qquad (11.5.19)$$

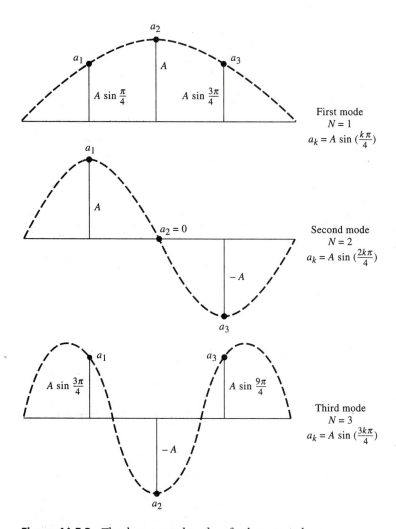

Figure 11.5.2 The three normal modes of a three-particle array.

The general type of motion is a linear combination of all the normal modes. This can be expressed as

$$q_k = \sum_{N=1}^{n} A_N \sin\left(\frac{N\pi k}{n+1}\right) \cos(\omega_N t - \epsilon_N) \tag{11.5.20}$$

in which the values of A_N and ϵ_N are determined from the initial conditions.

Suppose we look at the case where the number of masses on the string is very large. A real string is an aggregate of a very large number of very closely spaced atoms.

Let n increase, but at the same time, let the spacing d between neighboring particles decrease, such that the length of the string $L = (n + 1)d$ is held constant. Thus, for $N \ll n$, the argument $N\pi/[2(n + 1)]$ of the sine term in Equation 11.5.17 is small. So we have approximately

$$\omega_N = \omega_0 \left(\frac{N\pi}{n+1} \right) \tag{11.5.21}$$

but for the transverse oscillations we have

$$\omega_0 = \left(\frac{K}{m} \right)^{1/2} = \left(\frac{F}{md} \right)^{1/2} \tag{11.5.22}$$

and substituting this into Equation 11.5.21 gives

$$\omega_N \approx \left(\frac{F}{m/d} \right)^{1/2} \frac{N\pi}{(n+1)d} \tag{11.5.23a}$$

But $(n + 1)d = L$, the total length of the string, and m/d is its mass per unit length μ (the linear mass density). Thus, we have approximately

$$\omega_N = N \frac{\pi}{L} \left(\frac{F}{\mu} \right)^{1/2} \qquad (N = 1, 2, \ldots) \tag{11.5.23b}$$

In particular,

$$\omega_1 = \frac{\pi}{L} \left(\frac{F}{\mu} \right)^{1/2} \tag{11.5.23c}$$

and $\omega_N = N\omega_1$. The normal frequencies are integral multiples of the lowest, or fundamental, frequency ω_1. Remember, this is only an approximation, but for $N \ll n$ it is an exceedingly good one.

Let us now examine the displacements of the particles under these conditions. What might we guess? They ought to very closely approximate the vibration of a real string. For the Nth mode the displacement of the kth particle is given by Equation 11.5.19. Instead of denoting the particle by its k value, however, let us denote it by its distance down the string from the fixed end $x = kd$. Hence,

$$\frac{kN\pi}{(n+1)} = \frac{N\pi(kd)}{(n+1)d} = \frac{N\pi x}{L} \tag{11.5.24}$$

Replacing the argument of the sine term in Equation 11.5.19 by this factor permits us to rewrite that Equation as

$$q_N(x,t) = A \sin \left(\frac{N\pi x}{L} \right) \cos \omega_N t \qquad (N = 1, 2, \ldots)$$

$$= A \sin \left(\frac{2\pi x}{\lambda_N} \right) \cos 2\pi f_N t \tag{11.5.25a}$$

where we have defined the wavelength λ_N and the frequency f_N by

$$\lambda_N = \frac{2L}{N} \qquad f_N = \frac{\omega_N}{2\pi} \tag{11.5.25b}$$

The meaning of these terms applied to wave motion along a continuous medium is made more precise in the next section. Equation 11.5.25b expresses the displacement of any point along a continuous string when it is vibrating in its Nth mode. It represents a standing wave of wavelength λ_N. Each vibrational mode consists of an integral number of half-wavelength units constrained to fit within the length L such that the endpoints are nodes; that is, they do not vibrate. The meaning of the term's fundamental frequency—first, second, third, and so on harmonics of something like a vibrating violin string—should also be clear. No wonder the early Pythagoreans had such a high regard for integers.

In the next section, we treat the preceding situation directly as a continuous medium instead of one made up of a large number of discrete masses. We develop a differential "wave" equation governing the motion of the continuous medium, and we obtain standing wave solutions identical to the one shown earlier.

11.6 | Vibration of a Continuous System: The Wave Equation

Let us consider the motion of a linear array of connected particles in which the number of particles is made indefinitely large and the distance between adjacent particles indefinitely small. In other words, we have a continuous heavy cord or rod. To analyze this type of system it is convenient to rewrite the differential equations of motion of a finite system (Equation 11.5.10) in the following form:

$$m\ddot{q} = Kd\left[\left(\frac{q_{k+1} - q_k}{d}\right) - \left(\frac{q_k - q_{k-1}}{d}\right)\right] \tag{11.6.1}$$

in which d is the distance between the equilibrium positions of any two adjacent particles. Now if the variable x represents general distances in the longitudinal direction, and if the number n of particles is very large so that d is small compared with the total length, then we can write

$$\frac{q_{k+1} - q_k}{d} \approx \left(\frac{\partial q}{\partial x}\right)_{x=kd+d/2}$$

$$\frac{q_k - q_{k-1}}{d} \approx \left(\frac{\partial q}{\partial x}\right)_{x=kd-d/2} \tag{11.6.2}$$

Consequently, the difference between the two expressions in Equation 11.6.2 is equal to the second derivative multiplied by d, namely,

$$\frac{q_{k+1} - q_k}{d} - \frac{q_k - q_{k-1}}{d} \approx d\left(\frac{\partial^2 q}{\partial x^2}\right)_{x=kd} \tag{11.6.3}$$

The equation of motion can, therefore, be written

$$\frac{\partial^2 q}{\partial t^2} = \frac{Kd^2}{m}\frac{\partial^2 q}{\partial x^2} \qquad (11.6.4a)$$

or

$$\frac{\partial^2 q}{\partial t^2} = v^2 \frac{\partial^2 q}{\partial x^2} \qquad (11.6.4b)$$

in which we introduced the abbreviation

$$v^2 = \frac{Kd^2}{m} \qquad (11.6.4c)$$

Equation 11.6.4b is a well-known differential equation of mathematical physics. It is called the *one-dimensional wave equation*. It is encountered in many different places. Solutions of the wave equation represent traveling disturbances of some sort. It is easy to verify that a very general type of solution of the wave equation is given by

$$q = f(x + vt) \qquad (11.6.5a)$$

or

$$q = f(x - vt) \qquad (11.6.5b)$$

where f is any differentiable function of the argument $x \pm vt$. The first solution represents a disturbance that is propagating in the negative x direction with speed v, and the second equation represents a disturbance moving with speed v in the positive x direction. In our particular problem, the disturbance q is a *displacement* of a small portion of the system from its equilibrium configuration, (Figure 11.6.1). For the cord this displacement could be a kink that travels along the cord, and for a solid rod it could be a region of compression or of rarefaction moving along the length of the rod.

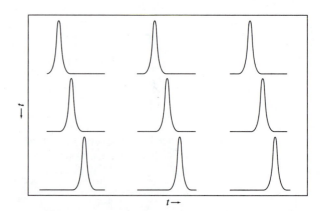

Figure 11.6.1   A sequence of pictures of a wave traveling to the right. The sequence was generated using *Mathematica's* Animate function.

Evaluation of the Wave Speed

In the preceding section we found that the constant K, for transverse motion of a loaded string, is equal to the ratio F/d, where F is the tension in the string. For the continuous string this ratio would, of course, become infinite as d approaches zero. If we introduce the linear density or mass per unit length μ, however, we have

$$\mu = \frac{m}{d} \qquad (11.6.6)$$

Consequently, the expression for v^2 (Equation 11.6.4c) can be written

$$v^2 = \frac{(F/d)d^2}{\mu d} = \frac{F}{\mu} \qquad (11.6.7a)$$

so that d cancels out. The speed of propagation for transverse waves in a continuous string is then

$$v = \left(\frac{F}{\mu}\right)^{1/2} \qquad (11.6.7b)$$

For the case of longitudinal vibrations, we introduce the elastic modulus Y, which is defined as the ratio of the force to the elongation per unit length. Thus, K, the stiffness of a small section of length d, is given by

$$K = \frac{Y}{d} \qquad (11.6.8)$$

Consequently, Equation 11.6.4c can be written as

$$v^2 = \frac{(Y/d)d^2}{\mu d} = \frac{Y}{\mu} \qquad (11.6.9a)$$

and again we see that d cancels out. Hence, the speed of propagation of longitudinal waves in an elastic rod is

$$v = \left(\frac{Y}{\mu}\right)^{1/2} \qquad (11.6.9b)$$

Sinusoidal Waves

In the study of wave motion, those particular solutions of the wave equation

$$\frac{\partial^2 q}{\partial t^2} = v^2 \frac{\partial^2 q}{\partial x^2} \qquad (11.6.10)$$

in which q is a sinusoidal function of x and t, namely,

$$q = A \frac{\sin}{\cos}\left[\frac{2\pi}{\lambda}(x + vt)\right] \qquad (11.6.11a)$$

$$q = A \frac{\sin}{\cos}\left[\frac{2\pi}{\lambda}(x - vt)\right] \qquad (11.6.11b)$$

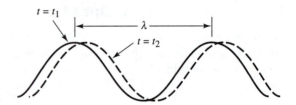

Figure 11.6.2   A traveling
sinusoidal wave.

are of fundamental importance. These solutions represent traveling disturbances, or
waves, in which the displacement, at a given point x, varies harmonically in time. The
amplitude of the wave is the constant A, and the frequency f is given by

$$f = \frac{\omega}{2\pi} = \frac{v}{\lambda} \tag{11.6.12}$$

Furthermore, at a given value of the time t, say $t = 0$, the displacement varies sinusoidally
with the distance x. The distance between two successive maxima, or minima, of the
displacement is the constant λ, called the *wavelength*. The waves represented by Equation
11.6.11a propagate in the negative x direction, and those represented by Equation 11.6.11b
propagate in the positive x direction, as shown in Figure 11.6.2. They are special cases of
the general type of solution mentioned earlier.

Standing Waves

Because the wave equation (Equation 11.6.4b) is linear, we can build up any number of
solutions by making linear combinations of known solutions. One possible linear combination of particular significance is obtained by adding two waves of equal amplitude
A that are traveling in opposite directions. In our notation such a solution is given by

$$q = A\sin\left[\frac{2\pi}{\lambda}(x + vt)\right] + A\sin\left[\frac{2\pi}{\lambda}(x - vt)\right] \tag{11.6.13}$$

By using the appropriate trigonometric identity and collecting terms, we find that the
equation reduces to

$$q = 2A\sin\left(\frac{2\pi}{\lambda}x\right)\cos\omega t \tag{11.6.14}$$

in which $\omega = 2\pi v/\lambda$. The amplitude of the resultant disturbance is $2A$. Note that this equation is identical to Equation 11.5.25a (only there the net amplitude was simply A), which
represents the motion of the loaded string in the limiting situation that the number of discrete masses approaches infinity, while their distance of separation approaches zero in such
a way that the total length of string remains constant. Again, Equation 11.6.14 represents
a *standing wave*. The amplitude of the displacement varies continuously with x. At $x = 0$,
$\lambda/2$, λ, $3\lambda/2$, . . . , the displacement of the string is always zero, because the sine term
vanishes at those points. These points of zero amplitude are the *nodes*. At $x = \lambda/4$, $3\lambda/4$,
$5\lambda/4$, ... , the amplitude of the vibrating string is a maximum. These points are the
antinodes. The distance between two successive nodes (or antinodes) is $\lambda/2$. These facts

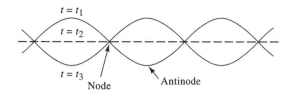

$t = t_1$

$t = t_2$

$t = t_3$

Node

Antinode

Figure 11.6.3 A standing sinusoidal wave.

are illustrated in Figure 11.6.3. Note again that there is a well-defined constraint on the values of allowable wavelengths λ. Because the endpoints of the string are fixed, we have as boundary conditions

$$q = 0 \qquad (x = 0, L) \qquad\qquad (11.6.15)$$

that our solution (Equation 11.6.14) must obey. The first condition at $x = 0$ is met automatically. The second boundary condition at $x = L$ is met if

$$L = N\left(\frac{\lambda}{2}\right) \qquad \lambda = \frac{2L}{N} \qquad\qquad (11.6.16)$$

An integral number of half wavelengths must fit within the length L if the endpoints are to be nodes. This is precisely the condition obtained previously for the normal modes of the loaded string.

Problems

11.1   A particle of mass m moves in one-dimensional motion with the following potential energy functions:

(a) $V(x) = \dfrac{k}{2}x^2 + \dfrac{k^2}{x}$

(b) $V(x) = kxe^{-bx}$

(c) $V(x) = k(x^4 - b^2x^2)$

where all constants are real and positive. Find the equilibrium positions for each case and determine their stability.

(d) Find the angular frequency ω for small oscillations about the respective positions of stable equilibrium for parts (a), (b), and (c), and find the period in seconds for each case if $m = 1$ g, and k and b are each of unit value in cgs units.

11.2   A particle moves in two dimensions under the potential energy function

$$V(x,y) = k(x^2 + y^2 - 2bx - 4by)$$

where k is a positive constant. Show that there is one position of equilibrium. Is it stable or unstable?

11.3   The potential energy function of a particle of mass m in one-dimensional motion is given by

$$V(x) = -\frac{k}{2}x^2$$

and so the force is of the antirestoring type

$$F(x) = kx$$

with $x = 0$ as a position of unstable equilibrium when k is a positive constant. If the initial conditions are $t = 0$, $x = x_0$, and $\dot{x} = 0$, show that the ensuing motion is given by an exponential "runaway"

$$x(t) = x_0(e^{\alpha t} + e^{-\alpha t})/2$$

where the constant $\alpha = \sqrt{k/m}$.

11.4 A light elastic cord of length $2l$ and stiffness k is held with the ends fixed a distance $2l$ apart in a horizontal position. A block of mass m is then suspended from the midpoint of the cord. Show that the potential energy of the system is given by the expression

$$V(y) = 2k[y^2 - 2l(y^2 + l^2)^{1/2}] - mgy$$

where y is the vertical sag of the center of the cord. From this show that the equilibrium position is given by a root of the equation

$$u^4 - 2au^3 + a^2u^2 - 2au + a^2 = 0$$

where $u = y/l$ and $a = mg/4kl$.

11.5 A uniform cubical block of mass m and sides $2a$ is balanced on top of a rough sphere of radius b. Show that the potential energy function can be expressed as

$$V(\theta) = mg[(a + b) \cos \theta + b\theta \sin \theta]$$

where θ is the angle of tilt. From this, show that the equilibrium at $\theta = 0$ is stable, or unstable, depending on whether a is less than or greater than b, respectively.

11.6 Expand the potential energy function of Problem 11.5 as a power series in θ. From this determine the stability for the case $a = b$.

11.7 A solid homogeneous hemisphere of radius a rests on top of a rough hemispherical cap of radius b, the curved faces being in contact. Show that the equilibrium is stable if a is less than $3b/5$.

11.8 Determine the frequency of vertical oscillations about the equilibrium position in Problem 11.4.

11.9 Determine the period of oscillation of the block in Problem 11.5.

11.10 Determine the period of oscillation of the hemisphere in Problem 11.7.

11.11 A small steel ball rolls back and forth about its equilibrium position in a rough spherical bowl. Show that the period of oscillation is $2\pi[7(b - a)/5g]^{1/2}$, where a is the radius of the ball and b is the radius of the bowl. Find the period in seconds if $b = 1$ m and $a = 1$ cm.

11.12 For an orbiting satellite in the form of a thin rod, show that the stable equilibrium attitude and period of oscillation are the same as those found in Example 11.2.2 for the dumbbell satellite.

11.13 In the system of two identical coupled oscillators shown in Figure 11.3.1, one oscillator is started with initial amplitude A_0, whereas the other is at rest at its equilibrium position, so that the initial conditions are

$$t = 0 \qquad x_1(0) = A_0 \qquad x_2(0) = 0 \qquad \dot{x}_1(0) = \dot{x}_2(0) = 0$$

Show that the amplitude of the symmetric component is equal to the amplitude of the antisymmetric component in this case and that the complete solution can be expressed as follows:

$$x_1(t) = \frac{1}{2}A_0(\cos\omega_a t + \cos\omega_b t) = A_0\cos\overline{\omega}t\ \cos\Delta t$$

$$x_2(t) = \frac{1}{2}A_0(\cos\omega_a t - \cos\omega_b t) = A_0\ \sin\overline{\omega}t\ \sin\Delta t$$

in which $\overline{\omega} = (\omega_a + \omega_b)/2$ and $\Delta = (\omega_b - \omega_a)/2$. Thus, if the coupling is very weak so that $K' \ll K$, then $\overline{\omega}$ will be very nearly equal to $\omega_a = (K/m)^{1/2}$, and Δ is very small. Consequently, under the stated initial conditions, the first oscillator eventually comes to rest while the second oscillator oscillates with amplitude A_0. Later, the system returns to the initial condition, and so on. Thus, the energy passes back and forth between the two oscillators indefinitely.

11.14 In Problem 11.13 show that, for weak coupling, the period at which the energy trades back and forth is approximately equal to $T_a(2K/K')$ where $T_a = 2\pi/\omega_a = 2\pi/(m/K)^{1/2}$ is the period of the symmetric oscillation.

11.15 Two identical simple pendula are coupled together by a very weak force of attraction that varies as the inverse square of the distance between the two particles. (This force might be the gravitational attraction between the two particles, for instance.) Show that, for small departures from the equilibrium configuration, the Lagrangian can be reduced to the same mathematical form, with appropriate constants, as that of the two identical coupled oscillators treated in Section 11.3 and in Problem 11.13. (*Hint: Consider Equation 11.3.9.*)

11.16 Find the normal frequencies of the coupled harmonic oscillator system (see Figure 11.3.1) for the general case in which the two particles have unequal mass and the springs have different stiffness. In particular, find the frequencies for the case $m_1 = m$, $m_2 = 2m$, $K_1 = K$, $K_2 = 2K$, $K' = 2K$. Express the result in terms of the quantity $\omega_0 = (K/m)^{1/2}$.

11.17 A light elastic spring of stiffness K is clamped at its upper end and supports a particle of mass m at its lower end. A second spring of stiffness K is fastened to the particle and, in turn, supports a particle of mass $2m$ at its lower end. Find the normal frequencies of the system for vertical oscillations about the equilibrium configuration. Find also the normal coordinates.

11.18 Consider the case of a double pendulum, Figure 11.3.7a, in which the two sections are of different length, the upper one being of length l_1 and the lower of length l_2. Both particles are of equal mass m. Find the normal frequencies of the system and the normal coordinates.

11.19 Set up the secular equation for the case of three coupled particles in a linear array and show that the normal frequencies are the same as those given by Equation 11.5.17.

11.20 A simple pendulum of mass m and length a is attached to a block of mass M that is constrained to slide along a frictionless, horizontal track as shown in Figure P11.20. Find the normal frequencies and normal modes of oscillation.

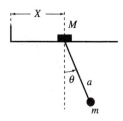

Figure P11.20

11.21 In Example 11.3.2
- **(a)** Find the normal modes of oscillation.
- **(b)** Relax the assumption that $2mg = kr$. Instead, let the support mass be M and the mass of the pendulum be m and assume that $\alpha = m/(m+M) \ll 1$. Now find the normal modes of oscillation and the normal mode frequencies.

11.22 Three beads of mass m, m, and $2m$ are constrained to slide along a frictionless, circular hoop. The two small masses are each connected to the large mass and to each other by springs of length a and force constants k and k', respectively. The masses are Shown in Figure P11.22 at their equilibrium positions, which are located at $120°$ angular separations. The largest mass is initially displaced $10°$ clockwise from its equilibrium position, and the other two are held fixed in place. The three masses are then simultaneously released from rest.
- **(a)** Find the normal frequencies and normal modes of oscillation.
- **(b)** Solve for the resulting motion of each mass.

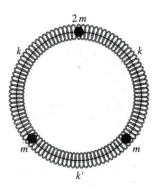

Figure P11.22

11.23 Find the matrix $\mathbf{A}$ that diagonalizes the $\mathbf{K}$ and $\mathbf{M}$ matrices in the case of the linear triatomic molecule of Example 11.4.1. Show that the ratio of their diagonal elements is equal to the eigenfrequencies of the normal modes of oscillation.

11.24 A triatomic molecule like hydrogen sulfide (H_2S) consists of two hydrogen atoms of mass m and one sulfur atom of mass M constrained by atomic bonding forces to assume the triangular configuration shown in Figure P11.24. Assume that the bonding forces can be approximated by springs whose force constant is k. When the three atoms are in their equilibrium configuration, the HS distance is $a = 1.67 \times 10^{-10}$ m and the H—S—H vertex angle is approximately $2\alpha = 90°$. Find the normal frequencies and normal modes of oscillation. Assume that the hydrogen atoms do not interact directly with each other.

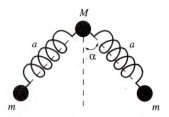

Figure P11.24

11.25 Two waves are traveling through a medium. Assume that the displacements from equilibrium of particles that make up the medium are given by the functions

$$q_1(x, t) = Ae^{i(\omega t - kx)}$$
$$q_2(x, t) = Ae^{i(\Omega t - Kx)}$$

whose real part represents the physical wave.
(a) Show that each of these functions are solutions of the wave equation.
(b) Assume that the frequencies and wave numbers differ by small amounts

$$\Omega = \omega + \Delta\omega \qquad K = k + \Delta k$$

Ignoring small differences of second order, show that the real part of the resultant wave function is given approximately by

$$Q(x,t) = q_1 + q_2 \approx 2\cos\left[\frac{(\Delta\omega)t - (\Delta k)x}{2}\right]\cos(\omega t - kx)$$

The resultant wave has the same frequency and wave number as the original wave, but it has a modulated amplitude (the wave number $k = 2\pi/\lambda$).
(c) Calculate the speed of propagation of the amplitude modulation (this speed is called the group speed v_g of the wave).

11.26 Illustrate the normal modes for the case of four particles in a linear array. Find the numerical values of the ratios of the second, third, and fourth normal frequencies to the lowest or first normal frequency.

11.27 A light elastic cord of natural length l and stiffness K is stretched out to a length $l + \Delta l$ and loaded with a number n of particles evenly spaced along its length. If m is the total mass of all n particles, find the speed of transverse and of longitudinal waves in the cord.

11.28 Work Problem 11.27 for the case in which, instead of being loaded, the cord is heavy with linear mass density μ.

11.29 In section 3.9, we showed how Fourier series could be used to represent a periodic function. Here we wish to apply that analysis to a vibrating string. Assume that a string of length l of mass per unit length μ is stretched horizontally between two supports and held with tension F_0. Assume that the middle of the string is displaced a distance a (where a $\ll l$) in the vertical direction. When the string is released, it vibrates in a standing wave pattern. Use Fourier analysis to calculate the pattern of vibration as a function of time (Figure P 11.29), that is calculate the Fourier coefficients of the series needed to describe the motion.

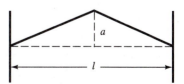

Figure P11.29

11.30 Find the solution to Problem 11.29 in terms of traveling waves.

Computer Problems

C 11.1 Consider a single pulse traveling down an infinitely long string. Assume that at $t = 0$, the shape of the pulse, or the vertical displacement of the string, is

$$y(x) = \frac{1}{1+x^2} \tag{1}$$

Analogous to the discussion of Fourier series in Section 3.9, this pulse can be thought of as a superposition of harmonic waves of differing wave numbers k. The infinite sum of Section 3.9, however, that approximates a repetitive function needs to be replaced here by an integral over an infinite number of harmonic waves, each one weighted by an appropriate amplitude function, that is,

$$y(x) = \int_0^\infty a(k) \cos(kx)\, dk \tag{2}$$

We use cosine functions because $y(x)$ is an even function of x. The amplitude function $a(k)$ is given by

$$a(k) = \frac{2}{\pi} \int_0^\infty y(x)\cos(kx)\, dx \tag{3}$$

- (a) Calculate $a(k)$ using Equation 3.
- (b) Substitute $a(k)$ into Equation 2, and show that it yields $y(x)$.
- (c) Integrate Equation 2 numerically for values of x ranging from 0 to 3, and show that the results agree with the exact values of $y(x)$. Assume that the speed of the pulse is given by $v = \omega/k = 1$. In such a case, the shape of the pulse is preserved as it travels down the string.
- (d) Write down an exact expression for $y(x,t)$ assuming that at $t = 0$, $y(x,0)$ is given by Equation 1.
- (e) Write down the appropriate integral expression for $y(x,t)$ using Equation 2.
- (f) Now assume that the pulse is traveling down a "dispersive" string, for which the wave velocity is not a constant but depends on the wave number of the wave. Assume that $\omega/k = 1 + 0.25\, k^2$. The many waves of differing k that are superimposed to make the traveling pulse change their phase relationship as each moves down the string. Thus, the shape of the pulse changes. To see this effect, modify the integral expression for $y(x,t)$ obtained in part (e) using the "dispersive" value of ω/k given above. Numerically integrate the resulting expression to obtain $y(x,t)$ for $t = 2.5, 5.0$, and 10.0 s. Pick a broad range of x about the location of the peak of the pulse at each of these times.
- (g) Plot these resultant waveforms, and compare them with $y(x,0)$. Comment on the result.

Appendix A

Units

Basic SI (Système International) Units

Unit	Symbol	Physical Quantity
meter	m	length
kilogram	kg	mass
second	s	time
ampere	A	electric current
Kelvin	K	temperature
mole	mol	amount of substance
candela	cd	luminous intensity

Derived SI Units (not a complete list)

Unit	Symbol	Physical Quantity	Equivalent
newton	N	force	$kg \cdot m/s^2$
joule	J	work or energy	$N \cdot m$
watt	W	power	J/s
pascal	Pa	pressure	N/m^2
volt	V	electric potential difference	W/A
couloumb	C	electric charge	$A \cdot s$
hertz	Hz	frequency	s^{-1}

SI Units of Other Physical Quantities

Physical Quantity	SI Unit
speed	m/s
acceleration	m/s^2
angular speed	$(rad)\, s^{-1}$
angular acceleration	$(rad)\, s^{-2}$
torque	$kg\, m^2/s^2$

Non-SI Units Converted to SI Units

Physical Quantity	SI Unit
Energy	
1 eV (electron volt)	1.6022×10^{-19} J
1 erg	10^{-7} J
1 BTU (British Thermal Unit)	1055 J
1 cal (calorie)	4.186 J
1 KWH (kilowatt-hour)	3.6×10^{6} J
Mass	
1 g (gram)	10^{-3} kg
1 u (atomic mass unit)	1.661×10^{-27} kg
1 eV/c^2	1.783×10^{-36} kg
1 lb (pound mass)	0.4536 kg
Force	
1 dyne	10^{-5} kg m/s^2
1 lb (pound force)	4.448 kg m/s^2
Length	
1 cm (centimeter)	10^{-2} m
1 km (kilometer)	10^{3} m
1 in (inch)	0.0254 m
1 ft (foot)	0.3048 m
1 yd (yard)	0.9144 m
1 mi (mile)	1609.3 m
1 AU (astronomical unit)	1.496×10^{11} m
1 ly (light-year)	9.46×10^{15} m
1 pc (parsec)	3.09×10^{16} m
Volume	
1 L (liter)	10^{-3} m^3
1 qt (quart)	0.9463×10^{-3} m^3
1 gal (gallon)	3.785×10^{-3} m^3
1 ft^3 (cubic foot)	0.02832 m^3
Angle	
1° (degree)	1.745×10^{-2} rad
1′ (arcminute)	2.909×10^{-4} rad
1″ (arcsecond)	4.848×10^{-6} rad
Time	
1 yr (year)	3.156×10^{7} s
1 d (day)	8.640×10^{4} s
1 hr (hour)	3600 s
1 min (minute)	60 s

Non-SI Units Converted to SI Units (Continued)

Physical Quantity	SI Unit
Power	
1 KW (kilowatt)	10^3 W
1 hp (horsepower)	745.7 W
Speed	
1 ft/s (foot per second)	0.3048 m/s
1 mph (mile per hour)	0.447 m/s

Prefixes for Multiplication by a Power of Ten

Name	Symbol	Factor	Name	Symbol	Factor
hecto	h	10^2	centi	c	10^{-2}
kilo	k	10^3	milli	m	10^{-3}
mega	M	10^6	micro	μ	10^{-6}
giga	G	10^9	nano	n	10^{-9}
tera	T	10^{12}	pico	p	10^{-12}
peta	P	10^{15}	femto	f	10^{-15}
exa	E	10^{18}	atto	a	10^{-18}
zetta	Z	10^{21}	zepto	z	10^{-21}

Appendix B

Complex Numbers

The quantity

$$z = x + iy$$

is said to be a *complex number* if x and y are real and $i = \ z =$. The *complex conjugate* is defined as

$$z^* = x - iy$$

The *absolute value* $|z|$ is given by

$$|z|^2 = zz^* = x^2 + y^2$$

The following are true

$$z + z^* = 2x = 2 \operatorname{Re} z$$
$$z - z^* = 2y = 2 \operatorname{Im} z$$

Exponential Notation

$$z = x + iy = |z| \, e^{i\theta} = |z| \, (\cos\theta + i \sin\theta)$$
$$z^* = x - iy = |z| \, e^{-i\theta} = |z| \, (\cos\theta - i \sin\theta)$$

where

$$\tan\theta = \frac{y}{x}$$

(For a proof of the relation $e^{i\theta} = \cos\theta + i\sin\theta$ see under Series Expansions in Appendix D.)

Circular and Hyperbolic Functions

The following relations are often useful

$$\cos\theta = \frac{e^{i\theta} + e^{-i\theta}}{2}$$

$$\sin\theta = \frac{e^{i\theta} - e^{-i\theta}}{2i}$$

$$\cosh\theta = \frac{e^{\theta} + e^{-\theta}}{2} \qquad \text{(hyperbolic cosine)}$$

$$\sinh\theta = \frac{e^{\theta} - e^{-\theta}}{2} \qquad \text{(hyperbolic sine)}$$

$$\tanh\theta = \frac{\sinh\theta}{\cosh\theta} = \frac{e^{\theta} - e^{-\theta}}{e^{\theta} + e^{-\theta}} \qquad \text{(hyperbolic tangent)}$$

Relations Between Circular and Hyperbolic Functions

$$\sin i\theta = i\sinh\theta$$
$$\cos i\theta = \cosh\theta$$
$$\sinh i\theta = i\sin\theta$$
$$\cosh i\theta = \cos\theta$$

Derivatives

$$\frac{d}{d\theta}\sin\theta = \cos\theta \qquad\qquad \frac{d}{d\theta}\sinh\theta = \cosh\theta$$

$$\frac{d}{d\theta}\cos\theta = -\sin\theta \qquad\qquad \frac{d}{d\theta}\cosh\theta = \sinh\theta$$

Trigonometric Identities

$$\cos^2\theta + \sin^2\theta = 1$$

$$1 + \tan^2\theta = \sec^2\theta$$

$$1 + \cot^2\theta = \csc^2\theta$$

$$\sin(\theta \pm \phi) = \sin\theta\cos\phi \pm \cos\theta\sin\phi$$

$$\cos(\theta \pm \phi) = \cos\theta\cos\phi \mp \sin\theta\sin\phi$$

$$\tan(\theta \pm \phi) = \frac{\tan\theta \pm \tan\phi}{1 \mp \tan\theta\tan\phi}$$

$$\sin 2\theta = 2\sin\theta\cos\theta$$

$$\cos 2\theta = \cos^2\theta - \sin^2\theta$$

$$\tan 2\theta = \frac{2\tan\theta}{1 - \tan^2\theta}$$

$$\sin^2\frac{\theta}{2} = \tfrac{1}{2}(1 - \cos\theta)$$

$$\cos^2\frac{\theta}{2} = \tfrac{1}{2}(1 + \cos\theta)$$

$$\tan^2\frac{\theta}{2} = \frac{1 - \cos\theta}{1 + \cos\theta}$$

$$\sin\theta + \sin\phi = 2\sin\left(\frac{\theta + \phi}{2}\right)\cos\left(\frac{\theta - \phi}{2}\right)$$

$$\cos\theta + \cos\phi = 2\cos\left(\frac{\theta + \phi}{2}\right)\cos\left(\frac{\theta - \phi}{2}\right)$$

$$\tan\theta \pm \tan\phi = \frac{\sin(\theta \pm \phi)}{\cos\theta\,\cos\phi}$$

Hyperbolic Identities

$$\cosh^2\theta - \sinh^2\theta = 1$$

$$\tanh^2\theta + \operatorname{sech}^2\theta = 1$$

$$\coth^2\theta - \operatorname{csch}^2\theta = 1$$

$$\sinh(\theta \pm \phi) = \sin\theta\cos\phi \pm \cosh\theta\sinh\phi$$

$$\cosh(\theta \pm \phi) = \cosh\theta\cos\phi \pm \sinh\theta\sinh\phi$$

$$\tanh(\theta \pm \phi) = \frac{\tanh\theta \pm \tanh\phi}{1 \pm \tanh\theta\tanh\phi}$$

$$\sinh 2\theta = 2\sinh\theta\cosh\theta$$

$$\cosh 2\theta = \cosh^2\theta + \sinh^2\theta$$

$$\tanh 2\theta = \frac{2\tanh\theta}{1 + \tanh^2\theta}$$

$$\sinh^2\frac{\theta}{2} = \tfrac{1}{2}(\cosh\theta - 1)$$

$$\cosh^2\frac{\theta}{2} = \tfrac{1}{2}(\cosh\theta + 1)$$

$$\tanh^2\frac{\theta}{2} = \frac{\cosh\theta - 1}{\cosh\theta + 1}$$

$$\sinh\theta + \sinh\phi = 2\sinh\left(\frac{\theta + \phi}{2}\right)\cosh\left(\frac{\theta - \phi}{2}\right)$$

$$\cosh\theta + \cos h\phi = 2\cosh\left(\frac{\theta + \phi}{2}\right)\cosh\left(\frac{\theta - \phi}{2}\right)$$

$$\tanh\theta + \tanh\phi = \frac{\sinh(\theta + \phi)}{\cosh\theta\,\cosh\phi}$$

Appendix C

Conic Sections

A *conic section* is a curve that is the locus of a point that moves in such a way that the ratio of its distance from a fixed point to its distance from a fixed line is a constant. This ratio is called the *eccentricity*, the fixed point is the *focus*, and the fixed line is the *directrix* of the curve. An example of a conic section is shown in Figure C.1a. The four possible conic sections, parameterized by their eccentricity, are shown in Figure C.1b

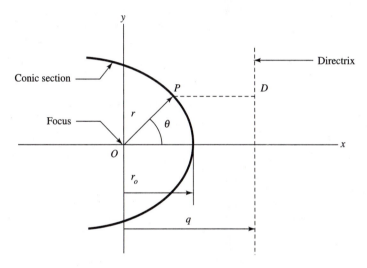

Figure C.1a   A conic section.

General Equation of a Conic Section (Figure C.1a) *where the focus is at the origin O:*

Cartesian coordinates: $(1 - \varepsilon^2)\, x^2 + 2\varepsilon^2 qx + y^2 = \varepsilon^2 q^2$

Polar Coordinates: $r = \dfrac{\varepsilon q}{1 + \varepsilon \cos \theta}$ or $\dfrac{1}{r} = \dfrac{1 + \varepsilon \cos \theta}{r_0(1 + \varepsilon)}$

where $\varepsilon = \dfrac{\overline{OP}}{\overline{PD}}$ and $r_0 = \dfrac{\varepsilon q}{1 + \varepsilon}$

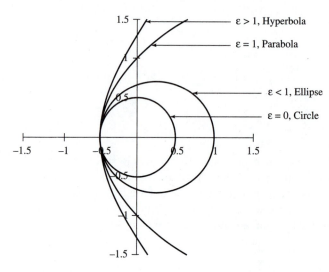

Figure C.1b   The four conic sections.

Equation of a Circle

For a circle: $\varepsilon = 0$, $q = \infty$, and $\varepsilon q = R$

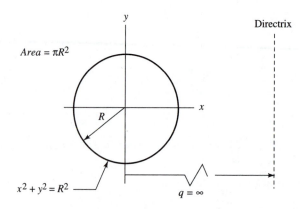

(**Note:** In the figures of the curves that follow (the ellipse, parabola, and hyperbola), the origin of the coordinate system does not coincide with the focus. The equation given above for these curves in polar coordinates applies only when the focus and origin of the coordinate system do coincide.)

Equation of an Ellipse

For an ellipse: $a^2 - b^2 = c^2$, $\varepsilon = c/a$, $b = a\sqrt{1 - \varepsilon^2}$, $q = \pm b^2/c$

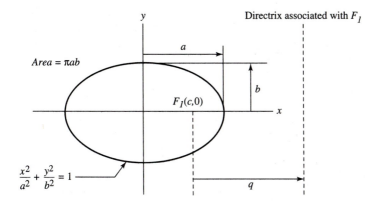

(**Note:**
(a) Only one focus, F_1, located at $(c,0)$ is shown. There is a second, symmetrically disposed focus F_2, located at $(-c,0)$ with its own symmetrically disposed directrix.
(b) If the origin of the coordinate system is positioned to coincide with F_1, then the polar equation of the ellipse may be used to relate the semi-major axis, a, to the eccentricity, ε:

$$a = \frac{r_0}{1 - \varepsilon} \quad \text{where } r_0 \text{ is the minimum distance of } r \text{ (see Figure C.1a).)}$$

Equation of a Parabola

For a parabola: $\varepsilon = 1$, $q = -2c$

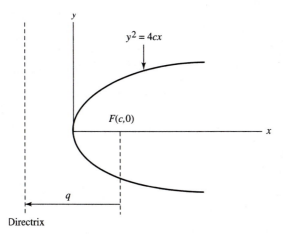

Equation of a Hyperbola

For a hyperbola: $a^2 + b^2 = c^2$, $\varepsilon = c/a$, $b = a\sqrt{\varepsilon^2 - 1}$, $q = \pm b^2/c$

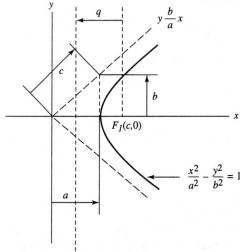

Directrix associated with F_1

(**Note:** Only one focus, F_1, located at $(c, 0)$ is shown. There is a second, symmetrically disposed focus F_2, located at $(-c, 0)$ with its own symmetrically disposed directrix.)

Appendix D

Series Expansions

Taylor's Series

$$f(x+a) = f(a) + xf'(a) + \frac{x^2}{2!}f''(a) + \cdots + \frac{x^n}{n!}f^n(a) + \cdots$$

$$f(x) = f(0) + xf'(0) + \frac{x^2}{2!}f''(0) + \cdots + \frac{x^n}{n!}f^n(0) + \cdots$$

where

$$f^n(a) = \left. \frac{d^n}{dx^n}f(x)\right|_{x=a}$$

Often-Used Expansions

$$e^x = 1 + x + \frac{x^2}{2!} + \cdots + \frac{x^n}{n!} + \cdots$$

$$\sin x = x - \frac{x^3}{3!} + \frac{x^5}{5!} - \cdots$$

$$\cos x = 1 - \frac{x^2}{2!} + \frac{x^4}{4!} - \cdots$$

$$\sinh x = x + \frac{x^3}{3!} + \frac{x^5}{5!} + \cdots$$

$$\cosh x = 1 + \frac{x^2}{2!} + \frac{x^4}{4!} + \cdots$$

$$\ln(1+x) = x - \frac{x^2}{2} + \frac{x^3}{3} - \cdots \qquad |x| < 1$$

$$\tan x = x + \frac{x^3}{3} + \frac{2}{15}x^5 + \cdots \qquad |x| < \frac{\pi}{2}$$

Complex Exponential

Setting $x = i\theta$ in the expansion for e^x gives

$$e^{i\theta} = 1 + i\theta + \frac{i^2\theta^2}{2!} + \frac{i^3\theta^3}{3!} + \cdots + \frac{i^n\theta^n}{n!} + \cdots$$

Because $i = \sqrt{-1}$

$$i^n = \begin{array}{l} +1: n = 0, 4, \ldots \\ -1: n = 2, 6, \ldots \\ +i: n = 1, 5, \ldots \\ -i: n = 3, 7, \ldots \end{array}$$

then

$$e^{i\theta} = \left(1 - \frac{\theta^2}{2!} + \frac{\theta^4}{4!} - \cdots\right) + i\left(\theta - \frac{\theta^3}{3!} + \frac{\theta^5}{5!} - \cdots\right)$$

$$= \cos\theta + i\sin\theta$$

from the series for the cosine and sine.

Binomial Series

$$(a + x)^n = a^n + na^{n-1}x + \frac{n(n-1)}{2!}a^{n-2}x^2 + \cdots + \binom{n}{m}a^{n-m}x^m + \cdots$$

where the binomial coefficient is

$$\binom{n}{m} = \frac{n!}{(n-m)!m!}$$

The series converges for $|x/a| < 1$.

Useful Approximations

For small x, the following approximations are often used

$$e^x \approx 1 + x$$

$$\sin x \approx x$$

$$\cos x \approx 1 - \frac{1}{2}x^2$$

$$\sqrt{1 + x} \approx 1 + \frac{1}{2}x$$

$$\frac{1}{1+x} \approx 1 - x$$

$$\frac{1}{1-x} \approx 1 + x$$

The last three are based on the binomial series, and the list can be extended for other values of the exponent:

$$(1 + x)^n = 1 + nx + \frac{1}{2}n(n-1)x^2 + \cdots$$

Appendix E

Special Functions

Elliptic Integrals

The elliptic integral of the *first kind* is given by the expressions

$$F(k, \phi) = \int_0^\phi \frac{d\phi}{(1-k^2 \sin^2 \phi)^{1/2}}$$

$$= \int_0^x \frac{dx}{(1-x^2)^{1/2} (1-k^2x^2)^{1/2}}$$

and the elliptic integral of the *second kind* by

$$E(k, \phi) = \int_0^\phi (1-k^2 \sin^2\phi)^{1/2} \, d\phi$$

$$= \int_0^x \frac{(1-k^2x^2)^{1/2}}{(1-x^2)^{1/2}} \, dx$$

Both converge for $|k| < 1$. They are called *incomplete* if $x = \sin\phi < 1$, and *complete* if $x = \sin\phi = 1$. The complete integrals have the following series expansions:

$$F(k) = F\left(k, \frac{\pi}{2}\right) = \frac{\pi}{2}\left(1 + \frac{k^2}{4} + \frac{9}{64}k^4 + \cdots\right)$$

$$E(k) = E\left(k, \frac{\pi}{2}\right) = \frac{\pi}{2}\left(1 - \frac{k^2}{4} - \frac{9}{64}k^4 - \cdots\right)$$

Gamma Function

The gamma function is defined as

$$\Gamma(n) = \int_0^\infty x^{n-1}e^{-x} \, dx$$

For any value of n

$$n\Gamma(n) = \Gamma(n+1)$$

If n is a positive integer

$$\Gamma(n) = (n-1)!$$

Special values

$$\Gamma\left(\tfrac{1}{2}\right) = \sqrt{\pi}$$

$$\Gamma(1) = 1$$

$$\Gamma\left(\tfrac{3}{2}\right) = \tfrac{1}{2}\sqrt{\pi}$$

$$\Gamma(2) = 1$$

Integrals expressible in terms of gamma functions

$$\int_0^1 \frac{dx}{\sqrt{1-x^n}} = \frac{\sqrt{\pi}}{n}\frac{\Gamma(1/n)}{\Gamma\left[\left(\tfrac{1}{2}\right) + (1/n)\right]}$$

$$\int_0^1 (1-x^2)^n x^m \, dx = \frac{\Gamma(n+1)\Gamma[(m+1)/2]}{2\Gamma[(2n+m+3)/2]}$$

Appendix F

Curvilinear Coordinates

We consider a general orthogonal system of coordinates u, v, and w with unit vectors $\mathbf{e}_1$, $\mathbf{e}_2$, and $\mathbf{e}_3$. The volume element is

$$dV = h_1 h_2 h_3 \, du \, dv \, dw$$

and the line element is

$$d\mathbf{r} = \mathbf{e}_1 h_1 \, du + \mathbf{e}_2 h_2 \, dv + \mathbf{e}_3 h_3 \, dw$$

The gradient, divergence, and curl are as follows:

$$\nabla f = \operatorname{grad} f = \frac{\mathbf{e}_1}{h_1} \frac{\partial f}{\partial u} + \frac{\mathbf{e}_2}{h_2} \frac{\partial f}{\partial v} + \frac{\mathbf{e}_3}{h_3} \frac{\partial f}{\partial w}$$

$$\nabla \cdot \mathbf{Q} = \operatorname{div} \mathbf{Q} = \frac{1}{h_1 h_2 h_3} \left[\frac{\partial}{\partial u}(h_2 h_3 Q_1) + \frac{\partial}{\partial v}(h_3 h_1 Q_2) + \frac{\partial}{\partial w}(h_1 h_2 Q_3) \right]$$

$$\nabla \times \mathbf{Q} = \operatorname{curl} \mathbf{Q} = \frac{1}{h_1 h_2 h_3} \begin{vmatrix} h_1 \mathbf{e}_1 & h_2 \mathbf{e}_2 & h_3 \mathbf{e}_3 \\ \dfrac{\partial}{\partial u} & \dfrac{\partial}{\partial v} & \dfrac{\partial}{\partial w} \\ h_1 Q_1 & h_2 Q_2 & h_3 Q_3 \end{vmatrix}$$

The h functions for some common coordinate systems are listed as follows.

Rectangular Coordinates: *x, y, z*

$$h_x = 1 \qquad h_y = 1 \qquad h_z = 1$$

Cylindrical Coordinates: *R, φ, z*

$$x = R \cos \phi \qquad y = R \sin \phi$$

$$h_R = 1 \qquad h_\phi = R \qquad h_z = 1$$

Spherical Coordinates: *r, θ, φ*

$$x = r \sin \theta \cos \phi \qquad y = r \sin \theta \sin \phi \qquad z = r \cos \theta$$

$$h_r = 1 \qquad h_\theta = r \qquad h_\phi = r \sin \theta$$

Parabolic Coordinates: *u, v, θ*

$$x = uv \cos \theta \qquad y = uv \sin \theta \qquad z = \tfrac{1}{2}(u^2 - v^2)$$

$$h_u = h_v = \sqrt{u^2 + v^2} \qquad h_\theta = uv$$

Example: The curl in spherical coordinates is

$$\text{curl } \mathbf{Q} = \begin{vmatrix} \mathbf{e}_r & r\mathbf{e}_\theta & r\sin\theta\,\mathbf{e}_\phi \\ \dfrac{\partial}{\partial r} & \dfrac{\partial}{\partial \theta} & \dfrac{\partial}{\partial \phi} \\ Q_r & rQ_\theta & r\sin\theta\,Q_\phi \end{vmatrix} \dfrac{1}{r^2 \sin\theta}$$

Appendix G

Fourier Series

To find the coefficients of the terms in the trigonometric expansion

$$f(t) = \frac{a_0}{2} + \sum_{n=1}^{\infty} [a_n \cos(n\omega t) + b_n \sin(n\omega t)]$$

multiply both sides of the equation by $\cos(n'\omega t)$ and integrate over the interval $-\pi/\omega$ to $+\pi/\omega$:

$$\int_{-\pi/\omega}^{\pi/\omega} f(t) \cos(n'\omega t)\, dt = \frac{a_0}{2} \int_{-\pi/\omega}^{\pi/\omega} \cos(n'\omega t)\, dt + \sum_{n=1}^{\infty} \left[a_n \int_{-\pi/\omega}^{\pi/\omega} \cos(n'\omega t) \cos(n\omega t)\, dt + \right.$$

$$\left. + b_n \int_{-\pi/\omega}^{\pi/\omega} \cos(n'\omega t) \sin(n\omega t)\, dt \right]$$

Now if n' and n are integers, we have the general formulas

$$\int_{-\pi/\omega}^{\pi/\omega} \cos(n'\omega t)\, dt = 2\pi/\omega \qquad n' = 0$$

$$= 0 \qquad n' \neq 0$$

$$\int_{-\pi/\omega}^{\pi/\omega} \cos(n'\omega t) \cos(n\omega t)\, dt = \pi/\omega \qquad n' = n$$

$$= 0 \qquad n' \neq n$$

$$\int_{-\pi/\omega}^{\pi/\omega} \cos(n'\omega t) \sin(n\omega t)\, dt = 0 \qquad \text{for all } n' \text{ and } n$$

Thus, for a given n', all of the definite integrals in the summation vanish except the one for which $n' = n$. Consequently we can write

$$a_n = \frac{\omega}{\pi} \int_{-\pi/\omega}^{\pi/\omega} f(t) \cos(n\omega t)\, dt \qquad \text{for } n = 0,1,2,\ldots$$

Similarly, if the equation for $f(t)$ is multiplied by $\sin(n'\omega t)$ and integrated term by term, we use the general formula

$$\int_{-\pi/\omega}^{\pi/\omega} \sin(n'\omega t) \sin(n\omega t)\, dt = \pi/\omega \qquad n' = n$$

$$= 0 \qquad n' \neq n$$

in addition to the ones shown previously. As before, all of the definite integrals vanish except for $n' = n$, and so we get

$$b_n = \frac{\omega}{\pi} \int_{-\pi/\omega}^{\pi/\omega} f(t) \sin(n\omega t) \, dt \qquad n = 1, 2, \ldots$$

Because the period $T = 2\pi/\omega$, the limits of integration can also be expressed as $-T/2$ to $T/2$. For more detailed information concerning continuity conditions, integrability, and so on, the reader should consult a text on Fourier series, such as R. V. Churchill, *Fourier Series and Boundary Value Problems*, McGraw-Hill, New York, 1963.

Appendix H

Matrices

A *matrix* $\mathbf{A}$ is an array of elements a_{ij} arranged thusly

$$\mathbf{A} = \begin{pmatrix} a_{11} & a_{12} & \cdots & a_{1j} & \cdots & a_{1m} \\ a_{21} & a_{22} & \cdots & a_{2j} & \cdots & a_{2m} \\ \vdots & \vdots & & \vdots & & \vdots \\ a_{i1} & a_{i2} & \cdots & a_{ij} & \cdots & a_{im} \\ \vdots & \vdots & & \vdots & & \vdots \\ a_{n1} & a_{n2} & \cdots & a_{nj} & \cdots & a_{nm} \end{pmatrix}$$

If $n = m$, it is called a *square* matrix. Unless stated otherwise, we consider only square matrices in this appendix. A *symmetric* matrix is one such that $a_{ij} = a_{ji}$. If $a_{ij} = -a_{ji}$, it is *antisymmetric*.

The sum of two matrices is defined as

$$(\mathbf{A} + \mathbf{B})_{ij} = a_{ij} + b_{ij}$$

The product of two matrices is defined as

$$(\mathbf{A}\,\mathbf{B})_{ij} = a_{i1}b_{1j} + a_{i2}b_{2j} + \cdots = \sum_{k} a_{ik}b_{kj}$$

The product $\mathbf{AB}$ is not, in general, equal to $\mathbf{BA}$. If $\mathbf{AB} = \mathbf{BA}$, the two matrices are said to *commute*. A *diagonal matrix* is one whose nondiagonal elements are zero, $a_{ij} = 0$ for $i \neq j$. The *identity* matrix[1] is a diagonal matrix with all diagonal elements equal to unity,

$$\mathbf{1} = \begin{pmatrix} 1 & 0 & 0 & \cdots & 0 \\ 0 & 1 & 0 & \cdots & 0 \\ 0 & 0 & 1 & \cdots & 0 \\ \cdot & \cdot & \cdot & \cdots & \cdot \\ 0 & 0 & 0 & \cdots & 1 \end{pmatrix}$$

From the definition of the product, it is easily shown that

$$\mathbf{A1} = \mathbf{1A}$$

The *inverse* $\mathbf{A}^{-1}$ of a matrix $\mathbf{A}$ is defined by

$$\mathbf{AA}^{-1} = \mathbf{1} = \mathbf{A}^{-1}\mathbf{A}$$

---

[1] This should not be confused with the inertia tensor defined in Chapter 9.

The *transpose* $\tilde{\mathbf{A}}$ of a matrix $\mathbf{A}$ is defined as

$$(\tilde{\mathbf{A}})_{ij} = (\mathbf{A})_{ji}$$

For two matrices $\mathbf{A}$ and $\mathbf{B}$, $(\widetilde{\mathbf{AB}}) = \tilde{\mathbf{B}}\tilde{\mathbf{A}}$.

The determinant of a matrix is the determinant of its elements,

$$\det \mathbf{A} = \begin{vmatrix} a_{11} & a_{12} & \cdots \\ a_{21} & a_{22} & \cdots \\ \cdot & \cdot & \cdots \end{vmatrix}$$

The determinant of the product of two matrices is equal to the product of the respective determinants,

$$\det \mathbf{AB} = \det \mathbf{A} \det \mathbf{B}$$

It can be shown that the inverse of a matrix $\mathbf{A}$ is given by the formula

$$\mathbf{A}^{-1} = \begin{pmatrix} \dfrac{\det \mathbf{A}_{11}}{\det \mathbf{A}} & \dfrac{\det \mathbf{A}_{21}}{\det \mathbf{A}} & \cdots \\[2ex] \dfrac{\det \mathbf{A}_{12}}{\det \mathbf{A}} & \dfrac{\det \mathbf{A}_{22}}{\det \mathbf{A}} & \cdots \\[2ex] \cdots & \cdots & \cdots \end{pmatrix}$$

where the matrix $\mathbf{A}_{ij}$ is the matrix left after the ith row and jth column have been removed from the matrix $\mathbf{A}$.

Matrix Representation of Vectors

A matrix with one row, or one column, defines a *row vector,* or *column vector,* respectively. If $\mathbf{a}$ is a column vector, then $\tilde{\mathbf{a}}$ is the corresponding row vector,

$$\mathbf{a} = \begin{pmatrix} a_1 \\ a_2 \\ \vdots \\ a_n \end{pmatrix} \qquad \tilde{\mathbf{a}} = (a_1, a_2, \ldots, a_n)$$

For two column vectors $\mathbf{a}$ and $\mathbf{b}$ with the same number of elements, the product $\tilde{\mathbf{a}}\mathbf{b}$ is a scalar, analogous to the dot product,

$$\tilde{\mathbf{a}}\mathbf{b} = (a_1, a_2, \ldots) \begin{pmatrix} b_1 \\ b_2 \\ \vdots \end{pmatrix} = a_1 b_1 + a_2 b_2 + \cdots$$

Two vectors $\mathbf{a}$ and $\mathbf{b}$ are *orthogonal* if $\tilde{\mathbf{a}}\mathbf{b} = 0$.

Matrix Transformations

A matrix $\mathbf{Q}$ is said to *transform* a vector $\mathbf{a}$ into another vector $\mathbf{a}'$ according to the rule

$$\mathbf{a}' = \mathbf{Qa} = \begin{pmatrix} q_{11} & q_{12} & \cdots \\ q_{21} & q_{22} & \cdots \\ \cdot & \cdot & \cdots \\ \cdot & \cdot & \cdots \\ \cdot & \cdot & \cdots \end{pmatrix} \begin{pmatrix} a_1 \\ a_2 \\ \cdot \\ \cdot \\ \cdot \end{pmatrix} = \begin{pmatrix} q_{11}a_1 & + & q_{12}a_2 & + & \cdots \\ q_{21}a_1 & + & q_{22}a_2 & + & \cdots \\ \cdot & & \cdot & & \cdots \\ \cdot & & \cdot & & \cdots \\ \cdot & & \cdot & & \cdots \end{pmatrix}$$

The transpose of $\mathbf{a}'$ is then

$$\tilde{\mathbf{a}}' = \tilde{\mathbf{a}}\tilde{\mathbf{Q}} = (a_1, a_2, \ldots) \begin{pmatrix} q_{11} & q_{21} & \cdots \\ q_{12} & q_{22} & \cdots \\ \cdot & \cdot & \cdots \end{pmatrix}$$

$$= (q_{11}a_1 + q_{12}a_2 + \cdots, q_{21}a_1 + q_{22}a_2 + \cdots, \cdots)$$

A matrix $\mathbf{Q}$ is said to be *orthogonal* if $\tilde{\mathbf{Q}} = \mathbf{Q}^{-1}$. It defines an *orthogonal transformation*. It leaves $\tilde{\mathbf{a}}\mathbf{b}$ unchanged, because $\tilde{\mathbf{a}}'\mathbf{b}' = \tilde{\mathbf{a}}\tilde{\mathbf{Q}}\mathbf{Q}\mathbf{b} = \tilde{\mathbf{a}}\mathbf{Q}^{-1}\mathbf{Q}\mathbf{b} = \tilde{\mathbf{a}}\mathbf{b}$.

The transformation defined by the matrix product $\mathbf{Q}^{-1}\mathbf{AQ}$ is called a *similarity transformation*. The transformation defined by the product $\tilde{\mathbf{Q}}\mathbf{AQ}$ is called a *congruent transformation*.

If the elements of $\mathbf{Q}$ are complex, then $\mathbf{Q}$ is called *Hermitian* if $q_{ij}^* = q_{ji}$, that is, $\tilde{\mathbf{Q}}^* = \mathbf{Q}$. If $\tilde{\mathbf{Q}}^* = \mathbf{Q}^{-1}$, then $\mathbf{Q}$ is called a *unitary* matrix, and the transformation $\mathbf{Q}^{-1}\mathbf{AQ}$ is called a *unitary transformation*.

Eigenvectors of a Matrix

An *eigenvector* $\mathbf{a}$ of a matrix $\mathbf{Q}$ is a vector such that

$$\mathbf{Qa} = \lambda\mathbf{a}$$

or

$$(\mathbf{Q} - \mathbf{1}\lambda)\mathbf{a} = 0$$

where λ is a scalar, called the *eigenvalue*. The eigenvalues are found by solving the *secular equation*

$$\det (\mathbf{Q} - \mathbf{1}\lambda) = \begin{vmatrix} q_{11} - \lambda & q_{12} & \cdots \\ q_{21} & q_{22} - \lambda & \cdots \\ \cdot & \cdot & \cdots \end{vmatrix} = 0$$

which is an algebraic equation of degree n (the number of rows or columns or order of the matrix).

If the matrix $\mathbf{Q}$ is diagonal, then the eigenvalues are its elements.

Consider two different eigenvectors $\mathbf{a}_\alpha$ and $\mathbf{a}_\beta$ of a symmetric matrix $\mathbf{Q}$. Then

$$\mathbf{Qa}_\alpha = \lambda_\alpha\mathbf{a}_\alpha$$

$$\mathbf{Qa}_\beta = \lambda_\beta\mathbf{a}_\beta$$

where λ_α and λ_β are the eigenvalues. Multiply the first, in the forward direction, by $\tilde{\mathbf{a}}_\beta$ and the second, transposed, by $\mathbf{a}_\alpha$, from the backward direction. Then

$$\tilde{\mathbf{a}}_\beta \mathbf{Q} \mathbf{a}_\alpha = \lambda_\alpha \tilde{\mathbf{a}}_\beta \mathbf{a}_\alpha$$

$$\tilde{\mathbf{a}}_\beta \tilde{\mathbf{Q}} \mathbf{a}_\alpha = \lambda_\beta \tilde{\mathbf{a}}_\beta \mathbf{a}_\alpha$$

But if $\mathbf{Q}$ is symmetric, then $\tilde{\mathbf{Q}} = \mathbf{Q}$, so the two expressions on the left are equal. Hence

$$(\lambda_\beta - \lambda_\alpha) \tilde{\mathbf{a}}_\beta \mathbf{a}_\alpha = 0$$

If the eigenvalues are different, then the two eigenvectors must be orthogonal.

Reduction to Diagonal Form

Given a matrix $\mathbf{Q}$, we seek a matrix $\mathbf{A}$ such that

$$\mathbf{A}^{-1} \mathbf{Q} \mathbf{A} = \mathbf{D}$$

where $\mathbf{D}$ is diagonal. Now

$$\mathbf{D} - \lambda \mathbf{1} = \mathbf{A}^{-1} \mathbf{Q} \mathbf{A} - \lambda \mathbf{1} = \mathbf{A}^{-1} (\mathbf{Q} - \lambda \mathbf{1}) \mathbf{A}$$

Hence, the eigenvalues of $\mathbf{Q}$ are the same as those of $\mathbf{D}$, namely, the elements of $\mathbf{D}$. Let λ_k be a particular eigenvalue, found by solving the secular equation $\det (\mathbf{Q} - \lambda \mathbf{1}) = 0$. Then the corresponding eigenvector $\mathbf{a}_k$ satisfies the equation

$$\mathbf{Q} \mathbf{a}_k = \lambda_k \mathbf{a}_k$$

which is equivalent to n linear homogeneous algebraic equations

$$\sum_j q_{ij} a_{jk} = \lambda_k a_{ik} \qquad (i = 1, 2, \ldots, n)$$

These may be solved for the ratios of the a's to yield the components of the eigenvector $\mathbf{a}_k$. The same procedure is repeated for each eigenvalue in turn. We then form the matrix $\mathbf{A}$ whose columns are the eigenvectors $\mathbf{a}_k$, that is, $(\mathbf{A})_{ik} = a_{ik}$. Thus, the matrix $\mathbf{A}$ must satisfy

$$\mathbf{Q} \mathbf{A} = \mathbf{A} \begin{pmatrix} \lambda_1 & 0 & \cdots & \\ 0 & \lambda_2 & \cdots & \\ \cdot & \cdot & \cdots & \\ 0 & 0 & & \cdot\cdot\lambda_n \end{pmatrix} = \mathbf{A} \mathbf{D}$$

so that $\mathbf{A}^{-1} \mathbf{Q} \mathbf{A} = \mathbf{D}$ as required. This method can always be done if $\mathbf{Q}$ is symmetric and the eigenvalues are all different.

Application to Oscillating Systems

For a system with n degrees of freedom, the generalized displacement vector is

$$\mathbf{q} = \begin{pmatrix} q_1 \\ q_2 \\ \vdots \\ q_n \end{pmatrix}$$

In matrix notation the kinetic and potential energies (defined in Sections 11.3 and 11.4 of the text) take the compact forms

$$T = \frac{1}{2}\tilde{\mathbf{q}}\mathbf{M}\dot{\mathbf{q}} \qquad V = \frac{1}{2}\tilde{\mathbf{q}}\mathbf{K}\mathbf{q}$$

in which

$$\mathbf{M} = \begin{pmatrix} M_{11} & M_{12} & \cdots \\ M_{21} & M_{22} & \cdots \\ \cdot & \cdot & \cdots \end{pmatrix}$$

$$\mathbf{K} = \begin{pmatrix} \kappa_{11} & \kappa_{12} & \cdots \\ \kappa_{21} & \kappa_{22} & \cdots \\ \cdot & \cdot & \cdots \end{pmatrix}$$

We note that both $\mathbf{M}$ and $\mathbf{K}$ are symmetric matrices. The differential equations of motion of the system given by Equation 11.4.6 can then be written as the matrix equation (11.4.7)

$$\mathbf{M}\ddot{\mathbf{q}} + \mathbf{K}\mathbf{q} = 0$$

If a harmonic solution of the form

$$\mathbf{q} = \mathbf{a}\cos\omega t$$

exists, then

$$\ddot{\mathbf{q}} = -\omega^2 \mathbf{q}$$

Consequently

$$(-\mathbf{M}\omega^2 + \mathbf{K})\mathbf{a} = 0$$

A nontrivial solution requires the secular determinant to vanish

$$\det(-\mathbf{M}\omega^2 + \mathbf{K}) = 0$$

or

$$|-M_{ij}\omega^2 + \kappa_{ij}| = 0$$

The roots give the normal frequencies, and the associated eigenvectors define the normal modes. For further reading, see any of the first seven titles under the heading *Advanced Mechanics* in Selected References.

Appendix I
Software Tools: *Mathcad* and *Mathematica*[1]

We presented many examples in the text in which we used the software tools available in either *Mathcad* or *Mathematica* to solve some of the otherwise intractable mathematical problems that frequently arise in physics. Other similar software packages exist, most notably, *Maple,* that we could have used as well. Our choice of *Mathcad* and *Mathematica* should not be taken as an endorsement for those two products. They are simply the ones with which we are most familiar. The avenues of analysis that mathematical software tools have opened up to the practicing scientist are virtually unlimited, and we have managed to explore only a few of them. Given this, it ought to come as no surprise that in this edition of the text we place greater emphasis than ever before on the use of such inexpensive and readily available tools for solving problems numerically.

Alternatively, we could have conscripted some business spreadsheet such as Microsoft *Excel,* Borland's *Quattro Pro* or Lotus *1-2-3* for duty as a scientific analysis engine. Typically, though, business spreadsheets are much more cumbersome to use this way than are any of the mathematical software packages already mentioned. Consequently, we have not described here how they might be used in scientific applications. Instead, we contend that serious students of physics ought to use one of the excellent computational tools that have been designed solely for the purpose of solving problems relevant to their own discipline. Knwledge of how to use such tools is an increasingly essential weapon that any practicing physicist should have in his or her analysis arsenal.

Space prevents us from attempting a complete discussion of how to use these tools. The companies that create them distribute massive tomes with the product that describe their use. Most of these accompanying manuals do an adequate and sometimes excellent job as a reference, but many suffer deficiencies common to most technical user manuals: frequently they are poorly organized and poorly written; almost invariably they are loaded with undefined cryptic jargon; critical information is sometimes buried away in unsuspected locations; and rarely do they perform well as a tutorial for the novice user. The *Mathcad* manual defies this convention: it is well organized and well written, jargon terms are usually defined before they are used; critical material is easy to find, and a novice user can get "up and running" without too much pain. On the other hand, the *Mathematica* manual is afflicted with most of the aforementioned ills. Indeed, its writers seem to have worked hard at rendering their otherwise excellent product unuseable. In spite of its manual, *Mathematica* itself proved to be an excellent computational tool and one well

---

[1] The interested reader can order *Mathcad* or *Mathematica* or related products on the Internet at the following respective addresses:

(1) http://www.mathsoft.com/ (2) http://www.store.wolfram.com/

A-24

worth the pain of learning how to use. In fact, it might be the most powerful tool currently available in the marketplace, if, as we suspect, that computational power is proportional to the cost of a product.

Fortunately, to fill the niche created by the inadequacy of most technical manuals, many independent companies have arisen that specialize in writing *How to . . . blah blah* or *blah blah . . . Software for Dummies* books. The entire body of such books is devoted to the presentation of examples that emphasize both how to choose the proper tool and how to use it to solve a particular problem. In each example, the authors take great pain to describe the precise details of each step taken in the problem-solving process. It is well worth having such a book in one's library to make intelligent use of the relevant software product. Our presentation here pales into insignificance in this regard—a deficiency for which we offer no apology given the primary mission of this text. Nonetheless, in what follows (see also Section 2.5) we present—what is admittedly only a minimal aid to the student—two worksheets that we set up in *Mathcad* and *Mathematica* to illustrate their use in solving problems. We urge the serious student to purchase a mathematical software package (and possibly a reference tutorial distributed by an independent vendor as well) and use it to try and solve the computer problems given at the end of each chapter.

Quick Plots Using *Mathcad*

It is frequently desirable to create a plot of numerical data or a function to visualize relationships between variables. It almost seems as though *Mathcad* were created especially for this purpose. Ease of generating plots is one of its strongest features, and for that reason alone, it is worth the price of its purchase.

To create a plot in *Mathcad*, simply

- Define the *x* and *y* variables.
- Type an @ key to open up a graphics region.
- Type the *x* and *y* variable names in the placeholder adjacent to the *x* and *y* axes.
- Move the cursor outside the graphics region, and *Mathcad* creates the plot.

An example of this process is shown in the following worksheet, which was set up to create a phase space plot for the simple harmonic oscillator (see Section 3.5). Notice that each step in the worksheet has a brief, accompanying comment that points to it. Such text regions can be created anywhere in a *Mathcad* worksheet, simply by typing the" key, followed by the text. Describing each step in a calculation in this or some similar way is a process well worth emulating. You never know when you might wish to resurrect a worksheet that you created sometime in the remote past and the presence of explanatory text quite often means the difference between looking at an algorithm that is comprehensible or one that instead bears a close resemblance to Egyptian hieroglyphics.

Phase Space plot for the simple Harmonic oscillator

$T := 4 \cdot \pi$ $\omega := \dfrac{2 \cdot \pi}{T}$ ← Define period, angular frequency

$j := 1 \ . \ . \ . \ 4$ $A_j := j$ ← Define 4 amplitudes, 1-4

i := 0 . . . 1000 $t_1 := \dfrac{i \cdot T}{1000}$ ← Divide period into 1000 time
                                                intervals

$x_{j,i} := A_j \cdot \sin(\omega \cdot t_i)$ ← Calculate x-coordinate of point

$v_{j,i} := \omega \cdot A_j \cdot \cos(\omega \cdot t_i)$ ← Calculate y-coordinate of point

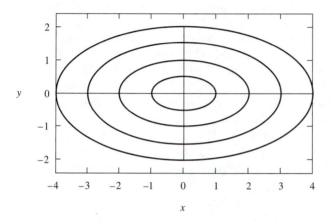

Figure 3.5.1   Phase-space plot for the simple harmonic oscillator ($\omega_0 = 0.5$ s^{-1}). No damping force ($\gamma = 0$ s^{-1}).

Coupled Differential Equations

The differential equations of motion of a particle expressed in component form quite frequently contain coupling terms that render the analytic solution at best difficult and at worst impossible. Typically, such terms are caused by forces arising from the motion of the particle in one dimension that affect its motion in another. Usually, we can find closed form, that is, analytic solutions to such coupled differential equations only for a limited number of cases in which either the circumstances that lead to the coupling prove to be fortuitous or the mathematical approximations that represent the forces are suitably contrived. In most cases, we must resort to numerical techniques to solve such equations. The coupled equations of motion governing the trajectory of the baseball discussed in Example 4.3.2 are an example of the latter kind; they can only be solved numerically.

We used *Mathematica* to solve Example 4.3.2. The example has the additional virtues that it not only illustrates how to solve coupled differential equations but also how to

- Create plots
- Create data tables
- Create interpolation functions to estimate values between data points
- Find roots of equations or locations where two lines intersect
- Find the extremum of a function.

The *Mathematica* worksheet solution is shown at the end of this appendix. The steps should be self-explanatory because each one is accompanied by a liberal dose of commentary. One particular point, made when the solution of Example 4.3.2 was discussed in the text, should be reiterated here: the process of solution is iterative. The variables θ_0 and $vMph_0$ are fixed during each iteration. Initially they were chosen to be $\theta_0 = 40°$ and $vMph_0 = 130$ mph. The values actually shown in the worksheet are the final ones that emerged from the iterative problem solving process.

The iterative steps carried out in the solution of the problem include all those in the following subsections:

- **Numerical solution of trajectory with quadratic air resistance**
- **Find range of baseball**

To begin, the steps in these subsections were executed sequentially with $vMph_0$ held fixed at 130 mph. Each trial used a different value of θ_0 ranging from 35° to 42° in 1° increments. The range of the baseball was found for each of these trials, and the resulting range versus θ_0 values were loaded into the table, *thetaData*. This data was then used to find the value of θ_0 at which the maximum range was obtained. This calculation was carried out in the subsection

- **Find optimum angle to launch baseball**

Next, the steps in the above two subsections were repeated again, only this time holding θ_0 fixed at the optimum value just found (39°) and changing the value of $vMph_0$ from 130 mph to 155 mph in 5 mph increments. Again the range of the baseball was calculated for each of these trials, and the resulting range versus $vMph_0$ values were loaded into the table, *RvsTheta*. Interpolation of this data was used to find the value of $vMph_0$ required to achieve a range of 565 ft (172.16 m). This calculation was carried out in the subsection

- **Find initial velocity for $R_{Mick} = 172.16$ m**

EXAMPLE 4.3.2

Calculate the Trajectory of a Baseball in Flight Subject to Air Resistance Proportional to the Square of Velocity.

Variables

$vMph_0$;   Initial velocity (mph)
v_0;   Initial velocity (ms^{-1})
u_0, w_0;   Initial velocity (x, z) components
θ_0;   Initial elevation angle (radians)
R_{mick};   Micky Mantle's range (565 ft = 172.16 m)
TofF;   Time of flight (s)
g;   Acceleration due to gravity (9.8 ms^{-2})
γ;   Air resitance factor (0.0055 m^{-1})

```
vMph₀ = 143.23;
v₀ = 0.447 vMph₀;
u₀ = v₀ Cos[θ₀];  w₀ = v₀ Sin[θ₀];
θ₀ = 39 (pi/180);
```

```
R_mick = 172.16;
TofF = 9;
g = 9.8;
γ = 0.0055;
```

Analytic Solution of Trajectory—no Air Resistance

$$z_1[t\_] = w_0 t - \frac{g}{2} t^2; \quad x_1[t\_] = u_0 t;$$

Numerical Solution of Trajectory with Quadratic Air Resistance

Call numerical differential equation solver NDSolve.
Plot results using ParametricPlot inhibit display.
Show the plot with curves labelled.

```
sol = NDSolve[
    {z''[t] == -g - γ(z'[t]² + x'[t]²)^0.5 z'[t], x''[t] == -γ(z'[t]² + x'[t]²)^0.5 x'[t],
    z[0] == 0, z'[0] == w_0, x[0] == 0, x'[0] == u_0}, {z, x}, {t, 0, TofF}]

trajectory = ParametricPlot[{{x_1[t], z_1[t]}, {Evaluate[{x[t], z[t]}/.sol]}},
    {t, 0, TofF}, Compiled -> False, PlotRange -> {{0, 420}, {0, 90}},
    PlotStyle -> {Thickness[0.005]}, AxesLable ->{"x(m)", "Height(m)"},
    PlotLabel -> " Baseball trajectories", DisplayFunction → Identity];

trajectory = Show[trajectory, Graphics[Text["no drag", {310, 80}]],
    Graphics[Text["Quadratic drag", {200, 50}]]];

show[trajectory, DisplayFunction → $DisplayFunction]

{{z → InterpolatingFunction[{{0., 9.}}, <>], x → InterpolatingFunction [{{0., 9.}}, <>]}}
```

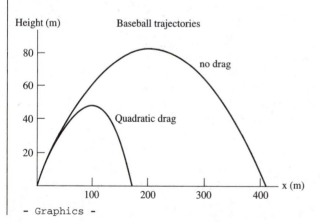

- Graphics -

Find Range of Baseball

Calculate time for ball to hit ground, that is, find root t_0 of $z[t]$
Calculate range in meters: xDistance = $x[t_0]$
Calculate range in feet: xFeet

```
Tzero = FindRoot[Evaluate[z[t] == 0/. sol], {t, 7.0}]
xDistance = x[t/. Tzero]/. sol
xFeet = xDistance 3.2808
{t → 6.23961}

{172.155}

{564.807}
```

Find Optimum Angle to Launch Baseball

Use plot to estimate optimum elevation angle for launching baseball at maximum range.
(Initial velocity guess: $vMph_0$: 130 mph).
Define data table {θ_0, Range} and interpolation function, RvsTheta.
Call Plot with interpolation function as argument.
Find angle θ_0 for which range is a maximum (Call FindMinimum).

```
thetaData = {{35, 154.974}, {36, 155.467}, {37, 155.816}, {38, 156.022},
    {39, 156.087}, {40, 156.014}, {41, 155.802}, {42, 155.455}};

RvsTheta = Interpolation[thetaData];

Plot[RvsTheta[x], {x, 35, 42}, PlotStyle -> {Thickness[0.005]}, AxesOrigin->
    {35, 155}, AxesLabel -> {"θ₀(degrees)", "Range(m)"}, PlotLabel->" Range vs θ₀"]
FindMinimum[-RvsTheta[x], {x, 39}]
```

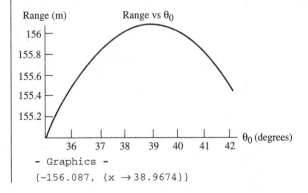

```
- Graphics -
{-156.087, {x → 38.9674}}
```

Find Initial Velocity of Baseball for R_{mick} = 565 ft (172.16 m)

Use plot to find initial velocity of baseball required to achieve desired range, R_{mick} (elevation angle θ_0 set to optimum value). Load data table (rangeData) with range versus initial velocity v_0.

Call ListPlot with data input as arguments {v_0, range}.

```
rangeData = {{156.087, 130}, {162.284, 135},
   {168.331, 140}, {174.23, 145}, {179.983, 150}, {185.595, 155}};
VvsR = ListPlot[rangeData, PlotJoined->True, AxesOrigin->{156.1, 130},
   AxesLabel->{"Range(m)", "v₀(mph)"}, PlotStyle->{Thickness[0.005]}]
```

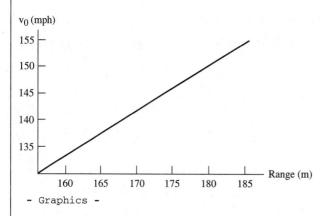

```
- Graphics -
```

Find initial velocity for R_{mick} = 172.16 m

Define interpolation function for velocity versus range data.
Find optimum speed by interpolation (Call Interpolation).
Call Line primitive to define horizontal and vertical line intercepts.

```
initSpeed = Interpolation[rangeData];
speed = initSpeed[Rmick]

143.231

rangeIntercept = Line[{{Rmick, 130}, {Rmick, 145}}];
speedIntercept = Line[{{156, speed}, {174.0, speed}}];
```

Redisplay Graph Showing Range and Corresponding Initial Velocity as Intercepts

Call Show for redisplay.

```
Show[VvsR, Graphics[{Dashing[{0.025, 0.025}], speedIntercept}],
  Graphics[{Dashing[{0.025, 0.025}], rangeIntercept}], AxesOrigin-> {156, 130},
  AxesLabel->{"Range(m)", "v₀(mph)"}, PlotLabel->" v₀ vs Range"]
```

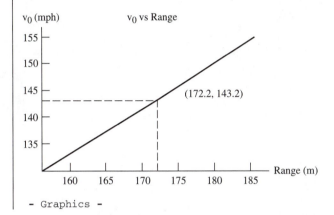

```
- Graphics -
```

Answers to Selected Problems

CHAPTER 1

1.1 (a) $\sqrt{6}$, (b) $3\mathbf{i} + \mathbf{j} - 2\mathbf{k}$, (c) 1, (d) $\mathbf{i} - \mathbf{j} + \mathbf{k}$

1.3 $\cos^{-1}\sqrt{5/14} = 53.3°$

1.7 $q = 1$ or 2

1.17 $b\omega(\sin^2\omega t + 4\cos^2\omega t)^{1/2}$, $2b\omega$, $b\omega$

1.27 $\mathbf{v} = v\hat{\boldsymbol{\tau}}$, $a = \dot{v}\hat{\boldsymbol{\tau}} + \hat{\mathbf{n}}v^2/\rho$, $|\mathbf{v} \times \mathbf{a}| = \mathbf{v} \cdot \mathbf{a}_n = vv^2/\rho = v^3/\rho$

CHAPTER 2

2.1 (a) $\dot{x} = (F_0/m)t + (c/2m)t^2$, $x = (F_0/2m)t^2 + (c/6m)t^3$
(b) $\dot{x} = (F_0/cm)(1 - \cos ct)$, $x = (F_0/c^2m)(ct - \sin ct)$
(c) $\dot{x} = -(F_0/cm)(1 - e^{ct})$, $x = -(F_0/c^2m)(ct - e^{ct} + 1)$

2.3 (a) $V = -F_0x - (c/2)x^2 + C$, (b) $V = (F_0/c)e^{-cx} + C$
(c) $V = -(F_0/c)\sin cx + C$

2.9 (a) 541, (b) 87, (c) 454

CHAPTER 3

3.1 6.43 m/s, 2.07×10^4 m/s^2

3.3 $x(t) = 0.25\cos(20\pi t) + 0.00159\sin(20\pi t)$ in meters

3.5 $[(\dot{x}_2^2 - \dot{x}_1^2)/(x_1^2 - x_2^2)]^{1/2}$, $[(x_1^2\dot{x}_2^2 - x_2^2\dot{x}_1^2)/(\dot{x}_2^2 - \dot{x}_1^2)]^{1/2}$

3.19 (a) $T = 2\pi(l/g)^{1/2} \times 1.041$, (b) g comes out to be about 8% low,
(c) $B/A = 0.0032$

CHAPTER 4

4.1 (a) $\mathbf{F} = -c(yz\mathbf{i} + xz\mathbf{j} + xy\mathbf{k})$, (b) $\mathbf{F} = -2(\alpha x\mathbf{i} + \beta y\mathbf{j} + \gamma z\mathbf{k})$
(c) $\mathbf{F} = ce^{-(\alpha x + \beta y + \gamma z)}(\alpha\mathbf{i} + \beta\mathbf{j} + \gamma\mathbf{k})$, (d) $\mathbf{F} = -cnr^{n-1}\mathbf{e}_r$

4.3 (a) $c = \frac{1}{2}$, (b) $c = -1$

4.7 $\theta = \sin^{-1}\left(-\dfrac{gb}{V_0^2}\right)$

4.21 It leaves at height $h = b/3$

CHAPTER 5

5.1 Up: 150 lb, Down: 90 lb

5.3 1.005 mg, about 5.7°

5.5 (a) $g/6$ forward, (b) $g/3$ toward rear

5.9 $(V_0^2/\rho)\mathbf{i}' + [(V_0^2/b) + (V_0^2 b/\rho^2)]\mathbf{j}$

CHAPTER 6

6.1 About 2×10^{-9}

6.3 About 1.4 h

6.23 $\psi = \pi[(1 + c)/(1 + 4c)]^{1/2}$, where $c = \rho 4\pi a^3/3M_{sun}$

6.25 $a > (\epsilon/k)^{1/2}$

6.29 $\psi = 180.7°$ for orbits near Earth.

6.31 $\theta = -30°$

CHAPTER 7

7.1 $\mathbf{r}_{cm} = (\mathbf{i} + 2\mathbf{j} + 2\mathbf{k})/3$, $\mathbf{v}_{cm} = (3\mathbf{i} + 2\mathbf{j} + \mathbf{k})/3$, $\mathbf{p} = 3\mathbf{i} + 2\mathbf{j} + \mathbf{k}$

7.5 Direction: downward at an angle of 26.6° with the horizontal, Speed: 1.118 v_0

7.7 Car: $v_0/2$, Truck: $v_0/8$. Both final velocities are in the direction of the initial velocity of the truck.

7.15 Proton: $v'_x = v'_y = 0.558\, v_0$, Helium: $v'_x = 0.110\, v_0$, $v'_y = -0.140\, v_0$

7.17 Approximately 57.3°

7.27 $-mg = m\dot{v} + V\dot{m}$, $m_f/m_p = \exp(1/k - g/(kv_c) \cdot m_0/\dot{m}) - 1$, $m_f/m_p = 77$

CHAPTER 8

8.1 (a) $b/3$ from center section, (b) $x_{cm} = y_{cm} = 4b/3\pi$, where lamina is in xy-plane, (c) $x_{cm} = 0$, $y_{cm} = 3b/5$, (d) $x_{cm} = y_{cm} = 0$, $z_{cm} = 2b/3$, (e) $b/4$ from base

8.3 $a/14$ from center of large sphere

8.5 $(31/70)ma^2$

8.11 $2\pi(2a/g)^{1/2}$, $2\pi(3a/2g)^{1/2}$

8.15 $g(m_1 - m_2)/(m_1 + m_2 + I/a^2)$

8.19 $v_0 t - \frac{1}{2} gt^2 (\sin\theta + \mu\cos\theta)$
$(2v_0^2/g)(\sin\theta + 6\mu\cos\theta)/(2\sin\theta + 7\mu\cos\theta)^2$

CHAPTER 9

9.1 (a) $I_{xx} = \dfrac{m}{3}a^2$, $I_{yy} = \dfrac{4m}{3}a^2$, $I_{zz} = \dfrac{5m}{3}a^2$

$I_{xz} = I_{yz} = 0$, $I_{xy} = -\dfrac{m}{3}a^2$

(b) $\frac{2}{15}ma^2$, (c) $\mathbf{L} = (ma^2\omega/6\sqrt{5})(\mathbf{i} + 2\mathbf{j})$, (d) $T = \frac{1}{15}ma^2\omega^2$

9.3   (a) Inclination of the l-axis is $\frac{1}{2}\tan^{-1}1 = 22.5°$
      (b) Principal axes in the xy-plane are parallel to the edges of the lamina.

9.9   (a) 1.414 s, 0.632 s; (b) 1.603 s, 0.663 s

9.13 $\alpha - \tan^{-1}[(I/I_s)\tan\alpha] \approx \alpha(I_s - I)/I_s = 0.00065$ arc sec

9.17 $S > \left[\dfrac{128\,ga}{b^4}\left(\dfrac{a^2}{3} + \dfrac{b^2}{16}\right) \right]^{1/2} \approx 2910$ rps

CHAPTER 10

10.3 $\ddot{x} = (\frac{5}{7})g\sin\theta$

10.5 $m_1 : -(\frac{5}{11})g$, $m_2 \colon (\frac{7}{11})g$, $m_3 \colon (\frac{3}{11})g$, $m_4 \colon -(\frac{5}{11})g$

10.7 $x = x_0\cosh\omega t - (g/2\omega^2)\sin\omega t + (g/2\omega^2)\sin\omega t$

10.9 $F_x = m(\ddot{x} - 2\omega\dot{y} - \omega^2 x)$, $F_y = m(\ddot{y} + 2\omega\dot{x} - \omega^2 y)$, $F_z = m\ddot{z}$

10.23 $U(r) = \dfrac{l^2\sin^2\alpha}{2r^2} + gr\cos\alpha$, where $l = r^2\dot{\phi} = constant$

10.27   (a) $\dot{\theta} = p_\theta/ml^2$, $\dot{p}_\theta = -mgl\sin\theta$
      (b) $\dot{x} = p_x/(m_1 + m_2)$, $\dot{p}_x = g(m_1 - m_2)$
      (c) $\dot{x} = p_x/m$, $\dot{p}_x = mg\sin\theta$

CHAPTER 11

11.1   (a) $x = k^{1/3}$ stable
      (b) $x = 1/b$ unstable
      (c) $x = 0$ unstable, $x = \pm b/\sqrt{2}$ stable
      (d) $(3k/m)^{1/2}$, 3.628 s; $2b(k/m)^{1/2}$, 3.14 s for parts (a) and (c), respectively.

11.9 $2\pi a[5/3g(b - a)]^{1/2}$

11.11   2.363 s

11.17 $\omega = (k/m)^{1/2}\,\dfrac{(5\pm\sqrt{17})^{1/2}}{2}$

11.27 $v_{long} = \sqrt{\dfrac{k}{m}(l + \Delta l)}$

       $v_{tran} = \sqrt{\dfrac{k}{m}(l + \Delta l)\Delta l}$

Selected References

Mechanics

Barger, V., and Olsson, M., *Classical Mechanics*, McGraw-Hill, New York, 1973.
Becker, R. A., *Introduction to Theoretical Mechanics*, McGraw-Hill, New York, 1954.
Lindsay, R. B., *Physical Mechanics*, Van Nostrand, Princeton, NJ., 1961.
Rossberg, K., *A First Course in Analytical Mechanics*, Wiley, New York, 1983.
Rutherford, D. E., *Classical Mechanics*, Interscience, New York, 1951.
Slater, J. C., and Frank, N. H., *Mechanics*, McGraw-Hill, New York, 1947.
Smith, P., and Smith, R. C., *Mechanics*, John Wiley & Sons, New York, 1990.
Symon, K., *Mechanics*, 3rd ed., Addison-Wesley, Reading, Mass., 1971.
Synge, J. L., and Griffith, B. A., *Principles of Mechanics*, McGraw-Hill, New York, 1959.

Advanced Mechanics

Chow, T. L., *Classical Mechanics*, John Wiley & Sons, New York, 1995.
Corbin, H. C., and Stehle, P., *Classical Mechanics*, Wiley, New York, 1950.
Desloge, E., *Classical Mechanics* (two volumes), Wiley-Interscience, New York, 1982.
Goldstein, H., *Classical Mechanics*, 2nd ed., Addison-Wesley, Reading, Mass., 1980.
Hauser, W., *Introduction to the Principles of Mechanics*, Addison-Wesley, Reading, Mass., 1965.
Landau, L. D., and Lifshitz, E. M., *Mechanics*, Pergamon, New York, 1976.
Marion, J. B., and Thornton, S. T., *Classical Dynamics*, 5th ed., Brooks/Cole—Thomson Learning, Belmont, CA, 2004.
Moore, E. N., *Theoretical Mechanics*, Wiley, New York, 1983.
Wells, D. A., *Lagrangian Dynamics*, Shaum, New York, 1967.
Whittaker, E. T., *Advanced Dynamics*, Cambridge University Press, London and New York, 1937.

Mathematical Methods

Burden, R. L., and Faires, J. D., *Numerical Analysis*, Brooks Cole Publ., Pacific Grove, CA, 1997.
Churchill, R. V., *Fourier Series and Boundary Value Problems*, McGraw-Hill, New York, 1963.
Grossman, S. I., and Derrick, W. R., *Advanced Engineering Mathematics*, Harper Collins Publ., New York, 1988.
Jeffreys, H., and Jeffreys, B. S., *Methods of Mathematical Physics*, Cambridge University Press, London and New York, 1946.
Kaplan, W., *Advanced Calculus*, Addison-Wesley, Reading, Mass., 1952.
Margenau, J., and Murphy, G. M., *The Mathematics of Physics and Chemistry*, 2nd ed., Van Nostrand, New York, 1956.
Mathews, J., and Walker, R. L., *Methods of Mathematical Physics*, W. A. Benjamin, New York, 1964.
Press, W. H., Teukolsky, S. A., Vetterling, W. T., and Flannery, B. T., *Numerical Recipes*, Cambridge University Press, New York, 1992.
Wylie, C. R., Jr., *Advanced Engineering Mathematics*, McGraw-Hill, New York, 1951.

Chaos

Baker, G. L., and Gollub, J. P., *Chaotic Dynamics*, Cambridge University Press, New York, 1990.
Hilborn, R. C., *Chaos and Nonlinear Dynamics*, Oxford University Press, New York, 1994.

Tables

Dwight, H. B., *Mathematical Tables*, Dover, New York, 1958.
Handbook of Chemistry and Physics, *Mathematical Tables*, Chemical Rubber Co., Cleveland, Ohio, 1962 or after.
Pierce, B. O., *A Short Table of Integrals*, Ginn, Boston, 1929.

Index

A

Acceleration, 33
   centrifugal, 208–209, 221, 290
   centripetal, 43, 193, 194, 221, 222
   Coriolis, 193, 196, 208, 209, 290, 295
   in rectangular coordinates, 31–36
   transverse, 43, 193
   uniform, 60–63
   velocity and, 36–39
α-Centauri system, 293
Adams, John Couch, 242
Addition, commutative law of, 12
Air gyroscope, 401–402
Air resistance, linear, projectile motion
    and, 161–162
   projectile motion and, 158–161
Airy, Sir George, 242
Alpha particles, scattering of, 264–269
Angle(s), apsidal, 262
   Eulerian, 391–397
   of kinetic friction, 63
Angular frequency, 87
Angular momentum, conservation of, 226–229
   general theorem concerning, 344–345
   and kinetic energy, 278–283, 361–371
   principle of conservation of, 279
   of rigid body in laminar motion, 344–347
   spin, 406
Angular momentum vector, 365–367
Antinodes, 508–509
Aphelion, 234
Aphelion points, 262
Apocenter, 234
Apogee, 234
Approximations, A-12
   successive, 125–127
Apsidal angle, 262
Apsides and apsidal angles, for nearly
    circular orbit, 262–264
Apsis(es), 262
Aristotle, 48, 49
Armstrong, Lance, 409–410, 412
Associative law, 12
Asteroids, Trojan, 295–302
Astronomical unit, 291
Atmosphere, 66
Atomic clock, cesium, 4–5

Atwood's machine, 433
   double, 434–435
Axis(es), fixed, body constrained
    to rotate about, 382–383
   principle, determination of, 375–377, 378–381

B

Balancing, dynamic, 374–375
Baliani, Giovanni, 157
Ballistocardiogram, 82–83
Baseball, trajectory of, 164–167, A-27 – A-31
"Baseball bat theorem," 354–356
Basis vectors, 13
Bernoulli, Daniel, 499
Bernoulli, Johann, 417, 418, 449, 499
Binary stars, 285–288, 2929, 499
Binding energy, of diatomic hydrogen molecule, 68
Binomial series, A-12
Black holes, 285–288
Blueshift, 54
Body, falling, 206, 221, 421
   rolling, 348, 349–350
Borelli, Giovanni, 220–221
Bovard, Alexis, 241–242
Brahe, Tycho de, 225
Brillovin, Leon, 499
Bullet, deflection of, 206–207
   speed of, 305–306
Burnout, time of, of rocket, 315

C

Cartesian coordinate system, 28, 156
Cartesian coordinates, 427–429, A-7
   scalar equations, 145
Cartesian unit vectors, 13–14
Center-of-mass coordinates, 306–312
Center of oscillation, 340–341
Center of percussion, 354–356
Centimeter, 5
Centrifugal acceleration, 208–209
Centrifugal force, 197, 221
Centrifugal potential, 257
Centripetal acceleration, 43, 193, 194
Centripetal force, 221
cgs system, 265
Challis, James, 242
Chaotic motion, 129–135

Circle, 235
    equation of, A-8
Circular disc, moment of inertia of, 331, 334
Circular function, A-5
Circular hoop, 330
Circular motion, 34–35, 220
Circular orbits, stability of, 260–261
Coefficient(s), of restitution, 303, 304
    of sliding friction, 62
    static, 62
    of static friction, 351–352
    of transformation, 25
Collision(s), 303–306
    direct, 303–305
    elastic, 303
    endoergic, 303
    impulse in, 305
    oblique, and scattering, 306–312
    of rigid bodies, 354–356
    totally inelastic, 304
Comet(s), 240
    orbit of, 234, 253–254, 260
Complex exponential, A-11 – A-12
Complex numbers, A-4
Compound pendulum, 338–344
Computer software packages, 75, A-24 – A-31
Cone, of variable density, center of
        mass of, 326–328
Configuration space, 453–454, 475–477
Conic sections, 233, 234, 235, A-7 – A-8
Conservative systems, Lagrange's equations of
        motion for, 430–431
Constraint(s), generalized forces of, 418, 444–448
    holonomic, 425–426
    nonholonomic, 426
    smooth, energy equation for, 176–177
Continuous system, vibration of, 505–509
Contour plot, 293–294, 299
Coordinate system(s), accelerated, and inertial
        forces, 184–189
    change in, 25–30
    fixed, rotation of rigid bodies and, 391–397
    rotating, 189–201
Coordinate(s), Cartesian, 427–429
    center-of-mass, 306–312
    curvilinear, A-15
    cylindrical, 39–40, 208, A-15
    generalized, 423–426, 438–444
        generalized momenta conjugate to, 438–439
        kinetic and potential energy calculated in
            terms of, 426–429
    normal, 472–493
        equations of motion in, 484–485
    parabolic, A-16
    plane polar, 36–39
    of point on Earth, 425
    rectangular, A-15
        velocity and acceleration in, 32–33
    spherical, 40–43, A-15
Copernicus, Nicolaus, 218–219
Coressio, Giorgio, 50

Coriolis acceleration, 193, 196, 208, 209
Coriolis force, 197, 202, 204, 207, 211, 295, 299
Cosines, laws of, 18–19
Cosmic microwave background, 53, 54
Coulomb's law, 265
Coupled harmonic oscillators. See Harmonic
        oscillator(s), coupled
Critical damping, 98–99
Crosetti, Frank, 166
Cross product, 19–23
    of two vectors, 21
Curl, 250, A-15, A-16
Curve, directrix of, A-7
Curvilinear coordinates, A-15
Cycloid, 36
    constrained motion on, 178–179
Cyclotron, 176
Cygnus X-1, 286, 287
Cylinder, moment of inertia of, 331, 336, 337
    rolling, rotational motion of, 346, 347
        without slipping, 352–353
    stability of, 468–469
Cylindrical coordinates, 39–40, 208, A-15

D
da Vinci, Leonardo, 3
D'Alembert, Jean LeRond, 417
D'Alembert's principle, 418–419, 449–455
Damping factor, 97
Deep Space I rocket, 317, 318
del Monte, Marquis and General, 157–158
Del operator, 151–156
de'Medici, Antonio, 158
Descartes, René, 48–49
Diagonal form, reduction to, A-22
Differential scattering cross section, 267
Dimensional analysis, determining
        relationships by, 7–9
    of equations, 5–7
Dimensions, 5–7
Displacements, virtual, 449
Distributive law, 12–13
Divergence, A-15
Doppler Effect, 54
Doppler shift, 54
Dot product, 16
Dugas, René, 323
Dunaway, Donald, 166
Dyad product, 366

E
Earth, atmosphere of, 66
        high-pressure system in, 197
    coordinates of point on, 425
    free precession of, 396–399
    gravity of, 222
    Moon and, 220–221, 23, 241, 292–293, 294,
            300–301
    rotation of, effects of, 201–207
    surface of, plumb bob hanging above, 202, 203
Earth satellite, launching of, 314

Eccentricity, A-7
Effective potential. *See* Potential, effective
Eigenfrequency, 481–482, 495
Eigenvalue, 481
Eigenvector(s), 481, 482, 487–488, 495
  of matrix, A-21 – A-22
Einstein, Albert, 1, 2, 54, 57, 184, 264, 418
Electric field, motion of charged
     particle in, 173–174
Electrical-mechanical analogs, 123–124
Electromagnetic theory, electric field and, 174
Ellipse(s), definition of, 233, 235
  eccentricity of, 233
  equation of, A-9
  law of, 225, 229–238
Ellipsoid(s), 388, 390
  Poinsot, 383–384, 385
Elliptic integrals, 341–343, A-13
Energy, harmonic motion and, 93–96
  kinetic. *See* Kinetic energy
  and laminar motion, 350
  of particle, 63–64
  potential. *See* Potential energy
     of simple pendulum, 95
  spin, 403
Energy balance condition, 306–307
Energy equation(s), 64, 176–177, 251, 401–407
Equation(s), of circle, A-8
  differential, coupled, *Mathematica* to
     solve, A-26 –A-31
    of orbit, 231
  dimensional analysis of, 7–9
  of ellipse, A-9
  energy, 64, 176–177, 251, 401–407
  energy balance, 306–307
  Euler's, 381–383, 384–390, 435–436
    for free rotation of rigid body, 384–390
  Hamilton's, 455–460
  of hyperbola, A-10
  Lagrange's, applications of, 431–438
  of motion for conservative systems, 430–431
  of motion, Hamilton's canonical, 455–456
    in normal coordinates, 484–485
    for restricted three-body problem, 289–290
    of rigid body, 381–383
  of parabola, A-9
  scalar, in Cartesian coordinates, 145
  for translation and rotation about
     fixed axis, 329–330
  van der Pol, 127–128
  vector form of, motion of particle, 144–145
  wave, 505–509
    one-dimensional, 506
Equilibrium, 465–467
  potential energy and, 465–469
  stable, 466–467
    oscillation of system and, 469–472
Equipotentials, 153
Escape speed, 66
Euclidean space, 2
Euler, Leonhard, 391, 418

Eulerian angles, 391–397
Euler's equations, for free rotation of rigid
     body, 384–390, 435–436
  of motion of rigid body, 381–383
Euler's identity, 100, 116
Euler's theorem for homogenous
     functions, 455
Exchange operation, 487, 488
Expansions, A-11
  series, A-11
Exponential, complex, A-11 – A-12
Exponential notation, A-4

F
Fall, vertical, through fluid, 71–74
Falling body, 206, 221, 421
Falling chain, attached to disc, 333
Falling disc, kinetic energy of, 447
Feynman, Richard, 454
Field, central, nearly circular orbits in, 260–261
    orbit in, energy equation of, 251
    particle in motion of, 457–458
    potential energy in, 250
  force. *See* Force field(s)
  gravitational, 244–245, 247
  inverse square, orbital energies in, 251–257
  inverse square repulsive, motion in, 264–269
Find minimum function, 301–302
Fluid, vertical fall through, 71–77
Fluid resistance, and terminal velocity, 69–74
Foot, 5
Force(s), 10
  central, 218, 227
  centrifugal, 197, 221
  centripetal, 221
  conservative, 64, 146–150
  constant external, 90–93
  Coriolis. *See* Coriolis force
  fictious, 185
  generalized, 449–455
    of constraint, 418, 444–448
  gravitational, between uniform sphere
     and particle, 223–225
  of gravity, 222
  impressed, 58
  impulsive, 305
  inertial, 185
    accelerted coordinate systems and, 184–217
  inverse-square law of, 156
    effective potential for, 257–259
  linear repulsive, 250
  linear restoring, 84–93
  mass and, 57–58
  moment(s) of, 22–23
  nonsinusoidal driving, 135–139
  and position, 63–68
  retarding, 78
  separable type, 156–167
  transverse, 197
  velocity dependent, 69–74
Force field(s), 146–150

central, single particle in, 432
conservative, 146
equipotential contour curves, 151–152
nonconservative, 147
Foucault, Jean, 214, 361
Foucault pendulum, 212–214
Fourier series, 135–138, A-17 – A-18
Fourier's theorem, 136
Free fall, 65, 92
Frequency, angular, 87
Friction, coefficient of, 62
static, coefficient of, 351–352
Frictionless pivot, thin rod suspended from, 356–357
Frisbee, precession of, 396
Functions, special, A-13 – A-14

G
Galactic rotation curve, 243
Galileo, 28, 48, 49–50, 51, 54–55, 157, 160, 219, 220, 222, 295
Galle, Johann, 242
Gamma function, A-13 – A-14
Gauss, Karl Friedrich, 418
Geiger, Hans, 265
Generalized forces, 449–455
Geodesics, 418
Gradient, A-15
Gram, 5
Gravitation, 218–274
universal, Newton's law of, 219–220
universal constant of, 219–220
universality of, 241–244
Gravitational field, 244–245
potential energy in, 244–250
uniform, projectile in, 157–158
Gravitational field intensity, 247
Gravitational force field, acting on planet, 263
Gravitational forces, between uniform sphere and particle, 223–225
Gravitational mass unit, 291
Gravitational potential, 244–250
Gravitational singularity, 292–293
Gravity, of Earth, 222
forces of, 222
and height, 65–66
mass and, 222
Gyration, radius of, 336–337
Gyrocompass, 407–409
Gyroscope, 400–401, 406
air, 401–402

H
Hall, Asaph, 263–264
Halley, Edmond, 226
Halley's Comet, 234
Hamilton, Sir William Rowan, 418
Hamiltonian function, 455–460
Hamilton's equations, 455–460
Hamilton's variational principle, 418, 419–423, 426, 430–431, 446, 452
Harmonic law, 225, 238–244

Harmonic motion, 83, 84–93, 113–124
damped, 96–105
energy considerations in, 93–96
forced, 113–124
Harmonic oscillator(s), 90–93, 104–113, 431
coupled, 472–493
displacement of, 473–474
linear array of, vibration of, 498–505
loaded string of, vibration of, 498–505
vibration of, 498–505
critically damped, 109–110, 111
damped, energy considerations for, 101–102
frequency of, 104
quality factor of, 102–105
simple, 106–108
driven, damped, 116–120, 131–135
linear isotropic, 168
nonisotropic, 170–171
one-dimensional, 456–457
overdamped, 111–113
simple, 84–88
no damping force, 106–108
three-dimensional isotropic, 167–170
energy considerations in, 171–173
two-dimensional isotropic, 167–170
underdamped, 108–109, 110
weakly damped, 109, 110
Height, gravity and, 65–66
maximum, 66
Helix, 175, 176
Hemisphere, solid, center of mass of, 325
Hertz, Heinrich, 88
Hertz (unit), 88
Hipparchus, 225
Hooke, Robert, 82, 226
Hooke's law, 86, 94, 125, 179
Hoop, circular, 330
Horizontal motion, with resistance, 70–71, 163–164
Huygens, Christian, 179, 275
Hyperbola, 235, 266
equation of, A-10
Hyperbolic function, A-5
Hyperbolic identities, A-6
Hyperfine transition, 4–5

I
Impact parameter, 266
Impulse, coefficient of restitution and, 303–304, 305
in collisions, 305
ideal, 305
involving rigid bodies, 354–356
rotational, 354
specific, of rocket engine, 315
Impulsive forces, 305
Inertia, 51
law of, 48–49, 50
moment(s) of, 281, 328–338, 361–371
Inertial forces, accelerated coordinate systems and, 184–217
Inertial frame of reference, 51–53, 55–56
Inertial reference systems, 51–53

Initial conditions, 483–484
Integrals, elliptic, 341–343, A-13
  line, 146
Invariable line, 394
Inverse-square field, orbital energies in, 251–257
Inverse-square law, 220–222, 232, 258
Inverse time unit, 291
Ion rockets, 317–319

J
Jupiter, 220–221, 295–298, 299–300
  Great Red Spot, 295

K
Kelvin, Lord, 264
Kepler, Johannes, 219
Kepler's laws, 221, 225–244
Kilogram, 5
Kinetic energy, 63–68, 282–283
  angular momentum and, 278–283, 361–371
  calculation of, in terms of generalized
      coordinates, 426–429
  of falling disc, 447
  rotational, of rigid body, 370–371
Kinetic friction, coefficient of, 62
Krypton, 4

L
L5 colony, 298–300
Lagrange, Joseph Louis de, 289, 293, 417
Lagrange multipliers, 444–448
Lagrange's equations, applications
      of, 431–438, 479
  of motion, for conservative
      systems, 418, 430–431
Lagrangian mechanics, 417–464, 484–488, 497, 501
Lagrangian points, 293–295, 301–302
  diagonalization of, 486
  invariance of, 487–488
Lamina, plane, perpendicular-axis
      theorem for, 334–335
  semicircular, center of mass of, 326
  square, moment of inertia of, 367–368
Latus rectum, 233
Lawrence, Ernest, 176
Law(s), associative, 12
  commutative, of addition, 12
  of cosines, 18–19
  Coulomb's, 265
  distributive, 12
  of ellipses, 225, 229–238
  of equal areas, 225
  harmonic, 225, 238–244
  Hooke's, 86, 94, 125, 179
  of inertia, 48–49, 50
  inverse-square, 220–222, 232
  of inverse-square force, 156
      effective potential for, 257–260
  Kepler's, 221, 225–244
  Kepler's third, 221
  of motion. See Motion, laws of

Newton's, of universal gravitation, 219–220, 222
  of sines, 203
Leibniz, Gottfried Wilhelm von, 144, 218, 417
Length, unit of, 3–4
Leverrier, Urbain Jean, 241–242, 263
Light, velocity of, 4
Linear momentum, 58–60
  center of mass and, 275–278
  conservation of, 277, 306
Linear motion, of triatomic molecule, 496–498
Linear restoring force, 84–93
Lines of nodes, 391, 399
Lissajous figure, 171, 173
Low earth orbit rocket, 315, 316

M
Mach, Ernst, 1
Magnetic field, motion of charged
      particle in, 174–176
  static, 174
Magnetic induction, 174
Magnitude, vector expressed by, 13
Mantle, Mickey, 164–166, 167
Marsden, Ernest, 265
Martin, Billy, 166
Mass, of attracting object, forces
      of attraction and, 222–223
  center of, and linear momentum, 275–278
      of rigid body, 323–328
  and force, 57–58
  and gravity, 222
  point. See Particle(s)
  reduced, 283–288
  unit of, 5
  variable, motion of body with, 312–319
Mass unit, "gravitational," 291
Mathcad, 75, 76–78, 112, 113, 128–129, 132, 297,
      A-24 – A-26
Mathematica, and Maple, 75
  quick plots using, A-25 – A-26
Mathematica, 75, 164, 166, 211, 296–298, 301–302,
      326–328, 388–389, A-24 – A-31
  FindMinimum, 301–302
  NDSolve, 296–298, 388–389, A-28
  ParaMetricPlot, 298, A-28
  to solve coupled differential
      equations, A-26 – A-31
Matrix multiplication, 26
Matrix(ices), A-19 – A-23
Maupertius, Pierre-Louis-Moreau de, 454–455
Mazzoleni, Marc'antonio, 157
Mechanics, Lagrangian, 417–464
  Newtonian, 47–81, 185, 264
  quantum, 454
Meter, 5
  definition of, 3–4
Milky Way galaxy, 53
Minkowski, Hermann, 2
Molecule, triatomic, linear motion of, 496–498
Moment(s), of force, 22–23
  of inertia, 281, 328–330, 361–371

calculation of, 330–338
    diagonalizing matrix of, 378–381
    and products of, 361–371
    of inertia tensor, 364
    principal, of rigid body, 371–373
Momentum, angular. *See* Angular momentum
    generalized, 438–444
    linear, 58–60, 276
        center of mass and, 275–278
        conservation of, 277, 306
Moon, and Earth, 220–221, 223, 241,
        292–293, 294, 300–301
Morse function, 67
Morse potential, 84–86
Motion, of body with variable mass, 312–319
    chaotic, 129–135
    of charged particles, in electric and magnetic
            fields, 173–176
    circular, 34–35, 220
    constants of, 88
    constrained, on cycloid, 178–179
        of particle, 176–179
    equations of. *See* Equation(s), of motion
    general, of rigid bodies, 410–411
    geometric description of, 383–384
    Hamilton's canonical equations of, 455–456
    harmonic. *See* Harmonic motion
    horizontal, with resistance, 70–71, 163–164
    in inverse-square repulsive field, 264–269
    isochronous, 179
    Lagrange's equations of, for conservative
            systems, 418, 430–431
    laminar, of rigid bodies, 344–353
    laws of, 47–60, 186, 210–211, 222,
            225–244, 276, 418, 420, 450
    linear, of triatomic molecule, 496–498
    multifrequency, 472–473
    with no slipping, 349
    of particle, 60, 94
        in central field, 457–458
        in three dimensions, 144–183
    periodic, 82
    planar, 323–360
    planetary, 49, 218
        Kepler's laws of, 225–244
    projectile, 33, 156–167, 204–212
    radial, limits of, 257–260
    rectilinear, 60–63
        of particle, 47–81
    retrograde, 218
    of rigid body(ies), Euler's equations of, 381–383
        in three dimensions, 361–416
    of rocket, 312–319
    simple harmonic, 83, 89–90
    steady-state, 114
    three-dimensional, potential energy
            function, 151–156
    of top, 397–401
    transient, 114
    turning points of, 64
    of two interacting bodies, 283–288
    uniform circular, 89–90

Multi-stage rockets, 315–317
Multiplication, by power of ten, A-3

N
NDSolve function, 296–298, 388–389, A-28
Neptune, discovery of, 242
Newcomb, Simon, 263
Newton, Sir Isaac, 1–2, 47, 51, 144, 218, 219, 220,
        221–222, 225, 264, 499
Newtonian mechanics, 47–81, 185, 264
Newton's law of universal gravitation, 219–220
Newton's laws of motion, 47–60, 186, 210–211, 418,
        420, 450
NIntegrate function, 328, 388
Nodal point of vibration, 477
Nodes, 508
    lines of, 391, 399
Noninertial frame of reference, 54
Noninertial reference systems, 184–217
Nonlinear oscillator, 125–135
Nonsinusoidal driving force, 135–139
Null vector, 12
Numerical integration, 326–328
Nutation, 401–402

O
One length unit, 291
Orbit(s), apocenter of, 234
    in central field, energy equation of, 251
    circular, 239
        stability of, 260–261
    of comet, 253–354
    differential equation of, 231
    energies of, in inverse-square field, 251–257
        nearly circular, apsides and apsidal angles
            for, 262–264
                in central field, 260–261
    energy equation of, 251
    pericenter of, 234
    reentrant, 262
    turning points of, 403–404
Orthogonal transformations, 28
Oscillating systems, dynamics of, 465–514
    matrices and, A-22 – A-23
Oscillation(s), 82–143, 343
    amplitude, at resonance peak, 120
    antisymmetric mode of, 477–479, 480, 491
    "breathing" mode of, 477–479, 480
    center of, 340–341
    of orbiting system, 471–472
    symmetric mode of, 477, 491, 492
    of system with one degree of freedom, 469–472
Oscillator(s), harmonic. *See* Harmonic oscillator(s)
    nonlinear, 125–135
    pulse-driven, 138
    quality factor of, 103–105
    self-limiting, 127–129, 130
    simple harmonic, 84–88
    undamped, 115, 117, 122
    van der Pol, 128
Overdamping, 98

P
Parabola, 160, 174, 235
    equation of, A-9
Parabolic coordinates, A-16
ParametricPlot function, 298, A-28
Particle(s), 5
    alpha, scattering of, 264–269, 310
    areal velocity of, 227
    charged, motion in electric field, 173–174
        motion in magnetic field, 174–176
    collisions of. *See* Collision(s)
    connected by rod, 424
    constrained, motion of, 176–179
    in free-fall, variation of coordinate
        of, 419–420
    motion of, 60, 94
        in three dimensions, 144–183
    position vector of, 31–36
    rectilinear motion of, 47–81
    in rotating coordinate system, 196–201
    single, in central force field, 432
    sliding, on movable inclined plane, 436–438
        on smooth sphere, 176–177
    systems of, dynamics of, 275–322
    test, 147–148
    total energy of, 63–64
    and uniform sphere, 223–225
Patterson, Red, 166
Pendulum, attached to movable support, 423–424,
        439–440, 492–493
    compound, 338–344
    conical, 442
    double, 488–492
    Foucault, 212–214
    physical, 338–344, 347
    simple, 92–93, 95, 126–127, 131, 281, 427
    spherical, 212–214, 440–444
    suspended, 186–187
        Newton's second law for, 186
    swinging, 423–424
    upside-down, 341–342
Percussion, center of, 354–356
Pericenter, 234
Perigee, 234
Perihelion, 234
Perihelion points, 262
Period doubling, 133, 134
Periodic motion, 82
Periodic pulse, 137–139
Periodicity, 82–83
Perpendicular-axis theorem, for plane
        lamina, 334–335
Phase difference, 115–116, 122–123
Phase space, 106–113
Physical pendulum, 338–344
Planar motion, 323–360
Plane, inclined, body rolling down, 348, 349–350
        movable, particle sliding down, 436–438
    of lamina, 334–335
Plane polar coordinates, 36–39
Planets, 240, 285

gravitational force field acting on, 263
    motion of, 49, 218
        Kepler's laws of, 225–244
Plato, 1
Plumb line, 201–204
Poincaré section, 133–134
Poinsot ellipsoid, 383–384, 385
Point mass. *See* Particle(s)
Polar coordinates, A-7
Pollio, Vitruvius, 3
Position, forces and, 63–68
Position vector, 31–36
Potential, centrifugal, 257
    effective, 257–260, 291–295, 403–405
    gravitational, 244–250
Potential energy, 63–68, 246, 247
    calculation of, in terms of generalized
        coordinates, 426–429
    and equilibrium, 465–469
    in general central field, 250
    in gravitational field, 244–250
Potential energy function, 64, 84, 151–156,
        468, 470
Pound, 5
Power of ten, multiplication by, A-3
Precession, free, of Earth, 396–397
    gyroscopic, 400–401
    nutational, 404–406
    of thin disc, 396
Primary (secondary, tertiary), 288
Principle axes, determination of, 375–377
Product(s), cross, 19–23
    dot, 16
    dyad, 366
    scalar, 15–19
    triple, 23–25
    vector, 19–22
Projectile, deflection of, 206–207
    motion of, 33, 204–212
        with linear air resistance, 161–162
        with no air resistance, 158–161
        in rotating cylinder, 207–212
        in uniform gravitational field, 157–158
Pseudo-vector, 392
Ptolemy, 218
Pythagorean theorem, 233

Q
Quadratic form, 468
Quality factor, 103–105, 121, 122
Quantum mechanics, 454

R
Radius, of gyration, 336–337
Rayleigh, Lord, 127, 418
Rectangular coordinates, 32–33, A-15
Rectilinear motion, 60–63
Redshift, 54
Reduced mass, 283–288
Reduction, to diagonal form, A-22
Reference, frames of, 52, 293

Reference systems, noninertial, 184–217
Resonance, 113–124
  amplitude of oscillation, 120
  sharpness, 120–121
Restitution, coefficient of, 303–304
  and impulse, 305
Restricted three-body problem, 288–302
Right-hand rule, 21
Rigid body(ies), 275
  center of mass of, 323–328, 346–347
    definition of, 323
  free rotation of, 383–384, 435–436
    with axis of symmetry, 384–390
    Euler's equations for, 384–390
    with three principal moments, 371–373
  general motion of, 410–411
  impulse and collisions involving, 354–356
  laminar motion of, 344–353
  mechanics of, 323–360
  motion of, Euler's equations of, 381–383
    in three dimensions, 361–416
  parallel-axis theorem for, 335–336
  principal axis of, 371–381
  rotation of, about arbitrary axis, 361–371
    about fixed axis, 329–330
    relative to fixed coordinate system, 391–397
    with three principal moments, 387–390
  rotational kinetic energy of, 370–371
  symmetric, motion of, 397–401
  symmetry of, 324–325
Ring, thin, potential and field of, 248–250
rkfixed function, 76, 128, 129
Rocket(s), 315–319
  Deep Space I, 317, 318
  ion, 317–319
  low earth orbit, 315, 316
  motion of, 312–319
  multi-stage, 315–317
  specific impulse of engine of, 315
  time of burnout, 315
Rod, thin, moment of inertia of, 330
    suspended from frictionless pivot, 356–357
Rolling wheel, 35–36, 194–196, 410–411, 412
Rotating cylinder, projectile motion in, 207
  rotational motion of, 346, 347
Rotation, of Earth, 201–207
  of rigid bodies. *See* Rigid body(ies)
Rotational impulse, 354
Runge-Kutta technique, 76, 128, 297
Ruth, Babe, 165–166
Rutherford, Ernest, 264, 265
Rutherford scattering formula, 267–268

S
Satellite, orbiting, 273
  boosting of, 255
  Earth, launching of, 314
  oscillation of, 471–472
  period of, 240–241
  speed of, 236

Saturn V, 316, 318
Saturnian moons, 300
Scalar, 9
  multiplication of, 11–12, 15
Scalar product, 15–19
Scalar triple product, 23–24
Scattering, of alpha particles, 267
  oblique collisions and, 306–307
Schrödinger, Erwin, 454, 465
Second, 5
Seismograph, 118–1206
Semicircle, center of mass of, 326
Series expansions, A-11 – A-12
Shell, cylindrical, moment of inertia of, 330
  hemispherical, center of mass of, 325–326
  spherical, gravitational field of, 223–224, 225
    moment of inertia of, 332–333
    potential of, 247–248
SI system of units, 4, A-1
Sidereal day, 472
Sidereal month, 241
Silly Putty, 304
Sines, law of, 203
Sinusoidal waves, 507–508
Sleeping top, 400
Slipping, and laminar motion, 351–352
Solar system, 49, 53
Space, 1, 2
  configuration, 453–454, 475–477
  Euclidean, 2
  force-field free region of, 422–423
  measurement of, 2–5
Space cone, 394–395
Spacecraft, 237–238, 254–257
Speed, 32–33
  of bullet, determination of, 305–306
  escape, 66
  terminal, 72, 73, 74
  and time, 42–43
  of waves, evaluation of, 507
Sphere, moment of inertia of, 332
  and particle, gravitational forces
    between, 223–225
Spherical body, stability of, 468–469
  uniform, particle attraction by, 225
Spherical coordinates, 40–43, A-15
Spherical pendulum, 212–214, 440–444
Spherical shell, gravitational field of, 223–224, 225
  uniform, potential of, 247–248
Spheroid, prolate, 386
Spin, 398
Spin energy, 403
Spring, restoring force of, 86
Spring constant, 86
Stability, 465–469
  attitude, 471–472
  of circular orbits, 260–261
Standing waves, 508–509
Stars, binary, 285–288, 292
Statics, 450

Steady precession, 399–400
Steady rolling, 411
Steady-state motion, 114
Stobbs, Chuck, 165, 167
Stokes' theorem, 150
Sun, 226, 227, 234, 239, 243, 253, 264, 285
Superposition principle, 83, 135, 138
Symmetry, axis of, free rotation of rigid
      body with, 384–390
Synchronous locking, 295
Synodic month, 241

T
Talleyrand-Perigord, Charles
      Maurice de, 3
Taylor series, 467
Taylor's expansion, 150, A-11
Tensors, 364
Terminal speed, 72, 73, 74
Terminal velocity, 71–74
Test particle, 147–148
Thomson, J.J., 264
Thomson atom, 264, 265
Three-body problem, restricted, 288–302
Time, 1, 2
      characteristic, 71, 74
      measurement of, 2–5
      speed and, 42–43
      unit of, 4–5
Time units, 291
Top, motion of, 397–401, 406–407
      sleeping, 400
      symmetric, 398
Torque(s), 22–23
Tour de France, 409–410
Trajectory, of baseball, 164–167, A-27 – A-31
      horizontal, of golf ball, 163–164
Transformation(s), coefficients of, 25
      congruent, 486
      matrix, 25–30, A-21
      orthogonal, 28
Transient, 114
Transverse acceleration, 43, 193
Transverse force, 197
Triatomic molecule, linear motion of, 496–498
Trigonometric identities, A-6
Trimble, Joe, 166
Triple products, 23–25
Trojan asteroids, 295–302
Turning points, of motion, 64

U
Ullrich, Jan, 409–410
Underdamping, 98, 99–100
Uniform acceleration, 60–63
Uniform spherical body, particle
      attraction by, 225
Unit coordinate vectors, 13, 16
Unit tensor, 365
Unit vectors, 13, 37, 40–41, 191

Units, 2–6, A-1 – A-2
Universal constant of gravitation, 219–220

V
van der Pol, B., 127
Variational principle, of Hamilton, 418, 419–423,
      426, 430–431, 446, 452
Vector(s), 1–46
      acceleration, velocity and, 38
      addition of, 11
      angular momentum, 365–367
      angular velocity, 190–191
      basis, 13
      in Cartesian coordinates, 9, 10
      Cartesian unit, 13–14
      components of, 9–10
      concept of, 9
      constant, 279
      cross product of two, 21
      derivative of, 30–31
      dot product between two, 15, 16
      equality of, 10, 11
      magnitude of, 13
      matrix representation of, A-20
      multiplication by scalar, 11–12
      normal mode, 495
      null, 12
      position, 31–36
      rotating, 89–90
      subtraction of, 12
      unit, 13, 37, 40–41
      unit coordinate, 13, 16
      velocity, 362
Vector algebra, 10–15
Vector product, 19–22
Vector triple product, 24–25
Velocity, 32–33
      and acceleration, 36–39
            in cylindrical coordinates, 39–40
            in spherical coordinates, 40–43
      and acceleration vectors, 40
      angular, as vector quantity, 190–191
      areal, of particle, 227
      in rectangular coordinates, 31–39
      terminal, 71–74
            fluid resistance and, 69–74
Velocity-dependent forces, 69–74
Velocity vector, 362
Vibrating systems, general
      theory of, 493–498
Vibration(s), 82
      of continuous system, 505–509
      of coupled harmonic
            oscillators, 498–505
      longitudinal, 501
      nodal point of, 477
      transverse, 500
Virgo cluster, 53
Virtual displacements, 449
Virtual work, 449

W
Wave(s), displacement of, 506
    sinusoidal, 507–508
    speed of, evaluation of, 507
    standing, 508–509
Wave equation, 505–509
    one-dimensional, 506
Wavelength, 508

Wheel, crooked, balancing of, 376–377
    rolling, 35–36, 194–196, 410–411, 412
    rotating, 43, 190, 194–195
White dwarfs, 285–288
Work, 17
    virtual, 449
Work principle, 145–146
Wren, Christopher, 226